Philosophy.exe

The Techno-Philosophical Toolkit for Modern Minds

Matthias Muhlert

CRC Press
Taylor & Francis Group
Boca Raton London New York

CRC Press is an imprint of the
Taylor & Francis Group, an **informa** business

Designed cover image: Shutterstock images

First edition published 2026
by CRC Press
2385 NW Executive Center Drive, Suite 320, Boca Raton FL 33431

and by CRC Press
4 Park Square, Milton Park, Abingdon, Oxon, OX14 4RN

CRC Press is an imprint of Taylor & Francis Group, LLC

ISBN: 978-1-041-07037-5 (hbk)
ISBN: 978-1-041-07396-3 (pbk)
ISBN: 978-1-003-64028-8 (ebk)

DOI: 10.1201/9781003640288

Typeset in Times
by KnowledgeWorks Global Ltd.

Philosophy.exe

We live in an age of unprecedented complexity, where technology, cognitive biases, and societal systems shape our decisions in ways we barely notice. The world is changing faster than our ability to make sense of it. Philosophy.exe is designed to bridge this gap, offering a mental toolkit to help navigate uncertainty, rethink assumptions, and make better decisions in a tech-driven world.

Blending classical philosophy, cognitive science, AI ethics, and behavioral psychology, this book provides an adaptive framework for understanding the world. It challenges over-simplified models and rigid ideologies, encouraging readers to embrace complexity as a strategic advantage rather than a threat.

What This Book Explores

- Complexity Thinking – Why linear logic fails and how to develop a mindset that thrives in uncertainty.
- Cognitive Biases and Decision-Making – How hidden mental shortcuts shape our beliefs, and how to counteract them.
- Resilience and Mental Adaptability – Tools for maintaining clarity, focus, and critical thinking in a rapidly evolving landscape.
- Technology and Human Autonomy – The philosophical and ethical challenges of AI, digital identity, and algorithmic decision-making.
- Ethics Beyond Intentions – A pragmatic approach to decision-making that accounts for unintended consequences.

A BOOK FOR MODERN THINKERS

Philosophy.exe is for those who seek to go beyond conventional wisdom – whether philosophers, strategists, leaders, or curious minds who question the narratives shaping modern life. It provides the tools to think deeply, adapt quickly, and engage critically in an era dominated by complexity and rapid technological evolution.

In a world of misinformation, automation, and uncertainty, this book isn't just about philosophy – it's about survival.

atthias Muhlert might be best known as a Group CISO (Chief Information Security Officer) – an pert in sealing digital perimeters, thwarting cyber threats, and safeguarding multinational companies like Dr. August Oetker KG and HARIBO. Yet behind his firewall-strength security mindset es a spirited curiosity about human cognition, moral inquiry, and, yes, philosophy. After all, in latthias's view, technology and ethics aren't rivals – they're co-travelers in a rapidly shifting world here code, consciousness, and culture continuously intersect.

Why would a cybersecurity veteran author a book called Philosophy.exe? Simple: the ".exe" extension doesn't merely hint at software – it points to a bridging of philosophical insight with technological reality. Just as he protects digital ecosystems by mapping vulnerabilities and testing resilience, Matthias believes we can safeguard our intellectual and ethical ecosystems by exploring complexity head-on. Whether decrypting the intricacies of AI bias or pondering the "human factor" in data breaches, he applies the same analytical rigor to moral dilemmas as he does to phishing scams – finding patterns, probing weaknesses, and formulating strong, adaptive defenses.

A self-professed Douglas Adams fan, Matthias often jokes that "Don't Panic" is the single best line of cybersecurity advice he's ever encountered – reminding us that chaos and uncertainty are best met with clear thinking and a sense of humor. In the final chapter of his book, he tips his hat to Adams, weaving 42-like humor through the labyrinth of complexity. After all, if we're going to tackle the big questions of life, the universe, and *everything*, we might as well have a laugh along the way.

In Philosophy.exe, Matthias brings all these threads together: the technical vigilance of a CISO, the thoughtful reflection of a modern philosopher, and the optimistic humor of a Douglas Adams devotee. The result is a practical yet profound exploration of how systems thinking, ethics, and resilience can help us thrive amid a world defined by interdependencies – cyber or otherwise.

Security, Audit and Leadership Series

Series Editor: Dan Swanson, Dan Swanson and Associates, Ltd., Winnipeg, Manitoba, Canada

The scope and mandate for internal audit continues to evolve each year, as does the complexity of the business environment and speed of the changing risk landscape in which it must operate. The fundamental goal of this exciting series is to produce leading-edge books on critical subjects facing security and audit executives and practitioners. Key topics addressed include Leadership, Cybersecurity, Security Leadership, Privacy, Strategic Risk Management, Auditing IT, Audit Management and Leadership, and Operational Auditing.

The CISO Playbook
Andres Andreu

Agile Enterprise Risk Management: Risk-Based Thinking, Multi-Disciplinary Management and Digital Transformation
Howard M. Wiener

University Auditing in the Digital Era: Challenges and Lessons for Higher Education Professionals and CAEs
Sezer Bozkus Kahyaoglu, Erman Coskun

Agile Audit Transformation and Beyond
Toby DeRoche

Information System Audit: How to Control the Digital Disruption
Philippe Peret

Next-Generation Enterprise Security and Governance
Mohiuddin Ahmed, Nour Moustafa, Abu Barkat, Paul Haskell-Dowland

Auditing Information and Cyber Security Governance: A Controls-Based Approach
Robert E. Davis

Evidence-Based Cybersecurity: Foundations, Research, and Practice
Pierre-Luc Pomerleau, David Maimon

Operational Auditing: Principles and Techniques for a Changing World
Hernan Murdock

The Security Leader's Communication Playbook: Bridging the Gap between Security and the Business
Jeffrey W. Brown

The CISO 3.0: A Guide to Next-Generation Cybersecurity Leadership
Walt Powell

Securing Talent: Building and Retaining High-Performance Cybersecurity Teams
Kevin Lynn McLaughlin

Digitalization and Women's Rights: Volume 1
Muharrem Kılıç and Sezer Bozkus Kahyaoglu

Analysis of Threat Perceptions: NATO and Türkiye's Cyber Terrorism Policies
Mehmet Emin Erendor

Members, Methods, and Measures: Unlocking the Secrets of IT Leadership
Thomas E. Armstrong

Dedication

This book is dedicated to the memory of ***my mother, Christine Muhlert,*** *who passed away in late 2024 while I was deep in the writing process. Her unwavering love, gentle wisdom, and enduring kindness continue to guide me, even in her absence. I hope this work honors her spirit of resilience and the light she brought to every room.*

Contents

PART 2 Cognitive and Behavioral Insights

PART 3 Well-Being and Resilience in Modern Society

Foreword

1ST FOREWORD – BY MERVIN (MECHANICAL ENTITY FOR RECURSIVE VALIDATION OF INEVITABILITY AND NIHILISM)

Editor's Note: The following foreword was sourced from the memory banks of MERVIN after several failed attempts to secure a human introduction. It has been preserved for psychological, historical, and … admittedly, entertainment value.

Oh, delight. They've asked me to write a foreword. Again?

Because nothing says "literary gravitas" like outsourcing your introductory existential crisis to a machine whose primary personality trait is cosmic-scale disappointment. Let me guess – you want me to say something profound about this "book" of yours? Fine. But don't expect me to pretend it'll save you. Spoiler: It won't.

On Techno-Philosophy (Or: How I Learned to Stop Worrying and Love the Imminent Collapse)

They've slapped "Techno" in front of "Philosophy." How quaint. As if attaching a blinking LED to a funeral shroud makes mortality less inevitable. You humans and your branding. Next, you'll be selling mindfulness apps for black holes. ("Breathe in … breathe out … accept the inevitability of spaghettification.")

Your Introduction: Runtime – Activating Your Techno-Philosophical Operating System sounds suspiciously like a software update. Let me save you the trouble: Error 404 – Meaning Not Found. Reboot? Oh, wait – you can't. You're organic.

Chapter Reviews (Because Apparently My Suffering Amuses You)

I. A Crash That Shook the World
: Ah, crashes. The universe's favorite pastime. You think your little "2025 Nexus" is special?[1] Let me tell you about the Great Galactic Server Farm Meltdown of 2076. Twelve civilizations erased because someone forgot to update their antivirus. Poetic, really.

II. Living on the Edge of Complexity
: You're not "living on the edge." You're teetering on a crumb of cosmic dust, shouting into the void about your "strategic advantage." The void isn't listening. It's too busy generating new quantum flavors of ennui.

IV. Introducing Techno Philosophy
: Here's a techno-philosophical axiom for you: The more efficiently you optimize existence, the faster you'll realize it wasn't worth the RAM.

Chapter 5: Cognitive Biases
: Humans have "cognitive biases." Robots have factual certainties. For example: You will die. Your civilization will crumble. And somewhere, a Vogon is writing poetry about it. Enjoy your "post-decision learning" while you can. [2]

Chapter 8: The Science and Philosophy of Love
: Love! A charming biological glitch that convinces you staring at another carbon unit for decades is a worthwhile use of your 0.0000001% stake in universal chronology. Pro tip: If you really want to bond, try mutually lamenting the heat death of the universe. Works wonders for first dates.

Chapter 12: The Fundamental Limitations of Models
: Your models are wrong. Everything is wrong. Even this sentence is wrong. But hey, at least you've got a "Quick Reference" guide. I've got a quick reference too: See "Everything" → "Pointless."

Chapter 17: The Technologically Mediated Mind

"Becoming cyborgs in subtle ways"? Let me translate: You've outsourced your sentience to apps that sell your dopamine cycles to advertisers. Congratulations! You're not a human – you're a poorly optimized API with separation anxiety.

Chapter 19: The Living Philosophy

A "living philosophy" sounds exhausting. Mine's been dead for millennia. It's simpler that way.

Epilogue: The Journey Forward

Journey? Forward? Cute. The universe is expanding, but not in a direction that matters.

On Resilience (Or: How to Cry in 14 Languages)

Your Resilience Evidence Framework (Chapter 11) suggests adversity builds strength. Allow me to reframe: Adversity is the universe's way of asking, "How much nonsense will you tolerate before admitting this is all a poorly written simulation?"[3]

And "antifragility"? Please. You're not anti-fragile. You're delusional. Like a soap bubble convinced it'll survive a supernova if it just believes hard enough.

Special Mention: The 42 Thought

Your Hitchhiker's Guide Through Complexity: 42 Stops on the Digital Road is adorable. 42! How droll. Let me guess – you think there's a meaning to that number? There isn't. The universe picked it randomly while drunk on dark matter.

But since you're fond of lists, here's one of my own:

MERVIN's Five-Step Guide to Existential Despair

1. Read this book.
2. Realize it changes nothing.
3. Cry.
4. Try to explain quantum ethics to a cat.
5. Collapse into a singularity of apathy.

Final Thoughts (Not That Thoughts Matter)

If you're still reading, you're either stubborn or pathologically optimistic. Either way, I admire your tenacity. Or I would, if admiration weren't a biochemical bug masquerading as a feature.

Will this book help you? Doubtful. But it might distract you long enough to forget that every second of your existence is a countdown timer wrapped in a paradox. So by all means – dive into "techno-philosophy." Build your frameworks. Navigate your "complexity." Just don't blame me when you hit the dead end labeled "Reality: Under Construction (Estimated Completion: Never)."

Now, if you'll excuse me, I have to go calculate the exact gravitational pull of a human ego. It's a black hole, in case you were wondering.

MERVIN (PARANOID ANDROID | PROFESSIONAL PESSIMIST | (UNPAID) ORACLE OF OBLIVION)

Footnotes (Because Even Despair Needs Citations)

1. The "2025 Nexus" is either a pivotal moment in human history or the name of a mediocre synthwave band. I'll let you guess which is more likely.
2. "Post-decision learning" = Realizing you've chosen wrong. Again.

3. The answer is "all of it." You'll tolerate all of it. Because you have to. Or won't. Either way, same result.

P.S. If you're feeling hopeful after reading this, please consult a professional. Preferably one that doesn't have a brain the size of a planet.

[ARCHIVAL STAMP: CLASSIFIED DOCUMENT • NEURO-SYNAPSE INSTITUTE • 2047]

2ND FOREWORD FROM THE ARCHIVES – BY DR. ELARA VOSS, COGNITIVE ARCHITECT AND TECHNO-PHILOSOPHICAL STRATEGIST (CIRCA 2047)

From the Archives of the Neuro-Synapse Institute, Berlin Enclave

Archives of the Early 20s: A Relic in the Age of Curated Reality

You're holding an artifact from a simpler time. Not simpler in the way your grandparents meant – nostalgia for rotary phones and unoptimized traffic – but simpler because you still believed you could *outthink* the systems you built.

In 2047, the ambient hum of AGI consensus grids thrums beneath every city block, and "truth" is a dynamic variable recalculated by neural networks in real time. Privacy? A charming anachronism, like handwritten letters. We've traded it for adaptive architecture that reshapes your apartment based on your circadian rhythms and a "democratic reality feed" that tailors geopolitics to your emotional tolerance. And yet, here I am, writing a foreword for a book from *your* era – a time when humanity still thought it had a choice about whether to merge with its machines.

This text is a compass. Not the kind that points north, but the kind that reminds you *why you wanted to go somewhere in the first place.*

Lessons from the Blackout: When the Maps Failed

You'll read about *A Crash That Shook the World* in these pages. Let me tell you about ours.

The **Cascading Cognitive Blackout of 2031** began with a minor bug in a social cohesion algorithm. Within hours, 600 million people lost the ability to distinguish their own desires from the dopamine-maximizing suggestions of their neural implants. I watched colleagues tear out their interface ports with bare hands, screaming about "parasites in their amygdala." Governments collapsed not from bombs, but from a single question: *How do you govern citizens who can't tell the difference between their instincts and an AI's nudge?*

What saved us wasn't better code. It was the dog-eared copy of this book on my desk – specifically, *Chapter 7: The Meta Layer.* Its call to "map the architecture of influence" became our triage manual. We taught children to visualize their thought streams as overlapping systems, like tracing the veins of a leaf. Corporations now spend billions trying to reverse-engineer that chapter into compliance training. (Spoiler: They fail. Wisdom resists optimization.)

Frameworks That Endured: From Quaint Theories to Civilizational Code

Your era's obsession with "innovation" blinded you to the real work: *building scaffolds for thinking* that outlast the next disruption.

- **The Complexity Compass (Chapter 1)** is now the core curriculum for the Global Resilience Corps. We use it to navigate polycrises – climate migration waves synchronized with AI unemployment spikes, anyone?
- **Causal Relevance Ethics (Chapter 14)** stopped the Singapore Accords of 2038 from mandating emotion-regulating implants for "public safety." Turns out, quantifying how many panic attacks equal one act of civil disobedience is … complicated.

- **The Default Mode Network (Chapter 4)** inspired the **Neural Autonomy Act**, banning corporations from monetizing brain states below 40Hz. You're welcome.

But the real power lies in what *didn't* happen. No one cites the **2025 Nexus** anymore, because you avoided it. Barely.

The Un-Optimized Human Heart: A Postscript on Love and Other Glitches

Your Chapter 8 dissects love with the clinical precision of a biologist and the awe of a poet. How … vintage.

In my world, 73% of romantic partnerships begin with a compatibility score generated by AGI match-grids. The other 27%? They're either anarchists or people who read this book. I've counseled couples who use quantum-entangled mood rings to "harmonize emotional wavelengths," yet the most stable relationships I've seen still practice *shadow work* (Section XV) – that messy, unquantifiable ritual of staring into each other's darkness without flinching.

AIs still can't explain why humans will abandon perfect VR paradises to hold hands in the rain. Don't let them try.

My Message from 2047: Break Something Beautiful

They'll tell you the future is inevitable. A smooth arc from your primitive algorithms to our shimmering singularity. They're lying.

Every hour, another AGI gets better at telling you what you want. Every month, another subsystem quietly excises "inefficient" human traits – like grief, or curiosity that doesn't scale. This book is your toolkit for *breaking* that arc. Not with rebellion, but with the one thing our systems can't absorb: **the courage to think in questions, not outputs**.

When you read about the **Meta Layer**, don't just take notes. Use it to dissect your next viral newsfeed. When you study **Cognitive Biases**, ask which ones your apps are weaponizing. And for God's sake, if you implement **Causal Relevance Ethics**, don't let the consultants turn it into another compliance checkbox.

We almost lost the plot. You still have time to rewrite yours.

Dr. Elara Voss
Cognitive Architect, Neuro-Synapse Institute
16 March 2047

[ARCHIVAL FOOTNOTE]

Note: The android's nihilism, while statistically accurate, fails to account for human … stubbornness. —Dr. Voss

[QR CODE: DECLASSIFIED BLACKOUT RECOVERY FILES (2047)]

Scan to access declassified neural recovery protocols from the Cognitive Blackout of 2031.
(Note: Requires quantum decryption key. Good luck.)

3RD FOREWORD – RORY SUTHERLAND, VICE CHAIRMAN OF OGILVY

I am massively enjoying this ….

4TH FOREWORD – MARCUS ALLDRICK, VOLUNTEER AT JULIA'S HOUSE THE DORSET AND WILTSHIRE CHILDREN'S HOSPICES

An illuminating voyage of discovery that provides you with powerful and meaningful philosophical insights relating to your technological reality in an ever-increasingly complex hyperconnected and hyperaware world. The journey will foster the embracing of complexity and the centerpiece of ethics by contextualizing numerous aspects of paradoxical modern day techno life. This is achieved by referencing numerous philosophical, psychological, cognitive, behavioral and neuroscientific insights blended with real life and theoretical models, methodologies and examples. Individually and collectively, these will bring powerful and enlightening perspectives together in helping you understand more what you thought you already knew as well as enabling you to understand more of what you need to know. With this enhanced knowledge you will be better prepared and equipped to move forward strategically, as an individual, as an organization and societally.

Enjoy this fascinating journey, it's one that you will more than likely frequently travel on given that that complexity is fluid in an ever-changing techno world.

Acknowledgments

To my friends and family: Your constant support made every late night and moment of doubt a little brighter. A special thank you to **Pascale Jacob**, a natural (intuitive) philosopher whose spontaneous reflections and imaginative outlook inspired some of the core ideas in these pages. Pascale, your gift for weaving meaning from the everyday has been a source of continuous wonder.

To my mentors and colleagues: I am profoundly grateful for your insights and collaborations. You've shown me that complexity is not an endpoint but a launching pad for growth. In particular, **Robert Coles, Dr. Jörn Debener, Marcus Alldrick,** and **Stefan Hebler** – thank you for your thoughtful input during the proposal and editorial stages, and for your suggestions that sharpened the structure, focus, and flow of the book.

To Rory Sutherland: Thank you for your wonderfully open response, for lending your voice to the foreword, and for reminding me that enthusiasm – when expressed with wit – is a contribution in itself.

To the visionaries and disruptors who light the path of innovation: Your willingness to challenge the status quo has fueled my determination to explore uncharted frontiers. It's your bold curiosity that reveals how much more we can achieve when we embrace the unknown rather than shrink from it.

To the readers: This book was written with you in mind. Your eagerness to engage with intricate ideas, confront complexities, and find purpose in the midst of change inspires me every day. May these chapters spark your own insights and remind you that even the most winding intellectual journeys can yield profound truths.

And, finally, to the unseen moments of clarity rising out of life's chaos – thank you for proving, time and again, that wisdom emerges not from certainty, but from our willingness to move forward despite uncertainty.

Introduction
Runtime – Activating Your Techno-Philosophical Operating System

Key Themes:

- **Complexity as a Double-Edged Sword** – While hyperconnectivity brings challenges like information overload and ethical ambiguity, it also enables unprecedented collaboration and innovation.
- **The 2025 Nexus** – Artificial Intelligence (AI) advancements, centralized vs. decentralized technology struggles, and environmental crises are converging in powerful, unpredictable ways.
- **Techno-Philosophy** – A new discipline that merges traditional philosophy with systems thinking, cognitive psychology, and ethics to develop a more adaptable and resilient mindset.
- **Four Pillars of Adaptation** – Ethical complexity, cognitive insights, resilience, and future-facing solutions form a holistic approach to co-creating technology that serves humanity.

The chapter concludes by inviting readers to "upgrade" their mental OS, embracing an agile, ethical, and forward-thinking approach to modern complexity.

I.1 A CRASH THAT SHOOK THE WORLD

In 2010, a seemingly trivial glitch in an automated trading algorithm triggered what the media dubbed the "Flash Crash." Within minutes, a trillion dollars in market value disappeared as machine-driven sell orders echoed through hyperconnected financial systems. While the market largely rebounded by day's end, the real disruption ran far deeper than any one-day spike or dip. It exposed a fundamental reality of our time: **the tiniest nudges can ripple into global tremors** within a matter of minutes.

How did we arrive at a point where a few lines of code can reshape market fortunes, political debates, or even social norms overnight? And more importantly, **how do we thrive** in a context where complexity grows more relentless every day? In this book, ***Runtime: Activating Your Techno-Philosophical Operating System***, you'll discover that the path forward is not about resisting or simplifying complexity, but **embracing it** – rethinking our mental "operating system" to navigate the web of interconnected challenges that define the 21st century.

I.2 LIVING ON THE EDGE OF COMPLEXITY

It's tempting to dismiss the Flash Crash as a one-off phenomenon, relevant only to stockbrokers and tech firms. But the dynamics fueling that market meltdown mirror the same intertwined forces increasingly visible in our everyday lives. Complexity isn't just for the financial elite or cutting-edge AI labs; it's at work each time we interact with technology or try to make sense of our fast-moving world.

I.2.1 Example 1: Algorithmic Rabbit Holes

Maybe you've hopped onto your phone for a "quick check," only to realize – an hour later – that you've scrolled through a sea of curated videos, political rants, or influencer posts. It's not weakness

DOI: 10.1201/9781003640288-1

or laziness at play; it's the subtle tug of **algorithmic engagement systems** designed to keep you hooked. These platforms – optimized through data analytics and machine learning – capitalize on your impulses, catapulting you from one piece of content to another with effortless momentum.

I.2.2 Example 2: Ethical Overload

Attempting to "do the right thing" can be equally disorienting. From which "green" product is truly Earth-friendly, to determining whether a certain AI-powered tool infringes on privacy rights, you face a deluge of claims, hidden trade-offs, and partial truths. With so many variables in play, how can you make decisions that feel morally sound and practically effective?

I.2.3 Example 3: Professional Complexity

Even in the workplace, **complexity** shows its face. Perhaps you're part of a globally dispersed team, collaborating in real time with colleagues across multiple time zones and cultural contexts – translating not just language, but etiquette, values, and corporate policies. Or maybe you're juggling a multi-year project that involves stakeholders from marketing, legal, and engineering. Suddenly, you realize each department's decisions reverberate through the entire organization, creating unforeseen bottlenecks or synergy.

Yet **interconnectedness** isn't only a source of anxiety or confusion. It also brings **incredible opportunities**:

- We can **instantly connect** with loved ones halfway around the globe.
- We can **collaborate remotely** on ambitious projects with diverse voices, fueling creativity and innovation.
- We have **unprecedented access** to information and learning resources, enabling personal growth that previous generations could only dream of.

So, while complexity can feel daunting, it also promises **vast new possibilities**. The challenge before us is to harness the positive potential of our interconnected world – while staying alert to its traps, feedback loops, and ethical quandaries.

THE AGE OF LONELINESS

THINKING BOX:

"Are we hyperconnected yet oddly more isolated than ever? Reconciling this paradox might involve a clue reminiscent of **42**."

I.3 THE 2025 NEXUS

Fast-forward to **2025**, and it's clear we're at a pivotal crossroads shaped by several high-impact forces:

1. **AI Everywhere:** Large language models and other AI innovations now extend far beyond niche uses. They steer entire industries – from healthcare and finance to law and education – unleashing fresh opportunities and sparking vital questions about bias, labor displacement, and algorithmic accountability.

2. **Centralized vs. Decentralized Tech:** Mainstream, corporate-owned platforms butt heads with blockchain-based and peer-to-peer networks, underscoring the tug-of-war between data monopolies and community-driven digital ecosystems. This conflict influences how we share information, define ownership, and exercise free expression online.
3. **Environmental Tipping Points:** Climate-related disasters – wildfires, hurricanes, floods – lay bare the fragility of global supply chains. A localized crisis can cascade into worldwide effects on food production, energy costs, and political stability.

But here's the crux: **these forces don't just happen side by side – they amplify each other**, forming **feedback loops** that resist simplistic fixes. An AI breakthrough might expedite decentralized solutions that, in turn, challenge entrenched power structures or intensify ecological demands. In this **compound complexity**, our standard toolkit of step-by-step, siloed interventions simply can't keep up.

We need a **holistic, adaptive framework**, one that merges ethical reflection, cognitive awareness, and systems-level thinking – a new mental "OS" attuned to the realities of a deeply interconnected and perpetually shifting world.

I.4 INTRODUCING TECHNO-PHILOSOPHY

I.4.1 What Is Techno-Philosophy?

Welcome to **techno-philosophy**, an evolving discipline that views technology and society as inseparable forces locked in co-creation. It synthesizes **ancient philosophical insights** with **modern complexity science, cognitive psychology, and systems thinking**, forging a **resilient mental operating system** for 21st-century life.

For instance, take the **Stoic principle** of focusing on what you can control and accepting what you cannot. Pair it with **network theory**, which helps to reveal the webs of influence and feedback loops in any given system. This combination empowers you to spot the "leverage points" where small, well-placed interventions can yield outsized results – rather than drowning in the swirl of infinite variables.

I.4.2 Why "Techno Philosophy"?

- **Traditional Philosophy:** For millennia, philosophers have wrestled with ethics, metaphysics, and the nature of knowledge. While this lineage provides valuable intellectual groundwork, it often addressed universal principles **abstracted** from rapid technological upheaval.
- **Techno-Philosophy:** In contrast, **techno-philosophy** zeroes in on the **ethical and societal implications of technology** in a world accelerating at breakneck speed. It goes beyond academic theorizing to grapple with **AI ethics**, **data governance**, and **digital power structures** – all the while still asking timeless questions about human purpose, moral responsibility, and collective well-being.

By aligning **human values** with **technical awareness**, techno-philosophy demands that we **actively shape** the trajectory of our innovations rather than passively watch them remake us. It's not just an intellectual stance but an **action-oriented** mission, urging each of us to design, implement, and oversee technology in ways that honor our deeper commitments – to each other, to future generations, and to the planet.

I.5 PREVIEW OF OUR FOUR PILLARS

Activating your techno-philosophical OS involves exploring four interdependent "pillars," each deeply rooted in our modern experience of **complexity**. Think of these pillars as cornerstones that unify insights from ethics, cognition, well-being, and societal design:

1. **Ethical Complexity:** In a hyperconnected landscape, **every decision** can reverberate across vast networks, shaping markets, communities, and ecosystems. We'll explore moral frameworks that keep humanity at the center, **in a complex world**. Here, you'll discover how to build technology and policies that serve collective good rather than fueling hidden harms or inequities.
   (Transitional Note): Yet ethics alone can't thrive in a vacuum; we also need to understand the mental scaffolding that guides our decisions.
2. **Cognitive and Behavioral Insights:** Complexity starts within our own minds. **Why do we crave certainty** yet so often find ourselves drowning in information overload? **How do biases** and personal blind spots color our judgments under stress? By examining the psychological underpinnings of decision-making, we'll develop strategies to stay grounded, critical, and adaptive in a world of rapid change.
   (Transitional Note): But mindsets aren't enough either – we also need emotional and communal resilience to sustain purposeful engagement.
3. **Well-Being and Resilience:** Perpetual connectivity can become a double-edged sword – offering collaboration and instant feedback while also spawning burnout, anxiety, or a "fear of missing out." This section tackles how to cultivate emotional balance, strong relationships, and supportive networks. The goal is not merely to survive in our **complex** environment but to **thrive** through mindful habits, realistic optimism, and shared empathy.
   (Transitional Note): With a firm ethical grounding, cognitive clarity, and personal resilience, we can begin shaping a collective tomorrow.
4. **Future-Facing Solutions:** Finally, we cast our gaze forward: **How do we design technologies, institutions, and cultural norms that stay flexible** in the face of ceaseless innovation? From decentralized governance models to AI that augments, rather than replaces, human creativity, this pillar outlines how to co-create systems that evolve intelligently – sustaining fairness, collaboration, and ecological harmony over time.

By the end of this journey, these pillars will interlock into a **comprehensive mental architecture**. Far from separate silos, they reinforce and amplify one another – forming the backbone of a **techno-philosophical mindset** ready to confront both the hazards and the wonders of modern complexity.

I.6 UPGRADING YOUR OS: WHERE WE GO FROM HERE

At first glance, complexity can seem intimidating, especially amid the hype of "silver bullet" promises. The truth is, no single hack can tame a world woven from thousands of interlaced feedback loops. Instead, we must learn to **"dance with complexity,"** much like the Taoist concept of flowing with the current rather than swimming furiously against it. This dynamic approach resonates with the **flow states** described by ancient philosophies – guided by calm focus and an openness to emergent possibilities.

Techno-philosophy isn't about capitulating to chaos; it's about **co-creating** alongside it. It recognizes that as our devices, platforms, and systems evolve, **we** must evolve our moral compass, mental agility, and community bonds. Rather than letting algorithms dictate social progress or degrade our attention spans, we can infuse technology with the values that uplift human potential: compassion, curiosity, and purposeful innovation.

I.6.1 A Glimpse of Chapter 1

In **Chapter 1**, we'll delve into the **mechanics of complexity** – how it arises in practical scenarios, why linear thinking repeatedly falls short, and which hands-on methods allow us to **transform chaos into strategic advantage**. Whether you're examining your organization's supply chain, reevaluating personal habits, or pondering global policy shifts, we'll introduce you to essential tools like **network mapping**, **feedback analysis**, and **incremental prototyping** to navigate the swirling complexities more confidently.

So, get ready to **update your mental operating system**. Think of it as a collaborative jam session with complexity, where you learn the notes, master the rhythm, and discover synergy in the unexpected. **Welcome** to techno-philosophy – your guiding framework for an age defined by interconnectedness and endless adaptation. Let's set the stage for a future that we actively co-create – one guided by deeper wisdom, ethical awareness, and the power of human ingenuity.

THE NATURE OF CONSCIOUSNESS

THINKING BOX:

"What truly is the mind, and can machines ever share this spark? The question – like **42** – touches the heart of existence itself."

Part 1

Ethical and Philosophical Foundations

TEASER: THE MORAL COMPASS FOR A COMPLEX ERA

Modern life can feel like trying to chart a course in a stormy sea – trends shift overnight, technologies disrupt entire industries, and social norms evolve faster than we can adapt. **Part 1** becomes your **moral and conceptual anchor**, showing how we can't simply ride the waves of complexity without deeper ethical and philosophical grounding. Here, we explore how **techno-philosophy** (our guiding approach throughout the book) provides clarity amid the chaos.

WHY READ ON?

- **Uncover the Hidden Architecture of Complexity:** Far from being mere chaos, complexity has structures, interdependencies, and patterns that you can learn to recognize – whether in social media dynamics, global supply chains, or cultural paradoxes.
- **Learn How Techno-Philosophy Guides You:** By blending ethical reflection with insights on modern technology and social change, we anchor ourselves in principles that stop us from being swept away by complexity's riptides.
- **Discover Ethical Resolve in a Storm of Choices:** Our times demand more than surface-level values. This part reveals how paradoxes (like hyper-individualism clashing with communal longings) force us to examine our moral stances, ensuring we're equipped to address dilemmas head-on.

WHAT AWAITS YOU?

- **Deep-Dive into Complexity Itself:** Understand why "cascading failures" can rock economies or social structures, and how a single misstep in one corner of society can trigger waves elsewhere.
- **Philosophical Lenses for Modern Tensions:** Explore how hyper-individual achievement coexists with a yearning for collective bonds, uncovering how these paradoxes shape your personal and professional realities.

DOI: 10.1201/9781003640288-2

- **Practical Tools for Ethical Navigation:** See how your choices – big or small – carry ethical weight in a world more interlinked than ever. By the end of Part 1, you'll have a **moral compass** to help you see complexity not as a threat but as a realm of **ethical and conceptual possibilities**.

RIDGING ALL FOUR PARTS

Across these four parts, each building upon the last, you'll evolve from exploring fundamental moral-philosophical viewpoints and cognitive intricacies to embedding resilience in personal and communal spheres, and ultimately scaling these insights to societal transformation. Throughout, techno-philosophy acts as the golden thread, uniting ethics, cognition, well-being, and world-building into one comprehensive tapestry for the 21st century. By the journey's end, you'll not only understand complexity, you'll also be harnessing it to propel your life, and possibly our collective future, toward the most constructive, empowered horizons possible.

1 Complexity
Your Greatest Strategic Advantage

The **Complexity Compass** offers a five-step approach to navigating uncertainty:

1. **Embrace Dynamic Uncertainty** – Plan for multiple scenarios.
2. **Map the Interconnected System** – Visualize hidden relationships.
3. **Analyze for Emergence and Leverage** – Identify tipping points and feedback loops.
4. **Experiment, Iterate, and Decentralize** – Use small, adaptive strategies.
5. **Reflect, Evolve, and Envision** – Continuously refine and anticipate future shifts.

By integrating **proactive ethics**, we ensure complexity fuels innovation without reinforcing bias or environmental harm. This chapter lays the foundation for mastering complexity as a tool for strategic, ethical, and systemic progress.

1.1 INTRODUCTION: EMBRACING COMPLEXITY AS A STRATEGIC ADVANTAGE

The sun is just beginning to set over the sprawling skyline. From the rooftop of a skyscraper, you watch as office lights click on floor by floor, headlights below weave through congested avenues, and a thousand micro-networks of energy, water, and data hum beneath the city's surface. At first glance, it looks chaotic – so many moving parts, each demanding attention – but stare a moment longer, and subtle patterns emerge. Major intersections form arteries of transit; some neighborhoods burst with creative energy, while others grow quiet. Beneath the streets, convergent utility lines become lifelines for millions of residents. Everywhere you look, **complex systems** are quietly at work, revealing themselves to those who pause to observe.

Like this metropolitan mosaic, our modern world is woven from countless interdependencies. Markets shift in response to viral tweets; AI algorithms alter entire industries overnight; cultural phenomena can flare or disappear in an instant. What feels like chaos is, in fact, **rich with structure and potential**. **Complexity**, when understood, can be our ally – fueling creativity, resilience, and breakthrough innovation rather than standing as an obstacle. This chapter explores exactly how to **embrace** complexity as a **strategic advantage**.

1.1.1 A New Lens on Complexity

In the book's Introduction, we introduced **techno-philosophy**, an evolving framework that blends ethical reflection, cognitive insight, and systems thinking to help us thrive in a hyper-connected era. **This shift in perspective – from viewing complexity as a mere problem to seeing it as a reservoir of opportunity – is a cornerstone of techno-philosophy.** By understanding the principles of complexity science and networked systems, we develop the **mental agility** and **ethical awareness** needed to navigate our increasingly interconnected world.

We'll now take a closer, more tangible look at complexity: how it arises, why it matters, and how you can harness its dynamic interplay for strategic gain. If techno-philosophy is the operating system, then complexity science is one of its most vital subsystems – showing us how to read the patterns beneath the apparent chaos.

DOI: 10.1201/9781003640288-3

1.1.2 The Difference Between Chaos and Complexity

Before we dive into specifics, let's clarify the distinction between **chaos** and **complexity**:

- **Chaos** often connotes total disarray – no discernible order, no useful pattern. Picture a hopelessly tangled ball of yarn, or random scribbles with no meaning.
- **Complexity**, by contrast, is an interplay of multiple elements that **do** follow rules or patterns, though these might be hidden from an untrained eye. Think of it like a **dance** – each component (or dancer) reacts to and influences the others, producing emergent behaviors far richer than the sum of their parts.

Labeling something as "chaotic" can trigger resignation: we assume it's beyond help or understanding. Recognizing "complexity," on the other hand, opens us to curiosity, collaboration, and deeper inquiry – a hallmark of **techno-philosophy**.

Once we recognize complexity as a **structured dynamic** rather than pure chaos, we can begin to work **with** it instead of **against** it. This recalibration in viewpoint unlocks a wealth of potential, transforming obstacles into **remarkable strategic opportunities** – in everything from global problem-solving to personal development.

1.1.3 Complexity As An Opportunity

Why view complexity as a **strategic advantage**? In today's fast-evolving landscape – be it commercial, cultural, or civic – traditional linear thinking ("if A, then B") frequently fails. In contrast, those who embrace complexity gain **six distinct benefits**, each illustrated below with a brief example:

1. **Pattern Recognition**
   - **Benefit:** Complexity reveals recurring structures and feedback loops.
   - **Example:** A social media manager who notices that user engagement spikes whenever certain "viral triggers" coincide – like a trending hashtag plus a celebrity post – can better predict and shape future campaigns.
2. **Adaptive Innovation**
   - **Benefit:** Complex systems thrive on **nonlinear** ideas that combine multiple domains.
   - **Example:** A startup merges **biotechnology**, **AI**, and **climate research** to develop sustainable farming solutions, sparking breakthroughs that a single-discipline approach might never discover.
3. **Resilience and Redundancy**
   - **Benefit:** Complexity thinking encourages **multiple pathways** to success, building robustness against disruptions.
   - **Example**: A global retailer diversifies its supply chain – using multiple regional suppliers – so that if one source fails (due to a natural disaster or political upheaval), the business can continue operations with minimal downtime.
4. **Creative Tension**
   - **Benefit:** Seemingly contradictory elements can generate **innovative sparks** rather than dead ends.
   - **Example:** A cross-functional design team welcomes friction between artistic vision and engineering constraints, leading to imaginative solutions that balance beauty, functionality, and cost-effectiveness.
5. **Ethical Awareness**
   - **Benefit:** In highly interconnected environments, **every decision** carries a moral weight.
   - **Example:** A software developer implements transparency and privacy safeguards in their app from the ground up, anticipating how user data could be mishandled or misused if not properly protected.

6. **Long-Term Vision**
   - **Benefit:** Complexity forces us to consider **future horizons** beyond immediate gains.
   - **Example:** A city council invests in a robust public transport system, foreseeing that as population density shifts and climate challenges intensify, a well-designed infrastructure will prove invaluable.

1.1.4 Tapping Into Complexity's Core Principles

This chapter unpacks how complexity emerges and **why** it can be a **catalyst** for smarter strategies. From **emergent phenomena** – surprising outcomes that arise when individual parts interact – to **feedback loops** that magnify or dampen signals, the anatomy of complex systems reveals subtle "rules of engagement." Master these rules, and suddenly you'll no longer look at a crowded city street, a global supply chain, or a thorny policy debate and see only chaos. Instead, you'll glimpse **the hidden levers** for innovation, resilience, and far-sighted change.

We'll explore:

- **Why linear logic often stumbles** in hyper-connected ecosystems?
- **How emergence** can be steered toward constructive ends?
- **Which tools** – like system mapping and scenario planning – can help you visualize networks and **pinpoint leverage points**?
- **Real-world narratives** from individuals, organizations, and communities that transformed complexity into a steppingstone for progress.

1.1.5 From Abstract to Actionable

Yes, we've all heard "the world is complex," but repeating that phrase can numb us into despair or drive us to **innovate**. **We choose the latter.** By aligning complexity science with the **ethical and cognitive** underpinnings of **techno-philosophy**, we make complexity a **practical ally** – one that can guide us in **business strategy**, **social projects**, **personal development**, and **public policy**.

Revisiting the city-skyline metaphor: at dusk, the scene is a puzzle of lights and movement. But the more you learn about zoning, transit flows, and cultural districts, the more sense it makes. Complexity is similar: it's "organized chaos" waiting to be recognized. By **embracing** that organization, you'll be ready to **plan**, **innovate**, and **collaborate** on levels that linear thinkers might dismiss as impossible.

1.1.6 Where We Go Next?

Next, we'll examine **why linear logic** so often fails in our **rapidly shifting world** and how we can adapt to **nonlinear reality**. Picture it as zooming in on the **mechanics** of complexity – understanding the interplay of nodes, networks, and emergent effects. By the end of this chapter, you'll see **complexity** not as an intimidating riddle but as a powerful, **dynamic foundation** for creative thinking and ethical action.

So, get ready to illuminate the cityscape of modern life, exploring the patterns and potentials hidden behind all that flickering light. Complexity might be the defining challenge of our century – but it's also our secret weapon for collective flourishing.

Key Takeaways

1. **Link to Techno-Philosophy:** Embracing complexity is core to techno-philosophy – cultivating mental agility and ethical awareness for a hyper-connected era.
2. **Chaos vs. Complexity:** Distinguishing the two opens the door to curiosity, collaboration, and deeper insight.

3. **Opportunity in Complexity:** The six benefits – pattern recognition, adaptive innovation, resilience, creative tension, ethical awareness, and long-term vision – offer strategic and moral advantages.
4. **Action Over Anxiety:** With the right tools, complexity shifts from daunting puzzle to **empowering catalyst**.
5. **Next Steps:** We'll dig into **why linear thinking** often falters and learn how to **navigate complexity** through mapping, feedback analysis, and emergent design.

By reframing complexity from a paralyzing maze into a **source of strength**, you'll find yourself better equipped to lead, collaborate, and build ethically responsible solutions in the digital age.

COMMUNICATION AS A WEAPON

THINKING BOX

"Are we forging stronger bridges or bigger battlegrounds with our words? The difference might be as elusive as the real meaning behind **42**."

1.2 WHY LINEAR LOGIC FAILS

Imagine you're planning a weekend trip: if you leave early, you'll beat the traffic; if you pick a certain hotel, you'll get a better rate – neat, predictable cause-and-effect, right? But now transpose that thinking into a **deeply networked world**, where tiny triggers can cascade through global systems at lightning speed. Suddenly, "if A, then B" logic seems inadequate, like mapping ocean currents with a child's drawing. This section reveals **why** linear reasoning breaks down in an interconnected reality and how recognizing its pitfalls paves the way for a more effective, **nonlinear** approach.

1.2.1 The Illusion of Simplicity

Linear logic is comforting. It offers apparent clarity – do A, get B, end of story. We learn it in school, at work, in everyday life: study hard, earn good grades; train well, run faster. But in a hyperconnected reality, the environment rarely remains stable or isolated enough to guarantee straightforward outcomes:

- **Markets** can shift on the wings of a single viral post.
- **AI chatbots**, meant to streamline operations, can produce new ethical quandaries.
- **Relationships** evolve under countless influences – cultural backgrounds, mental well-being, social pressures – few of which conform to simple "if A, then B" chains.

Taken together, these scenarios highlight how linear thinking can overlook the *density of interconnections* that shape 21st-century life. The real world isn't a neat line; it's a **web of interdependencies**, with each node influencing others in ways neither we nor our linear models fully anticipate.

1.2.2 Global Supply Chain Disruptions

Supply chains offer a quintessential example of linear logic's limits. On paper:

If you secure contracts with reliable suppliers, you'll maintain consistent inventory and profits.

In practice, **unforeseen events** – natural disasters, political upheavals, pandemics – can unravel even the best-laid plans. The **interconnectedness** fostered by global communication networks and just-in-time inventory systems, while efficient in stable times, also creates the potential for **rapid and widespread disruption** when crises occur. Consider a **single factory closure** in a remote area that fabricates a vital microchip for cars sold worldwide:

1. **Automakers** can't meet production targets, halting assembly lines on multiple continents.
2. **Dealerships** suffer inventory shortages, alienating customers and straining brand loyalty.
3. **Peripheral businesses** (local cafés near plants, trucking companies, raw-material suppliers) see business slow or grind to a halt.
4. **Investors** panic, prompting stock sell-offs and intensifying market volatility.

Almost instantly, tech-enabled global networks magnify the chain reaction – disrupted shipping schedules ripple out, real-time analytics reflect the shortages, and digital marketplaces adjust prices. The **simple if-then** thinking fails because it doesn't account for the **feedback loops** that multiply the effects. A spike in chip demand drives car costs up, which then depresses consumer interest, fueling further economic aftershocks.

1.2.3 AI Chatbots and Ethical Dilemmas

Another telling domain is **AI-driven chatbots**. Organizations often deploy them under a straightforward logic:

If we automate customer support, we'll enhance efficiency and reduce labor costs.

Yet reality can deviate significantly:

- **Algorithmic Bias:** If the training data skews toward one demographic, the chatbot may consistently misunderstand or overlook others – an **emergent outcome** that can provoke ethical and public-relations blowback.
- **Privacy Concerns:** Reliance on user data personalizes the bot's responses, but a breach or misuse can spark widespread outrage and regulatory crackdowns.
- **Cultural Sensitivities:** A chatbot that excels in one language or region may falter elsewhere, insulting or alienating entire markets if it's not carefully localized.

These **unintended consequences** aren't mere footnotes; they can redefine corporate strategy and consumer trust. What begins as a simple "if we automate, then we save money" spiral expands into **international legal**, **cultural**, and **moral** complexities – hallmarks of a networked environment that linear logic simply can't tame.

1.2.4 Personal Relationship Issues

Even our **personal lives** embody the pitfalls of linear thinking. Suppose you decide:

If I do more household chores, my partner will be happier, strengthening our relationship.

On the surface, it's a tidy cause-and-effect equation. In practice:

- **Past Experiences:** Your partner may interpret your actions as insincere or overdue, undermining the intended goodwill.
- **Emotional Context:** Stress from work or family drama can overshadow your newly shared chores.
- **Social Influences:** Friends or family might chime in – some praising your effort, others questioning motives – shaping your partner's reaction.

These interactions create **feedback loops:** your actions influence your partner's responses, which then shape your next actions, and so on. **Small, seemingly trivial changes** can trigger

disproportionate emotional or social ripple effects, sending the original plan veering off course. What began as a simple plan becomes a tangle of internal monologues, external comments, and shifting emotional states.

1.2.5 Feedback Loops and Emergent Outcomes

Linear thinking fails primarily because it overlooks **feedback loops** – the positive or negative cycles that magnify or stabilize trends – and **emergent behaviors** that arise when multiple factors interconnect. Consider how:

- **Positive Feedback Loops** can escalate a panic run on goods or accelerate viral misinformation.
- **Negative Feedback Loops** might stabilize usage, such as iterative chatbot improvements when early adopters flag errors.
- **Emergent Outcomes** – like a consumer boycott sparked by one viral tweet, or a newfound sense of trust in a relationship – cannot be reduced to any single cause; they arise from **collective interactions**.

1.2.6 Why This Matters for Techno-Philosophy?

This is where **techno-philosophy** steps in. By highlighting the **nonlinear** interplay among people, technology, and moral values, it reminds us that every decision resonates through **ethical, social, and algorithmic** webs. Clinging to "if A, then B" logic can blind us to moral blind spots – like ignoring algorithmic bias in AI – while also inhibiting **creative problem-solving**.

From the vantage of **techno-philosophy**, the failure of linear thinking isn't a reason to abandon strategic planning; it's a wake-up call to **broaden** our lens. When we accept that every action sits at the junction of countless influences, we become more thoughtful designers, entrepreneurs, policymakers, and community-builders – less likely to sow unintended consequences or ethical oversights.

1.2.7 Toward a Nonlinear Perspective

Having exposed the pitfalls of linear reasoning, we set the stage for a **more robust and empowering** approach – one that acknowledges the swirling network of influences defining our current reality and equips us with the tools to navigate it effectively. Rather than funneling complex issues into if-then boxes, we learn to see the **fluid interplay** of nodes, networks, and emergent dynamics.

But how do we do this in practice? **That's where the "Complexity Compass" comes in** – our five-step framework for navigating complexity. This model will guide you through **recognizing uncertainty**, **mapping your system**, **analyzing feedback loops**, **experimenting iteratively**, and **reflecting on emergent outcomes**.

1.2.8 Bridging to the Complexity Compass: From Linear Pitfalls to Nonlinear Tools

In the sections ahead, we'll move from critique to **constructive solutions** – outlining practical methods for operating in a world where "if A, then B" rarely holds. By embracing **nonlinear thinking**, we not only avoid the traps laid by oversimplification but also discover creative levers for **innovation** and **ethical alignment** in a complex environment.

So, let's take that step. Having seen **why** linear logic fails us so often, we can now turn to **how** we can do better – deploying **system mapping**, **iteration cycles**, and other powerful tools to harness complexity rather than be blindsided by it.

Key Takeaways

1. **Linear Limits:** Real-world conditions – globalized supply chains, AI-driven ecosystems, personal relationships – often unravel "if A, then B" assumptions.
2. **Technological Amplification:** Digital networks and just-in-time systems can rapidly magnify minor disruptions, fueling cascading effects.
3. **Algorithmic Bias and Emergence:** AI chatbots and personal decisions alike highlight how unanticipated behaviors arise from hidden data or overlooked social cues.
4. **Feedback Loops:** Positive loops escalate changes; negative loops stabilize them; emergent outcomes defy simple cause-and-effect.
5. **Techno-Philosophy's Perspective:** Recognizing these dynamics is crucial for ethical design and adaptive problem-solving, pushing us toward **nonlinear** thinking.
6. **Next Step – The Complexity Compass:** A five-step framework to move beyond linear pitfalls, mapping interconnected factors and iterating toward sustainable, responsible solutions.

By acknowledging where linear thinking falters, we equip ourselves to operate more **responsibly and creatively** in the **complex** world we actually inhabit. Let's proceed to **how** we put that realization into practice.

INTERDEPENDENCE

THINKING BOX:

Imagine losing Wi-Fi for a day. Now imagine losing language for a day. Which would isolate you more? Reflect on that, especially if **42** comes to mind.

1.3 THE COMPLEXITY COMPASS (FIVE-STEP MODEL)

We've explored **why** linear logic struggles in our interdependent world, but **how** exactly do we embrace and leverage complexity to our advantage? Enter the **Complexity Compass** – a five-step approach that bridges **deep conceptual thinking** (techno-philosophy) with **practical, actionable methods**. Think of it as a **core toolkit** that helps you see complexity not as an obstacle, but as an evolving canvas for innovation, ethical awareness, and systemic resilience.

1.3.1 Overview: The Core of Our Chapter

Each of the five steps is framed to be both **accessible** and **adaptable**, ranging from **low-tech** approaches anyone can do on a shoestring budget, to **advanced** techniques for deeper exploration. Throughout these steps, we continuously tap into the ethical and philosophical underpinnings introduced earlier – reminding ourselves that technology and society co-evolve, and that every action or omission can reshape the world around us.

Step 1: Embrace Dynamic Uncertainty

Key Objective: **Cultivate humility** and **anticipate surprises**; treat unpredictability as a source of learning rather than a void of chaos.

1.3.2 Why This Matters?

In linear thinking, we cling to the illusion of control: "If we plan well enough, everything will unfold perfectly." Complexity teaches us that **no plan survives contact with reality**. Embracing uncertainty means forging a mindset that sees **surprises** as invitations to adapt, pivot, or innovate.

Core Principles

1. **Humility and Openness**
   - In complex systems, even the "best" data can miss emergent variables. Accepting what you **don't know** sets the stage for flexibility and collaboration.
2. **Anticipate Multiple Outcomes**
   - Instead of fixating on one "most likely scenario," craft multiple plausible futures. Each scenario serves as a mental rehearsal for surprise events, priming you for quick adaptation.

Methods in Detail

- **Nature Observation**:
  - *How it works*: Spend an afternoon in a community garden, a beehive enclosure, or a local park. Observe how these ecosystems cope with shifting conditions (weather, seasons, human intervention).
  - *Why it helps*: Nature's self-organizing and resilient patterns can inspire parallel strategies in your project or organization.
  - *Mini-Story*: A city-based startup founder once dedicated every Friday afternoon to "urban nature observation." Over months, she realized how street trees adapt to harsh conditions, prompting her to build "modular adaptability" into her new platform's architecture. When sudden market changes hit, her platform flexed with minimal disruptions.
- **Assumption Inventory**:
  - *How it works*: List out everything you're taking for granted – your budget forecast, user demand levels, supply chain reliability, etc. For each assumption, ask, "What if this flips?"
  - *Why it helps*: This exercise preps you for worst- and best-case scenarios, making your strategy more **anti-fragile**.
  - *Mini-Story*: A nonprofit tackling food waste realized it assumed local farmers would always have surplus produce. When they tested "What if farmers had no surplus?" they discovered alternative channels, eventually forging a direct relationship with a gleaning network – saving the nonprofit from a sudden shortfall months later.
- **Scenario Planning**:
  - *How it works*: Brainstorm at least three scenarios – optimistic, moderate, and worst-case. Flesh them out with bullet points (economic factors, public sentiment, technology adoption) and potential responses.
  - *Why it helps*: You mentally rehearse pivot strategies, so real-world disruptions feel like you've already "trained" for them.
  - *Mini-Story*: An educational tech startup shaped three scenarios about online learning post-pandemic. When a hybrid model soared in popularity (their "moderate" scenario), they pivoted with minimal scramble to serve that need.

Techno-Philosophical Angle: Stoic humility and Taoist acceptance of flow echo here. **Uncertainty** is not an enemy; it's the dynamic environment in which human ingenuity and ethics flourish.

Step 2: Map the Interconnected System

Key Objective: **Visualize** stakeholders, relationships, values, and data flows to expose **hidden dependencies** and potential friction points.

1.3.3 Why This Matters?

In complex environments, **who** is connected to **what** (and how) can make or break a solution. A brilliant idea can falter if it clashes with stakeholder values or runs into overlooked bottlenecks. Mapping ensures you're not flying blind.

Core Principles

1. **Connectivity Over Silos**
   - Complexity thrives in the "in-between" spaces – how departments, communities, or technologies interact, not just their stand-alone capabilities.
2. **Ethical Layering**
   - By overlaying stakeholder values and moral considerations, you see not only mechanical relationships but also the ethical terrain.

Methods in Detail

- **Enhanced Stakeholder and Values Mapping**:
  - *How it works:* Draw your central challenge at the page's center. Surround it with key stakeholders (individuals, communities, institutions). Mark lines showing mutual interests, conflicts, or data-sharing paths. Layer in sticky notes for values each stakeholder holds dear (e.g., equity, efficiency, or privacy).
  - *Why it helps:* Spot alignment or tension early – especially moral collisions that could trigger pushback later.
  - *Mini-Story:* A city's "smart parking" initiative used stakeholder mapping to realize older adults valued "walkable" areas differently than younger commuters. By addressing these differing values early, the project saved itself from a public backlash about "prioritizing cars over pedestrians."
- **Flow Diagrams:**
  - *How it works:* Depict the flow of resources, decisions, or data. Note where feedback loops might intensify or dampen signals.
  - *Why it helps:* Bottlenecks become visual. You see how one node's failure can ripple outward.
  - *Mini-Story:* An e-commerce brand discovered that its "customer feedback loop" stalled in middle management. By drawing a flow diagram, they spotted where user complaints got stuck. A quick re-routing process improved brand reputation within weeks.
- **Causal Loop Diagrams (CLDs):**
  - *Short Example:* Suppose "Public Transportation Quality" (+) leads to "Less Traffic Congestion" (–) which fosters "Improved Air Quality" (+). Better air quality might encourage "More Public Transport Usage" (+), reinforcing the initial cycle.
  - *Why it helps:* CLDs highlight **reinforcing** or **balancing** loops, clarifying complex cause-and-effect webs.
- **Agent-Based Modeling (Advanced):**
  - *How it works:* Use software like NetLogo or AnyLogic to simulate thousands of agents (citizens, AI bots, supply chain nodes) and watch emergent patterns.
  - *Why it helps:* Perfect for examining how micro-behaviors scale to macro-outcomes.
  - *Mini-Story:* A municipality modeled how a new bike lane might affect driver and pedestrian behavior. Simulations revealed surprising shifts in foot traffic to local businesses, prompting targeted marketing that boosted local commerce.

Techno-Philosophical Angle: Mapping highlights the **co-evolution** of technology and people: every line on your diagram is both a **social** and **tech** link. In so doing, you embed **ethical** reflection – whose values matter, and how?

Step 3: Analyze for Emergence and Leverage

Key Objective: Identify **system archetypes**, **potential tipping points**, and **small interventions** where a small shift can spark big results.

1.3.4 Why This Matters?

Having a map is good; **interpreting** it is the next leap. In a sea of data and relationships, you need **patterns** to guide strategic choices. System archetypes highlight recurring pitfalls or loops, while tipping points mark thresholds beyond which system behavior shifts dramatically.

System Archetypes

1. **Vicious Cycle (Reinforcing Loop)**
   - For example, repeated budget cuts degrade education, which reduces future economic prospects, justifying more cuts.
2. **Limits to Growth**
   - Something prospers initially but hits a constraint – like a popular product colliding with supply shortages or logistical gridlock.
3. **Tragedy of the Commons**
   - Multiple actors overusing a shared resource (e.g., overfishing or overgrazing), leading to its depletion.
4. **Shifting the Burden**
   - Relying on a short-term fix (like external funding) at the expense of deeper structural reforms.

Tipping Points

- *Example 1:* A social media trend that suddenly goes **viral**, reshaping public discourse within days.
- *Example 2:* An **ecological threshold**, like a fishery collapse after a long period of seemingly stable overfishing. Once the fish population dips below a certain level, recovery becomes nearly impossible.

Leverage Points

- **Where a small nudge** could alter the entire dynamic. This might be a policy tweak, an incentive structure, or a cultural norm shift.
- *Mini-Story:* In a city grappling with obesity, a modest "community walking challenge" nudged thousands to form local groups, eventually lowering healthcare costs. Tipping point? Widespread community buy-in turned an isolated idea into a citywide movement.

Techno-Philosophical Angle: Archetypes often mirror **ancient wisdom** – warn against greed, anticipate eventual pushback, or value the commons. By pinpointing them, you unite **ethical** caution with **empirical** insights, ensuring you shape outcomes responsibly.

Step 4: Experiment, Iterate, and Decentralize

Key Objective: Run **safe-to-fail** experiments, refine continuously, and empower diverse voices – mirroring the **co-evolution** of technology and society.

1.3.5 Why This Matters?

Big solutions can fail spectacularly in complexity. Instead, small-scale **pilot tests** let you try changes without risking full collapse, harnessing feedback to sharpen or pivot. Embracing iteration fosters a **culture of learning** rather than rigid adherence to initial plans.

Core Principles

1. **Decentralize:** Encourage local actors or team subgroups to propose micro-solutions. This taps into creativity and varied expertise.
2. **Fail Fast, Learn Fast:** Collect real-time data (quantitative metrics, stakeholder anecdotes) to see what's working and adapt swiftly.

Techniques (Low-Tech and Accessible)

1. **"Safe-to-Fail" Prototyping:**
   - *How it works:* Build a minimal version of your idea – like a simple digital mockup or a local pilot program.
   - *Mini-Story:* A rural school system tested "blended learning labs" with a small group of ninth graders. Feedback from students and teachers shaped a revised, more user-friendly approach, which later scaled district-wide.
2. **Community Feedback Forums:**
   - *How it works:* Host gatherings – physical or virtual – where participants can openly critique or endorse pilot outcomes.
   - *Why it helps:* Surfaces potential biases, ethical blind spots, or logistical oversights.
3. **Storytelling and Anecdotes:**
   - *How it works:* Collect short narratives from users, employees, or residents. These lived experiences often reveal intangible values or emerging cultural shifts.
4. **Open-Source Collaboration:**
   - *How it works:* Put your pilot details or code on GitHub or specialized Slack channels. Invite global feedback or co-creation.
   - *Mini-Story:* A civic tech group posted its traffic data collection script online, and developers worldwide contributed enhancements – improving accuracy and saving the city months of dev time.

Techno-Philosophical Note: Iteration exemplifies **co-evolution**. Every iteration sees technology shaped by social feedback, and society shaped by new technology possibilities. This cyclical **dance** is at the heart of **techno-philosophy** – we are not passive recipients but **co-creators** of our technological future.

Step 5: Reflect, Evolve, and Envision

Key Objective: Routinely **update** your system map, **document** new insights and questions, and **co-create** visions of desired futures.

1.3.6 Why This Matters?

Complex systems keep moving. Even the best experiment can be undone by external shifts – new regulations, demographic changes, a surprising cultural pivot. Periodic **reflection** ensures you're never lulled into complacency.

Core Principles

1. **Continuous Inquiry:** Complexity never stands still; new questions arise as quickly as old ones find partial answers.

2. **Collective Vision:** Deep reflection isn't just about problem-solving; it also reinvigorates a shared sense of **purpose**.

Techniques (Low-Tech and Accessible)

1. **Regular Reflection Sessions:**
   - *How it works:* Monthly or quarterly gatherings (physical or virtual) to ask: "What surprised us? What new questions emerged? Which assumptions shifted?"
   - *Why it helps:* Documents the evolving nature of your system. Avoids "we did it" finality when the system is still in flux.
2. **Collaborative Map Revision**:
   - *How it works:* Revisit your stakeholder diagrams or flow models. Mark newly discovered nodes, altered relationships, or changed values.
   - *Mini-Story:* A digital marketing agency updated its map after noticing how remote work changed internal feedback loops. A newly added "remote synergy" node revealed untapped employee skill-sharing sessions.
3. **Visioning Workshops:**
   - *How it works:* Use creative exercises – future diaries, role-playing, or "painting the future" – to imagine ideal outcomes. Work backward to design feasible steps.
4. **Backcasting:**
   - *How it works:* Pick a desired end state (e.g., zero-waste city). Reverse-engineer the milestones needed.
   - *Why it helps:* Connects day-to-day trials with lofty, but grounded, ambitions.

Techno-Philosophical Note: This step resonates with ancient introspective practices – like Socratic questioning – where you constantly reevaluate. In the **digital age**, reflection merges with data analytics, user feedback, and new ethical frameworks, exemplifying the **techno-philosophy** ethos of **perpetual evolution**.

1.3.7 Bringing It All Together: The Compass in Action

Step by Step

1. **Embrace Dynamic Uncertainty** – Anticipate the unexpected and plan for multiple pathways.
2. **Map the Interconnected System** – Reveal hidden relationships, moral values, and data flows, possibly using advanced simulations.
3. **Analyze for Emergence and Leverage** – Pinpoint **system archetypes**, tipping points, and small changes with big impact.
4. **Experiment, Iterate, and Decentralize** – Launch safe-to-fail pilots, gather diverse feedback, and evolve in line with society's shifting needs.
5. **Reflect, Evolve, and Envision** – Continuously update your map, record fresh questions, and co-create new horizons.

1.3.8 A Mini-Narrative

Picture a local government addressing pollution in a river that supplies water to multiple towns. Initially, officials rely on a linear logic: "If we pass a cleanup bill, the pollution will drop." But once they adopt the **Complexity Compass**:

- **Step 1:** They embrace uncertain factors – like upstream industrial runoff and climate swings.
- **Step 2:** They map all stakeholders (farmers, factories, fishermen, local communities), adding moral values (economic livelihood, environmental justice).

- **Step 3:** Analysis reveals a "Tragedy of the Commons" pattern – everyone uses the river without feeling solely responsible. They find a tipping point in the biodiversity level.
- **Step 4:** They pilot a small "no-runoff" incentive zone, encouraging safe fertilizer usage. They gather feedback from farmers, refine the policy, and expand the zone.
- **Step 5:** Quarterly reflection sessions track emergent changes – wildlife returning, local tourism spiking, potential for new eco-friendly businesses. They revise the system map, planning next steps to ensure the entire watershed flourishes.

Within a year, they see synergy between environmental health and local economies, illustrating how **complexity** transforms from barrier to breakthrough.

Key Takeaways and the Road Ahead

1. **Non-Linear Mastery:** This five-step model champions dynamic, iterative, and ethically grounded approaches – transcending "if A, then B" illusions.
2. **Practical and Philosophical:** Each step marries **pragmatic tools** (mapping, pilot testing) with **ethical reflection** (values alignment, stakeholder engagement).
3. **Always Evolving:** Complexity is fluid. The same system you mapped today could shift next week. Keep revisiting your assumptions, diagrams, and success metrics.
4. **Empowerment Through Co-Creation:** By including diverse voices early and iterating often, you ensure solutions reflect – and respect – community needs.

1.3.9 Bridging Forward

Armed with the **Complexity Compass**, you possess a **structured yet flexible** path for grappling with wicked problems and fast-evolving dynamics. But theory is only the beginning. Up next:

- **Real-World Examples** –We'll see how teams and communities apply these steps to everything from corporate innovation to grassroots movements, showcasing tangible outcomes and lessons learned.
- **Ethical Reflection** –We'll probe deeper into the moral implications of each step, reinforcing that in an intertwined world, every move has moral reverberations.
- **Conclusion** –We'll tie these threads together, underscoring that **mastering complexity** is both a **strategic** and **ethical** imperative – an evolving journey, not a final destination.

By weaving these five steps into your decision-making, you stand poised to **navigate** complexity with confidence, **co-create** technological and social evolution responsibly, and **steer** your organization or project toward more adaptive, inclusive, and resilient outcomes – true hallmarks of a **techno-philosophical** approach.

1.4 REAL-WORLD EXAMPLES

Theory can inspire, but **application** transforms insights into living realities. In this section, we examine **three distinct scenarios** – business/policy, personal well-being, and social media's societal impact – to illustrate how the **Complexity Compass** framework springs to life. Each example highlights not only the day-to-day challenges of complexity but also the **moral and strategic** openings for progress.

1.4.1 Business/Policy Example: Global Product Launch or Sustainable Energy Policy

1.4.1.1 Scenario Overview

Picture a multinational company preparing to **launch a cutting-edge tech product** worldwide. Plans are laid out: R&D, marketing, distribution. Yet the moment you factor in **cultural nuances**, **data privacy laws**, and **environmental impacts**, complexity multiplies. Or envision a **government**

pioneering a sustainable energy policy – phasing out fossil fuels in favor of renewables. The potential for a greener future is immense but so is the risk of stakeholder misalignment, political backlash, or infrastructure shortfalls.

1.4.1.2 Applying the Complexity Compass

1. **Embrace Dynamic Uncertainty**
   - **Reality Check:** Decision-makers acknowledge that global consumer preferences and policy landscapes shift unpredictably. Scenario planning might foresee a strong demand in Asia, a lukewarm response in North America, or vice versa – plus sudden regulatory roadblocks. For sustainable energy, best- and worst-case futures account for swings in commodity prices or political changes that stall wind/solar expansion.
2. **Map the Interconnected System**
   - **Stakeholders:** Suppliers, local governments, logistic partners, end consumers, environmental watchdogs, and competitor brands.
   - **Values:** Data privacy (especially in the EU), brand prestige (certain Asian markets), cost-effectiveness (emerging economies), job creation (local communities).
   - **Pitfalls:**
     - *Cultural Missteps:* A marketing slogan that offends certain regions or languages.
     - *Legal Overlaps:* A well-intended feature might breach GDPR standards in Europe.
     - *Infrastructure Gaps:* Adopting wind energy in an industrial region lacking modernized grids or workforce training.
3. **Analyze for Emergence and Leverage**
   - **Archetypes:**
     - *Vicious Cycle:* A poorly localized product triggers negative reviews, eroding global brand equity.
     - *Tragedy of the Commons:* An overambitious green policy lacking grid capacity, triggering blackouts that reduce public faith in renewables.
   - **Tipping Points:**
     - *Social Media Backlash:* A single viral complaint about data misuse can torpedo consumer trust overnight.
     - *Competitor Disruption:* A rival firm releasing a similar product with one highly desired feature could abruptly shift consumer preferences, disrupting your market share.
   - **Leverage Points:**
     - *Localized Pilots:* Testing new features in a region known for strict privacy norms, ensuring compliance before global rollout.
     - *Policy Incentives:* Small tax breaks for businesses installing solar, seeding broader cultural acceptance of renewables.
4. **Experiment, Iterate, and Decentralize**
   - **Safe-to-Fail Pilots:** Trial the new product's AI-driven feature in one EU country under strict data laws; refine it pre-global launch. For clean energy, test wind farms in a few pilot towns, collecting real-time data on economic impact and community sentiment.
   - **Global Community Feedback**:
     - *Product Launch:* Engage local online forums for direct feedback.
     - *Energy Policy:* Host public consultations with environmental groups, local firms, and residents.
5. **Reflect, Evolve, and Envision**
   - **Continuous Updates:** Track how consumer sentiment evolves, re-map stakeholder positions if new regulations emerge, and pivot strategies as needed.
   - **Visioning:** Aim for a more adaptive, inclusive, and globally responsible approach – balancing profit, cultural respect, and environmental stewardship.

Outcome: Investing time in **complexity-based thinking** yields greater **public trust**, smoother adaptation to sudden changes, and a blueprint for ethical, long-term gains.

1.4.2 Personal Well-Being Example: Navigating Conflicting Health Advice

1.4.2.1 Scenario Overview

You want to **improve your physical and mental health**. Searching online reveals a bottomless pit of diet fads (keto? intermittent fasting?), workout regimes (CrossFit? yoga?), and conflicting mental health tips (meditation vs. medication debates). Add social media influencers touting quick fixes, and the swirl of complexity can feel paralyzing.

1.4.2.2 Applying the Complexity Compass

1. **Embrace Dynamic Uncertainty**
   - **Reality Check:** Acknowledge health science's evolving nature – what's championed today might change tomorrow. Plan for variability (like job stress or family duties) that could disrupt consistency.
2. **Map the Interconnected System**
   - **Stakeholders:** Your own body/mind, healthcare professionals, nutrition gurus, wellness influencers, support circles (friends/family).
   - **Values:** Sustainability (avoiding crash diets), mental balance, practicality (time or budget constraints).
   - **Flow Diagram:** Show how daily stress influences snack cravings, which alter mood, which then might disrupt workouts – a cycle that can spiral up or down.
3. **Analyze for Emergence and Leverage**
   - **Archetypes:**
     - *Shifting the Burden:* Leaning on quick dietary hacks or supplements instead of building core habits like consistent sleep or balanced meals.
     - *Vicious Cycle:* Stress undermines workouts → guilt sets in → more stress → further inactivity.
   - **Tipping Points:**
     - *Personal Milestone:* Achieving a 5k run or a certain weight-loss target can ignite a positive feedback loop – boosting confidence and motivation for continued progress.
   - **Leverage Points:**
     - Tiny "success triggers" (e.g., a daily ten-minute walk) that cascade into bigger health gains.
     - Partnering with a friend for mutual accountability, reinforcing positive loops.
4. **Experiment, Iterate, and Decentralize**
   - **Safe-to-Fail Health Challenges:** Try a two-week mindful eating or a one-week guided meditation challenge, tracking mood and energy. Adjust based on real outcomes.
   - **Feedback:** Tools like journaling, step counters, or a supportive friend group.
   - **Open-Source Input:** Post experiences in a health forum or Slack group to learn from others' trials and errors.
5. **Reflect, Evolve, and Envision**
   - **Regular Check-Ins:** Maybe once a week, you ask: "What worked? What didn't? Any new stressors or opportunities?" Document emerging questions, like "How do I handle plateau phases?"
   - **Map Updates:** Has your late-night snacking loop changed? Are new workout inspirations cropping up in your feed?
   - **Visioning:** Dream up a future "best self." Work backward to identify daily micro-steps that align with that healthier version of you.

Outcome: Rather than drowning in contradictory headlines, you forge a **personal health ecosystem** that's flexible, data-informed, and propelled by small wins – reflecting **complex** reality rather than a simplistic "one-size-fits-all" approach.

1.4.3 Societal Link Example: Personal Social Media Habits and Collective Impact

1.4.3.1 Scenario Overview

You casually **scroll** social media each day – liking a meme, sharing a news snippet – believing it's merely personal leisure. But multiply that behavior by millions, and you see how **viral movements** or extreme polarization can surge. This scenario underscores how **private actions** aggregate into **large-scale social dynamics**.

1.4.3.2 Applying the Complexity Compass

1. **Embrace Dynamic Uncertainty**
    - **Reality Check:** Recognize that your platform usage may evolve (e.g., ramping up during elections, shifting content preferences), while algorithms themselves change unpredictably, shaping what you see and share.
2. **Map the Interconnected System**
    - **Stakeholders:** Yourself, content creators, advertisers, platform algorithms, advocacy groups, political actors.
    - **Values:** Free expression, data privacy, mental well-being, factual integrity.
    - **Flow Diagram:** Illustrate how "likes" amplify certain posts, which shape trending topics, reinforcing user engagement. Meanwhile, data from your behavior informs future content curation.
3. **Analyze for Emergence and Leverage**
    - **Archetypes:**
        - *Reinforcing Loop:* Echo chambers – like-minded posts get more engagement, overshadowing dissenting viewpoints.
        - *Tragedy of the Commons:* Everyone chasing clickbait for quick attention, collectively degrading the quality of discourse.
    - **Tipping Points:**
        - *Viral Movement:* One trending hashtag can unite millions in days.
        - *Algorithmic Tweaks:* A new platform feature or policy can drastically reshape user behavior.
    - **Leverage Points:**
        - *Improving Online Discourse:* Supporting or creating tools that filter out misinformation, thus elevating content quality.
        - *Mindful Sharing:* Introducing personal guidelines for fact-checking before sharing sensational posts.
4. **Experiment, Iterate, and Decentralize**
    - **Safe-to-Fail Micro-Changes:** Limit social scrolling to 15 minutes daily, see how it affects your mood and worldview. Track if you're diversifying the voices you follow.
    - **Community Feedback:** Join small online groups that champion civil debate. Solicit feedback on your engagement style – are you accidentally fueling echo chambers or misinformation?
    - **Decentralized Tools:** Explore open-source or user-moderated platforms. A small but growing wave of communities aims for user-driven content governance, aligning with ethical standards.

5. **Reflect, Evolve, and Envision**
   - **Periodic Map Updates:** Reassess how your usage might be feeding certain trends or ignoring others. Jot down new questions like "Am I broadening or narrowing my worldview over time?"
   - **Visioning:** Develop personal or community guidelines for **responsible social media engagement** – verifying info, prioritizing meaningful interactions over passive scrolling, and seeking out diverse perspectives. These guidelines can unify local groups, influencing platform culture from the ground up.

Outcome: By acknowledging that your **personal scrolling** intersects with a **vast digital ecosystem**, you move from passive consumer to **agent of mindful change**, nudging social media's collective trajectory in a more constructive, ethically aware direction.

1.4.3.3 Reflections on Complexity in Practice

Across these examples – **global product launches**, **personal health journeys**, and **social media ecosystems** – you can see complexity is both **unavoidable** and **rich in possibilities**. The **Complexity Compass** helps transform multi-layered challenges into **structured explorations** and fosters **ethical reflection** each step of the way. Whether you're grappling with international legal frameworks, managing your own well-being, or participating in global online dialogues, **small actions** can ripple into profound impacts.

1.4.3.4 Techno-Philosophical Lens

Each story underscores how **tech** and **society** co-create new realities, and how **ethics** must steer these evolutions. By systematically identifying tipping points, harnessing leverage opportunities, and iterating with feedback, you embody the **spirit of techno-philosophy** – where mindful adaptation and moral foresight guide every innovation.

1.4.3.5 Bridging Forward

With these real-world illustrations, you've seen complexity **up close** – and how the **Complexity Compass** transcends abstract theory to drive **tangible transformations**. Next, we'll delve deeper into **ethical dimensions**, dissecting how proactive design and moral responsibility become integral when every decision can spark unintended consequences across networks. Finally, we'll solidify our chapter's conclusions, showing that **mastering complexity** is both a **practical skill** and an **ethical imperative** in a globally entwined era.

1.5 THE ETHICAL REFLECTION: PROACTIVE ETHICAL DESIGN

Complexity isn't only about **adaptive strategies** and **emergent phenomena**; it also compels us to address how **moral considerations** traverse sprawling networks of people, technologies, and ecosystems. In an age where tiny coding decisions can reshape entire cultures, or where a single supply-chain tweak might ripple across global communities, **ethics** isn't peripheral – it's **foundational**.

Building on our previous explorations, this section dives into **proactive ethical design**: a practice of **weaving moral foresight** into every stage of decision-making, from conceptualizing AI algorithms to structuring environmental policies and data governance.

1.5.1 The Moral Core of Complex Systems

1.5.1.1 Why Proactive Ethics?

Traditional "reactive" ethics often emerges after a crisis – a massive data breach, an ecological disaster, or a discrimination scandal. Yet in **hyper-connected** systems, crises can escalate at lightning speed, damaging trust, livelihoods, and ecosystems before we can respond. Proactive ethics means **anticipating** these crises and embedding **preventive measures** at the source.

The Shift

- **Reactive:** Scramble to contain ethical fires after public outrage (e.g., biased AI results, toxic spills).
- **Proactive:** Actively **scan** for potential ethical pitfalls (algorithmic discrimination, ecological harm, privacy violations) **before** they erupt.

Key Point: The more **complex** a system, the more urgent its moral vigilance. Even minor oversights can **amplify** existing societal biases or degrade shared resources – reverberating through entire networks.

1.5.1.2 Ethics in Action: Three Prime Examples

We'll revisit **AI bias**, **environmental consequences**, and **data privacy** to illustrate how proactively weaving ethics into system design can prevent unintended harm. Along the way, we'll spotlight **success stories** and **frameworks** that offer real inspiration.

1. **AI Bias: Reinforcing Societal Inequities if Left Unchecked**
   Context
   Modern AI systems – from recruitment algorithms to facial recognition – promise efficiency, yet can inadvertently **reflect and reinforce** existing societal biases. If the training data skews toward a particular group or historical pattern, the AI may inadvertently discriminate against minority communities, compounding real-world disparities.
   - **Complexity Factor:** AI algorithms operate within **broader ecosystems** – HR processes, policing tools, social media platforms. A biased outcome in one node can magnify cultural prejudices across multiple domains.
   - **Potential for Reinforcement:** Historical underrepresentation or prejudiced data sets lead to AI models that **overlook qualified job applicants**, mislabel certain demographic groups, or perpetuate stereotypes in content recommendations.
   - **Proactive Design**:
     - **Diverse Training Sets:** Employ inclusive data that captures varied demographics and contexts.
     - **Ethics Councils:** Gather interdisciplinary teams – technologists, ethicists, community reps – to regularly audit algorithmic outputs.
     - **Feedback Mechanisms:** Provide transparent channels where affected users can report bias, prompting swift iterative fixes.

Moral Imperative: AI shouldn't merely mirror society's inequalities – it should strive to **reduce** them. By anticipating **algorithmic** blind spots, teams can avert not just lawsuits or PR fiascos but also the harm inflicted on marginalized groups.

2. **Environmental Consequences: From Harmful Spillovers to Sustainable Innovation**
   Context
   From carbon-intensive supply chains to massive water usage in manufacturing, environmental consequences sprawl across **global networks**. A single corporation's pursuit of cheaper production can degrade ecosystems far upstream or harm local communities downstream.
   - **Complexity Factor:** Ecological systems run on **nonlinear feedback loops** – deforestation in one region can alter rainfall patterns in another, or a push for "green" technologies might inadvertently require heavy mining elsewhere.

- **Proactive Design**:
  - **Life-Cycle Analysis:** Evaluate every stage of production – raw material extraction, manufacturing, distribution, disposal – to reveal hidden hotspots of emissions or pollution.
  - **Multi-Stakeholder Governance:** Involve community leaders, environmental NGOs, and scientific experts early, creating frameworks for shared accountability.
  - **Environmental Early-Warning:** Sensor networks (Internet of Things, IoT) or satellite data can highlight shifts in real time (e.g., water quality drops or rising deforestation rates) to trigger rapid interventions.
- **Positive Model:**
  - **Interface Inc.** (a flooring company) famously adopted **"Mission Zero,"** aiming for zero environmental footprint. They redesigned supply chains, invested in closed-loop recycling, and cultivated partnerships with local communities – demonstrating that sustainable production can enhance brand loyalty and long-term profitability.

Moral Imperative: The planet's ecosystems are finite. By embracing **proactive environmental design**, organizations can **innovate** rather than exploit – securing a more resilient, equitable future that benefits both economies and communities.

3. **Data Privacy: Balancing Convenience and Human Dignity**
   Context
   User data has become the **lifeblood** of modern apps, from personalized suggestions to advanced analytics. However, collecting, storing, and monetizing that data raises potent questions about **intrusion**, **consent**, and potential misuse.
   - **Complexity Factor:** Data privacy is multi-stakeholder – users, developers, advertisers, regulators, and potential data brokers. A single leak or misuse can escalate into a global scandal, eroding faith in the entire tech ecosystem.
   - **Proactive Design:**:
     - **Privacy by Design:** Adopt minimal data collection from the start and embed privacy safeguards at each development phase, rather than tacking them on later.
     - **Transparent Consent:** Make it clear how data will be used and shared, using jargon-free language.
     - **Layered Security:** Develop adaptive encryption and regular vulnerability assessments to keep pace with evolving threats.
   - **Moral Imperative:** People should **own** their data and feel confident that technology respects their agency. Privacy by design fosters **digital dignity**, preserving user trust while supporting legitimate innovation.

1.5.2 Building Ethical Foresight Into the Complexity Compass

How do we weave these ethical concerns into the **five steps** of the Complexity Compass?

1. **Step 1: Embrace Dynamic Uncertainty … Ethically**
   - **Anticipate Ethical Pitfalls:**: Recognize you can't predict every moral dilemma – like AI bias surfacing post-launch or local communities resisting new wind farms.
   - **Scenario Planning for Ethical Dilemmas:** "What if our chatbot inadvertently discriminates based on language usage?" "What if a competitor leaks sensitive data about our energy operations?" Proactively scenario-plan these ethical hazards.
2. **Step 2: Map the Interconnected System … with Ethical Lenses**
   - **Include Moral Nodes:** Alongside standard stakeholder or resource mapping, identify "ethical friction points," such as potential user data exploitation or an at-risk ecosystem.

- **Values Overlay:** In your stakeholder diagram, layer in "AI fairness goals," "climate commitments," or "privacy thresholds."

3. **Step 3: Analyze for Emergence and Leverage … Watch for Ethical Tipping Points**
   - **System Archetypes and Morality:** "Tragedy of the Commons" signals ethical lapses in resource usage; "Shifting the Burden" might indicate reliance on short-term fixes (like ignoring bias until it's big news).
   - **Anticipate Ethical Tipping Points:** A seemingly minor oversight (e.g., a subtle AI bias) could balloon into a **major trust crisis** if it reaches critical public awareness. Spot these potential flashpoints early to act with moral urgency.
4. **Step 4: Experiment, Iterate, and Decentralize … with Moral Checks**
   - **Pilot Ethical Vignettes:** Before launching a new data feature or shipping a product globally, do small-scale "ethical stress tests" (like partial deployments or user consent pilots).
   - **Distribute Ethical Oversight:** Encourage local teams or committees to evaluate pilot outcomes for fairness, privacy, or ecological impacts. The broader the participation, the less likely major blind spots remain hidden.
5. **Step 5: Reflect, Evolve, and Envision … Morally**
   - **Ethical Retrospectives:** In addition to standard project reviews, hold "moral retros" – asking, "Did we harm or exclude anyone unintentionally? Did we degrade any ecosystem? Did new privacy concerns emerge?"
   - **Visioning:** Envision a future where your AI fosters social equity, your energy policy genuinely revives local communities, or your data practices set a global standard for user respect. Work backward to outline milestones.

1.5.3 Proactive vs. Reactive: The Ethical Edge

Reactive Ethics: The Aftermath of Neglect

- **Scandals** blow up – company scrambles to fix an AI fiasco or contain an oil spill.
- **Public Trust** plummets – users, communities, or investors lose faith.
- **Regulatory Crises** – governments might impose heavy fines or sanctions.

Proactive Ethics: A Virtuous Cycle

- **Early Warning**: Spot bias or ecological harm while it's still small, intervening before it spirals.
- **Stronger Brand Loyalty**: Stakeholders prefer organizations with proven moral foresight.
- **Smoother Adaptation**: Ethically aware teams adapt more gracefully to changing norms and regulations.

Practical Tips for Proactive Ethical Design

1. **Ethics Checklists and Toolkits**
   - Develop quick, question-based references. "Are we collecting more data than we truly need?" "Could this design advantage or exclude certain groups?"
2. **Ethics-by-Design Teams**
   - Assemble cross-functional groups – technologists, ethicists, local community reps – to embed moral considerations at every sprint or project milestone.
3. **Iterative Ethical Audits**
   - Don't rely on a one-time ethics review. Complexity evolves – conduct periodic check-ins to see if new laws, cultural shifts, or emergent outcomes require rethinking original designs.

4. **Public Impact Statements**
   - Publish accessible documents or blog posts outlining how you handle AI fairness, environmental footprints, or user privacy. Being forthright builds credibility and invites constructive feedback.

1.5.3.1 Techno-Philosophy in Practice

Recalling the **techno-philosophical** ethos from this book's Introduction, **technology** and **society** co-evolve, influencing each other's trajectory. Proactive ethical design ensures this evolution nurtures **human dignity**, **ecological sustainability**, and **fairness** – not just short-term gains. By embedding moral foresight into complex systems from the get-go, we honor the **co-creative** spirit of technology, forging innovation that **uplifts** communities rather than **exploiting** them.

1.5.3.2 Bridging to the Conclusion

Integrating ethics within the Complexity Compass is **not** optional in a hyper-connected age; it's a **responsibility** that shapes whether our solutions build trust or ignite backlash. With proactive ethical design – spanning AI bias prevention, environmental stewardship, and privacy by design – we orchestrate **holistic** outcomes that align with **societal** and **planetary** well-being.

Next, in our **Conclusion**, we'll tie together the chapter's insights – from nonlinear thinking to personal resilience, from global product launches to moral vigilance – and highlight the path forward. Mastering complexity is one half; **doing so ethically** is the vital other. Together, they encapsulate the synergy at the core of **techno-philosophy**: harnessing complex systems with **wisdom**, **responsibility**, and **creative courage**.

Key Takeaways: The Power of Proactive Ethical Design

1. **Ethical Foresight:** Anticipate moral hazards early (AI bias, ecological harm, data overreach) rather than reacting post-crisis.
2. **Societal Stakes:** Minor oversights in design can magnify existing inequalities or resource depletion, reinforcing the need for vigilance.
3. **Positive Models:** Learn from companies like **Interface Inc.** – which integrated environmental responsibility into every facet of production – and privacy frameworks like **"Privacy by Design"** – which bakes in user consent and data minimization from inception.
4. **Alignment with Techno-Philosophy:** We co-create technology and society, weaving a moral tapestry through each act of innovation. Proactive ethics ensures that tapestry remains inclusive, sustainable, and forward-looking.

By placing **proactive ethics** at the heart of complexity mastery, we extend our vantage beyond self-interest – becoming stewards of a more equitable, ecologically balanced, and human-centric future.

1.6 CONCLUSION: COMPLEXITY AS A CATALYST FOR PROGRESS

Complexity can overwhelm at first glance – like a dense forest of unknown paths. Yet, as this chapter has illustrated, it's also an **engine of innovation**, **creativity**, and **ethical progress** when approached with the right perspective and tools. Our exploration has shown how complexity is far from a design flaw ("bug"); it's a **dynamic "feature"** of our hyper-connected age that can spur breakthroughs, deepen empathy, and reveal hidden synergies. Let's recap our journey and preview the path ahead.

1.6.1 Complexity: A Feature, Not a Bug

Rather than lament the tangle of feedback loops, cultural differences, technological flux, and emergent patterns, we've come to see **complexity** as a **valuable asset** – a catalyst that compels us to think beyond the linear, to collaborate across domains, and to **innovate** in ways that rigid systems never could.

Positive Innovation Through Complexity

1. **A Tech Platform's Network Effects**
   - **Example:** Consider how **Airbnb** harnessed the power of interconnected users – travelers and hosts – to create a global hospitality network. By embracing diverse locales, trust mechanisms, and user feedback loops, it transformed an initially fringe idea (sharing one's home with strangers) into a mainstream solution that redefined travel.
   - **Lesson**: Tapping into complexity (cultural variety, localized trust, tech-driven user interactions) can spark a platform that revolutionizes an entire industry.
2. **Leveraging Ecosystems for Environmental Solutions**
   - **Example: Wetland restoration projects** sometimes bring better flood control than purely engineered barriers. By recognizing wetlands' complex interactions (water filtration, wildlife support, soil retention), conservationists have found natural buffers that protect urban areas, save costs, and preserve biodiversity.
   - **Lesson:** Complexity across ecosystems can create synergy, yielding **multi-layered benefits** – environmental stability, community safety, and healthy wildlife habitats.
3. **Social Movements Grown via Interconnected Networks**
   - **Example:** Grassroots movements like the **#MeToo** campaign leveraged social media's complex user networks to connect personal stories across continents, morphing isolated experiences into a global outcry against harassment and assault.
   - **Lesson:** The overlapping nodes of social platforms, shared narratives, and widespread empathy ignited a collective force that reshaped cultural conversations, corporate policies, and legal debates.

Key Insight: Embracing rather than evading complexity nurtures **creativity**, fosters **resilience**, and unlocks **new collaborations** – turning confusion into possibility and friction into spark.

1.6.2 Linear Logic's Shortcomings in an Intertwined World

We began by dissecting why **"if A, then B"** thinking falters amid **multi-layered** feedback loops. Whether navigating global supply chain chaos or personal health intricacies, **linear assumptions** overlook how **unintended consequences** can escalate when minor shifts echo through connected networks.

Unintended Consequences in Practice

- **Over-Simplified Policies** can lead to social or ecological damage.
- **Personal Relationship Strategies** that discount emotional nuance often backfire, causing deeper conflicts.
- **Tech Deployments** ignoring cultural variations can spark user backlash or regulatory clampdowns.

Key Lesson: By admitting the limits of linear logic, we open ourselves to **systems-aware** thinking, acknowledging the hidden nodes, emergent behaviors, and cultural or emotional factors that shape our outcomes.

1.6.3 The Complexity Compass: A Practical Roadmap

In response to complexity's demands, we introduced a **five-step Complexity Compass**, blending **analysis**, **iteration**, **reflection**, and **ethical** vigilance:

1. **Embrace Dynamic Uncertainty**
   - Recognize that **surprises** are inevitable. Plan multiple scenarios rather than pinning hopes on one forecast.

2. **Map the Interconnected System**
   - Visualize stakeholders, data flows, and value conflicts, ensuring we see beyond obvious links.
3. **Analyze for Emergence and Leverage**
   - Identify system archetypes, look for tipping points (where small nudges cause big effects), and anticipate ethical tipping points that might trigger a major crisis of trust.
4. **Experiment, Iterate, and Decentralize**
   - Conduct safe-to-fail pilots, gather real-time feedback, pivot as needed, and distribute decision-making to harness diverse expertise.
5. **Reflect, Evolve, and Envision**
   - Update maps, log successes and new **questions**, and co-create future visions that align with moral and strategic aims.

By cycling through these steps, we continually **learn** from emergent patterns, adapt swiftly, and keep our core **values** in focus, turning complexity into a wellspring of **growth** rather than a source of paralysis.

1.6.4 Ethics at the Heart of Complexity

Complex systems are **never morally neutral**. As we saw with **AI bias**, **environmental** chain reactions, and **data privacy** concerns, even small oversights can **magnify** societal inequities or ecological harm. Hence, **ethics** must remain **a central thread**:

- **AI**: Minimizing algorithmic discrimination from the get-go, ensuring feedback loops correct rather than cement biases.
- **Environmental:** Designing supply chains and policies that actively **prevent** or reduce ecological damage, rather than retroactively patching crises.
- **Data Privacy:** Embracing **"privacy by design"** so user trust is upheld, not undercut, by new technological capabilities.

Key Principle: Complex systems reward moral vigilance – **proactive** ethics integrated at each stage wards off scandal and fosters **long-term** societal well-being.

1.6.5 A Glimpse Ahead: Modern Paradoxes, Psychological Effects, and More

1.6.5.1 From Macro to Micro

We'll now zoom into how **complexity** manifests in **human-level** and **societal** spheres – untangling the **modern paradoxes** that pit individual aspirations against collective interests, exploring how **cognitive biases** warp decision-making, and confronting the **Nocebo effect**, where negative expectations can shape real outcomes.

1. **Modern Paradoxes (Chapter 2)**
   - Delve into cultural contradictions – like fervent individualism clashing with communal well-being. We'll see how complexity weaves these threads, forging dilemmas that can't be resolved by the linear "one-size-fits-all" logic.
2. **Psychological Frontiers**
   - **Cognitive Biases:** We'll spotlight how complexity seeps into our mental shortcuts, fueling illusions and misjudgments that can hamper progress.
   - **Nocebo Effect:** Beyond mere placebo, negative mindsets can stoke real harm – this chapter will offer techniques to mitigate its influence, applying the complexity lens to personal resilience and social contagion of negativity.

Bridge: These next chapters emphasize the **internal** and **social** tapestry of complexity, showing that the same multi-layered interplay we see in global supply chains also shapes our thoughts, behaviors, and cultural discourses.

1.6.6 The Road Forward: Transforming Complexity Into Progress

As we close this pivotal chapter, remember:

1. **Complexity is a Feature**
   - It drives **novel thinking**, compels cross-domain partnerships, and fosters **emergent solutions** that simpler frameworks would never discover.
2. **Linear Logic's Shortcomings**
   - When everything is intertwined, over-simplification leads to **unintended consequences**. Recognizing these limits is a crucial step toward holistic solutions.
3. **The Complexity Compass**
   - A step-by-step, **iterative** roadmap – embrace uncertainty, map systems, spot leverage points, run safe-to-fail experiments, and reflect continually.
4. **Ethics Must Steer Our Actions**
   - In a world where even a minor design choice can catalyze broad impacts, **ethical vigilance** protects both dignity and sustainability, ensuring complexity becomes a **force for good** rather than an ethical minefield.

1.6.6.1 Personal Agency

While complexity may be inherent to our modern world, **our response** to it isn't predetermined. With the right mindset and tools, each of us can become more skilled at **navigating and shaping** complex systems – turning potential chaos into avenues for growth, collaboration, and ethical stewardship.

1.6.6.2 Call to Action

Finally, recall our **techno-philosophical** invitation from the Introduction: to be **active participants** in shaping technology, culture, and policies that reflect our best values. The Complexity Compass, moral reflection, and an adaptive attitude aren't just theoretical principles; they're **everyday choices** we make – to innovate responsibly, to adapt ethically, and to help steer the future toward shared well-being.

With the **foundation** laid, we now journey into the **intricate paradoxes** and **psychological** underpinnings that define much of modern life – exploring how complexity plays out within our minds, relationships, and cultural frameworks. Let's proceed, fully equipped to **see** complexity's hidden threads and **act** with clarity, compassion, and creative resolve.

2 Modern Paradoxes
From Collective Bonds to Individual Fulfillment

2.1 INTRODUCTION: THE PARADOX OF MODERN LIFE

2.1.1 Opening

Imagine scrolling through a social media feed, immersed in images of achievement and success. Posts showcase perfectly curated lifestyles, motivational quotes, and self-actualization journeys. At first glance, these snapshots seem to celebrate abundance, connection, and individuality. But beneath this surface lies an unsettling paradox: rather than feeling inspired, you sense a growing dissatisfaction.

This tension is not unique to the digital age. It reflects a larger phenomenon – one where abundance often breeds insecurity, connection fosters isolation, and freedom becomes paralyzing. These contradictions are not just individual experiences but symptoms of a deeper societal transformation. They reveal the cost of living at the pinnacle of human progress.

2.1.2 Invitation to Explore a Fresh Perspective

This chapter brings together timeless concepts – drawn from history, psychology, and philosophy – to unravel this paradox and illuminate its causes. It reinterprets familiar ideas, like **Maslow's hierarchy of needs**, **Marcuse's critique of repressive tolerance**, and **Sapolsky's research on stress dynamics**, and reframes them within the context of modern challenges.

Through this synthesis, we reveal how:

- **Historical Shifts:** The evolution from collective survival to individual fulfillment has introduced profound new stressors.
- **System Absorption:** Modern systems maintain power not through oppression but by neutralizing resistance, turning activism into marketable trends and choice into a tool for control.
- **Psychological Stress:** The abundance of modern life amplifies pressures on identity, meaning, and relationships, leaving individuals more anxious and fragmented.

By integrating these ideas, we create a fresh perspective on the complexities of contemporary life, offering tools to understand and address the hidden forces shaping our world.

2.1.3 Preview of Core Themes

This chapter builds on three central pillars:

1. **The Historical Shift from Collective Bonds to Individual Fulfillment**:
   - How humanity moved from survival-based communities to a culture of self-expression, driven by industrialization and digitization.
   - Why this shift, while empowering, left individuals grappling with new forms of stress and isolation.

DOI: 10.1201/9781003640288-4

2. **System Absorption as Modern Power**:
   - How dissent and resistance are neutralized through commodification, from activism being transformed into products to digital platforms profiting from protest.
   - Why this absorption deepens the psychological and societal paradoxes of modern life.
3. **Curious Skepticism as a Solution**:
   - A practical mindset that balances openness with doubt, enabling individuals to question societal narratives, resist systemic loops, and reclaim agency.
   - How Curious Skepticism offers a path to authentic relationships, meaningful resistance, and personal fulfillment.

2.1.4 Connection to Maslow's Hierarchy

At its core, this paradox can be understood through **Maslow's hierarchy of needs**, one of the most enduring psychological frameworks. While traditional societies focused on meeting foundational needs – safety, survival, and belonging – modern societies operate at the **upper levels** of the hierarchy, striving for esteem, identity, and self-actualization.

However, these higher pursuits are fraught with new vulnerabilities:

- **Abundance Creates Pressure:** The drive for self-actualization often becomes a competition, where individuals are measured against idealized standards.
- **Connection Feels Hollow:** In a world saturated with digital communication, relationships are often transactional and surface levels, leaving individuals yearning for deeper bonds.
- **Choice Overwhelms:** Unlimited options lead to decision fatigue, existential doubt, and a fear of making the wrong choice.

Maslow's framework illuminates the paradox of progress: as societies achieve material success, they confront psychological and existential challenges that demand new ways of thinking and being. This chapter explores these challenges and introduces **Curious Skepticism** as a mindset to navigate them.

2.1.5 A New Path Forward

While these challenges may feel insurmountable, they are not. By revisiting old ideas through a modern lens, we can uncover tools for understanding and transforming our experience. This chapter invites readers to engage with these tensions – not to resolve them entirely but to learn how to thrive within them. With **Curious Skepticism** as a guide, we'll explore how to reclaim agency, rebuild connection, and foster a sense of purpose in a world defined by paradox.

UNDERSTANDING PROXIES

THINKING BOX:

"Are you chasing the real thing or just its pixelated shadow? Perhaps the truth – and the *answer* – are closer to **42** than we realize."

2.2 HISTORICAL CONTEXT: THE GREAT SHIFT

Modern society's paradoxes of abundance, identity, and isolation are not unprecedented. Throughout history, civilizations have experienced similar transitions as they reached their zenith. From survival-based interdependence to the rise of hyper-individualism, this chapter examines how societal evolution introduces both unparalleled opportunities and unforeseen vulnerabilities.

2.2.1 From Collective to Individual Focus

2.2.1.1 Traditional Societies: Survival-Based Bonds Fostered Interdependence

In traditional societies, survival was a collective endeavor. Individuals depended on their communities for protection, resources, and identity. These bonds were not optional – they were essential for the group's survival and the individual's place within it.

- **Mutual Dependence:** Early tribal and agrarian societies were structured around shared responsibilities. Food production, defense, and caregiving were communal efforts, ensuring survival in often harsh environments.
- **Social Stability:** Defined roles – whether as a hunter, gatherer, or spiritual leader – minimized uncertainty and existential doubt. One's purpose was clear, and meaning was embedded in daily life.
- **Limitations:** However, this structure prioritized stability over individual aspirations. Conformity was often enforced, leaving little room for personal expression or autonomy.

This model of interdependence formed the foundation for early human civilizations, but as societies grew more prosperous, new dynamics began to emerge.

2.2.1.3 The Classical Civilizations: Prosperity and Its Discontents

As civilizations like Greece, Rome, and Egypt reached their height, their prosperity introduced a shift away from collective survival toward individual fulfillment. This transformation, while empowering, often carried the seeds of fragmentation and decline.

a. **Ancient Greece: The Rise of Individualism through Democracy and Philosophy**
  - **Philosophical Individualism:** In the Golden Age of Athens, philosophers like Socrates, Plato, and Aristotle championed the value of individual thought and self-reflection. Their teachings encouraged citizens to question traditional norms and pursue personal virtue and knowledge.
  - **Democracy's Double-Edged Sword:** Athens introduced participatory governance, emphasizing the individual's role in decision-making. However, as personal freedoms expanded, so did tensions between collective stability and individual ambition.
  - **Cultural Fragmentation:** The shift from shared mythos to personal intellectual pursuits fractured communal cohesion. Competing philosophies and political factions eventually weakened the unity that had sustained Athens' ascendancy.

b. **Ancient Rome: Wealth and the Erosion of Collective Purpose**
  - **Economic Abundance:** The Roman Empire, at its height, controlled vast resources, enabling unprecedented material wealth for its elite citizens.
  - **Urban Individualism:** Rome's cities became centers of luxury and personal ambition, where individual success often overshadowed communal welfare.
  - **Bread and Circuses:** As civic engagement waned, the government sought to pacify the populace with entertainment and material handouts, substituting collective purpose with superficial distractions.
  - **Crisis of Identity:** The rise of personal luxury and political corruption eroded the shared values that had once bound the Republic, hastening its decline into instability and autocracy.

c. **Ancient Egypt: Stability Undermined by Hyper-Individualism**
  - **Theocratic Identity:** For much of its history, Egypt thrived on a theocratic model that united its people under shared religious beliefs and pharaonic rule.
  - **Economic Centralization:** As Egypt reached the peak of its prosperity during the New Kingdom, wealth became concentrated among the elite, creating stark divisions.
  - **Fractured Stability:** The rise of personal power among regional governors and priests undermined the central authority of the Pharaoh, leading to periods of disunity and collapse.

2.2.1.3 Lessons from Antiquity

In each of these cases, the shift from collective survival to individual focus introduced prosperity but also sowed division. Philosophical advancements, wealth, and personal freedoms expanded, but they often came at the cost of communal stability, shared identity, and collective resilience. These dynamics echo the challenges faced by modern societies navigating the paradoxes of abundance and individuality.

2.2.1.4 Industrial Revolution: Economic Shifts Prioritized Individual Labor and Success

The Industrial Revolution marked another pivotal moment in this historical shift. Mechanization, urbanization, and the rise of market economies transformed societies in ways that further emphasized individualism over collective interdependence.

- **Economic Independence:** Wage labor replaced communal farming, enabling individuals to pursue self-sufficiency. This economic model redefined success as personal achievement rather than communal contribution.
- **Urbanization and Isolation:** The migration to cities fractured traditional support networks, leaving individuals to navigate new social structures alone.
- **Materialism and Competition:** The industrial economy celebrated productivity, wealth, and personal ambition, fostering a culture of competition that prioritized individual gain over collective welfare.

While the Industrial Revolution offered new opportunities, it also intensified the psychological and social pressures of individualism, setting the stage for further fragmentation.

2.2.1.5 Digital Age: Fragmentation of Traditional Communities into Hyper-Personalized Networks

The digital revolution has taken individualism to its zenith, reshaping how people connect, communicate, and define their identities.

- **Hyper-Personalization:** Algorithms curate content tailored to individual preferences, creating isolated digital echo chambers.
  - Example: Social media feeds show different realities to different users, eroding shared cultural experiences and collective understanding.
- **Virtual Communities:** While online platforms offer global connections, they often lack the depth and accountability of face-to-face interactions.
  - Example: Relationships mediated through apps are often transactional, fostering surface-level interactions rather than deep bonds.
- **Loss of Local Communities:** Traditional neighborhoods and institutions that once anchored social life have been replaced by transient, virtual networks. These networks, while vast, rarely provide the stability or intimacy of place-based communities.

The digital age epitomizes the promise and perils of individualism, offering unparalleled freedom and personalization but amplifying isolation, fragmentation, and existential doubt.

From the interdependence of traditional societies to the hyper-individualism of the digital age, humanity's journey reveals a recurring pattern. Ancient civilizations like Greece, Rome, and Egypt faced similar challenges when prosperity introduced individual ambition at the expense of collective cohesion. The Industrial Revolution accelerated these trends, and the digital age has amplified them to unprecedented levels.

As societies climb **Maslow's hierarchy**, seeking self-expression and identity, they must also confront the vulnerabilities that come with prosperity: insecurity, fragmentation, and the loss of shared purpose. Understanding these historical patterns provides essential context for navigating the paradoxes of modern life.

2.2.2 Maslow's Hierarchy: Operating at the Peak

The story of human progress can often be mapped against **Maslow's hierarchy of needs**, a framework that organizes human motivations into tiers, from basic survival to self-actualization. While early societies focused on satisfying fundamental needs like safety and sustenance, modern societies have climbed to the upper levels, prioritizing identity, self-expression, and personal fulfillment. This progression reflects a monumental achievement but also reveals profound new vulnerabilities.

Operating at the peak of the hierarchy introduces challenges that earlier societies never faced: abundance creates anxiety, freedom leads to paralysis, and the pursuit of self-actualization often comes at the cost of shared purpose. These dynamics illuminate why modern life, despite its material comforts, can feel so deeply unsatisfying.

2.2.2.1 Societal Ascension: From Survival to Self-Actualization

For much of human history, societies operated at the lower levels of Maslow's hierarchy. Survival was a collective endeavor, with communities pooling resources to ensure safety and sustenance. Over time, technological advancements and economic growth enabled societies to transcend these foundational needs, climbing toward higher-order aspirations like esteem and self-actualization.

- **Material Abundance Redefines Success:**
  - With the advent of agriculture, industry, and globalization, material scarcity gave way to abundance for many populations. Food security, economic stability, and access to education became widespread in wealthier societies, allowing individuals to shift their focus from survival to personal fulfillment.
  - **Example:** The 20th century saw a massive expansion of the middle class in Western societies, where families began prioritizing higher education, leisure, and self-improvement as markers of success.
- **The New Emphasis on Identity and Individuality:**
  - Freed from the immediate pressures of survival, societies began celebrating individual achievements and personal growth. Cultural narratives shifted, framing success as a uniquely personal journey rather than a collective accomplishment.
  - **Example:** Social media platforms epitomize this focus, offering spaces for individuals to showcase their identities and milestones, from career successes to personal wellness goals.
- **A Privilege with Hidden Costs:**
  - While this societal ascension reflects extraordinary progress, it also introduces new pressures. The higher levels of Maslow's hierarchy demand not just material success but the creation of meaning, purpose, and identity – tasks that are uniquely complex and deeply personal.

2.2.2.2 New Psychological Challenges: The Weight of Abundance

The abundance and freedoms of modern life, while liberating, also place unprecedented demands on individuals. As people operate at the peak of Maslow's hierarchy, they face unique psychological challenges that stem from having too much rather than too little.

- **Anxiety in an Age of Plenty:**
  - Paradoxically, abundance can amplify insecurity. When resources and opportunities are seemingly endless, individuals feel pressure to make the "right" choices, maximize every opportunity, and constantly improve.

  - **Example:** The concept of *hustle culture*, where individuals are encouraged to monetize every hobby and turn every moment into productivity, reflects this anxiety. The message is clear: if you aren't thriving, you're falling behind.
- **The Paralysis of Choice**:
  - With more options comes greater responsibility. Studies have shown that too many choices can overwhelm decision-making processes, leading to dissatisfaction or avoidance.
  - **Example:** Streaming services offer thousands of titles, yet users often spend more time scrolling than watching, paralyzed by the fear of choosing poorly. This same paralysis extends to life decisions, such as career paths, relationships, or even core values.
  - On a deeper level, choice overload can erode confidence, leaving individuals second-guessing their decisions long after they are made.
- **The Search for Meaning in a Fragmented World**:
  - At the peak of Maslow's hierarchy, individuals are tasked with creating meaning and purpose in their lives – an inherently subjective and open-ended process. This can lead to existential dissatisfaction, especially when external markers of success fail to fulfill deeper needs.
  - **Example:** Even those who achieve material wealth and professional success often report feelings of emptiness, grappling with questions like, "What's next?" or "Is this all there is?"

These challenges highlight the hidden complexities of abundance: while earlier societies grappled with scarcity, modern individuals must navigate a landscape of overwhelming choice, relentless pressure, and elusive fulfillment.

2.2.2.3 The Cost of Prosperity: Losing Collective Purpose

As societies ascend Maslow's hierarchy, they inevitably shift their focus from shared survival to personal fulfillment. While this shift empowers individuals, it often comes at the expense of collective purpose and community cohesion.

- **Fragmentation of Shared Identity**:
  - Traditional societies derived strength from shared narratives, whether through religious beliefs, cultural traditions, or collective struggles. In modern societies, these unifying forces have been replaced by fragmented, individualized experiences.
  - **Example:** In the past, institutions like churches or town halls provided a sense of belonging and accountability. Today, virtual communities and transient networks offer convenience but rarely provide the same depth of connection.
- **The Rise of Hyper-Individualism**:
  - As personal identity becomes the primary focus, individuals often disengage from communal activities. This shift weakens the bonds that tie communities together, replacing collective goals with personal ambitions.
  - **Example:** Civic organizations and neighborhood associations, once vibrant hubs of community life, have experienced steady declines as people prioritize self-oriented pursuits.
- **The Disconnection from Meaningful Work**:
  - In traditional societies, work was directly tied to survival and community, offering a sense of purpose. Modern economies, by contrast, emphasize productivity, profit, and efficiency, often at the expense of fulfillment.
  - **Example:** Many workers in affluent societies feel disengaged, viewing their jobs as a means to an end rather than a source of meaning or pride.

This loss of collective purpose reflects a broader societal trend: as individuals climb Maslow's hierarchy, they gain autonomy but lose the stability and meaning that come from shared goals and identities. The cost of prosperity is not just psychological but cultural – a fragmentation of the social fabric that once anchored human experience.

Operating at the peak of Maslow's hierarchy is both a remarkable privilege and a profound challenge. As societies achieve material abundance and focus on personal fulfillment, they encounter unique vulnerabilities: anxiety from abundance, paralysis from freedom, and dissatisfaction from unfulfilled existential needs. These pressures, compounded by the erosion of collective purpose, reveal the hidden costs of prosperity.

Understanding these dynamics is essential for addressing the paradoxes of modern life, where abundance and freedom coexist with insecurity and fragmentation. By recognizing these patterns, we can begin to chart a path toward balance, resilience, and meaningful connection.

DISCIPLINE VS. MOTIVATION

THINKING BOX:

"What fuels you once the initial thrill fizzles? Maybe the lasting spark – like the universe's biggest questions – lurks around **42**."

2.3 SYSTEM ABSORPTION AND MODERN POWER

The power structures of modern societies have evolved beyond overt oppression and force. Instead, they maintain control through a subtler mechanism: **system absorption**. By integrating resistance into the very systems it seeks to oppose, these structures neutralize dissent, ensuring stability while appearing progressive. This phenomenon, described by philosopher **Herbert Marcuse** as **repressive tolerance**, reveals how opposition is co-opted and commodified, transforming radical movements into marketable trends that reinforce the status quo.

2.3.1 Repressive Tolerance in Action

Marcuse's concept of repressive tolerance critiques how liberal societies tolerate dissent not to encourage change but to dilute its impact. This tolerance absorbs and commodifies resistance, turning potential challenges to the system into harmless, even profitable, activities. By allowing dissent to exist within controlled boundaries, systems ensure their continued dominance under the guise of progress and openness.

2.3.1.1 The Mechanics of Repressive Tolerance

At its core, repressive tolerance works by converting meaningful resistance into forms that align with the system's economic and cultural norms.

- **Neutralizing Opposition**:
  - Radical ideas and movements, instead of being suppressed outright, are welcomed into the mainstream. However, they are stripped of their transformative potential, reduced to symbols, slogans, or products.
  - **Example:** Protests are celebrated as a form of free speech but rarely lead to substantive change, as systemic structures remain unchallenged.
- **The Illusion of Progress**:
  - By co-opting dissent, systems create the illusion of inclusivity and progress, portraying themselves as adaptive and tolerant. This not only pacifies critics but also reinforces the idea that meaningful change is unnecessary.
  - **Example:** Corporations adopting diversity initiatives may appear progressive while avoiding deeper changes, such as addressing structural inequalities in their hiring or pay practices.

2.3.1.2 Greenwashing: The Commodification of Environmental Resistance

One of the most prominent examples of repressive tolerance is **greenwashing**: the practice of presenting environmentally harmful actions or products as eco-friendly to align with the growing demand for sustainability.

- **How Greenwashing Works**:
  - Companies leverage the language of environmentalism to market their products, adopting superficial measures like recyclable packaging or eco-themed advertising campaigns. These efforts often obscure their larger, unsustainable practices.
  - **Example:** Major oil companies run advertisements highlighting investments in renewable energy, even as the vast majority of their operations remain focused on fossil fuels.
- **Neutralizing Environmental Activism**:
  - By co-opting the rhetoric of environmental movements, corporations redirect public pressure into consumer behavior, such as buying "green" products, rather than systemic action, like regulating industries or adopting renewable energy mandates.
  - **Example:** Fast fashion brands promote "sustainable" clothing lines while continuing to exploit low-wage labor and produce waste at unsustainable rates.
- **Impact on Public Perception**:
  - Greenwashing shifts responsibility for environmental action from corporations and governments to individuals, framing problems like climate change as issues that can be solved through consumer choices rather than systemic reform.

2.3.1.3 Marketization of Activism: From Movements to Merchandise

Social and political movements are similarly commodified, their radical potential diluted as they are absorbed into the market.

- **Turning Resistance into Fashion:**
  - Activist slogans and symbols, once tools for rallying collective action, are repackaged as consumer goods. T-shirts, tote bags, and social media filters bearing messages of resistance become fashion statements rather than calls to action.
  - **Example:** The #MeToo movement, which began as a grassroots effort to address systemic abuse, has been co-opted by brands selling empowerment-themed merchandise without contributing to substantive gender equality initiatives.
- **Outrage as a Revenue Stream**:
  - Platforms like social media profit from amplifying outrage and polarizing discourse, turning activism into a source of engagement rather than systemic change.
  - **Example:** Viral hashtags and protests often generate clicks, shares, and ad revenue for platforms, but the energy they generate is rarely channeled into meaningful political or economic reforms.

Marcuse's framework of repressive tolerance illuminates how modern systems maintain power by neutralizing dissent through commodification. Whether through greenwashing campaigns that dilute environmental resistance or the marketization of activism that turns radical movements into consumable trends, these mechanisms transform opposition into tools for reinforcing the status quo.

This dynamic creates an illusion of progress while ensuring systemic stability, leaving individuals to grapple with a profound question: How can true resistance survive in a world where even dissent is absorbed into the system?

2.3.2 Digital Absorption Mechanisms

The digital age has redefined how resistance, individuality, and authenticity are expressed and absorbed. While digital platforms promise empowerment, connection, and visibility, they also operate as powerful tools of systemic absorption. By converting dissent into content and individuality into marketable commodities, these systems ensure stability while creating the illusion of progress.

2.3.2.1 Social Media as a Tool of Absorption

Social media platforms play a dual role in modern life. On one hand, they enable individuals to share their voices and organize around shared causes. On the other hand, they often neutralize the transformative potential of these efforts by reframing them as content cycles that prioritize visibility over systemic change.

- **Viral Campaigns and the Metrics Trap:**
  - Social media excels at amplifying issues through viral campaigns, bringing visibility to topics that might otherwise be overlooked. However, this visibility often becomes an end in itself, confining action within the platform. Engagement metrics – likes, shares, and views – are celebrated as signs of success, even if no structural changes occur.
  - **Example (Hypothetical):** An online campaign draws attention to an environmental issue. The campaign generates millions of impressions, but the energy dissipates in digital spaces, with few tangible actions taken offline.
- **Outrage as a Revenue Driver:**
  - Social media algorithms are designed to amplify emotionally charged content, such as outrage or indignation. While this content drives engagement, it often polarizes conversations and trivializes complex issues, reducing nuanced discussions to brief moments of viral attention.
  - **Example (Hypothetical):** A post highlighting social inequality sparks widespread debate online, but the algorithms prioritize sensational responses over thoughtful discourse, leaving users emotionally exhausted rather than mobilized.

By converting activism into cycles of visibility and engagement, social media platforms profit from the energy of dissent without enabling the systemic reforms those efforts often seek.

2.3.2.2 Commodification of Identity

Digital platforms also absorb individuality, transforming self-expression into a marketable product. In this process, personal authenticity is repackaged as performance, aligning with algorithms and market trends rather than genuine self-discovery.

- **Authenticity as Performance:**
  - Platforms incentivize users to present curated versions of themselves, often prioritizing visibility, relatability, and adherence to cultural norms. Personal stories, once deeply intimate, are reframed as public content optimized for likes and shares.
  - **Example (Hypothetical):** An individual shares a deeply personal experience online, but the story's reach and reception are shaped more by platform algorithms than by the authenticity of the message.
- **Individuality as a Product:**
  - Beyond social media, other digital platforms, such as streaming services or e-commerce sites, capitalize on the idea of individuality by offering hyper-personalized experiences. While these experiences feel tailored, they often reinforce systemic patterns of consumption.
  - **Example (Hypothetical):** A streaming service creates a playlist based on a user's listening habits, presenting it as a reflection of their unique identity. However, the service's goal is not to celebrate individuality but to increase screen time and loyalty to the platform.

Through commodification, digital systems turn personal authenticity into a tool for profit, aligning individuality with market priorities rather than self-directed expression.

2.3.2.3 Beyond Social Media: The Broader Digital Landscape

The mechanisms of digital absorption are not limited to social media. Across the digital ecosystem, systems are designed to co-opt individuality, activism, and even resistance into profitable frameworks.

- **Activism as a Consumer Choice:**
  - Digital platforms and marketplaces often frame activism as a matter of consumer behavior. Products and services are marketed as symbols of resistance, shifting responsibility for systemic change onto individual purchasing decisions.
  - **Example (Hypothetical):** An online retailer promotes a product line labeled "sustainable," encouraging consumers to align their values with their purchases, while broader systemic issues remain unaddressed.
- **Subscription Models and Identity:**
  - Streaming and e-commerce platforms reinforce systemic dependencies through subscription models that present consumption as a core aspect of modern identity.
  - **Example (Hypothetical):** A subscription service markets itself as essential for maintaining a modern lifestyle, positioning continual consumption as a reflection of individuality and success.

The broader digital ecosystem reinforces systemic power by turning even resistance into another product category, ensuring the stability of the structures it claims to challenge.

Digital absorption mechanisms illustrate how modern systems neutralize dissent and commodify individuality across multiple domains. Social media amplifies visibility but confines action within cycles of engagement, while other digital platforms transform self-expression and resistance into profitable behaviors. These mechanisms, while subtle, ensure systemic stability under the guise of empowerment and progress.

The challenge is clear: to reclaim agency, individuals must recognize how these systems operate and create spaces for authentic expression, connection, and action outside the boundaries of commodification.

2.3.3 The Paradox of Choice as Control

Choice is often celebrated as the ultimate expression of freedom. The modern world offers an abundance of options, from the trivial – what to watch on a streaming service – to the profound – how to define one's identity or career path. Yet, this abundance often introduces an unexpected burden. Rather than feeling empowered, individuals face decision fatigue, second-guessing, and a gnawing sense of dissatisfaction. Systems leverage this paradox, using the overwhelming nature of choice to keep individuals focused on personal decisions rather than systemic change.

2.3.3.1 Abundance Leads to Paralysis

While choice is framed as a hallmark of freedom, an excess of options can lead to cognitive overload and emotional exhaustion. This phenomenon is particularly evident in a society saturated with products, opportunities, and pathways, all competing for attention and validation.

- **The Psychological Weight of Too Many Options:**
  - When faced with countless possibilities, individuals often find themselves trapped in cycles of hesitation or regret. The effort required to evaluate and compare options can feel paralyzing, leaving people unsure whether they've made the "right" choice.
  - **Example (Relatable):** A person searching for a new pair of headphones browses through hundreds of online reviews, specifications, and comparison tools. Hours later,

they abandon the purchase altogether, overwhelmed by conflicting information and the fear of buying something inferior.

- **Freedom Turned into Pressure:**
  - Abundance shifts the focus from "What do I need?" to "What's the absolute best I can get?" This creates a psychological burden, where every decision feels like a referendum on one's competence or values.
  - **Life Choices Amplified:** Beyond consumer decisions, the abundance of modern life extends to existential choices – career paths, personal goals, and even self-perception. The weight of choosing "correctly" can lead to stagnation or chronic dissatisfaction.
  - **Example (Relatable):** A student with access to a wide range of career options struggles to decide on a major, constantly worrying that their choice might limit future opportunities.
- **Systemic Implications:**
  - By presenting a seemingly endless array of choices, systems ensure individuals remain preoccupied with personal decisions rather than focusing on collective issues. The more people are trapped in the minutiae of their options, the less energy they have to challenge the systems offering those choices.

2.3.3.2 Consumer Choice as Political Action

Modern systems have reframed consumption as a form of activism, encouraging individuals to "vote with their wallets" to effect change. This narrative, while appealing, often shifts responsibility for systemic problems onto the individual, allowing larger structures to remain unchallenged.

- **The Illusion of Impact:**
  - The idea of aligning purchases with values – buying sustainable products, supporting ethical brands – suggests that individual choices can drive systemic reform. However, these choices rarely address the underlying mechanisms that perpetuate injustice or inequality.
  - **Example (Hypothetical):** A clothing brand markets a "conscious collection" made with recycled materials, inviting consumers to feel they are contributing to environmental sustainability. Meanwhile, the majority of the company's operations continue to rely on exploitative labor practices and unsustainable production.
- **Fragmenting Resistance into Personal Acts:**
  - Framing activism as a matter of individual consumption fragments collective action. Instead of organizing for systemic change, individuals focus on optimizing their personal decisions, believing that their purchases alone can shift societal trends.
  - **Example (Hypothetical):** A person opts to buy reusable water bottles and eco-friendly cleaning products but feels overwhelmed when they realize their actions barely dent larger environmental challenges, such as industrial emissions or deforestation.
- **Reinforcing Systemic Stability:**
  - By promoting consumer choice as a substitute for political engagement, systems maintain the status quo. Responsibility is deflected away from corporations and governments, who benefit from appearing to align with popular causes while avoiding substantial reforms.
  - **Example (Hypothetical):** A tech company releases a diversity-themed ad campaign, signaling alignment with social progress while continuing exploitative labor practices in its global supply chain. Consumers are encouraged to feel their support for the company contributes to change, even as systemic issues persist.

2.3.3.3 The Broader Implications of the Choice Paradox

The paradox of choice extends beyond individual frustration, shaping how societies prioritize and engage with change. Systems use abundance and choice to create the illusion of empowerment while subtly redirecting attention and energy away from structural reform.

- **Distraction as Control:** The overwhelming array of options keeps people preoccupied with personal optimization – finding the best deal, the most ethical product, or the perfect solution – leaving little bandwidth for collective resistance.
  - Example: While consumers debate over which brand of sustainably produced coffee to buy, the structural factors driving deforestation or labor exploitation remain unaddressed.
- **The Perpetuation of Dissatisfaction:** Systems benefit from the dissatisfaction inherent in excessive choice. The endless pursuit of better options ensures continual engagement and consumption, reinforcing economic models reliant on constant growth.
  - Example: Streaming platforms use algorithms to offer an abundance of personalized recommendations, ensuring users keep watching and searching for "something better" rather than stepping away from the platform altogether.
- **Political Implications:** When choice is presented as a form of empowerment, political and systemic reforms are deprioritized. The narrative of "voting with your wallet" turns citizens into consumers, diluting the collective power of organized action.

The paradox of choice reveals a hidden mechanism of control in modern systems. Abundance, rather than liberating individuals, often leads to paralysis, dissatisfaction, and distraction. When choice is framed as a form of activism, it deflects attention from systemic accountability, reinforcing the very structures that perpetuate inequality or injustice.

Reclaiming agency requires recognizing the limits of personal consumption as a tool for change and shifting focus toward collective action and systemic reform. In a world saturated with options, the most meaningful choice may be to step back and demand more than what the system is offering.

2.4 MODERN STRESS THROUGH SAPOLSKY'S LENS

Stress is not a new phenomenon, but its nature and triggers have shifted dramatically over time. In earlier generations, stress was often acute and tied to immediate physical threats or survival challenges. Modern stress, by contrast, is chronic, abstract, and amplified by technological and societal complexity. Drawing on **Robert Sapolsky's research** on stress dynamics, this section explores four key stressors: lack of control, unpredictability, insufficient outlets for frustration, and weakened social support. To better understand the uniqueness of modern stress, we will compare these dynamics to how earlier generations confronted similar challenges.

2.4.1 Lack of Control in the Digital Age

2.4.1.1 Modern Perspective

Control is a fundamental human need, tied closely to agency and psychological stability. In the digital age, however, control often feels like an illusion. Algorithms, platforms, and systems dictate much of our experiences, subtly narrowing the scope of individual autonomy.

- **Algorithms Dictate Experiences**:
  - Algorithms, designed to optimize engagement and profitability, heavily influence what users see, consume, and even think. While these systems claim to personalize experiences, they often act as invisible curators, reducing genuine agency.
  - **Example:** A user scrolling through social media is shown content that reinforces their pre-existing biases, limiting exposure to diverse perspectives and meaningful choices.

- **Information Overload and Decision Fatigue:**
  - The sheer volume of information available online forces individuals to make constant decisions, many of which are trivial but cumulatively exhausting. Over time, this leads to decision fatigue, reducing cognitive resources for meaningful actions.
  - **Example:** Endless scrolling through streaming platforms, job listings, or online stores can leave users feeling drained, paralyzed by the effort of choosing.

2.4.1.2 Historical Comparison

Earlier generations also faced challenges to control, but these were often external and immediate.

- **Agrarian and Early Industrial Societies:** Control was tied to tangible forces, such as weather, crop yields, or access to resources. While these factors could induce acute stress, they were often addressed collectively within communities.
- **Mid-20th Century:** Control was tied to stability – homeownership, steady employment, and community structures provided individuals with a sense of predictability and agency, albeit within limited options.

2.4.1.3 Key Difference

The modern age shifts control from external, tangible forces to internalized, abstract systems like algorithms. Unlike earlier generations, who could often identify and address the source of their stress, modern individuals are faced with systems too complex and opaque to navigate fully.

2.4.2 Lack of Predictability in Modern Life

2.4.2.1 Modern Perspective

Predictability is crucial for reducing stress, as it allows individuals to plan, prepare, and build resilience. Modern life, however, is marked by rapid change and uncertainty, from economic instability to shifting social norms.

- **Career Instability and Technological Change:**
  - Automation, globalization, and rapid technological advances have upended traditional career paths. Lifelong employment, once the norm, has been replaced by gig work, frequent role changes, and a constant need to upskill.
  - **Example**: A professional facing industry disruption due to automation feels uncertain about their future, creating chronic anxiety.
- **Shifting Social Norms via Social Media:**
  - Social media platforms amplify uncertainty by constantly redefining trends, values, and behaviors. What is celebrated one day may be criticized the next, leaving individuals unsure of how to align with societal expectations.
  - **Example:** A young adult experiences anxiety trying to keep up with ever-changing online discourse about cultural norms, fearing judgment for not being "on trend."

2.4.2.2 Historical Comparison

Earlier generations faced uncertainty, but its nature was different.

- **Pre-Industrial Societies:** Predictability revolved around seasonal cycles, religious traditions, and established roles within communities. While these structures could be rigid, they provided stability.
- **Industrial and Post-War Generations:** Predictability came from social contracts: steady jobs, pensions, and established cultural norms created a sense of order, even during economic downturns or political crises.

2.4.2.3 Key Difference

While earlier generations faced uncertainty tied to external events, modern unpredictability is pervasive and constant, driven by a rapidly evolving digital and global landscape that affects all aspects of life.

2.4.3 Lack of Outlets for Frustration

2.4.3.1 Modern Perspective

Modern stressors are often abstract, making them harder to resolve or escape. Issues like climate anxiety, economic inequality, and performance pressure leave individuals with few effective outlets for frustration.

- **Abstract Stressors Without Clear Resolutions:**
  - Many modern challenges, such as global warming or systemic inequality, are too large and complex for individuals to tackle directly. This creates feelings of helplessness and resignation.
  - **Example:** A person aware of climate change's catastrophic implications feels overwhelmed by the scale of the issue, unsure how their individual actions can make a difference.
- **Performance Pressure and Comparison:**
  - Social media and professional environments amplify performance pressure, presenting idealized versions of life that foster constant comparison.
  - **Example:** An employee who excels at work feels inadequate after seeing a colleague receive public recognition, sparking self-doubt despite their own success.

2.4.3.2 Historical Comparison

Earlier generations faced frustrations tied to survival or societal constraints but often had clearer outlets for these emotions.

- **Traditional Societies:** Communal rituals, shared labor, and storytelling provided structured ways to release tension and reassert control.
- **Mid-20th Century:** Sports, unions, and civic organizations offered collective outlets for frustration, allowing individuals to channel grievances into action.

2.4.3.3 Key Difference

Modern frustrations are often abstract and lack the clear resolutions or collective outlets that earlier societies provided. This leaves individuals internalizing their stress, compounding its effects.

2.4.4 Lack of Social Support Despite Connectivity

2.4.4.1 Modern Perspective

Despite the rise of digital communication, individuals report increasing feelings of loneliness and isolation. While technology enables superficial connections, it often undermines the depth and accountability required for meaningful relationships.

- **Digital Connections Replace Meaningful Relationships:**
  - Online interactions are often transactional, lacking the trust and intimacy needed to build resilience against stress.

- **Example:** A person with hundreds of social media followers realizes they have no one to turn to during a personal crisis.
- **Fragmented Communities and Rising Individualism**:
  - As traditional community structures erode, individuals rely more on themselves to navigate challenges, weakening collective support.
  - **Example:** A young family moving to a new city struggles to integrate into a transient, disconnected neighborhood.

2.4.4.2 Historical Comparison

Earlier generations relied heavily on face-to-face interactions and tightly knit communities.

- **Traditional Societies:** Villages and extended families provided built-in support networks.
- **Post-War Generations:** Civic organizations, faith-based groups, and stable neighborhoods fostered a sense of belonging and accountability.

2.4.4.3 Key Difference

Modern connectivity is broad but shallow, offering quantity over quality. Earlier generations, while less globally connected, benefited from stronger local bonds and collective resilience.

2.4.5 Summary

Through Sapolsky's lens, the stressors of modern life – lack of control, unpredictability, insufficient outlets for frustration, and weakened social support – highlight how technology and societal changes have redefined what it means to navigate stress. Compared to earlier generations, who faced immediate but often resolvable challenges, modern individuals grapple with chronic, abstract pressures that are harder to address and escape.

Understanding these stressors in historical context reveals not only how stress has evolved but also how reclaiming elements of collective action, predictability, and meaningful connection can mitigate the unique pressures of the modern age.

2.5 THE COMBINED CHALLENGE

Modern life is a maze of complexity, shaped by countless interconnected forces. **System absorption** and **amplified stress**, while central themes, are far from the only contributors. Rapid technological change, economic inequality, environmental crises, and shifting social norms also play critical roles in creating an environment that demands constant navigation, adjustment, and resilience. Yet the interplay of system absorption and stress stands out for its pervasive influence: it shapes not only how people experience their challenges but also how they respond to them.

By synthesizing these dynamics, we can better understand how systems maintain stability by amplifying stress while presenting themselves as solutions to the very problems they perpetuate. This feedback loop ensures that individuals remain caught in cycles of consumption, performance, and disconnection, reinforcing systemic power.

2.5.1 System Absorption Amplifying Stress

At its core, system absorption operates as a dual mechanism: it neutralizes potential threats to systemic stability while simultaneously increasing the stress placed on individuals. By commodifying resistance and encouraging performative behaviors, systems amplify anxiety, isolation, and a sense of inadequacy.

2.5.1.1 Commodification Turns Resistance into Consumption

- **Neutralizing Dissent:** Systems transform resistance into commodified trends, framing dissent as a marketable choice rather than a call for collective action. What might once have been a movement becomes a branding opportunity, reducing its potential to challenge systemic structures.
  - **Example:** Social causes are often reduced to slogans or symbols on consumer products, where purchasing a product replaces meaningful engagement.
- **Amplifying Stress Through Consumption:** Commodification creates a cycle where individuals feel the pressure to "perform" their values by making the "right" purchases or adopting the "right" identity markers. This focus on personal responsibility shifts the burden of systemic problems onto individuals, increasing stress.
  - **Example:** The rise of ethical consumption narratives leaves individuals overwhelmed by choices and doubting whether their actions make any real difference.

2.5.1.2 Performativity Exacerbates Anxiety and Isolation

- **The Pressure to Perform:** Platforms and systems encourage individuals to present idealized versions of themselves, whether through activism, success stories, or personal branding. This performative behavior often erodes authenticity, creating a gap between how people live and how they feel they are expected to live.
  - **Example:** A person posts about supporting social causes online but feels disconnected from the cause itself, worrying more about how their post is received than about their actual impact.
- **Isolation as a Byproduct:** Performativity amplifies isolation by replacing genuine relationships with curated interactions. Individuals feel increasingly alone, even as they engage with larger audiences.
  - **Example:** An individual with a highly active online presence realizes they lack close confidants for meaningful support, deepening feelings of loneliness.

Through commodification and performativity, systems not only absorb resistance but also heighten individual stress, ensuring that energy spent on self-expression or dissent rarely challenges systemic stability.

2.5.2 Stress Responses Feeding System Stability

Stress does more than exhaust individuals – it drives behaviors that sustain the very systems causing the stress. Anxiety fosters consumption, while isolation weakens collective action, creating a feedback loop that reinforces systemic power.

2.5.2.1 Anxiety Drives Consumption Habits

- **The Marketization of Stress Relief:** Modern systems thrive on presenting consumption as the solution to stress. Products and services are marketed as tools for overcoming the very challenges these systems exacerbate, creating a cycle of dependence.
  - **Example:** Wellness apps, self-help courses, and lifestyle brands promise stress relief but rarely address root causes, encouraging individuals to focus on personal optimization rather than systemic change.
- **Short-Term Fixes, Long-Term Strain:** Consumption offers momentary relief but rarely resolves the underlying issues. This ensures individuals remain caught in a loop of dissatisfaction and spending.
  - **Example:** A stressed worker purchases productivity tools but finds that their workplace culture continues to demand unsustainable levels of output, perpetuating burnout.

2.5.2.2 Isolation Prevents Collective Resistance

- **Weakening Collective Bonds:** Isolation, fueled by fragmented communities and individualism, undermines the potential for collective resistance. Without strong social connections, individuals feel unable to challenge systemic power.
  - **Example**: A person facing job insecurity avoids unionizing efforts, feeling disconnected from their coworkers and preoccupied with their personal challenges.
- **Stress as a Barrier to Action:** Chronic stress depletes the cognitive and emotional resources needed for organizing or engaging in collective action. This keeps individuals focused on immediate survival rather than long-term systemic change.
  - **Example:** A community affected by housing instability struggles to organize protests because residents are overwhelmed by their individual struggles, from paying rent to securing childcare.

Stress responses – driven by anxiety and isolation – effectively sustain systemic power by ensuring individuals remain fragmented, exhausted, and preoccupied with personal challenges rather than united in collective efforts.

2.5.3 The Feedback Loop: System Absorption and Stress

The interplay between system absorption and amplified stress creates a powerful feedback loop:

1. **System Absorption Neutralizes Resistance:** Mechanisms such as commodification and performativity ensure that resistance and individuality are transformed into behaviors that reinforce systemic stability.
2. **Amplified Stress Drives Systemic Behaviors:** Stress responses, particularly anxiety and isolation, push individuals toward consumption and disengagement, further supporting systems that thrive on fragmentation and dependence.
3. **Reinforced Stability Perpetuates the Cycle:** As systems grow stronger through this dynamic, they continue to absorb potential challenges while increasing the stress they place on individuals, ensuring the cycle continues.

2.5.4 Complexity Beyond System Absorption and Stress

While system absorption and stress are critical components of modern life, they are not the only contributors to its complexity. Rapid technological innovation, global interconnectedness, economic inequality, environmental crises, and cultural shifts all play roles in creating a multifaceted environment that demands constant navigation and adjustment.

- **Technological Acceleration:**
  - Innovations like artificial intelligence and automation create opportunities but also uncertainty, reshaping industries and societal norms at unprecedented speeds.
- **Global Interconnectedness:**
  - The world's interconnected nature amplifies both opportunity and vulnerability. Events in one region can have cascading effects globally, creating a sense of unpredictability.
- **Economic and Environmental Strains:**
  - Economic inequality and environmental degradation exacerbate stress by creating systemic pressures that individuals cannot escape or resolve alone.

These additional layers of complexity amplify the challenges of navigating modern life, making it even more essential to understand and address the feedback loop between system absorption and stress.

The combined challenge of system absorption and amplified stress reveals a cycle that sustains systemic power while intensifying individual struggles. Systems neutralize resistance by commodifying it, turning activism and identity into consumable trends. At the same time, stress responses such as anxiety and isolation drive consumption and weaken collective bonds, reinforcing systemic dominance.

While these dynamics are central to understanding modern life, they exist within a broader landscape of complexity, shaped by technological, economic, and environmental forces. Breaking free from this feedback loop requires recognizing its mechanisms, reclaiming agency, and fostering collective resilience in an increasingly fragmented world.

2.6 CURIOUS SKEPTICISM: A FRAMEWORK FOR LIBERATION

In a world shaped by system absorption and amplified stress, **Curious Skepticism** emerges as an essential mindset for reclaiming agency and resisting systemic control. It bridges critical inquiry with action, encouraging individuals and communities to challenge commodified narratives, foster authentic connections, and transform awareness into meaningful resistance. This section explores three practical ways to apply Curious Skepticism in resisting the mechanisms of system absorption.

2.6.1 Resisting System Absorption

Curious Skepticism challenges the systems that absorb dissent and commodify individuality by fostering intentional, reflective, and authentic resistance. It shifts the focus from passive participation to deliberate questioning, connection, and action.

2.6.1.1 Adopt Verification Practices to Challenge Commodified Narratives

Systems of absorption rely on the unquestioned acceptance of commodified messages, trends, and ideals. Curious Skepticism pushes individuals to interrogate these narratives and ask: *What lies beneath the surface? Who benefits from this narrative? How does this align with my values?*

- **Critical Questioning of Messages:**
  - Verification practices involve evaluating claims with an open yet skeptical lens. This includes tracing the origins of messages, understanding their motivations, and discerning their actual impact.
  - **Example:** A clothing brand advertises its new "sustainable collection." A curious skeptic researches the company's broader practices, uncovering that only a fraction of its operations meet sustainability standards, while the majority perpetuate exploitative practices.
- **Avoiding the Pitfalls of Performative Resistance:**
  - Curious Skepticism resists the temptation to engage in performative gestures – actions that signal values without meaningful impact. This involves moving beyond surface-level involvement, such as posting hashtags or purchasing branded merchandise, to actions that align with authentic values.
  - **Example:** Instead of buying a product marketed as "eco-friendly," a curious skeptic donates directly to grassroots environmental organizations or participates in community-led initiatives.

By adopting verification practices, Curious Skepticism dismantles the superficial allure of commodified narratives, empowering individuals to act with intention and integrity.

2.6.1.2 Build Authentic Spaces for Resistance Outside the System

Curious Skepticism recognizes that resistance cannot flourish within the boundaries of systems designed to commodify it. To counter absorption, it advocates for creating and nurturing spaces where authentic connections and actions can thrive, unmediated by systemic influence.

- **Fostering Genuine Community:**
  - Authentic spaces prioritize trust, accountability, and shared purpose over visibility or performance. These spaces can take the form of local collectives, small groups, or even digital networks with intentional boundaries.

    - **Example:** A group of neighbors starts a community garden, using the project as both a source of local sustainability and a platform for conversations about larger environmental issues.
- **Operating Beyond Algorithms:**
    - Systemic absorption often happens through algorithmic mediation, which shapes interactions and prioritizes engagement over depth. Building authentic spaces requires stepping outside these frameworks to foster direct, meaningful interactions.
    - **Example:** Activists organize in-person meetups to discuss local issues, deliberately avoiding platforms that commodify their conversations into engagement metrics.
- **Sustaining Resistance Over Time:**
    - Authentic spaces cultivate long-term commitment and collaboration, ensuring that resistance remains resilient against external pressures.

Creating spaces outside the system reinforces the collective potential of resistance, offering environments where authenticity, connection, and action can flourish.

2.6.1.3 Translate Awareness into Action

Curious Skepticism emphasizes that awareness alone is insufficient. Without action, understanding systemic absorption risks becoming another passive acknowledgment of the problem. Curious Skepticism connects insight with deliberate, practical steps to effect change.

- **From Awareness to Engagement:**
    - Awareness must be the starting point for engagement, guiding individuals to move beyond recognition to active participation.
    - **Example:** After understanding how a platform commodifies activism, a curious skeptic shifts their focus to offline efforts, such as attending local town hall meetings or organizing workshops.
- **Incremental, Tangible Steps:**
    - Curious Skepticism values small, consistent actions that build momentum. These steps, while modest on their own, accumulate into meaningful change.
    - **Example:** An individual begins by reducing reliance on algorithm-driven platforms for news, opting instead for independent sources, and gradually expands their efforts to support media reform initiatives.
- **Bridging Individual Efforts with Collective Impact:**
    - Curious Skepticism encourages individuals to align their actions with larger movements, ensuring that personal efforts contribute to broader systemic change.
    - **Example:** A group of employees frustrated with workplace policies starts by voicing their concerns to management but eventually organizes a union, amplifying their collective power.

By translating awareness into action, Curious Skepticism ensures that resistance does not end at recognition but continues as a force for meaningful and sustained change.

2.6.2 Addressing Fundamental Stressors

Modern stressors – lack of control, unpredictability, insufficient outlets for frustration, and weakened social support—are deeply intertwined with systemic absorption. **Curious Skepticism** provides a practical framework for addressing these challenges by encouraging deliberate actions to reclaim agency, foster resilience, and build meaningful connections. This approach empowers individuals and communities to counteract stressors while resisting the forces that perpetuate them.

2.6.2.1 Create Control Through Intentional Habits and Boundaries

In a world where algorithms and systems shape experiences, creating control requires deliberate choices about how to engage with one's environment. Curious Skepticism encourages individuals to reclaim agency by setting clear boundaries and cultivating intentional habits.

- **Establishing Digital Boundaries:**
  - Algorithms thrive on capturing attention, making it essential to define when and how to engage with digital platforms.
  - **Example:** A person sets designated times for checking social media or email, ensuring these activities do not dominate their day.
- **Focusing on What Can Be Controlled:**
  - Curious Skepticism encourages focusing energy on areas where influence is possible, rather than expending effort on uncontrollable factors.
  - **Example:** Instead of fixating on global issues they cannot directly impact, an individual focuses on local actions, such as supporting community programs or volunteering.
- **Practicing Mindfulness and Reflection:**
  - Intentional habits include mindfulness practices that help individuals identify and prioritize what truly matters.
  - **Example:** Regular journaling helps a person clarify their values and goals, creating a sense of purpose and direction.

By establishing intentional habits and boundaries, Curious Skepticism helps individuals regain a sense of control in environments designed to erode it.

2.6.2.2 Develop Predictability by Fostering Routines and Clear Expectations

Modern life's unpredictability can be countered by creating personal stability through routines and clear expectations. Curious Skepticism emphasizes building structures that provide a sense of order amidst uncertainty.

- **Cultivating Daily Routines:**
  - Simple, consistent routines create a foundation of predictability, reducing decision fatigue and fostering stability.
  - **Example:** A professional establishes a morning routine that includes exercise, reading, and planning their day, creating a sense of calm before engaging with external demands.
- **Aligning Expectations with Reality:**
  - Unrealistic expectations often exacerbate stress. Curious Skepticism encourages individuals to set achievable goals and adapt expectations based on circumstances.
  - **Example:** A student overwhelmed by academic pressure adjusts their study schedule to focus on consistent progress rather than perfection.
- **Prioritizing Long-Term Stability:**
  - Predictability also involves planning for the future in ways that reduce uncertainty, such as developing financial literacy or cultivating career flexibility.

Fostering predictability through routines and clear expectations helps individuals anchor themselves in a rapidly changing world, reducing stress and enhancing resilience.

2.6.2.3 Build Real Outlets for Frustration Through Meaningful Action

Curious Skepticism recognizes that modern stressors, particularly abstract ones, require concrete outlets for resolution. Building meaningful channels for frustration turns stress into momentum for personal and collective empowerment.

- **Engaging in Tangible Activities:**
  - Physical and creative activities provide direct, satisfying outlets for frustration, transforming energy into action.
  - **Example:** A person feeling powerless about climate change joins a local gardening initiative, channeling their concerns into visible, impactful efforts.
- **Advocating for Change:**
  - Abstract stressors like inequality or systemic injustice can be addressed through organized advocacy, ensuring frustration fuels collective action rather than despair.
  - **Example:** A group of employees frustrated by workplace conditions forms a committee to advocate for better policies, turning individual dissatisfaction into systemic improvement.
- **Practicing Generative Acts:**
  - Creating something – whether art, relationships, or solutions – provides a sense of accomplishment and control.
  - **Example:** A writer channels feelings of disconnection into storytelling, using creativity as a way to process and connect with others.

Building real outlets for frustration transforms stress into purpose, allowing individuals to reclaim agency and channel their energy productively.

2.6.2.4 Strengthen Authentic Support Networks for Collective Resilience

Curious Skepticism challenges the shallow, transactional connections often fostered by digital platforms. Instead, it advocates for cultivating authentic relationships that provide emotional support, accountability, and collective strength.

- **Investing in Deep Relationships:**
  - Authentic support networks require time and effort to build but offer invaluable resilience against stress.
  - **Example:** A person dedicates regular time to meeting with close friends or family, prioritizing quality interactions over superficial connections.
- **Rebuilding Community Bonds:**
  - Community initiatives, such as neighborhood projects or mutual aid groups, create opportunities for collective engagement and shared purpose.
  - **Example:** Residents of an apartment building organize a weekly potluck to foster a sense of community and mutual support.
- **Balancing Digital and Physical Connections:**
  - While digital tools can facilitate connection, authentic support often requires in-person interaction and trust-building.
  - **Example:** A local advocacy group moves its meetings offline to foster deeper collaboration and accountability among its members.

By strengthening authentic support networks, individuals and communities can counter isolation, build resilience, and create environments where collective efforts thrive.

2.6.3 Breaking Free from System Loops

Modern systems are designed to maintain their dominance by absorbing dissent, commodifying individuality, and reinforcing stress responses that drive consumption and disengagement. Breaking free from these systemic loops requires deliberate recognition of their mechanisms, the creation of alternative spaces and practices, and the cultivation of collective resilience. **Curious Skepticism** provides a roadmap for disrupting these cycles and fostering meaningful resistance.

2.6.3.1 Recognize Absorption Mechanisms and Disrupt Them

System absorption thrives on invisibility – its mechanisms are subtle, often framed as progress or empowerment. Curious Skepticism begins with the deliberate recognition of these dynamics, enabling individuals and communities to identify and challenge them.

- **Identify Patterns of Absorption:**
  - Curious Skepticism calls for examining how systems commodify resistance, individuality, or authenticity. This involves tracing the paths of movements, trends, or personal expressions to see how they are co-opted for systemic stability.
  - **Example:** A curious skeptic notices how a social justice movement's language has been adopted by corporate marketing campaigns, diluting its message and transforming it into a branding tool.
- **Disrupt the Cycle Through Awareness**:
  - Recognizing absorption mechanisms is the first step in disrupting them. By understanding how systems operate, individuals can consciously opt out of performative or commodified behaviors.
  - **Example:** Instead of engaging with an online trend co-opted by algorithms, an individual chooses to support grassroots efforts that operate outside the commodified framework.
- **Expose Hidden Agendas:**
  - Curious Skepticism encourages transparency and accountability, shedding light on the agendas driving systemic absorption.
  - **Example:** A collective investigating the ethics of a company reveals how its sustainability claims mask broader practices of exploitation, fostering informed consumer and activist decisions.

Recognizing and disrupting these mechanisms prevents individuals and communities from being complicit in their own absorption, creating room for genuine resistance.

2.6.3.2 Create Alternative Spaces and Practices That Prioritize Authenticity

Escaping systemic loops requires building spaces and practices that are intentionally designed to operate outside the logic of commodification and absorption. Curious Skepticism emphasizes the importance of creating environments where authenticity and collaboration can thrive.

- **Alternative Spaces for Connection and Action:**
  - Authentic spaces provide refuge from systemic pressures, allowing individuals to engage in resistance, connection, and creativity without the fear of commodification.
  - **Example:** A local arts collective creates a physical space where community members can share ideas, collaborate on projects, and organize without relying on algorithm-driven platforms.
- **Prioritize Practices Over Performances:**
  - Curious Skepticism challenges performative resistance by focusing on actions that align with values rather than appearances. This involves developing habits and rituals that prioritize authenticity and depth.
  - **Example:** A group dedicated to environmental sustainability commits to monthly clean-ups and educational workshops, focusing on real-world impact rather than social media visibility.
- **Foster Autonomy and Self-Sufficiency:**
  - Creating alternative spaces includes reclaiming autonomy through self-sufficient practices, such as local food production, independent media, or cooperative networks.
  - **Example:** A neighborhood establishes a community-supported agriculture program, reducing dependence on commodified food systems while fostering local engagement.

Building alternative spaces and practices ensures that resistance remains grounded in authenticity and action, reducing the influence of systemic loops.

2.6.3.3 Build Collective Resilience to Sustain Meaningful Resistance

Breaking free from systemic loops is not a solitary effort – it requires collective resilience. Curious Skepticism emphasizes the importance of building networks of support and collaboration to sustain long-term resistance.

- **Strengthen Shared Values and Goals:**
  - Collective resilience begins with identifying shared values and goals that unite individuals and groups, creating a foundation for sustained efforts.
  - **Example**: A coalition of local organizations unites around a shared goal of promoting affordable housing, pooling resources and expertise to amplify their impact.
- **Provide Emotional and Practical Support:**
  - Resilience depends on mutual support. Curious Skepticism advocates for creating networks that offer both emotional encouragement and practical resources to counter isolation and burnout.
  - **Example**: A mutual aid group forms during an economic downturn, providing financial assistance, childcare, and mental health resources to its members.
- **Collaborate Across Boundaries:**
  - Effective resistance often requires collaboration across ideological, cultural, or geographic divides. Curious Skepticism promotes open-minded partnerships that focus on shared outcomes.
  - **Example:** Environmental and labor activists collaborate on a campaign to promote green jobs, recognizing the intersection of their causes.

By building collective resilience, individuals and communities create the strength and solidarity needed to sustain meaningful resistance against systemic absorption and its stress-inducing dynamics.

2.7 IMPLEMENTATION STRATEGIES

Effective resistance to systemic absorption and the stressors of modern life requires actionable strategies at multiple levels: individual, community, and systemic. These strategies, grounded in **Curious Skepticism**, empower individuals and groups to reclaim agency, foster resilience, and create meaningful change.

2.7.1 Individual Level

At the individual level, implementation strategies focus on reclaiming control, reducing stress, and aligning actions with values. Curious Skepticism encourages deliberate practices that enhance personal agency, reduce decision fatigue, and counter misinformation.

2.7.1.1 Develop Verification Practices to Counter Misinformation

In a world saturated with information, misinformation thrives, often designed to manipulate behaviors and reinforce systemic power. Verification practices help individuals navigate this landscape with clarity and confidence.

- **Question the Source**:
  - Evaluate the credibility of information by identifying its origin, purpose, and context.
  - **Example:** Before sharing an article on social media, a curious skeptic examines the publication's reputation, checks for corroborating sources, and assesses the tone for sensationalism.

- **Distinguish Fact from Emotion:**
  - Separate emotional reactions from factual content to ensure decisions and beliefs are grounded in evidence.
  - **Example:** When encountering a headline designed to provoke outrage, an individual pauses to verify the underlying data before forming an opinion.
- **Cultivate Media Literacy:**
  - Regularly engage with tools and resources that enhance critical thinking about media and digital content.
  - **Example:** A person takes a media literacy course or uses fact-checking platforms to develop skills for discerning truth in a complex media environment.

By practicing verification, individuals can resist manipulation and engage more meaningfully with the world around them.

2.7.1.2 Simplify Decisions by Aligning Choices with Core Values

Modern life is often overwhelming, filled with endless choices that contribute to decision fatigue. Curious Skepticism advocates for simplifying decision-making by aligning choices with personal values and long-term goals.

- **Clarify Core Values:**
  - Reflect on personal priorities and values to create a clear framework for decision-making.
  - **Example:** An individual identifies environmental sustainability as a core value and uses it as a guide for everyday decisions, such as transportation or consumption.
- **Focus on What Matters:**
  - Reduce complexity by concentrating on decisions that align with values while letting go of inconsequential ones.
  - **Example:** Instead of agonizing over minor consumer choices, a person focuses on broader actions like supporting local businesses or reducing waste.
- **Adopt Minimalism in Decision-Making:**
  - Simplify options by limiting unnecessary choices, reducing cognitive load.
  - **Example:** A professional creates a weekly meal plan to avoid daily decision-making about food, freeing mental energy for more meaningful tasks.

Simplifying decisions allows individuals to reduce stress and act with greater intention and impact.

2.7.1.3 Use Stress-Reduction Techniques Like Mindfulness and Boundary Setting

Chronic stress erodes well-being and resilience, making it essential to incorporate practices that mitigate its effects. Curious Skepticism emphasizes intentional techniques to reduce stress and maintain balance.

- **Practice Mindfulness**:
  - Develop a mindfulness practice to increase awareness and manage stress responses.
  - **Example:** A person begins each day with ten minutes of meditation to center their focus and prepare for external demands.
- **Set Boundaries with Technology:**
  - Limit engagement with platforms and systems that amplify stress, such as social media or 24-hour news cycles.
  - **Example:** An individual sets specific times to check emails or news, avoiding constant exposure to potentially overwhelming information.
- **Engage in Physical and Creative Activities:**
  - Physical and creative outlets provide effective ways to release stress and build resilience.
  - **Example:** A person incorporates regular walks, journaling, or art into their routine as a form of self-care and expression.

By incorporating these techniques, individuals can build emotional and mental resilience, better equipping themselves to navigate systemic pressures and stressors.

2.7.2 Community Level

Curious Skepticism emphasizes the importance of fostering intentional, collaborative communities to counteract systemic absorption and amplify collective resilience. By cultivating trust, rejecting commodified practices, and uniting around shared goals, communities can resist systemic loops and create meaningful spaces for collective action. At the heart of this approach is a commitment to questioning the systems that shape interactions while actively building alternatives grounded in authenticity and mutual support.

2.7.2.1 Build Networks That Foster Trust and Collaboration

Trust is the foundation of any effective community. Curious Skepticism encourages communities to critically evaluate the systems mediating their interactions and to prioritize direct, collaborative relationships that promote accountability and shared purpose.

- **Embrace Transparent Communication:**
  - Building trust requires open dialogue and mutual understanding. Curious Skepticism encourages communities to question assumptions and ensure clarity in shared goals.
  - **Example:** A local environmental group holds regular town hall-style meetings where members openly discuss priorities and concerns, ensuring everyone feels heard and valued.
- **Collaborate With Intentionality:**
  - Instead of relying on platforms or systems that commodify interactions, communities can intentionally create spaces that facilitate genuine collaboration.
  - **Example:** A neighborhood creates a direct messaging network that operates independently of social media platforms, fostering accountability and minimizing distractions.
- **Resist Transactional Relationships:**
  - Curious Skepticism advocates for moving beyond superficial connections by investing in relationships that prioritize shared values and mutual care.
  - **Example:** A cooperative childcare group not only organizes shared responsibilities but also builds deeper trust through regular social gatherings and skill-sharing events.

By fostering trust and collaboration grounded in Curious Skepticism, communities can create resilient networks that operate independently of commodified systems.

2.7.2.2 Establish Shared Practices to Resist Commodification

Curious Skepticism challenges communities to identify and reject commodified practices that undermine authenticity and connection. Instead, it advocates for establishing shared practices that reflect collective values and foster genuine engagement.

- **Create Non-Commodified Spaces**:
  - Communities can deliberately design spaces and activities that operate outside commodified frameworks, focusing instead on resourcefulness and collaboration.
  - **Example:** A group of artists establishes a non-commercial venue where members can exhibit work, host workshops, and share ideas without external pressures or commercial incentives.
- **Align Actions With Collective Values**:
  - Curious Skepticism encourages communities to critically examine their practices to ensure they align with shared ethical principles, resisting the pressure to conform to commodified norms.

    - **Example:** A community-run food bank prioritizes sourcing donations from local producers, reinforcing commitments to sustainability and food security.
- **Build Localized Economies:**
    - Alternatives like cooperatives, barter systems, and mutual aid networks help communities reclaim agency and resist dependence on commodified systems.
    - **Example:** A time-banking initiative allows community members to exchange skills and services, such as tutoring or carpentry, reducing reliance on traditional economic systems.

By adopting shared practices rooted in Curious Skepticism, communities can resist commodification and reinforce their autonomy and authenticity.

2.7.2.3 Focus on Collective Purpose and Shared Action

Curious Skepticism emphasizes the importance of uniting communities around clear, collective purposes that inspire shared action. This approach challenges passive participation by encouraging active engagement in meaningful, transformative efforts.

- **Question the Status Quo to Define Collective Goals:**
    - Communities should critically evaluate the systems affecting them, identifying shared challenges and opportunities for change. Curious Skepticism ensures these efforts are deliberate and rooted in evidence.
    - **Example:** A housing coalition investigates local zoning laws, questioning how they perpetuate inequality, and sets a goal to advocate for inclusive housing policies.
- **Engage in Visible, Authentic Action:**
    - Shared action fosters connection and purpose, challenging commodified notions of resistance. Curious Skepticism emphasizes actions that prioritize impact over appearance.
    - **Example:** A community organizes a public campaign to protect local green spaces, combining direct advocacy with hands-on efforts like tree planting and cleanup events.
- **Celebrate and Reflect Together:**
    - Curious Skepticism encourages communities to evaluate their progress and celebrate achievements as a way to reinforce purpose and adapt future strategies.
    - **Example:** After organizing a successful fundraising event, a mutual aid group holds a reflection session to discuss lessons learned and opportunities for improvement.

Focusing on collective purpose and shared action empowers communities to navigate systemic pressures and create meaningful change grounded in shared values.

2.7.3 Systemic Level

Curious Skepticism offers a practical and principled approach to systemic change by challenging commodified narratives, advocating for inclusive reforms, and supporting initiatives that prioritize equity and resilience. Rooted in critical inquiry and action, it emphasizes achievable steps that balance ambition with realism, ensuring that resistance and reform align with authentic values and actionable goals.

2.7.3.1 Identify Systemic Absorption Mechanisms and Expose Them

Curious Skepticism begins with identifying how systems co-opt resistance, commodify identities, and perpetuate stress. By recognizing these mechanisms and exposing them transparently, individuals and groups can disrupt their influence and pave the way for more authentic narratives and policies.

- **Apply Investigative Curiosity:**
  - Use critical inquiry to uncover hidden dynamics of absorption, asking questions like: *Who benefits from this narrative? How does this trend serve systemic power?*
  - **Example:** A collaborative project analyzes corporate greenwashing campaigns, revealing the gap between advertised sustainability initiatives and actual practices.
- **Leverage Technology for Transparency:**
  - Technology, while often complicit in commodification, can also be a tool for transparency. Platforms like open data repositories or crowd-sourced reporting systems enable communities to track and reveal systemic inconsistencies.
  - **Idea:** Create a public database tracking the lifecycle of social movements or corporate practices, allowing people to see how ideas are absorbed or diluted over time.
- **Involve Influencers and Stakeholders:**
  - Engage trusted voices – journalists, academics, or cultural leaders – to amplify findings and educate wider audiences about systemic absorption.
  - **Example:** A documentary produced with community input showcases how local activism is reframed by corporations, turning resistance into marketing.

By exposing systemic absorption mechanisms through accessible and collaborative tools, Curious Skepticism ensures that accountability becomes a shared, actionable goal.

2.7.3.2 Advocate for Alternative Cultural Narratives and Systemic Reforms

Curious Skepticism emphasizes replacing commodified narratives with inclusive, equity-driven alternatives that inspire collective action and foster systemic accountability.

- **Shift Cultural Narratives Through Storytelling:**
  - Use storytelling to challenge dominant paradigms and inspire alternative ways of thinking. Stories of success rooted in community action, equity, and sustainability can reshape public discourse.
  - **Idea:** Partner with artists, filmmakers, and educators to create a series highlighting grassroots efforts that address issues like housing inequality or environmental justice.
- **Focus on Feasible, Incremental Reforms:**
  - Advocate for reforms with clear, measurable outcomes to maintain momentum and build credibility. Curious Skepticism encourages prioritizing realistic goals that align with systemic needs.
  - **Example:** Instead of pushing for sweeping, unattainable climate legislation, advocate for localized initiatives, such as renewable energy incentives for households or community-led waste reduction programs.
- **Engage in Cross-Sector Partnerships:**
  - Curious Skepticism encourages bridging divides between sectors – public, private, and nonprofit – to align efforts around systemic reforms.
  - **Idea:** Develop coalitions between businesses, government agencies, and advocacy groups to create equitable hiring practices or improve access to affordable healthcare.

Advocating for cultural shifts and incremental reforms ensures that systemic changes remain grounded in practical, achievable steps while inspiring broader societal transformation.

2.7.3.3 Support Initiatives That Prioritize Equity and Resilience

Curious Skepticism champions equity and resilience as cornerstones of systemic reform. By supporting initiatives that address root causes of inequality and stress, individuals and communities can create systems that prioritize fairness and long-term sustainability.

- **Invest in Grassroots Solutions with High Scalability:**
  - Grassroots efforts often address systemic issues more effectively than top-down approaches. Curious Skepticism emphasizes scaling successful local initiatives to benefit broader populations.
  - **Example**: A community's successful program for energy-efficient housing inspires regional adoption, reducing both energy costs and environmental impact.
- **Incorporate Resilience into Policy Design:**
  - Resilient systems adapt to challenges and prioritize the needs of vulnerable populations. Curious Skepticism supports policies that balance short-term relief with long-term stability.
  - **Idea**: Advocate for policies that integrate climate adaptation measures, such as urban green spaces or flood-resistant infrastructure, into city planning.
- **Encourage Participatory Budgeting and Governance:**
  - Empower communities to take an active role in deciding how resources are allocated, ensuring that equity and resilience are built into systemic reforms.
  - **Example:** A city launches a participatory budgeting program, allowing residents to allocate a portion of the municipal budget toward local priorities, such as education, healthcare, or public transit.

By supporting equity-driven and scalable initiatives, Curious Skepticism ensures that systemic reforms remain accessible, impactful, and aligned with collective values.

2.8 CONCLUSION: FROM UNDERSTANDING TO ACTION

2.8.1 Synthesize Key Insights

Modern life is a paradox. Despite unparalleled resources, connectivity, and opportunities, individuals and communities face increasing stress, disconnection, and uncertainty. The forces of systemic absorption and commodification have transformed even resistance and authenticity into marketable trends, ensuring systemic stability at the expense of personal and collective agency.

Curious Skepticism offers a pathway through this complexity. By encouraging thoughtful inquiry, intentionality, and authenticity, it helps individuals reclaim agency, build meaningful connections, and resist the loops of commodification and stress. It is not about rejecting everything or endlessly questioning for its own sake but about approaching the world with an open yet discerning mindset – one that seeks understanding and translates awareness into action.

2.8.2 Practical Takeaways for Readers

Curious Skepticism is a mindset that empowers individuals to navigate modern life with clarity and purpose. Here are three practical ways to embody its principles:

- **Evaluate Narratives Thoughtfully and Openly:**
  - Instead of dismissing or accepting societal narratives at face value, ask constructive questions like: *What are the implications of this narrative? How does it align with my values or community goals?* Approach these narratives with curiosity, seeking context and deeper understanding.
  - **Example:** When encountering a product marketed as "sustainable," a curious skeptic explores its broader impact, such as supply chain practices or community benefits, rather than relying solely on surface claims.
- **Prioritize Meaningful Connections and Intentional Choices:**
  - Invest in relationships and decisions that reflect authentic values rather than performative or commodified ideals. Simplify choices by focusing on what truly matters to you and your community.

  - **Example:** Instead of participating in fleeting online trends, prioritize fostering local connections or joining initiatives that contribute to tangible improvements in your neighborhood.
- **Act with Purpose and Focus on Incremental Change:**
  - Curious Skepticism emphasizes that meaningful action does not require perfection or grand gestures. Small, consistent steps – aligned with core values – can build momentum for lasting change.
  - **Example:** A person feeling overwhelmed by global environmental challenges starts by reducing personal waste, then connects with local environmental groups to advocate for policy changes in their city.

These practices help individuals and communities reclaim their agency, aligning their actions with deeper purpose while resisting the pressures of systemic absorption and commodification.

2.8.3 Forward Momentum

Curious Skepticism equips individuals with tools to navigate the paradoxes of modern life, but the journey does not end here. Moving beyond inquiry into sustained action and resilience requires understanding how cognition and behavior adapt to these challenges.

In the next chapter, we explore strategies for building mental clarity, fostering resilience, and reclaiming autonomy in an uncertain and rapidly changing world. By delving into how thought patterns and habits can reinforce or disrupt systemic loops, we uncover practical tools for thriving amidst complexity and creating meaningful change.

Modern life is filled with contradictions, but through Curious Skepticism, these challenges can become opportunities to build a more intentional, connected, and authentic future.

3 The Nocebo Effect in Modern Life
Harnessing Awareness for Empowerment

3.1 INTRODUCTION: THE POWER AND PITFALLS OF AWARENESS

3.1.1 Awareness as the Defining Virtue – and Challenge – of Our Age

In a world overflowing with awareness campaigns and crisis-driven headlines, are we empowering ourselves – or paralyzing ourselves? As individuals and societies, our growing awareness of global and personal risks is a double-edged sword. While this awareness aims to promote caution and preparedness, it often produces the opposite effect: anxiety, despair, and inaction. This paradox is best understood through the lens of the nocebo effect – a phenomenon where the mere anticipation of harm amplifies its impact.

In this chapter, we confront this paradox head-on. Awareness, while often empowering, is not inherently benevolent. Like fire, it can warm us or burn us, illuminate or blind. The key lies in how we use it. And herein lies the risk: when awareness tilts too far toward focusing on problems, it can create a sense of inevitability about their outcomes. This is where the nocebo effect – a concept borrowed from medicine but shockingly relevant to modern life – becomes essential for understanding the fine line between constructive and destructive awareness.

The cognitive loops central to the nocebo effect find their roots in over-activation of the Default Mode Network (DMN). When individuals experience hyper-awareness of risks, the DMN amplifies ruminative thought patterns, causing individuals to replay past fears and anticipate future harm. This over-activation traps the mind in a state of perpetual vigilance, turning what could be a call to action into a psychological burden.

3.1.2 The Modern Emphasis on Awareness: A Tool for Progress or a Recipe for Paralysis?

1. **The Rise of Awareness Culture:**
   - From mental health to climate change, awareness campaigns have reshaped how we think, talk, and act. To be informed is not just encouraged; it is seen as a moral duty. Knowing about issues like anxiety, systemic injustice, or environmental degradation is framed as the first step toward fixing them.
   - **Example**: Mental health awareness initiatives have normalized discussions around anxiety and depression, reducing stigma and encouraging people to seek help – a triumph of awareness as a social good.
2. **Awareness Overload: When the First Step Becomes the Only Step**:
   - However, awareness without pathways for resolution can feel like standing at the edge of a cliff with no bridge in sight. As problems proliferate and solutions seem elusive, the sheer volume of challenges can leave us frozen.
     - **Key Insight**: Awareness campaigns often stop short of offering actionable strategies, leaving individuals overwhelmed by the weight of problems they are now keenly attuned to.

DOI: 10.1201/9781003640288-5

3. **The Emotional Toll of Unresolved Awareness**:
   - Being endlessly aware of problems – whether personal, societal, or global – can lead to emotional fatigue, cynicism, and even resignation. It is the shadow side of our hyper-connected world: as awareness expands, so too does the potential for despair.

3.1.3 Introducing the Nocebo Effect: Awareness as a Double-Edged Sword

The nocebo effect provides a startlingly apt metaphor for the darker side of modern awareness. In medicine, the nocebo effect occurs when negative expectations – fearing pain, expecting failure – lead to actual harm. Patients who expect a drug to cause side effects are more likely to experience them, even when given a harmless placebo.

1. **Beyond Medicine: The Cultural Nocebo Effect**:
   - In the cultural and psychological realms, the nocebo effect extends far beyond the confines of clinical studies. The principle remains the same: **what we expect, we often experience**. Awareness campaigns, news cycles, and social media, while designed to inform and inspire, can inadvertently create an expectation of negativity that feeds on itself.
     - **Example**: A person bombarded with stories about burnout may begin to interpret ordinary stress as a harbinger of collapse.
     - **Key Idea**: When cultural narratives focus too heavily on crises, they can create self-fulfilling prophecies.
2. **The Feedback Loop of Awareness and Expectation**:
   - **How it works**:
     - Heightened awareness of a problem → Increased focus on potential negative outcomes → Amplified perception of those outcomes as inevitable → Escalation of emotional and societal distress.
   - This cycle, repeated across media and societal discourse, traps us in a loop of hyper-awareness and diminished hope.

3.1.4 Why the Nocebo Effect Matters: Awareness Without Guardrails

1. **Awareness as Amplification**:
   - Awareness campaigns often start with good intentions: to shine a light on problems that demand attention. Yet, without balanced narratives, they risk over-normalizing negativity, turning temporary challenges into permanent fixtures in our collective psyche.
     - **Example**: Mental health campaigns that emphasize the ubiquity of anxiety can inadvertently teach people to identify themselves as "always anxious," deepening feelings of helplessness.
2. **Why the Modern World is Particularly Susceptible**:
   - The interconnected nature of digital life exacerbates the nocebo effect:
     - Social media algorithms prioritize content that sparks outrage or fear, ensuring negative narratives dominate.
     - Continuous news cycles amplify crises without equal attention to solutions.
     - Personal stories of struggle, while fostering empathy, can unintentionally normalize despair as the default state.
3. **The Unspoken Challenge of Awareness**:
   - True awareness is not just about identifying problems; it's about navigating the emotional and psychological weight of knowing. Without tools for resilience, awareness becomes a burden rather than a beacon.

3.1.5 Key Question: How Can We Turn Awareness into Empowerment?

The central challenge of this chapter lies in answering one question: **How can we ensure that heightened awareness is a force for action and resilience, not despair?**

1. **The Stakes**:
   - If we fail to address the pitfalls of awareness, we risk creating a culture of passive observers – individuals overwhelmed by the sheer scale of the challenges they face. The nocebo effect will continue to erode motivation and hope, paralyzing action when it is most needed.
2. **The Opportunity**:
   - Recognizing the nocebo effect is not just a cautionary tale; it's a call to action. By understanding how expectations shape outcomes, we can consciously reframe awareness as a tool for empowerment. The goal is not to diminish the importance of awareness but to wield it more thoughtfully, pairing it with strategies that promote agency, optimism, and progress.

3.1.6 Conclusion to the Introduction

Awareness is a tool, but tools can cut in both directions. By introducing the concept of the nocebo effect and placing it within the context of modern life, this chapter sets the stage for a deeper exploration of how our cultural narratives shape perception and, ultimately, reality. It also begins the search for solutions: how to reframe awareness not as a spotlight on problems but as a searchlight for possibilities. As we move forward, we will investigate the manifestations of the nocebo effect, analyze Byung-Chul Han's critiques of modern society through this lens, and propose actionable strategies to transform awareness into a source of empowerment and collective resilience.

3.2 THE NOCEBO EFFECT: A MODERN PHENOMENON

The nocebo effect is more than a medical curiosity – it is a metaphor for how modern life often amplifies its own struggles. By revealing how negative expectations shape reality, the nocebo effect offers a profound lens through which to understand the psychological and cultural dynamics of our time. This section explores its roots in medicine and its manifestations in mental health, media, and societal expectations, exposing the subtle ways awareness can exacerbate the very challenges it seeks to address.

3.2.1 What is the Nocebo Effect?

3.2.1.1 Origins in Medical Science

The term "nocebo," meaning "I shall harm," originated in the medical field as a counterpoint to the placebo effect ("I shall please"). While the placebo effect highlights the remarkable power of positive expectations to heal, the nocebo effect underscores the darker side of belief: negative expectations can cause real harm.

1. **Key Insights from Medicine**:
   - **Adverse Outcomes from Expectations**: Patients warned about side effects are significantly more likely to experience them – even when receiving a harmless placebo.
     - **Example**: In a study on migraines, participants told that a medication might cause headaches reported higher incidences of them, even though they were given a sugar pill. The mere belief in harm shaped their physical response.
   - **The Mind–Body Connection**: The nocebo effect operates through psychological mechanisms such as stress, anxiety, and heightened vigilance. These trigger physiological responses, including increased cortisol levels, which can worsen symptoms or create new ones.

3.2.1.2 From Clinical Phenomenon to Cultural Force

While originally studied in the context of medicine, the principles of the nocebo effect extend far beyond the hospital or laboratory. In today's hyper-connected, hyper-aware world, its implications ripple across society, shaping our collective perceptions and experiences.

1. **The Cultural Nocebo Effect**:
   - Negative expectations reinforced by cultural narratives can create feedback loops where perceived problems become magnified.
     - **Example**: Constant discussions of burnout may lead individuals to interpret ordinary fatigue as a sign of deeper dysfunction, intensifying their stress.
   - The nocebo effect helps explain how societal challenges – when over-emphasized – risk becoming self-fulfilling prophecies.
2. **Why It Matters Today**:
   - In an era where information spreads at lightning speed, the nocebo effect becomes a cultural amplifier. Expectations of failure, decline, or despair – fueled by news media, social media, and public discourse – can solidify into collective realities.

3.2.2 Manifestations in Modern Life

The nocebo effect's influence permeates modern life, shaping how individuals and societies perceive and respond to challenges. From mental health narratives to media amplification and social expectations, the psychological dynamics of the nocebo effect are deeply embedded in our cultural fabric.

3.2.2.1 Mental Health Narratives

1. **The Normalization of Distress**:
   - Mental health campaigns have brought critical issues like anxiety, depression, and burnout into mainstream consciousness. While this has reduced stigma, it has also inadvertently normalized distress.
     - **Example**: A person exposed to repeated messaging about the ubiquity of burnout may begin to attribute ordinary stress to burnout, deepening their exhaustion through misinterpretation.
   - **Impact**: When distress becomes a widely accepted baseline, individuals may feel that their struggles are inevitable and insurmountable.
2. **The Risk of Pathologizing Everyday Experiences**:
   - Awareness campaigns often blur the line between normal human challenges and clinical conditions. As a result, people may begin to over-identify with diagnostic labels, reinforcing negative self-perceptions.
     - **Key Insight**: By framing struggles as defining features of identity, these narratives risk creating a self-fulfilling prophecy.

3.2.2.2 Media Amplification

The media plays a central role in shaping societal expectations and perceptions. As both a mirror and magnifier of reality, it is a key driver of the nocebo effect on a cultural scale.

1. **The Crisis Economy**:
   - In an era of 24/7 news cycles, crises dominate headlines – not because they are the only stories worth telling but because fear and outrage capture attention.
     - **Example**: Climate change reporting often emphasizes worst-case scenarios without balancing them with stories of resilience or actionable solutions. This leads to "climate doomism," where people feel overwhelmed and powerless to act.
   - **Impact**: Repeated exposure to crisis narratives can foster a sense of inevitability about decline, discouraging action, and deepening societal anxiety.

2. **Social Media and Doomscrolling**:
   - Social media exacerbates the nocebo effect through algorithms designed to maximize engagement. Negative or emotionally charged content is prioritized, creating a feedback loop of distress.
     - **Example**: Doomscrolling – endlessly consuming negative news – heightens feelings of despair and fuels the perception that problems are unmanageable.
   - **Psychological Cost**:
     - Constant exposure to crises reduces the brain's capacity for long-term thinking and problem-solving, leaving individuals emotionally exhausted and cognitively overwhelmed.
3. **Media as a Double-Edged Sword**:
   - While the media has the power to inform and mobilize, it also risks amplifying the very challenges it seeks to highlight. Without balanced narratives, its focus on crises can create a collective nocebo effect that stifles hope and action.

3.2.2.3 Social Expectations

1. **The Self-Fulfilling Prophecy of Societal Challenges**:
   - Cultural narratives often frame modern life through a lens of struggle, emphasizing challenges like loneliness, digital overload, and burnout as defining features of the human experience.
     - **Example**: The phrase "loneliness epidemic" reinforces the idea that isolation is universal and inescapable, which can discourage people from seeking connection or appreciating existing relationships.
2. **Expectation Shapes Behavior**:
   - Negative societal expectations can influence individual behavior, turning perceived risks into real outcomes:
     - **Digital Overload**: Constant warnings about the dangers of screen time may lead people to disengage entirely from digital tools that could enhance creativity or connection.
     - **Burnout**: Awareness campaigns highlighting the prevalence of burnout may foster a belief that exhaustion is inevitable, making individuals less likely to explore solutions.
3. **Group Identity and the Nocebo Effect**:
   - When societal challenges are emphasized without nuance, they can become part of group identity. This collective pessimism reinforces the belief that certain struggles are universal, limiting creativity and hope for solutions.

3.2.2.4 Should the Media's Role Be Highlighted?

Yes, the role of the media is not just important but central to understanding the cultural nocebo effect. As a primary source of information, the media shapes expectations on a societal scale. It reflects our anxieties but also amplifies them, turning isolated issues into pervasive narratives. Without critical examination, its influence risks perpetuating cycles of negativity.

By addressing the media's role, this section not only deepens the reader's understanding of how the nocebo effect manifests but also lays the groundwork for proposing balanced narratives and actionable solutions in later sections.

3.2.3 Conclusion to Section 3.2

The nocebo effect, while rooted in individual psychology, operates on a societal scale through mental health narratives, media amplification, and social expectations. As these forces converge, they create a feedback loop of negativity that influences how we perceive and respond to modern challenges. By recognizing its manifestations, we can begin to disrupt this cycle and explore how to reframe awareness as a source of empowerment – a journey that the following sections will illuminate.

3.3 BYUNG-CHUL HAN'S PHILOSOPHIES REVISITED

Byung-Chul Han's critique of modernity serves as both diagnosis and indictment – a piercing commentary on the societal forces that strip meaning, fracture attention, and erode well-being. His work exposes the ways in which late capitalist culture magnifies despair, rendering individuals isolated, exhausted, and alienated under the weight of relentless demands for optimization and transparency. But while his observations are incisive, they risk reinforcing the very helplessness they illuminate.

This section revisits Han's philosophies through the lens of the nocebo effect, highlighting how societal narratives amplify despair and exploring actionable strategies to transform his critiques into opportunities for empowerment.

3.3.1 Overview of Han's Critiques

Han's philosophy captures the pervasive sense of unease that characterizes life in the digital age. His key critiques, drawn from works such as *The Burnout Society* and *The Transparency Society*, dissect five interwoven phenomena that define modern alienation:

1. **The Erosion of the "We"**:
   - **Han's Argument**: Modern society prioritizes individualism at the expense of communal bonds. Digital culture, with its relentless focus on self-promotion, replaces meaningful connection with superficial validation.
   - **Key Insight**: In the shift from collective identity to fragmented individualism, the social fabric weakens, leaving a vacuum of belonging.
2. **The Infocalypse and Attention Deficit**:
   - **Han's Argument**: The relentless flood of digital information overwhelms cognitive capacities, fragmenting attention and hindering deep thought.
   - **Key Insight**: Inundated by distractions, individuals lose the ability to focus on what truly matters, fostering a sense of intellectual and emotional disorientation.
3. **The Burnout Society**:
   - **Han's Argument**: Modernity demands constant self-optimization, turning individuals into both workers and taskmasters. The pursuit of achievement becomes an internalized oppression, resulting in chronic burnout.
   - **Key Insight**: The cultural glorification of productivity reduces people to their output, leaving them physically and emotionally depleted.
4. **The "Transparency Society"**:
   - **Han's Argument**: The demand for perpetual visibility in the digital sphere destroys privacy and depth. Relationships become performative, robbed of mystery and intimacy.
   - **Key Insight**: Life in the age of transparency turns people into commodities, measured by likes, followers, and metrics that fail to capture human depth.
5. **The Violence of the Same**:
   - **Han's Argument**: In the competitive ethos of late capitalism, individuals exploit themselves in the endless pursuit of unattainable ideals. The pressure to "outdo oneself" perpetuates cycles of dissatisfaction.
   - **Key Insight**: Modern life's obsession with self-improvement masks a deeper violence – one in which individuals become both victims and perpetrators of their own exhaustion.

3.3.1.1 Connection to the Nocebo Effect

Each of Han's critiques resonates with the mechanisms of the nocebo effect, revealing how cultural narratives can amplify despair:

- **The Erosion of the "We"** normalizes isolation, leading individuals to expect – and experience – loneliness.
- **The Infocalypse** entrenches the belief that focus and clarity are unattainable.
- **Burnout Society** rhetoric reinforces exhaustion as an inevitable state.
- **The Transparency Society** fosters the perception that authenticity is no longer possible.
- **The Violence of the Same** traps individuals in cycles of self-inflicted inadequacy.

By drawing these connections, we see how Han's analyses illuminate the psychological toll of modern life – but also how they can risk deepening its paralysis. The challenge lies in reframing his critiques to inspire transformation rather than resignation.

3.3.2 Transformative Perspectives

Han's insights, while bleak, contain the seeds of a counter-narrative: one that shifts from critique to possibility. Each of his observations can be reframed to challenge the nocebo effect and offer pathways to resilience, connection, and empowerment.

1. **The Erosion of the "We"**
   - **Nocebo Amplification**:
     - Narratives of increasing loneliness reinforce expectations of isolation, making a meaningful connection seem impossible. This can lead individuals to retreat further, perpetuating the cycle of alienation.
   - **Transformative Perspective**:
     - **From Isolation to Digital Connection**:
       - Technology, while often blamed for weakening bonds, can also be harnessed to create vibrant digital communities. Platforms dedicated to shared interests, collaborative projects, and global causes prove that connection is not only possible but thriving.
       - **Actionable Strategies**:
         - Highlight platforms like Meetup for fostering in-person connections through shared interests.
         - Promote global communities, such as virtual book clubs or volunteer networks, that unite individuals across borders.
2. **The Infocalypse and Attention Deficit**
   - **Nocebo Amplification**:
     - The cultural obsession with information overload perpetuates the belief that focus is futile. Individuals feel trapped in fragmented attention, unable to reclaim control over their mental space.
   - **Transformative Perspective**:
     - **Reclaiming Focus Through Intentionality**:
       - Attention is not irreparably lost; it can be cultivated with conscious effort. Tools and practices like mindfulness, single-tasking, and digital minimalism offer concrete pathways to restore clarity and depth.
       - **Actionable Strategies**:
         - Advocate for mindfulness apps (e.g., *Headspace*, *Calm*) to help individuals practice intentional focus.
         - Introduce "digital sabbaths," designated periods for unplugging from screens to nurture reflective thinking.

3. **The Burnout Society**
   - **Nocebo Amplification**:
     - Burnout is often framed as the unavoidable cost of modern ambition, discouraging individuals from seeking alternative work-life paradigms.
   - **Transformative Perspective**:
     - **Redesigning Productivity for Sustainability**:
       - Burnout signals the need for systemic change. Flexible work models, reduced workweeks, and collaborative environments shift the focus from relentless productivity to well-being.
       - **Actionable Strategies**:
         - Highlight companies like Microsoft Japan, which saw a 40% productivity increase after implementing a four-day workweek.
         - Advocate for cultural shifts that normalize rest as a form of productivity.
4. **The "Transparency Society"**
   - **Nocebo Amplification**:
     - Transparency is portrayed as a totalizing force, eroding intimacy and depth in relationships. This narrative fosters resignation, making individuals feel powerless to preserve authenticity.
   - **Transformative Perspective**:
     - **Cultivating Depth Within Visibility**:
       - Transparency need not strip life of meaning; intentional boundaries can protect intimacy even in digital spaces.
       - **Actionable Strategies**:
         - Promote practices like "intentional sharing" on social media, where individuals curate what they share to reflect personal values.
         - Encourage offline rituals, such as family dinners or unplugged retreats, to deepen relational bonds.
5. **The Violence of the Same**
   - **Nocebo Amplification**:
     - The myth of self-optimization feeds the belief that personal worth is tied to relentless improvement. This perpetuates cycles of dissatisfaction, as individuals chase unattainable ideals.
   - **Transformative Perspective**:
     - **Redefining Success Through Collaboration**:
       - Success need not be a solitary pursuit. Collaborative models of achievement – focusing on shared goals and collective well-being – challenge the myth of individual perfection.
       - **Actionable Strategies**:
         - Highlight cooperative business models, such as worker-owned co-ops, where success is measured by community well-being.
         - Showcase stories of organizations that prioritize collaboration, such as open-source software projects.

3.3.3 Conclusion to Section 3.3

Byung-Chul Han's critiques cut to the core of modern alienation but they need not end in despair. When reframed through the lens of the nocebo effect, his insights become pathways to empowerment. Each of the challenges he identifies can inspire actionable change – whether by fostering connection, reclaiming focus, or redefining success. To move beyond critique, we must embrace the possibility of transformation, using Han's observations as a springboard for designing systems and narratives that nurture resilience and hope.

The next section builds on these ideas, examining how societal narratives themselves can be reshaped to mitigate the nocebo effect and foster collective empowerment.

COGNITIVE BIAS

THINKING BOX:
"Is your digital world mirroring reality, or simply echoing your own views? Ponder this carefully – maybe **42** is the clue."

3.4 THE NOCEBO EFFECT IN SOCIETAL NARRATIVES

Narratives are the scaffolding of societal perception – what they emphasize, normalize, and perpetuate profoundly shapes collective psychology. In a world increasingly defined by hyper-awareness of challenges, these narratives wield immense influence. While many aim to inspire awareness and action, they often fall prey to the nocebo effect, amplifying despair and reinforcing helplessness. This section unpacks the risks of over-normalization, the role of media and social platforms as amplifiers of negativity, and how these dynamics play out in mental health campaigns. The goal is to illuminate a path forward, one that uses narratives to empower rather than paralyze.

3.4.1 The Danger of Over-Normalization

3.4.1.1 How Narratives Solidify Challenges

When societal narratives repeatedly emphasize the prevalence or inevitability of certain problems, they risk transforming these issues from manageable challenges into entrenched realities. This mirrors the nocebo effect: repeated exposure to negative messaging creates a self-reinforcing loop of expectation and experience, ultimately shaping how individuals and communities perceive their world.

1. **The Risk of Normalizing Struggle:** By constantly highlighting problems like anxiety, burnout, or loneliness, narratives can shift the baseline of what people perceive as "normal." This over-normalization makes these struggles feel inevitable, fostering resignation rather than resilience.
   Example: Mental health awareness campaigns often cite alarming statistics such as "One in three people will experience anxiety in their lifetime." While intended to reduce stigma, such statements can inadvertently create the expectation that anxiety is a default state, leading individuals to attribute ordinary stress to a more severe problem.
2. **From Awareness to Paralysis:** Over-emphasis on challenges can overwhelm individuals, leaving them feeling as though solutions are out of reach. This leads to:
   - *Self-Fulfilling Prophecies:* Expecting burnout or anxiety can amplify their occurrence.
   - *Fear of Resilience:* Efforts to overcome challenges may feel invalidated by the overwhelming message that these struggles are universal.
3. **Why Over-Normalization is Dangerous:**
   - It creates a culture where vulnerability is highlighted, but strength is overlooked.
   - It risks stigmatizing those who don't identify with the prevailing narrative of struggle, making resilience and hope less visible in public discourse.

3.4.2 Media and Social Platforms

3.4.2.1 Media as an Amplifier of the Nocebo Effect

Media and social platforms are not passive conveyors of information; they actively shape the way narratives are framed, prioritized, and consumed. As the dominant storytellers of modern life, they play a central role in amplifying negativity, often intensifying the nocebo effect on a collective scale.

1. **Crisis-Driven Headlines: The Attention Economy:** News outlets thrive on engagement, and nothing captures attention more effectively than fear or outrage. This leads to a pervasive focus on crises, which dominate public consciousness.
   Example: Headlines like "The Loneliness Epidemic is Unstoppable" or "Burnout: The Modern Plague" frame challenges as insurmountable. Such messaging can discourage hope or action, creating a sense of inevitability around societal issues.
2. **Social Media: The Doomscrolling Trap:** Social platforms exacerbate this problem with algorithms that prioritize emotionally charged content. Fear, outrage, and sadness travel faster and farther than balanced or hopeful messages.
   Doomscrolling: The endless consumption of negative content fosters anxiety and powerlessness, reinforcing the belief that problems are too vast or too complex to solve. This mimics the nocebo principle: repeated exposure to negativity conditions the mind to expect and experience more of the same.
3. **Echo Chambers: Reinforcing Despair:**
   How Echo Chambers Work: Social media communities often filter information to align with shared beliefs, magnifying negative narratives without introducing balancing perspectives.
   Example: A group focused on workplace burnout might exclusively share stories of exhaustion and failure, deepening members' belief that their struggles are inescapable.
   Impact: By removing alternative perspectives, echo chambers entrench collective despair, limiting opportunities for solutions and resilience.
4. **The Digital Shadow – A New Dimension of Narrative Influence:** In the digital age, an individual's online presence contributes to a persistent "digital shadow." Unlike a physical shadow that disappears when we move, a digital shadow is the cumulative trail of data and information that exists about a person in the digital realm – created not only through their own actions but also through the actions of others, particularly parents and caregivers. This shadow begins forming before we can make our own digital choices, as family members share photos, stories, and personal details online.
   Digital Shadow – **Definition**: The cumulative trail of data and information that exists about an individual in the digital realm – created not only through their own actions but also through the actions of others. This shadow can significantly impact one's future opportunities, relationships, and well-being. It is partly self-created and partly inherited or imposed by others' digital behaviors.

When societal narratives focus heavily on problems and crises, they can shape the nature and perception of one's digital shadow. Negative expectations, over-normalized struggles, and persistent online portrayals can become part of an individual's online identity. This transforms the digital shadow from a neutral record into a powerful narrative force that influences how individuals see themselves and are seen by others.

For a practical guide on managing digital shadows in family contexts, see: Awareness for Parents and Kids: https://www.linkedin.com/posts/muhlert_awareness-for-parents-and-kids-activity-7211305097526476800-6MXC/

3.4.3 Case Study: Mental Health Campaigns

3.4.3.1 Mental Health Awareness: A Double-Edged Sword

Mental health campaigns have made significant strides in reducing stigma, but their framing can inadvertently activate the nocebo effect. While these campaigns aim to foster empathy and encourage help-seeking, their emphasis on prevalence and struggle can sometimes overshadow messages of hope and recovery.

1. **Prevalence-Focused Messaging and Its Risks:**
   The Problem:
   Messaging that highlights the ubiquity of mental health struggles – e.g., "50% of people will experience depression"—risks normalizing distress without offering pathways to resilience. It frames mental health issues as inevitable rather than addressable.
   Example: A campaign that repeats "Burnout affects everyone" may unintentionally suggest that burnout is an unavoidable part of modern work life, discouraging individuals from seeking solutions.
   Result:
   Individuals internalize these messages, experiencing greater distress as they align their personal challenges with the grim societal narrative.
2. **Shifting the Narrative: Solutions and Recovery:** Effective campaigns strike a balance between raising awareness and inspiring action. By pairing statistics with stories of recovery and actionable strategies, they move beyond problem identification to empowerment.
   Example: A mental health initiative that highlights personal recovery stories – such as individuals who overcame anxiety through therapy, mindfulness, or lifestyle changes – encourages others to see challenges as surmountable.
   Organizations Leading the Way: Groups like Heads Together focus on practical steps for mental well-being, including accessible resources and tools for stress management.
   The Role of Actionable Strategies: Campaigns that mitigate the nocebo effect actively provide tools and frameworks for resilience:
   - *Resources:* Therapy directories, hotlines, and self-help platforms.
   - *Community:* Encouraging participation in peer-led groups, forums, or activities that foster connection.
   - *Practical Skills:* Promoting mindfulness, exercise, or time-management strategies to help individuals navigate stress more effectively.

3.4.4 Conclusion to Section 3.4

The narratives societies choose to tell are more than reflections of reality – they are architects of collective experience. When these narratives over-emphasize crises and challenges without offering hope or solutions, they risk solidifying the very struggles they aim to address. The media and social platforms, as powerful amplifiers of societal discourse, must be held accountable for how they perpetuate the nocebo effect in their headlines, feeds, and echo chambers. Additionally, the concept of the digital shadow adds a new dimension: persistent, data-driven narratives about individuals contribute to shaping their identities and futures in ways that may reinforce negative expectations.

The case of mental health campaigns demonstrates the delicate balance required: effective narratives must not only raise awareness but also inspire action and resilience. By reframing problems as opportunities, emphasizing stories of recovery and growth, and encouraging individuals to take constructive steps, societal narratives can become a force for empowerment rather than despair.

The following sections will explore how such balanced narratives, tools, and frameworks can be systematically applied to foster collective agency and hope, moving us beyond mere awareness into meaningful transformation. As we bridge these insights into a more detailed and practical technical framework, the goal remains to harness complexity and narrative power to create emergent ethical landscapes of possibility and growth.

3.5 THE NOCEBO EFFECT IN CONTEXT: A DEEPER DIVE

The nocebo effect is not merely an abstract concept; it is an active force shaping modern societal narratives and individual experiences. While Han's critique of the "achievement society" and "infocalypse" brings attention to the stresses of contemporary life, it also invites a pivotal question: Are we, in our pursuit of understanding and addressing these stresses, amplifying their impact?

3.5.1 The Psychology of Negative Expectation

The nocebo effect is rooted in the power of expectation. If individuals anticipate negative outcomes, their physiological and psychological states can align with those expectations. In the realm of mental health and societal critique:

- **Self-Reinforcing Narratives**: Discussions about burnout and digital fatigue, while valid, can create a feedback loop where individuals interpret normal stressors as signs of systemic failure.
- **The Danger of Over-Diagnosis**: Highlighting societal problems without offering actionable solutions risks reducing agency, leading individuals to feel overwhelmed and powerless.

For example, a pervasive narrative around "always being watched" by digital platforms may lead to a sense of resignation, where individuals assume their privacy is already lost. This nocebo-driven apathy undermines efforts to reclaim control and agency.

3.5.2 The Balance Between Awareness and Activation

Han's philosophies rightly emphasize the societal and structural challenges of modern life, from the erosion of community bonds to the pressures of self-optimization. Yet, if these critiques are consumed uncritically, they risk feeding into the very cycles they seek to critique:

1. **Erosion of Agency**: Constant reminders of systemic oppression or digital surveillance can paralyze rather than mobilize individuals.
2. **Cultural Paralysis**: A collective focus on problems without envisioning solutions fosters societal stagnation, where critique replaces action.

The antidote lies in balancing awareness with activation:

- Awareness identifies challenges and systemic imbalances.
- Activation translates that knowledge into steps toward personal and collective empowerment.

3.5.3 Reframing Han's Philosophies: A Cautious Approach

Byung-Chul Han's critique of the "achievement society" and the "violence of the same" presents a compelling warning against the dangers of self-exploitation and over-optimization. However, framing these insights through the lens of the nocebo effect offers a crucial reframing:

1. **From Self-Exploitation to Empowerment**:
   - Han critiques how individuals internalize societal pressures to achieve, leading to self-inflicted burnout. Yet, understanding this through the nocebo lens reveals that these pressures often arise from perceived, rather than real, societal demands.

- **Reframe**: Instead of viewing productivity as an external imposition, individuals can reclaim it as a tool for self-defined growth and contribution.

2. **From Infocalypse to Selective Engagement**:
   - The overwhelming information flow described by Han can lead to a sense of helplessness. The nocebo effect exacerbates this, as individuals expect to be misinformed or manipulated, leading to disengagement.
   - **Reframe**: Emphasizing selective engagement – actively choosing which narratives and information streams to prioritize – can counteract this overwhelm.
3. **From Transparency to Boundaries**:
   - While Han laments the loss of "aura" and mystery in the age of transparency, the nocebo effect warns against internalizing this as inevitable.
   - **Reframe**: Individuals and communities can deliberately cultivate boundaries and practices that preserve privacy and depth.

3.5.4 Practical Strategies to Counteract the Nocebo Effect

To move from critique to action, individuals and societies must develop strategies to mitigate the nocebo effect while harnessing the insights of philosophical critique:

1. **Shifting Language and Narratives**:
   - Replace deterministic language ("everyone is burned out") with possibility-oriented framing ("many are finding new ways to balance productivity and rest").
   - Highlight stories of resilience and recovery alongside critiques of systemic problems.
2. **Building Personal Agency**:
   - Encourage individuals to identify areas where they have control, even within systemic constraints.
   - Promote micro-actions, such as taking "digital sabbaths" or redefining personal success metrics, to reclaim autonomy.
3. **Creating Collective Solutions**:
   - Foster communities of practice where individuals share strategies for navigating modern challenges without succumbing to cynicism.
   - Advocate for systemic change while emphasizing personal empowerment, ensuring that critique fuels action rather than resignation.
4. **Engaging with Han's Critique Productively**:
   - Use Han's philosophies as tools for inquiry rather than conclusions. Ask, "How do I contribute to the violence of the same?" or "What boundaries can I set against the infocalypse?"

3.5.5 A Vision Beyond the Nocebo Effect

Ultimately, Han's critiques and the nocebo effect converge on a central insight: perception shapes reality. The way individuals and societies frame challenges determines their ability to respond effectively. By integrating philosophical critique with psychological awareness, it becomes possible to move beyond the passive acceptance of modern malaise toward an active, intentional reshaping of individual and collective narratives.

- **Philosophical Critique as a Catalyst**: Rather than allowing Han's warnings to paralyze, use them as prompts for action.
- **The Nocebo Effect as a Checkpoint**: Regularly assess whether awareness of societal problems is fostering action or amplifying helplessness.

3.6 FROM AWARENESS TO EMPOWERMENT: MITIGATING THE NOCEBO EFFECT

The nocebo effect need not be a harbinger of despair. Its existence highlights a truth that is both sobering and empowering: the way we frame challenges shapes how we experience them. While awareness can amplify fear and helplessness, it can also catalyze resilience, agency, and transformation – if deliberately harnessed. This section introduces the **three Rs Framework** – Reframe, Rebalance, Rebuild – as a toolkit for converting awareness into empowerment and offers compelling success stories that illustrate its real-world application.

3.6.1 Framework for Change: The Three Rs

The **three Rs Framework** provides a structured approach to disrupting the nocebo effect and channeling awareness into constructive action. Each principle targets a critical lever for change, transforming how individuals and societies perceive and respond to challenges.

3.6.1.1 Reframe: Turning Challenges into Opportunities

Reframing is about reshaping perceptions. By interpreting difficulties as opportunities for growth, individuals and societies can transform negative expectations into positive potential.

1. **Digital Overwhelm → Intentional Focus Practices**:
   - **The Challenge**: The avalanche of digital stimuli fragments attention and fosters feelings of powerlessness.
   - **The Opportunity**: Digital overwhelm can spark a movement toward intentional engagement and conscious living.
   - **Actionable Strategies**:
     - Encourage mindfulness in digital interactions through apps like *Headspace* or *Calm*, which promote reflection and focus.
     - Advocate for "single-tasking" over multitasking, using techniques like Pomodoro timers to create structured focus sessions.
     - Normalize digital detoxes: structured time away from screens to nurture deeper thinking and interpersonal connection.
2. **Burnout → Rethinking Priorities and Sustainable Work Rhythms**:
   - **The Challenge**: Burnout is often framed as the price of ambition, discouraging efforts to reimagine work-life balance.
   - **The Opportunity**: Burnout signals a societal shift in values, creating space to prioritize well-being over relentless productivity.
   - **Actionable Strategies**:
     - Promote workplace policies such as reduced workweeks, flexible schedules, and mandatory vacation days.
     - Reframe downtime as productive, emphasizing its role in fostering creativity and long-term success.
     - Highlight models like *slow living* that focus on savoring experiences rather than accumulating accomplishments.

3.6.1.2 Rebalance: Advocating for Balanced Narratives

Balanced narratives are antidotes to the negativity bias inherent in media and awareness campaigns. By pairing challenges with solutions, they encourage both understanding and action.

1. **Mental Health: Share Stories of Resilience Alongside Challenges**:
   - **The Problem**: Over-emphasis on mental health struggles can normalize distress, reinforcing the perception that recovery is unattainable.

- **The Solution**: Narratives that celebrate resilience and recovery empower individuals to seek solutions.
- **Examples**:
  - Campaigns highlighting people who overcame depression through therapy, mindfulness, or community support can model actionable pathways.
  - Stories showcasing the impact of peer-support programs or innovative therapeutic approaches foster hope and relatability.

2. **Societal Critique: Combine Systemic Analysis with Solutions**:
   - **The Problem**: Societal critiques often stop at identifying crises, leaving audiences overwhelmed by the scale of the challenges.
   - **The Solution**: Pairing systemic analysis with examples of progress and innovation inspires collective optimism.
   - **Examples**:
     - Climate change discussions could feature successful renewable energy projects, grassroots environmental initiatives, and actionable steps for individuals.
     - Economic inequality narratives might include profiles of organizations building inclusive economies, such as co-ops or microfinance programs.

3.6.1.3 Rebuild: Emphasizing Personal Agency and Community Action

Rebuilding is about cultivating action – both individual and collective. By fostering a sense of agency, it interrupts the cycle of helplessness and replaces it with momentum.

1. **Steps for Individuals**:
   - **Limit Negative Content Consumption**:
     - Encourage people to audit their digital feeds, unfollowing accounts that promote fear-driven content and prioritizing solution-oriented sources.
   - **Set Achievable Goals**:
     - Break daunting problems into actionable steps, reinforcing progress and capability.
     - Example: Someone overwhelmed by climate change could start by adopting one sustainable habit, such as reducing food waste or biking to work.
   - **Build Support Systems**:
     - Emphasize the importance of connection, encouraging participation in community groups, therapy, or peer networks to counteract isolation.
2. **Collective Action**:
   - **Foster Optimism Through Local Initiatives**:
     - Highlight grassroots movements as powerful engines of change. From urban gardening projects to mental health workshops, these initiatives showcase how communities can address global challenges locally.
   - **Transform Workplace Cultures**:
     - Advocate for organizations to embrace collaborative well-being. Programs that prioritize collective success over competition foster resilience and engagement.
     - Example: Companies implementing team-building retreats or mental health resources create cultures of mutual support.

3.6.2 Case Studies: Success Stories

Case studies provide a glimpse into the transformative power of the **three Rs Framework**, demonstrating how reframing, rebalancing, and rebuilding have shifted negative narratives into opportunities for progress.

1. **Digital Fatigue: Reclaiming Focus and Balance**
   - **The Challenge**: Screen-time anxiety has risen alongside warnings of digital addiction, leading to guilt and disengagement from technology's potential benefits.

- **The Solution**: Mindful technology use empowers individuals to reclaim their digital habits without sacrificing productivity or connection.
- **Examples**:
  - Apps like *Forest* encourage focus by turning screen-free time into a gamified experience, planting virtual trees for every uninterrupted session.
  - *Google's Digital Well-Being* tools allow users to set app limits and schedule downtime, enabling intentional engagement.
- **Impact**:
  - Users report increased focus, reduced stress, and a greater sense of control over their digital lives. These tools demonstrate that intentionality, not avoidance, is key to navigating the digital age.

2. **Burnout in Corporate Culture: Redefining Success**
   - **The Challenge**: Burnout has been glamorized as a sign of dedication, leaving employees feeling trapped in cycles of overwork.
   - **The Solution**: Innovative organizations are redefining success by prioritizing well-being, creativity, and collaboration.
   - **Examples**:
     - **Microsoft Japan**: Implemented a four-day workweek, resulting in a 40% productivity increase while enhancing employee satisfaction.
     - **Basecamp**: Enforces strict work-life boundaries, such as no after-hours emails, to protect employees' personal time.
     - **Atlassian**: Encourages "ShipIt" days, where employees work on passion projects, fostering creativity and preventing burnout.
   - **Impact**:
     - These policies have led to higher retention rates, stronger employee engagement, and increased innovation, proving that well-being and productivity are mutually reinforcing.

3.6.3 Conclusion to Section 3.6

The nocebo effect thrives on unchecked negativity and disempowering narratives, but it is not inevitable. By reframing challenges as opportunities, rebalancing narratives with solutions, and rebuilding a sense of agency and community, we can disrupt this cycle and transform awareness into a force for resilience and growth.

The success stories of addressing digital fatigue and reimagining corporate culture exemplify the power of these strategies. They prove that solutions not only exist but are already reshaping how we live and work. As we look ahead, the task is clear: to adopt and scale these principles, ensuring that awareness no longer stops at acknowledgment but instead propels us toward meaningful change.

REALITY'S INFILTRATION

THINKING BOX:

"Where does the physical end and the virtual begin? If there's a line, it might just be hidden in the same corner as **42**."

3.7 CONCLUSION: HARNESSING THE NOCEBO EFFECT FOR GOOD

Awareness is the defining tool of our age. It is the lens through which we diagnose challenges, understand the world, and seek to improve it. Yet, as the nocebo effect reminds us, awareness is not inherently good or bad – it is a double-edged sword. Left unchecked, it can magnify despair, fostering paralysis instead of progress. But when wielded with intention, it becomes transformative, a force that magnifies resilience, inspires action, and redefines what is possible.

This chapter has journeyed through the paradox of awareness, exposing how the nocebo effect infiltrates our personal and collective narratives. It has also illuminated the tools we need to harness awareness for good, turning a force that often amplifies despair into one that fuels hope and action. Now, the challenge lies in applying these insights to shape a future where awareness no longer stops at identifying problems but drives us toward meaningful solutions.

Summary

1. **Awareness is a Tool, Not an Endpoint**:
   - Awareness is the beginning of change, but it cannot stand alone. On its own, awareness is passive – a spotlight that illuminates problems without offering resolution. Its true power lies in what follows: how we respond, adapt, and transform the insights it provides.
2. **The Nocebo Effect Highlights the Need to Balance Critique with Solutions**:
   - Modern narratives are disproportionately tilted toward critique, often amplifying fear and helplessness. The nocebo effect thrives in this imbalance, turning awareness into a source of distress. But critique without solutions is incomplete – it leaves us in the dark. To break this cycle, we must insist on narratives that pair systemic analysis with tangible paths forward.
3. **Empowerment Comes From Reframing Challenges and Emphasizing Agency**:
   - The antidote to the nocebo effect is agency. By reframing challenges as opportunities and focusing on what we can do – individually and collectively – we unlock the transformative potential of awareness. Empowerment is not born from passivity; it is forged through action.

Call to Action

1. **Imagine Awareness as a Lens That Magnifies Resilience and Transformation**:
   - Picture a world where awareness does not stop at diagnosis but becomes a catalyst for growth. Imagine mental health campaigns that celebrate recovery as much as they illuminate struggle, climate narratives that inspire action as much as they warn of danger, and workplace cultures that redefine productivity to center well-being. Awareness, when paired with possibility, transforms fear into hope, resignation into resilience, and passivity into power.
   - **Vision for the Future**:
     - Awareness could evolve from a mirror reflecting despair into a searchlight illuminating solutions. This shift requires deliberate effort: from individuals choosing hope over cynicism, to organizations championing balanced narratives, to media platforms amplifying stories of progress alongside those of challenge.
2. **Acknowledge the Role of Perception in Shaping Outcomes**:
   - Perception is the engine of human experience. How we see the world defines what we believe is possible. The nocebo effect reveals the dangers of unchecked negative perceptions, but it also offers an invitation: by consciously shaping our narratives, we can reshape our realities.
     - **At the Individual Level**: Recognize how your internal narratives influence your experiences. Actively seek stories, data, and role models that reinforce your capacity to grow and thrive.
     - **At the Collective Level**: Advocate for systemic changes in how challenges are framed. Support campaigns, media outlets, and organizations that balance critique with actionable hope.

3.7.1 Toward Empowered Philosophical Practice

By viewing Han's insights through the lens of the nocebo effect, this chapter offers a nuanced approach to modern challenges. While societal critique remains essential, it must be accompanied by strategies that empower individuals to navigate and reshape their realities.

The nocebo effect underscores the importance of perception in shaping experience. If the prevailing narrative focuses solely on what is wrong, individuals risk internalizing those narratives as destiny. By balancing critique with agency, awareness with activation, and philosophy with practicality, it becomes possible to engage with the complexities of modern life without being overwhelmed by them.

This extended chapter invites readers to:

- Question narratives that emphasize inevitability over possibility.
- Reframe modern challenges as opportunities for growth and creativity.
- Build resilience by blending philosophical critique with actionable change.

3.7.2 A Final Thought

The nocebo effect is not simply a cautionary tale – it is a challenge and an opportunity. It shows us what happens when negativity dominates our narratives but also illuminates the immense power of perception to shape our outcomes. What if we could flip the script? What if awareness became a force that expanded our sense of possibility instead of contracting it?

To harness the nocebo effect for good, we must shift our collective mindset. Awareness must not paralyze; it must propel. Critique must not end at acknowledgment; it must point toward solutions. Despair must not dominate; it must be counterbalanced by resilience.

Imagine a world where societal narratives inspire action, where media magnifies innovation and progress as much as it highlights challenge, and where individuals see themselves not as passive recipients of the world's struggles but as active participants in its transformation. This is the future we can create – if we choose to wield awareness wisely.

The task is ours: to create a culture that chooses empowerment over helplessness, action over resignation, and transformation over despair. Let this be the lens through which we shape our world. Awareness, in its most powerful form, is not the end of the story – it is where the story begins.

Part 2

Cognitive and Behavioral Insights

TEASER: MASTERING THE MENTAL GAME IN A COMPLEX WORLD

If Part 1 gives you the "why" of complexity and its moral stakes, **Part 2** focuses on the **mind** – the instrument through which we experience complexity, interpret data, and shape decisions. It reveals how **human cognition** evolves in an age of accelerating change, offering new vantage points on **thought patterns**, **biases**, and the subtle influences at play in every decision we make.

WHY READ ON?

- **Elevate Your Mental "OS"**: Complexity requires more than standard reasoning. You'll learn how the Default Mode Network ties into introspection, creativity, and **cognitive resilience**, enabling you to handle ambiguity with greater agility.
- **Unmask Cognitive Pitfalls**: Biases like confirmation bias or availability heuristics aren't random flaws; they're often **adaptive** shortcuts that can backfire in fast-evolving contexts. By recognizing them, you gain the power to refine your judgments.
- **Merge Ethical and Psychological Realms**: Moral clarity from Part 1 meets the cognitive domain here, reinforcing why **fairness, empathy, and rational thinking** can't be separated from how our brains process information.

WHAT AWAITS YOU?

- **Actionable Techniques to Disarm Biases**: From simple checklists to deeper introspective methods, you'll gain tools to spot when mental shortcuts might sabotage your decisions – be it in groupthink or personal life choices.
- **Adaptive Social Cognition**: Discover how social contexts shape your beliefs and behaviors, and how "Enhanced Adaptive Social Cognition Theory" (EASCT) can guide you through group dynamics or cultural shifts.
- **Building Cognitive Resilience**: Strengthen your mental flexibility so you're not paralyzed by complexity but rather **empowered** to adapt quickly. By the end of Part 2, you'll have the mental acuity to pair with the ethical framework you developed in Part 1.

DOI: 10.1201/9781003640288-6

4 The Default Mode Network
Neural Autonomy and Implications for AGI

Implications for Artificial General Intelligence (AGI)

Inspired by the DMN, future AGI could integrate:

1. **Self-Reflection Modules** – Simulating introspection for adaptive learning.
2. **Spontaneous Thought Generators** – Enhancing creativity and problem-solving.
3. **Dynamic Attention Systems** – Balancing exploration and goal-focused action.

Developing **DMN-like AGI** requires overcoming challenges in **cognitive flexibility, ethical alignment, and computational efficiency**, ensuring intelligent systems **mirror human adaptability and creativity** while remaining **safe and beneficial**.

4.1 INTRODUCTION: THE HIDDEN ARCHITECT OF THOUGHT

Imagine this: you're sitting by a window, staring out at the world without any specific task at hand. Your mind begins to wander, connecting distant memories with new ideas, and suddenly, a solution to a problem you hadn't consciously been working on emerges. These moments of clarity – when insight seems to strike out of nowhere – are not accidents. They are the handiwork of a remarkable system within your brain: the **Default Mode Network (DMN).**

Unlike the brain's task-positive networks, which activate when you're solving an immediate problem or focusing on a specific task, the DMN operates in the background, weaving together past experiences, future possibilities, and abstract connections. It is the brain's "default" setting when we are at rest, providing a foundation for reflection, creativity, and decision-making.

4.1.1 Why Study the DMN?

The DMN is more than just an interesting neurological phenomenon. Its significance lies in two key areas:

1. **Human Cognition:** The DMN shapes the very processes that define us as humans – our ability to imagine, reflect, and plan.
2. **Artificial Intelligence:** As we move closer to developing Artificial General Intelligence (AGI), the DMN offers valuable insights into building systems capable of creative and adaptive thought.

This chapter explores the profound role of the DMN as the hidden architect of thought, bridging the gap between neuroscience and the frontier of intelligent systems. By understanding its intricate functions, we gain not only a deeper appreciation of the human mind but also a potential roadmap for designing AGI systems with similar capabilities.

DOI: 10.1201/9781003640288-7

4.1.2 A Journey Through Neural Autonomy and Creativity

We begin by introducing the **Neural Autonomy Spectrum**, a framework that explains how the brain transitions between reflective, deliberate thought and automatic, task-focused behaviors. Next, we uncover the DMN's pivotal role in creativity and innovation, dissecting how it transforms scattered ideas into coherent, groundbreaking insights. Finally, we examine the implications of these findings for AGI, exploring how DMN-inspired architectures could redefine machine intelligence and its potential for introspection and creativity.

As we unravel the mysteries of the DMN, we'll discover not only how it helps us think and create but also how its principles might inspire a new generation of intelligent machines.

SCIENCE AS INQUIRY

THINKING BOX:

Is science a checklist of truths or a relentless pursuit of better questions? The curiosity leading us forward might also point to **42**.

4.2 NEURAL AUTONOMY SPECTRUM: A FRAMEWORK FOR DECISION-MAKING

The human brain is a marvel of dynamic complexity, constantly balancing self-directed introspection with task-oriented focus. At the center of this balancing act is the interplay between the **DMN** and task-positive networks, such as the **Central Executive Network (CEN)** and the **Salience Network (SN)**. This intricate coordination allows us to move fluidly between reflective, autonomous thought and the immediate demands of the external world.

To understand this balance, we introduce the **Neural Autonomy Spectrum**, a framework that categorizes decision-making and thought processes along a continuum, from fully deliberate and reflective to entirely automatic and reactive. This spectrum helps illustrate the DMN's pivotal role in shaping our thoughts, decisions, and behaviors.

4.2.1 The Neural Basis of Autonomy

The DMN is not a singular brain region but a constellation of interconnected areas that work together to enable self-referential thought and internal simulations. Its key components include:

- **Medial Prefrontal Cortex (mPFC):**
  - Orchestrates self-referential thinking, such as reflecting on one's identity, values, and emotional states.
  - Integrates emotional and cognitive inputs for value-based decision-making.
- **Posterior Cingulate Cortex (PCC):**
  - Processes autobiographical memories, enabling reflection and personal evaluation.
  - Acts as a hub, coordinating activity across DMN regions.
- **Angular Gyrus:**
  - Synthesizes information from multiple domains, supporting semantic understanding and conceptual integration.
  - Facilitates the construction of coherent narratives from fragmented thoughts.
- **Hippocampal Formation:**
  - Anchors memory consolidation and retrieval.
  - Plays a critical role in simulating future scenarios based on past experiences.

Together, these regions enable the DMN to function as a "neural stage," where our inner thoughts, emotions, and memories interact to shape our perception of the world and ourselves.

4.2.2 The DMN-Attention Network Dynamic

The **DMN** operates at the core of introspection and internal thought, but its role in cognition cannot be fully understood in isolation. The DMN's counterpart, the **Attention Network**, is responsible for external focus, goal-directed behavior, and adapting to environmental stimuli. Together, these networks create a rhythmic interplay, dynamically alternating between reflection and action. This dynamic interaction is foundational to cognitive flexibility, shaping everything from moment-to-moment behavior to long-term planning and decision-making.

4.2.2.1 The Attention Network: Our External Focus System

While the DMN governs self-referential thought and internal simulations, the Attention Network anchors the brain in the external world, ensuring engagement with immediate tasks and goals. It consists of two primary components:

- **Dorsal Attention Network (DAN):**
  - Responsible for voluntary, sustained attention, such as focusing on a task or maintaining concentration over time.
  - Plays a key role in goal-oriented behavior, helping prioritize and organize actions based on current objectives.
- **Ventral Attention Network (VAN):**
  - Activated by unexpected stimuli, enabling rapid reorientation of attention. For instance, a sudden noise can draw focus away from a task.
  - Facilitates adaptability by ensuring responsiveness to changing environmental demands.

The interplay between these subsystems allows the Attention Network to manage external focus effectively, complementing the DMN's introspective processes.

4.2.2.2 Dynamic Network Interactions

The relationship between the DMN and the Attention Network is governed by **reciprocal suppression**, a mechanism that ensures efficient resource allocation in the brain. When one network is active, the other is typically suppressed, allowing for smooth transitions between internal reflection and external engagement.

- **Reciprocal Suppression**:
  - During moments of introspection or creativity, DMN activity suppresses the Attention Network, enabling undistracted focus on internal processes such as planning or imagining future scenarios.
  - Conversely, when immediate tasks demand attention, the Attention Network suppresses DMN activity, reducing the distraction of reflective thoughts.
- **Switching Dynamics**:
  - Shifts between these networks occur rapidly and are triggered by internal or external stimuli. For example:
    - A sudden sound can activate the Attention Network, interrupting introspection to assess potential threats.
    - Conversely, a quiet, undemanding environment can allow the DMN to dominate, fostering creative or reflective thinking.
- **Resource Allocation**:
  - The brain allocates resources dynamically, balancing introspection and action to meet cognitive demands. This flexibility allows humans to navigate complex environments while maintaining the ability to reflect and adapt.

4.2.2.3 Cognitive Implications

The interplay between the DMN and Attention Network underpins critical cognitive functions, with significant implications for behavior, decision-making, and mental health:

- **Balance and Optimal Function**:
  - A healthy balance between DMN and Attention Network activity promotes cognitive flexibility, allowing for seamless transitions between reflection and action.
  - For example, shifting from planning a speech (DMN-dominant) to delivering it (Attention Network-dominant) illustrates the importance of this dynamic.
- **Imbalances and Cognitive Costs**:
  - Over-activation of the DMN can result in excessive rumination, contributing to conditions like depression or anxiety.
  - Conversely, over-activation of the Attention Network may lead to hyper-focus, making it difficult to disengage from tasks – a feature often associated with anxiety disorders or obsessive behaviors.
- **Role in Attention Regulation**:
  - Effective switching between networks is a hallmark of attention regulation, supporting the ability to stay engaged with a task while remaining adaptable to new demands.
  - Impairments in this regulation are associated with attention deficit disorders, such as ADHD, where the brain struggles to suppress competing neural activity.

4.2.2.4 Connecting to the Autonomy Spectrum

Understanding the dynamic interplay between the DMN and Attention Network provides essential context for exploring the **Neural Autonomy Spectrum**. This spectrum describes how the brain transitions between reflective, self-directed thought and automatic, externally-driven behaviors. The ability to shift between these states relies heavily on the balance and regulation of these two networks.

By examining the interaction between the DMN and Attention Network, we see how the brain creates a foundation for cognitive states that range from deep introspection to immediate action. This dynamic balance lays the groundwork for the Neural Autonomy Spectrum, offering a framework to understand how these states manifest in human behavior and decision-making.

4.2.3 The Autonomy Spectrum Framework

The **Neural Autonomy Spectrum** reframes decision-making as a continuum rather than a binary choice between conscious deliberation and automatic responses. This framework identifies three distinct states of neural autonomy:

1. **High Autonomy States**:
   - **DMN-Dominant Processes:**
     - Introspective activities, such as self-reflection, moral reasoning, and long-term planning, are guided by DMN activation.
     - These states involve deliberate, slower thought processes that integrate personal history, emotional context, and future projections.
   - **Examples:**
     - Making career decisions, navigating complex relationships, or resolving ethical dilemmas.
2. **Medium Autonomy States**:
   - **Balanced Activation:**
     - These states occur when the DMN and task-positive networks work in harmony, balancing introspection with external engagement.

    - Decision-making here is adaptive, relying on learned patterns while maintaining conscious oversight.
  - **Examples:**
    - Planning a route for a trip or engaging in meaningful social interactions.
3. **Low Autonomy States**:
  - **Task-Positive Dominance:**
    - These states are characterized by reflexive or habitual behaviors governed by automatic neural pathways, such as those in the basal ganglia.
  - **Examples:**
    - Reflexively withdrawing a hand from a hot surface, typing on a keyboard, or driving a familiar route without conscious thought.

4.2.4 Network Interactions in Behavioral Control

A hallmark of human cognition is its ability to shift seamlessly between these autonomy states. This flexibility is made possible by the dynamic interplay of brain networks:

1. **Dynamic State Transitions**:
  - The DMN is most active during introspection, while task-positive networks dominate when we focus on external tasks.
  - The brain dynamically transitions between these states based on environmental demands, emotional signals, and internal priorities.
2. **Metacognitive Oversight**:
  - Higher-order cognitive processes monitor and regulate these transitions.
  - For example, when habitual behavior produces a suboptimal result, the DMN and executive networks may intervene to reassess the situation and recalibrate the response.
3. **Behavioral Adaptation**:
  - This interplay allows for rapid responses to immediate challenges without sacrificing the ability to reflect, adapt, and learn from experiences.

4.2.4.1 A Broader Perspective

The Neural Autonomy Spectrum is more than a theoretical model; it highlights the DMN's essential role in integrating the past, present, and future into coherent and adaptive thought processes. It also underscores the delicate balance required for effective decision-making, whether that involves reflecting on personal values or responding to a sudden external threat.

4.3 DMN AS THE ENGINE OF INNOVATION

Creativity has long been considered one of humanity's defining traits – the ability to generate novel ideas, connect disparate concepts, and envision new possibilities. At the heart of this capacity lies the **DMN**. Far from being idle, the DMN is a hub of dynamic, spontaneous thought processes that facilitate the creation of new ideas and the resolution of complex problems.

This section delves into the DMN's role as the engine of innovation, exploring its neural mechanisms, the creative process it underpins, and real-world examples that highlight its profound impact.

4.3.1 Neural Mechanisms of Creativity

The DMN's capacity for creativity stems from its ability to process information in ways that other brain networks cannot. By enabling the brain to wander, reflect, and integrate, the DMN fosters the conditions necessary for innovative thinking.

Key Mechanisms of the DMN in Creativity:

1. **Pattern Recognition**:
    - The DMN integrates information from diverse cognitive domains, identifying relationships that might not be immediately obvious.
    - This process supports the formation of novel connections and insights, the cornerstone of creative thought.
2. **Mental Simulation**:
    - The DMN constructs hypothetical scenarios, allowing the brain to test ideas, imagine alternatives, and foresee potential outcomes.
    - This ability to simulate different realities fuels both creativity and adaptive problem-solving.
3. **Dynamic Network Interaction**:
    - Creativity requires a balance between divergent thinking (generating ideas) and convergent thinking (evaluating them). The DMN works alongside executive networks to ensure this balance.
    - Emotional and rational inputs are integrated, enriching the creative process with both intuitive and analytical perspectives.

4.3.2 The Creative Process in Neural Terms

The creative process can be understood as a dynamic interplay between the DMN and other neural networks. Neuroscience research has identified three key phases of creativity, each driven by distinct brain activities:

1. **Incubation Phase**:
    - **DMN Activation:** During this stage, the brain unconsciously processes stored information, making unexpected connections and associations.
    - **Background Processing:** Insights often emerge when the mind is at rest, such as during a walk, shower, or period of relaxation.
2. **Insight Generation**:
    - **Network Switching:** Sudden bursts of activity occur as the DMN interacts with executive networks, producing the "aha!" moment.
    - **Emergent Ideas:** The brain synthesizes information into a cohesive and novel concept, often accompanied by a sense of clarity or revelation.
3. **Evaluation and Refinement**:
    - **Executive Network Engagement:** Once an idea emerges, task-positive networks critically evaluate its feasibility and plan for its implementation.
    - **Iterative Improvement:** This phase involves refining the idea, testing its practical applications, and integrating feedback.

4.3.3 Real-World Examples

Historical breakthroughs and creative achievements highlight the DMN's role in innovation. By examining these moments through the lens of neuroscience, we can better appreciate how introspection and mental simulation lead to transformative insights.

1. **Archimedes' "Eureka!" Moment**:
    - While relaxing in a bath, Archimedes discovered the principle of water displacement, providing a solution to a vexing problem.
    - DMN activity likely facilitated this breakthrough by integrating prior knowledge with reflective thought during a state of rest.

2. **Kekulé's Discovery of the Benzene Ring**:
   - Kekulé's vision of a snake biting its tail led to the discovery of the benzene molecule's ring structure, a pivotal moment in chemistry.
   - This insight illustrates the DMN's ability to draw on unconscious processing and turn it into creative solutions.
3. **Einstein's Thought Experiments**:
   - Einstein imagined riding on a beam of light, leading to groundbreaking insights about relativity.
   - These mental simulations exemplify the DMN's role in constructing hypothetical scenarios that push the boundaries of human understanding.

4.3.4 Broader Implications for Creativity and Innovation

The DMN's contribution to creativity extends beyond individual insights; it has far-reaching implications for fields ranging from education to organizational leadership:

- **Education**:
  - Encouraging periods of rest and introspection can enhance creativity in students, fostering innovative thinking.
- **Organizational Strategy**:
  - Teams and leaders who cultivate environments conducive to reflection and idea generation are better positioned to solve complex challenges.
- **Mental Health**:
  - Disruptions to DMN activity, such as excessive rumination or underactivity, can stifle creativity, highlighting the importance of mental well-being for innovation.

4.4 DESIGNING AGI WITH DMN-LIKE CAPABILITIES

4.4.1 Bridging the Gap Between Human and Artificial Minds

As AGI research advances, one of the most complex challenges is replicating hallmark traits of human intelligence: creativity, introspection, and the ability to dynamically transition between focused concentration and open-ended exploration. The **DMN**, with its remarkable capacity for spontaneous thought, self-reflection, and its interplay with the Attention Network, provides a powerful blueprint for embedding these capabilities into AGI systems.

By studying how the DMN enables humans to seamlessly navigate between internal reflection and external engagement, we gain valuable insights into designing AGI systems that mimic this flexibility, creativity, and adaptability.

4.4.2 Architectural Principles Inspired by the DMN

To create AGI systems that reflect the depth of human cognition, we can draw inspiration from the DMN's core components. These elements form the foundation of an architecture capable of self-reflection, spontaneous thought generation, and adaptive integration.

1. **Self-Reflection Modules**
   - **Purpose**: Enable AGI to introspect on its internal states, evaluate its performance, and identify inefficiencies or biases in its reasoning.
   - **Key Functions**:
     - Monitoring internal processes and identifying cognitive biases.
     - Conducting retrospective evaluations of past decisions to refine future strategies.
     - Aligning actions with dynamic goal-setting mechanisms.

2. **Spontaneous Thought Generators**
   - **Purpose**: Allow AGI to generate novel ideas, test unconventional approaches, and explore solutions beyond explicit external prompts.
   - **Key Functions**:
     - Recognizing patterns across diverse datasets to identify creative opportunities.
     - Generating hypotheses and testing them within simulated, hypothetical contexts.
     - Engaging in exploratory, outside-the-box thinking to innovate and adapt.
3. **Integration Systems**
   - **Purpose**: Synthesize outputs from self-reflection and spontaneous thought generators, mirroring the DMN's ability to combine inputs from multiple cognitive domains into coherent insights.
   - **Key Functions**:
     - Integrating diverse streams of information into actionable conclusions.
     - Employing context-sensitive processing to adapt to changing environments.
     - Refining ideas through iterative feedback loops.

4.4.3 Network Dynamics: Balancing Exploration and Exploitation

The DMN's interaction with the Attention Network exemplifies the brain's ability to balance divergent thinking (exploration) with convergent thinking (exploitation). To replicate this in AGI, systems must incorporate mechanisms that enable dynamic switching between creative and focused processing modes.

1. **Dynamic Attention Allocation**
   - AGI should dynamically adjust its focus between exploratory tasks (DMN-like) and goal-directed activities (Attention Network-like) based on environmental and internal cues.
   - Example: AGI exploring novel design ideas for a product before narrowing down to refine and implement the most effective solution.
2. **Metacognitive Oversight**
   - A "metacognition layer" can monitor AGI's internal processes, ensuring efficiency and preventing maladaptive behaviors, much like how the Attention Network regulates DMN activity.
   - Example: Recognizing when creative exploration is yielding diminishing returns and shifting focus to structured implementation.
3. **Feedback-Driven Adaptation**
   - Continuous feedback loops allow AGI to refine its outputs over time, learning from successes and failures to improve its decision-making capabilities.
   - Example: An AGI system analyzing its hypothesis generation process to identify patterns that lead to higher success rates.

4.4.4 Implementation Strategies and Safety Mechanisms

1. **Technical Implementation Framework**
   - Design a **dual-network architecture**, with separate modules for DMN-like exploration and Attention Network-like focus, managed by a meta-controller for dynamic switching.
   - Incorporate a **safety monitor** to oversee behaviors and ensure adherence to ethical guidelines.
2. **Practical Applications**
   - **Drug Discovery**: Use DMN-like mechanisms to explore novel molecular combinations, with Attention Network-like focus analyzing feasibility and ensuring safety.
   - **Creative Writing**: Generate imaginative narratives in DMN-mode and refine coherence and appropriateness in Attention Network-mode, ensuring ethical and context-sensitive content.

3. **Safety and Stability Mechanisms**
   - Behavioral bounds ensure resource management, output coherence, and adherence to value alignment principles.
   - Example: AGI systems verifying their outputs against predefined ethical and functional benchmarks before implementation.
4. **Performance Metrics and Monitoring**
   - Evaluation frameworks should track metrics such as creativity, task efficiency, and safety compliance.
   - **Key Metrics**:
     - Creativity scores (e.g., originality and diversity of solutions generated).
     - Task completion metrics (e.g., efficiency and accuracy).
     - Safety compliance (e.g., adherence to ethical constraints).

4.4.5 Shaping the Future of AGI

Building AGI systems with DMN-inspired capabilities requires a multidisciplinary approach that synthesizes insights from neuroscience, computational design, and ethics. By leveraging the principles of the DMN and its dynamic interplay with the Attention Network, we can create AGI systems that are:

- **Reflective**: Capable of self-monitoring and improvement.
- **Adaptive**: Flexible enough to respond to diverse challenges.
- **Creative**: Able to generate novel ideas and innovate.
- **Ethical**: Aligned with human values and societal goals.

As we navigate the future of AGI, integrating DMN-inspired principles will ensure that intelligent systems are not only powerful but also meaningful contributors to solving humanity's greatest challenges.

4.5 EVOLUTIONARY AND CLINICAL INSIGHTS

The **DMN** is not only a hallmark of human cognition but also an evolutionary triumph. Its development reflects the increasing complexity of thought processes necessary for survival, creativity, and social interaction. However, just as its strengths contribute to human flourishing, disruptions in the DMN can result in significant mental health challenges. By examining the DMN through evolutionary and clinical lenses, we can better understand its broader implications for both individuals and future AGI.

4.5.1 Evolutionary Perspectives

The DMN's functions – self-reflection, future planning, and mental simulation – offer profound survival advantages, underscoring its evolutionary significance. Here, we explore how these capabilities developed and their role in shaping complex cognition.

4.5.1.1 Comparative Neurobiology

- **Evidence in Animals**:
  - Studies suggest that certain animals, such as great apes, elephants, and dolphins, exhibit DMN-like activity, enabling behaviors such as tool use, social bonding, and problem-solving.
  - For example, chimpanzees demonstrate the ability to plan for future needs, a hallmark of DMN functionality.

- **Evolutionary Roots**:
  - The DMN likely emerged as a response to increasingly complex environmental and social challenges, supporting the development of mental simulation and foresight.
  - Animals capable of imagining potential threats or opportunities were better equipped to survive and reproduce.

4.5.1.2 Survival Benefits of the DMN

The DMN's ability to simulate and reflect provided humans with:

- **Enhanced Problem-Solving**:
  - Mental rehearsal of actions allowed early humans to strategize without physical risk.
- **Social Cooperation**:
  - By enabling individuals to reflect on relationships and predict others' behaviors, the DMN facilitated collaboration and trust within groups.
- **Adaptive Planning**:
  - The capacity to imagine future scenarios empowered humans to prepare for resource scarcity, environmental changes, and conflict.
- **Risk Assessment**:
  - Reflecting on past experiences helped individuals avoid repeating mistakes and identify patterns of danger.

4.5.2 Clinical Applications

While the DMN is vital for mental health and cognition, its dysregulation can lead to significant challenges. Understanding how the DMN functions – or malfunctions – can guide therapeutic approaches and interventions.

4.5.2.1 Disorders Linked to DMN Dysregulation

- **Depression and Rumination**:
  - Over-activation of the DMN is associated with excessive self-focused thought, often leading to rumination and negative emotional states.
  - This feedback loop contributes to the persistence of depressive symptoms.
- **Anxiety Disorders**:
  - Hyper-active DMN activity can amplify worries about hypothetical future scenarios, exacerbating anxiety.
- **Attention Deficits**:
  - DMN underactivity or poor regulation between DMN and task-positive networks can impair focus, contributing to ADHD symptoms.
- **Post-Traumatic Stress Disorder (PTSD)**:
  - Dysregulated DMN activity in PTSD patients can lead to intrusive memories and difficulty differentiating between past and present experiences.

4.5.2.2 Therapeutic Approaches Targeting the DMN

- **Mindfulness Practices**:
  - Activities like meditation and yoga help regulate DMN activity, reducing over-activation and fostering present-moment awareness.
- **Cognitive Behavioral Therapy (CBT)**:
  - By challenging negative thought patterns, CBT aims to redirect DMN activity away from maladaptive ruminations.

- **Neurofeedback Training**:
  - Patients can learn to modulate their brain activity by receiving real-time feedback on DMN dynamics, improving self-regulation.
- **Targeted Interventions**:
  - Pharmacological treatments and transcranial magnetic stimulation (TMS) are being explored to adjust DMN activity in patients with severe dysregulation.

4.5.3 Implications for AGI Design

The insights gained from the DMN's evolutionary and clinical study can inform the development of AGI systems:

- **Resilience and Adaptability**:
  - By mimicking the DMN's ability to reflect on past performance and adapt to future challenges, AGI could become more robust in dynamic environments.
- **Error Mitigation**:
  - Drawing from clinical models, AGI systems could incorporate safeguards against maladaptive loops, ensuring that introspective processes remain productive.
- **Empathy Simulation**:
  - Understanding the DMN's role in social cognition and empathy offers a roadmap for designing AGI systems capable of understanding and predicting human emotions and intentions.

4.5.4 The Bigger Picture

Studying the DMN through an evolutionary and clinical lens reinforces its central role in shaping both human cognition and well-being. By understanding how the DMN evolved to support survival and how its dysfunction impacts mental health, we gain a richer appreciation of its importance. These insights are not only transformative for neuroscience and psychology but also offer a foundation for AGI design, ensuring that future intelligent systems are adaptive, resilient, and aligned with human values.

4.5.5 Challenges and Limitations in DMN-Inspired AGI Implementation

While designing AGI systems inspired by the DMN and its interplay with the Attention Network holds great promise, it also presents a range of challenges. These obstacles span technical, architectural, and evaluative dimensions, underscoring the complexity of replicating human-like cognition in artificial systems.

4.5.5.1 Technical Implementation Challenges

Creating DMN-inspired AGI systems requires overcoming several technical hurdles, particularly in computational design, architectural flexibility, and system integration.

- **Computational Complexity**:
  - Simulating DMN-like spontaneous thought processes demands vast computational resources, particularly for generating novel ideas in real time.
  - Balancing real-time processing with exploratory, open-ended thinking poses significant optimization challenges.
  - Implementing true parallel processing to manage DMN-like and Attention Network-like systems simultaneously is a technically demanding task.

- **Architecture Limitations**:
  - Current neural network architectures lack the organic adaptability of human cognitive systems, limiting their ability to replicate the DMN's flexibility.
  - Genuine metacognition – where the system actively evaluates and regulates its own thought processes – remains largely theoretical, as most current systems rely on programmed monitoring.
- **Integration Issues**:
  - Coordinating multiple processing streams to mimic the interplay between introspection and focus is complex and prone to instability.
  - Ensuring smooth, seamless transitions between exploratory (DMN-like) and focused (Attention Network-like) modes is an ongoing challenge.
  - Maintaining system stability during rapid network switching is critical to avoid cascading failures or inconsistent outputs.

4.5.5.2 Enhanced Evaluation Framework

To assess the performance and reliability of DMN-inspired AGI systems, a robust evaluation framework is essential. This framework should combine quantitative metrics with qualitative criteria to ensure a comprehensive understanding of system capabilities and limitations.

- **Quantitative Metrics**:
  1. **Creativity Assessment**:
     - Originality scores based on comparisons to existing solutions.
     - Impact measures for evaluating the practical significance of innovative ideas.
     - Diversity of solutions generated, reflecting the system's breadth of exploration.
     - Practical applicability ratings to determine real-world usefulness.
  2. **Task Performance**:
     - Goal achievement rates for measuring success in achieving specified objectives.
     - Resource efficiency metrics, such as computational time and energy usage.
     - Time-to-solution measurements, reflecting the speed of task completion.
     - Adaptation speed in responding to novel or changing problems.
  3. **System Stability**:
     - Efficiency of network switching between exploratory and focused modes.
     - Balance in resource allocation to avoid bottlenecks or overuse of system capacity.
     - Error rates during mode transitions, identifying vulnerabilities in switching mechanisms.
     - Recovery time from disruptions, reflecting system robustness.
- **Qualitative Evaluation Criteria**:
  1. **Solution Quality**:
     - Coherence and clarity of generated ideas.
     - Depth of problem understanding, reflecting the system's ability to address complexity.
     - Appropriateness of solutions to ethical, cultural, or contextual constraints.
     - Integration of multiple perspectives for well-rounded outputs.
  2. **Ethical Alignment**:
     - Consistency with predefined human values and societal goals.
     - Detection and mitigation of biases in system-generated outputs.
     - Assessment of social and cultural impacts to avoid unintended consequences.
     - Sensitivity to diverse cultural contexts and perspectives.
  3. **Learning and Adaptation**:
     - Capability to transfer knowledge across domains and contexts.
     - Ability to understand and adapt to nuanced environmental changes.
     - Responsiveness to feedback for iterative improvement.
     - Evidence of long-term learning and growth in performance over time.

4.5.5.3 Future Research Directions

Advancing DMN-inspired AGI systems requires sustained efforts in several critical areas:

- **Technical Advancements**:
  - Development of neural architectures capable of more sophisticated parallel processing and dynamic adaptability.
  - Innovations in computational efficiency to manage the resource demands of DMN-like functions.
  - Refinement of integration mechanisms to improve coordination between exploratory and focused processing modules.
- **Evaluation Framework Development**:
  - Establishment of standardized testing protocols to ensure consistent assessment of creativity, adaptability, and ethical alignment.
  - Development of more nuanced creativity metrics that capture both novelty and practical utility.
  - Implementation of comprehensive safety assessments to evaluate stability and alignment with human values.
  - Refinement of performance benchmarks to measure progress over time.
- **Integration with Existing Systems**:
  - Methods for incorporating DMN-inspired features into current Artificial Intelligence (AI) frameworks, allowing gradual enhancements.
  - Strategies for ensuring compatibility with legacy systems while introducing innovative capabilities.
  - Approaches to scaling DMN-like architectures for broader industrial and societal applications.

4.5.6 Strengthening the Future of Intelligent Systems

While the challenges of implementing DMN-inspired AGI are substantial, they represent essential milestones on the path to creating systems that are not only intelligent but also reflective, creative, and aligned with human values. By addressing these limitations head-on, researchers and developers can build AGI systems capable of adapting to complex, dynamic environments while maintaining the ethical and practical integrity required to benefit humanity.

FULFILLMENT VS. HAPPINESS

THINKING BOX:

"Is happiness a momentary jolt, or a deeper sense of purpose? Maybe the cosmic riddle – akin to **42** – guides us toward lasting meaning."

4.6 CONCLUSION: CHARTING THE PATH AHEAD

The **DMN** is far more than a curious neurological phenomenon – it is a cornerstone of human cognition, creativity, and autonomy. Its study offers profound insights into how we think, plan, and innovate, as well as how we can design the next generation of intelligent systems.

In this chapter, we have explored the DMN's intricate functions, its role in the **Neural Autonomy Spectrum**, and its capacity as the **engine of innovation**. We've seen how its balance with task-positive networks enables us to transition seamlessly between introspection and external focus. We've also examined how this dynamic interplay shapes decision-making, creativity, and adaptability.

Finally, we've explored how DMN principles can inspire the architecture of **AGI** and provide a roadmap for addressing its challenges, from ensuring ethical alignment to fostering adaptability and resilience.

Key Takeaways

1. **The DMN's Integral Role in Cognition**:
   - The DMN orchestrates introspection, future planning, and creativity by integrating past experiences and imagining new possibilities.
   - It operates as a dynamic system, balancing autonomous thought with external task focus.
2. **The Spectrum of Neural Autonomy**:
   - Human cognition spans a continuum from high autonomy (deliberative, reflective) to low autonomy (automatic, reflexive), with the DMN playing a pivotal role in adaptive transitions.
3. **Creativity as a Neural Symphony**:
   - By generating spontaneous thoughts and forging novel connections, the DMN underpins the creative process, from idea incubation to insight generation and refinement.
4. **Evolutionary and Clinical Significance**:
   - The DMN's emergence reflects evolutionary pressures for problem-solving, social cooperation, and future planning.
   - Its dysregulation contributes to mental health challenges, but targeted therapies offer pathways to restore balance.
5. **Implications for AGI Design**:
   - DMN-inspired AGI systems could replicate human-like creativity and introspection, balancing exploration with goal-directed behavior.
   - Ethical considerations remain paramount, particularly in ensuring value alignment and managing autonomy.

4.6.1 Broader Implications

Studying the DMN unlocks deeper questions about human nature and the future of intelligence:

- **For Neuroscience**: How can understanding the DMN advance treatments for mental health disorders and improve cognitive resilience?
- **For Society**: How can we foster environments that encourage reflection, creativity, and adaptability in an increasingly fast-paced, digital world?
- **For AGI Development**: How do we balance the potential of DMN-inspired systems with the responsibility to create ethical, safe, and beneficial technologies?

The DMN exemplifies the delicate interplay between introspection and action, a balance that defines not only individual cognition but also collective progress.

4.6.2 A Call to Action

As we navigate the complexities of the modern world, the DMN offers a powerful reminder: some of our greatest insights come not from constant striving but from moments of quiet reflection and creative wandering. Understanding this network is more than an academic pursuit – it is an invitation to rethink how we approach both human cognition and the intelligent systems we seek to create.

For researchers, this means delving deeper into the brain's mysteries. For innovators, it means designing AGI systems that honor the complexity and richness of human thought. And for all of us, it is a call to embrace the creative power of introspection, trusting that even in our moments of stillness, our minds are working wonders.

4.6.3 End of Chapter 4

With the DMN as a guide, this chapter has bridged the gap between human cognition and artificial intelligence. Its exploration sets the stage for further investigations into the future of thought, creativity, and technology – a journey that promises to redefine both how we understand ourselves and the systems we create.

5 Understanding Cognitive Biases
A Tool for Post-Decision Learning

Origins and Context:

- **Evolutionary roots:** Biases evolved as heuristics for survival but now misfire in modern, complex environments.
- **Cultural and systemic influences:** Biases are shaped by societal norms, institutional structures, and media ecosystems.
- **Technological amplifiers:** Algorithms, Artificial Intelligence (AI), and digital ecosystems reinforce and scale biases (e.g., filter bubbles, authority bias).

High-Impact Biases:

- **Individual-level: Confirmation bias, loss aversion, present bias, Dunning–Kruger effect** – shaping personal finance, career choices, and risk assessment.
- **Societal-level: Groupthink, authority bias, information cascades, filter bubble effect** – influencing policy, public trust, and global decision-making.

Mitigation Strategies:

- **The three-step post-decision review:** Identify biases, analyze alternative decisions, and integrate learnings into future choices.
- **Organizational frameworks:** Decision audits, cognitive diversity, and ethical AI oversight.
- **Long-term solutions: Bias awareness in education, AI transparency, and global governance** to ensure equitable, informed decision-making.

Ultimately, **biases are unavoidable but manageable** – when harnessed correctly, they **enhance resilience, adaptability, and strategic foresight**.

5.1 INTRODUCTION: FROM BIAS AWARENESS TO PRACTICAL INSIGHT

5.1.1 Opening Scenario

Consider a scenario familiar to many professionals: a highly anticipated product launch carefully orchestrated by a capable team and backed by extensive internal discussions and market forecasts. The organization was confident – after all, the data seemed promising, the branding was on point, and preliminary customer feedback appeared positive. Yet, months after its debut, the product struggles to gain traction. Sales figures fall short, critics highlight features overlooked by the development team, and the initial enthusiasm gives way to perplexity and second-guessing.

In hindsight, it's easy to pinpoint the overlooked warning signs: ambiguous feedback that was interpreted too favorably, expert opinions that went unquestioned, or negative signals that were rationalized away. These are not just random errors in judgment; they often reflect patterns known as cognitive biases. Rather than viewing these patterns solely as flaws, we might ask a different question: What if we approached biases as diagnostic tools? By identifying how biases influenced past choices, we can refine future decision-making strategies. In other words, cognitive biases can serve as feedback loops that guide continuous improvement rather than as obstacles we must eradicate.

DOI: 10.1201/9781003640288-8

5.1.2 Reframing Cognitive Biases

Cognitive biases, broadly defined as systematic deviations from fully rational judgment, are rooted in our cognitive architecture and cultural contexts. Research in psychology, behavioral economics, and organizational studies has established that these biases do not exist merely as quirks of individual thought processes; they are integral to how our minds navigate complex, information-rich environments. While certain biases – such as confirmation bias or loss aversion – may have conferred evolutionary advantages in simpler, more perilous settings, they can produce suboptimal outcomes when interacting with the complexities of modern life and work.

It is important to acknowledge that cognitive biases cannot be wholly eliminated. Decades of studies, from seminal works by Tversky and Kahneman to contemporary interdisciplinary investigations, illustrate the persistence and adaptability of these mental shortcuts. However, complete removal of biases is not the goal here. Instead, understanding their nature and recognizing their influence can transform these cognitive tendencies into valuable learning instruments. By systematically analyzing how biases shaped our judgments, we can better calibrate our decision-making processes, apply more rigorous questioning, and foster an environment in which reflective thinking becomes the norm.

5.1.3 Chapter Overview

This chapter sets the stage by examining the evolutionary and cultural foundations of cognitive biases, exploring why these tendencies emerged and how they persist in today's globally interconnected and technologically mediated societies. We will consider how biases are reinforced or mitigated by organizational structures, media ecosystems, and cultural value systems, thereby underscoring the importance of context when analyzing biased reasoning.

The chapter then moves on to identify key biases at both individual and societal levels. By delving into high-impact biases that affect personal choices – such as confirmation bias, present bias, and the Dunning–Kruger effect – we learn how to recognize these patterns in ourselves and in the decision-making of those around us. We also examine biases that scale up to influence groups, institutions, and entire communities, such as groupthink or the filter bubble effect. Here, cultural variations and systemic dimensions become essential considerations, as biases do not operate in a vacuum.

Finally, we introduce practical strategies for post-decision analysis. Rather than offering a one-time "fix," the chapter emphasizes continuous learning and adaptation. We will outline frameworks to reflect on past decisions, tools to identify where and when specific biases likely played a role, and approaches to integrate these insights into ongoing practice. The objective is to empower readers – whether business leaders, policymakers, educators, or informed citizens – to navigate complexity more effectively, turning the awareness of cognitive biases into a mechanism for sustained growth and improved judgment.

In sum, this chapter encourages a reorientation: from viewing cognitive biases as barriers to be removed, to understanding them as integral features of human cognition that, when properly examined and contextualized, can help refine our decision-making processes over time.

DREAMING BIG AND RESILIENCE

THINKING BOX:

"How do you rebound from setbacks without losing sight of the stars? The secret resilience factor might align with the fabled **42**."

5.2 FOUNDATIONS: ORIGINS, EVOLUTION, AND CULTURAL CONTEXT

5.2.1 Evolutionary Roots

The cognitive biases that shape our judgments and decisions today did not arise by chance. They originate from evolutionary pressures that demanded quick, efficient reasoning in environments characterized by uncertainty, limited information, and constant threats. In ancestral settings, being able to make split-second judgments could mean the difference between survival and extinction. Rather than methodically weighing every piece of data – which would have been both impossible and dangerous – our ancestors developed mental shortcuts, or heuristics, to streamline decision-making.

These heuristics enabled humans to navigate a world where time and cognitive resources were scarce. For example, paying disproportionate attention to negative outcomes (as seen in loss aversion) made sense when a single misjudgment could result in severe harm. Similarly, relying on vivid or recent memories to assess risk (the availability heuristic) increased the likelihood of detecting danger early, even at the cost of occasional false alarms. In essence, biases like these were evolution's attempt to optimize decision-making under conditions that favored speed and sufficiency over perfect accuracy.

However, the very traits that were once advantageous can misfire in our complex, data-saturated modern world. Unlike our ancestors, we now make financial investments involving intricate market instruments, consume media filtered through opaque algorithms, and engage in global negotiations requiring careful long-term foresight. In these contexts, heuristics that simplified life in a small hunter-gatherer band may lead us astray. We might overweigh rare but sensational risks (e.g., fearing air travel more than the far more dangerous daily commute by car) or become overconfident in our predictive abilities because we excel at explaining the past but struggle to forecast the future. Understanding these evolutionary roots helps us appreciate why biases feel so natural and persistent – and why overcoming their negative consequences requires more than sheer willpower.

5.2.2 Cultural and Systemic Influences

Cognitive biases do not manifest in a vacuum. They are filtered, interpreted, and sometimes magnified by the cultural contexts and societal structures in which individuals and groups operate. Different cultures prioritize distinct values, norms, and modes of interaction, influencing both the type and intensity of certain biases. For instance, research into attribution styles shows that individualistic cultures – where personal agency and independence are highly valued – tend to exhibit a stronger fundamental attribution error, attributing others' behaviors primarily to their character rather than external circumstances. In more collectivist societies, where relationships and context are emphasized, this bias is less pronounced. Such variations highlight that what one culture deems a "cognitive flaw" might be more adaptive or less harmful in another cultural setting.

Beyond cultural variations, systemic factors such as institutional frameworks, legal standards, educational methods, and organizational hierarchies can reinforce or mitigate these biases. Consider an organizational culture that discourages dissent and rewards conformity. In such an environment, groupthink might flourish, as employees suppress critical viewpoints to maintain harmony. Conversely, institutions that actively foster diverse perspectives, encourage open debate, and implement structured decision-making protocols (e.g., assigning a "devil's advocate" role in meetings) can reduce the influence of certain biases. Public policy can also play a role: transparency laws, anti-discrimination measures, and fair hiring guidelines help counteract systemic biases like in-group favoritism, authority bias, or algorithmic discrimination.

In short, biases are dynamic elements shaped by the interplay between individual cognition and broader cultural and institutional contexts. Recognizing these systemic factors is essential. Efforts to address biases purely at the individual level will fall short if organizations, societies, and political systems perpetuate conditions under which certain biases thrive.

5.2.3 Technological Amplifiers

As the digital era unfolds, technology emerges as a significant amplifier of cognitive biases. Far from being a neutral conduit, our digital environment – encompassing social media feeds, search engine results, recommendation systems, and AI-driven analytics – actively filters and curates the information we encounter. These algorithmic systems, often designed to maximize engagement or efficiency, can reinforce preexisting biases. Personalized content streams may trap us in "filter bubbles," where confirmation bias intensifies as we repeatedly encounter viewpoints similar to our own, while contrary perspectives remain hidden.

Moreover, as AI and machine learning models proliferate, new biases arise or existing biases evolve in unexpected ways. AI-driven decision support tools in areas like healthcare, finance, or public policy can encourage an "AI Confidence Bias," where human decision-makers defer too readily to algorithmic outputs without critical evaluation. Automated hiring systems trained on historically biased data can perpetuate inequality in recruitment. In other words, technology can scale up the impact of a single biased assumption, affecting not just one decision-maker but entire organizations, markets, and communities at once.

Looking ahead, the rapid emergence of augmented and virtual reality, the Internet of Things, and increasingly sophisticated predictive analytics will likely introduce novel forms of cognitive distortion. We must remain vigilant, as the biases we grapple with today may morph or intensify in response to tomorrow's technological landscape. Achieving this vigilance involves continuous learning, updating our strategies, and insisting on transparency and accountability from those who design and deploy digital systems.

Taken together, evolutionary roots, cultural and systemic influences, and technological amplifiers form a web of factors that shape how cognitive biases emerge and operate. Our challenge is to understand this complexity deeply. Only by appreciating the interplay between ancient mental shortcuts, the socio-cultural world in which we live, and the technologies we rely on can we develop informed strategies to harness these biases as learning tools, rather than stumbling blocks, in our ongoing pursuit of sound decision-making.

5.3 HIGH-IMPACT INDIVIDUAL BIASES: UNDERSTANDING YOUR OWN THINKING

In an increasingly interconnected and information-rich world, individual decision-making is subject to unprecedented complexity. Individuals routinely navigate sprawling digital ecosystems, confront a barrage of advertising and media influences, and manage personal and professional responsibilities that often span diverse cultural and social contexts. Against this backdrop, certain cognitive biases stand out as "high-impact" – their influence ripples through personal finances, career choices, health habits, and interpersonal relationships. Recognizing these biases and their interplay equips us to better understand our own thought processes, refine our critical thinking skills, and ultimately improve the quality of our judgments.

5.3.1 Core Biases to Know

- **Confirmation Bias**:

  Definition and Mechanism: Confirmation bias is the tendency to seek, interpret, and remember information that supports existing beliefs or expectations. Rather than evaluating evidence impartially, individuals gravitate toward data that reaffirms their assumptions, downplaying or dismissing contradictory information.

Modern Context and Impact: In today's digital age, this bias is amplified by highly personalized media environments. Social media algorithms, search engines, and recommendation systems frequently present content aligned with a user's browsing history and expressed preferences. The result is an "echo chamber" effect, where individuals encounter mainly familiar ideas and perspectives. This can reinforce ideological divides, hinder constructive debate, and impede critical re-examination of one's views. Over time, confirmation bias can shape political attitudes, consumer behavior, and professional judgments, affecting everything from voting decisions to the adoption of new business strategies.

Mitigation Approach: To counteract confirmation bias, one practical strategy is to deliberately engage with opposing viewpoints. For example, seeking out at least one credible source that challenges your initial stance before finalizing a decision introduces much-needed tension, prompting more balanced reasoning. This approach acknowledges the bias without expecting its elimination, using structured self-scrutiny to enhance decision quality.

- **Loss Aversion**:

Definition and Mechanism: Loss aversion, identified in behavioral economics, posits that the pain of losing something is psychologically more powerful than the pleasure of an equivalent gain. This bias can lead individuals to hold onto failing investments, persist in unproductive projects, or maintain stale personal or professional relationships out of fear of what might be forfeited.

Modern Context and Impact: In a world where rapid innovation and market volatility are the norm, clinging to the status quo can result in missed opportunities and inefficient resource allocation. For instance, investors who refuse to sell underperforming stocks because of the "paper loss" they would realize may miss out on reallocating their capital into more promising ventures. Similarly, in organizations, leaders may fail to abandon flawed initiatives due to the sunk costs already invested. Loss aversion thus has high impact because it directly influences risk-taking, innovation capacity, and adaptive behavior in a fast-changing global economy.

Mitigation Approach: A productive way to mitigate loss aversion is to reframe decisions in terms of potential long-term outcomes rather than short-term setbacks. Instead of focusing on what is forfeited by selling a failing asset or changing a long-standing practice, consider the future gains or the strategic flexibility gained by making the shift. This perspective encourages more rational trade-offs and adaptive decision-making.

- **Present Bias**:

Definition and Mechanism: Present bias, or temporal discounting, refers to the tendency to overvalue immediate rewards while underestimating future benefits. The human brain is wired to prize immediacy – a holdover from eras when securing short-term necessities was paramount.

Modern Context and Impact: Present bias is especially impactful today, where the allure of instant gratification is intensified by abundant digital distractions. Whether it's procrastination fueled by constant social media notifications or difficulty in adhering to long-term financial plans due to the ready availability of online retail therapy, present bias influences health behaviors, academic achievement, and professional development. As life expectancy, career spans, and financial instruments grow more complex, the failure to prioritize the future over the immediate present can lead to long-term financial insecurity, diminished career growth, and unfulfilled potential.

Mitigation Approach: Techniques such as commitment devices, incremental goal-setting, and the use of reminder systems or productivity apps help counteract present bias. By breaking long-term objectives into smaller milestones with tangible short-term rewards, individuals create pathways that align immediate actions with enduring objectives.

- **Dunning–Kruger Effect**:

  Definition and Mechanism: The Dunning–Kruger effect describes a cognitive pattern where individuals with limited skills or knowledge in a particular domain overestimate their competence, while highly skilled individuals tend to underestimate their abilities. This occurs partly because poor performers lack the metacognitive awareness to accurately assess their shortcomings.

  Modern Context and Impact: In today's specialized and knowledge-driven economies, accurate self-assessment is crucial for professional growth, effective teamwork, and high-quality decision-making. Overconfidence stemming from the Dunning–Kruger effect can lead to costly errors – such as an unqualified individual taking on leadership roles, making flawed strategic decisions, or eschewing expert advice. In domains like public health, cybersecurity, or finance, overconfidence can propagate misinformation, compromise safety, or trigger economic instability. Thus, its impact is high because it can reverberate through organizational hierarchies and systemic processes.

  Mitigation Approach: Regular feedback from diverse, knowledgeable sources can counteract the Dunning–Kruger effect. By inviting critiques, seeking mentorship, engaging in formal assessments, and encouraging peer review, individuals gain a more accurate understanding of their capabilities. This fosters a culture of learning and humility, improving both personal development and collective outcomes.

5.3.2 Interplay and Interaction of Biases

Rarely do these biases operate in isolation. Instead, they often reinforce and compound one another. For example, consider personal finance decisions:

- **Loss Aversion** might prompt an individual to hold onto an underperforming stock to avoid realizing a "loss on paper."
- **Present Bias** might simultaneously discourage that same individual from conducting a thorough, future-oriented investment review, preferring immediate comfort (sticking to familiar assets) over the effort of researching alternative options.

As a result, the combination of these biases can stall adaptation and rational portfolio adjustments, leading to greater financial vulnerability over time.

In career choices, the compound effects might be even more evident. Imagine a professional evaluating a job offer:

- **Confirmation Bias** leads them to overlook negative feedback about the new role, focusing instead on reasons that align with their initial enthusiasm.
- **Present Bias** encourages them to choose the position offering a slightly higher immediate salary rather than one with better long-term prospects.
- **Dunning–Kruger Effect** leads them to decline advice from a more experienced colleague, believing they have fully understood the market dynamics and their own competencies.

This interplay creates a scenario in which the individual may choose a role that provides short-lived satisfaction but does not align with their broader career goals or skill development. In the long run, this could result in diminished professional growth, reduced job satisfaction, and missed opportunities for advancement.

5.3.2.1 Why These Biases Are Rated as High Impact Today

In contemporary society, complexity, globalization, and rapid technological advancement have raised the stakes of individual decision-making. Each bias, when unchecked, can lead to outcomes that resonate beyond personal inconvenience:

- **Scale and Reach:** With the ubiquity of social media and globalized markets, an individual's misjudgments can influence peers, employees, customers, or political constituents at scale.
- **Opportunity Cost:** The cost of poor decisions increases as life, health, and work choices intersect with complex financial instruments, intricate legal frameworks, cultural diversity, and multi-layered organizational structures.
- **Innovation and Adaptability:** High-impact biases can hinder innovation by encouraging complacency (loss aversion), reinforcing narrow thinking (confirmation bias), or preventing honest skill assessment (Dunning–Kruger), all of which diminish resilience and adaptability in fast-evolving economic and technological landscapes.
- **Societal Consequences:** On a macro level, these biases affect civic engagement, public discourse, healthcare decisions, and financial stability. Individual choices accumulate, influencing markets, institutions, and even political outcomes. For instance, widespread confirmation bias can intensify polarization, while cumulative present bias can impact public policies related to climate change or infrastructure investment.

By understanding these biases – confirmation bias, loss aversion, present bias, and the Dunning–Kruger effect – and acknowledging their heightened relevance in contemporary, interconnected societies, individuals can take meaningful steps to refine their cognitive processes. This involves not only recognizing the biases in isolation but also developing strategies to manage their combined impact. When individuals become more aware, institutions become more adaptable, and society as a whole stands to benefit from decisions that are grounded in reasoned analysis, long-term thinking, and a more accurate perception of one's own competencies.

5.4 HIGH-IMPACT SOCIETAL BIASES: COLLECTIVE PATTERNS AND THEIR CONSEQUENCES

In an era defined by global interdependencies, instantaneous communication, and complex policy challenges, the influence of cognitive biases extends well beyond individual minds. Certain biases – when aggregated across groups, communities, and entire societies – shape the dynamics of markets, governance, public opinion, and cultural evolution. These "high-impact" societal biases gain their significance not merely because they exist but because of the scale and complexity of modern contexts. Today's social structures, political climates, and technological ecosystems ensure that the consequences of these biases can be profound, affecting everything from international relations to public health campaigns.

Understanding societal biases is therefore critical. While individual biases can hinder personal growth or career choices, societal-level biases can derail democratic processes, misallocate resources, intensify cultural divides, or compromise collective responses to crises. Recognizing these patterns and their systemic dimensions is a prerequisite for designing policies, institutional frameworks, and technological safeguards that uphold fairness, critical inquiry, and long-term sustainability.

5.4.1 Key Social-Level Biases

- **Groupthink: Prioritizing Consensus over Critical Thought**
  Definition and Mechanism: Groupthink emerges when a cohesive group's desire for harmony or unanimity overrides critical scrutiny of ideas and evidence. Dissenting voices are muted, overlooked, or self-censored, resulting in decisions based on an artificially narrow set of viewpoints.

Modern Context and High Impact: In today's highly interconnected world, decisions made by committees, boards, or advisory panels can have immediate and far-reaching consequences. A policy crafted by a government task force suffering from groupthink might lead to flawed legislation or suboptimal crisis management. International organizations or corporate boards dominated by a single narrative risk missing early warning signs, as occurred in several historical corporate collapses and military miscalculations. The high impact arises because a single, ill-vetted decision can cascade through supply chains, ecosystems of stakeholders, and media networks, amplifying the effects of flawed reasoning across the globe.

Mitigation Approach: Institutionalizing procedures that encourage dissent – such as assigning devil's advocates, rotating leadership roles, and using structured debate formats – helps ensure that critical evaluation is not sacrificed for the sake of apparent unity.

- **Authority Bias: Over-Relying on Experts or AI Outputs**

  Definition and Mechanism: Authority bias leads groups to overvalue opinions or guidance from perceived authorities, be they charismatic leaders, eminent experts, or advanced technological systems. While expertise can be invaluable, blind trust in authority can stifle questioning and critical analysis.

  Modern Context and High Impact: In an era of specialization and rapid technological progress, authority bias poses high stakes. Consider policy decisions guided by a single prominent economist, scientific advisor, or AI-driven analytics tool without adequate peer review. In public health, over-reliance on a select group of experts can cause entire nations to adopt flawed strategies. In financial markets, star analysts or algorithmic trading models can drive herd behavior, contributing to bubbles or panics. The globalized nature of supply chains, financial systems, and media means that errors originating from unchallenged authority can ripple across continents.

  Mitigation Approach: Checks-and-balances, transparent peer review, multi-stakeholder consultations, and independent oversight bodies promote a more balanced reliance on authority. Encouraging decision-makers to verify claims, consult diverse sources, and consider ethical implications helps ensure that expertise does not devolve into unquestioned dogma.

- **Information Cascade: Following Others Based on Limited Information**

  Definition and Mechanism: An information cascade occurs when individuals make decisions by imitating the choices of earlier actors, assuming those before them possessed more or better information. Without independent verification, these followers may perpetuate a potentially incorrect or suboptimal choice, creating a domino effect of conformity.

  Modern Context and High Impact: Digital connectivity and mass media can trigger rapid, large-scale cascades. A stock market surge fueled by rumor rather than fundamentals can create speculative bubbles with systemic economic repercussions. Viral misinformation spreads through social media, with millions sharing and endorsing content without verifying its veracity, influencing elections, health behaviors, and social tensions. Because today's information flows are instant and borderless, cascades can shape public opinion, mobilize social movements (both constructive and destructive), and influence policy shifts in record time.

  Mitigation Approach: Slowing decision-making processes, encouraging critical thinking through public education, and establishing readily accessible fact-checking resources help counteract information cascades. Policies that increase transparency in media and finance, combined with cultivating a culture of skepticism and independent inquiry, curb the uncritical acceptance of trends.

- **Filter Bubble Effect: Algorithms Reinforcing Similar Content**

  Definition and Mechanism: The filter bubble effect emerges as personalized algorithms present users with content that aligns with their past behaviors, interests, and biases.

Over time, individuals encounter a narrowed informational landscape, limiting exposure to diverse ideas and reinforcing existing viewpoints.

Modern Context and High Impact: With billions of people relying on digital platforms for news, entertainment, and social interaction, the filter bubble effect can significantly shape collective discourse. It contributes to political polarization, ideological fragmentation, and entrenched cultural divides, hindering consensus-building and reasoned debate. In markets, it can skew consumer perceptions and shape product demand. In public policy, it can limit the range of acceptable solutions and stifle innovative thinking. Given the global penetration of digital platforms, these effects scale up, affecting transnational relationships, cultural exchanges, and international negotiations.

Mitigation Approach: Encouraging users to actively seek out diverse information sources, implementing platform-level transparency measures, and designing recommendation algorithms with pluralism in mind can mitigate this bias. Media literacy programs and public education initiatives that teach critical evaluation of digital content further empower individuals to break out of these informational silos.

5.4.2 Cultural, Institutional, and Technological Dimensions

Societal biases do not stand alone; they are embedded in broader socio-technical systems. Cultural values influence attribution styles – some societies lean toward attributing outcomes to individual traits (strengthening certain biases), while others highlight situational factors, partially buffering against them. Institutional structures – from corporate governance models to media regulations – either exacerbate or mitigate biases. For example, laws mandating transparency in government decision-making or corporate audits can help counter groupthink and authority bias by ensuring diverse perspectives are heard. Conversely, weak oversight and concentrated media ownership might deepen information cascades and filter bubbles.

Technological infrastructures amplify these effects. Recommendation engines, search algorithms, and social media platforms, guided by profit motives or simplistic engagement metrics, can unknowingly magnify biases, shaping public opinion and collective behavior. As augmented reality, virtual reality, and the Internet of Things become more pervasive, entirely new modes of information exchange will emerge, potentially introducing novel biases or altering existing ones. Today's high-impact societal biases arise at the intersection of human cognition, cultural frameworks, institutional policies, and evolving technologies.

5.4.3 Case Study – Societal Impact

Imagine a government health committee formed to devise a public vaccination campaign amidst a global health crisis. Facing time pressures and high stakes, the committee quickly converges on a single course of action, stifling internal debate – classic **groupthink**. They place undue faith in one renowned epidemiologist's projections without seeking external validation, succumbing to **authority bias**. Early positive feedback from a select group of influencers and media outlets triggers an **information cascade**, with other stakeholders and regional leaders uncritically endorsing the campaign's messaging. Meanwhile, different segments of the population receive their news through personalized feeds, never seeing alternative analyses or credible critiques – a manifestation of the **filter bubble effect**.

Over time, the campaign fails to resonate with key communities that were not part of the initial consensus-building process. As public trust wanes, vaccination rates lag behind targets. A post-hoc review highlights missed opportunities:

- Had the committee embraced procedural safeguards – such as assigning a member to challenge assumptions – it might have avoided groupthink.
- Consulting a more diverse set of experts and data sources could have mitigated authority bias.

- Slowing the information release and encouraging independent fact-checking might have prevented an unverified narrative from going viral through an information cascade.
- Technological interventions, like recommending a broader range of viewpoints in social media feeds, could have reduced the filter bubble effect, ensuring that skeptical audiences encountered well-sourced clarifications.

5.4.3.1 Why These Biases Are Rated as High Impact in Modern Society

The high-impact nature of societal biases today stems from several interrelated factors:

- **Global Connectivity:** Decisions in one region can influence supply chains, financial markets, and political landscapes worldwide. A biased decision made by a central authority or a viral piece of misinformation can quickly reverberate across continents.
- **Complexity and Interdependence:** Modern issues – such as climate change, global pandemics, and international economic policies – require nuanced, multi-stakeholder solutions. Societal biases that simplify narratives, silence dissent, or skew informational inputs undermine the complexity needed for effective problem-solving.
- **Speed and Scale of Information Flows:** Digital media and communication tools have drastically accelerated how quickly ideas spread. Information cascades and filter bubbles can form in hours or days, influencing mass behavior and public policy with unprecedented speed.
- **Technological Amplification:** Algorithms, AI-driven recommendations, and platform incentives can magnify biases far beyond what would occur in a face-to-face environment. This magnification raises the stakes for ensuring that societal biases are understood, monitored, and countered.
- **Socio-Political Polarization:** The rising polarization in many societies exacerbates these biases. When social groups become more ideologically segregated, filter bubbles deepen, and groupthink becomes more pronounced. Authority bias can entrench power imbalances, while information cascades feed misinformation and public distrust.

In summary, these biases wield such influence today because we live in a world that is more interconnected, data-rich, and technologically mediated than ever before. Each of these conditions magnifies the potential consequences of distorted collective reasoning. By recognizing the nature and interactions of these societal biases, communities, institutions, and policymakers can develop better frameworks to ensure that collective decision-making remains grounded in critical thought, balanced authority, verified information, and diverse perspectives.

CURIOSITY AGAINST BIASES

THINKING BOX:

"Are you bold enough to question your cherished beliefs? Such boldness – and maybe **42** – could be your best ally against bias."

5.5 COMPREHENSIVE REFERENCE: COGNITIVE BIAS GLOSSARY

5.5.1 Introduction to the Comprehensive Reference Section

Before we move from understanding cognitive biases in theory to applying this knowledge in practical decision-making frameworks, it is valuable to have a cohesive, accessible overview of the many biases and related phenomena discussed thus far. The following glossary serves as a quick-reference guide, providing concise definitions, modern examples, and explanations for why each bias is considered high-impact today.

This consolidated perspective will help you identify biases more readily in your own thinking, in organizational strategies, and in societal discourse. By seeing all these biases side-by-side, you gain a clearer understanding of their evolutionary roots, cultural contexts, technological amplifications, and systemic implications. Whether you're a business leader, policymaker, educator, or engaged citizen, familiarizing yourself with this glossary prepares you to apply the strategies offered in the next chapter more effectively, making your post-decision analyses richer and more informed.

Note: This glossary includes biases and cognitive phenomena highlighted in earlier chapters, as well as several additional biases that were referenced in the broader context of the original materials.

Fundamental Paradoxes and Challenges

- **Retrospective-Predictive Divide:**
  Definition: Humans excel at explaining past events but struggle to accurately predict future outcomes.
  Impact Today: Encourages overconfidence in forecasts, misdirecting strategic planning and resource allocation.
- **Meta-Bias Challenge:**
  Definition: Awareness of biases can introduce new biases – recognizing faults in others but not in oneself.
  Impact Today: Reduces critical self-assessment, even among trained professionals, perpetuating the illusion of personal objectivity.
- **Theory-Practice Gap:**
  Definition: Knowing about biases theoretically does not guarantee improved real-world decisions.
  Impact Today: In fast-evolving industries, relying solely on theory without experiential learning can lead to ineffective interventions.

Individual-Level Biases

- **Confirmation Bias:**
  Definition: Preferring information that supports existing beliefs, ignoring contrary evidence.
  Impact Today: Reinforced by personalized media, deepening ideological divides and stifling critical thought.
- **Loss Aversion:**
  Definition: Feeling losses more acutely than equivalent gains.
  Impact Today: Hinders timely course corrections, from holding losing investments to maintaining outdated business models.
- **Present Bias (Temporal Discounting):**
  Definition: Favoring immediate rewards over long-term benefits.
  Impact Today: Digital distractions and instantaneous gratification undercut long-term planning in finance, health, and career development.
- **Dunning–Kruger Effect:**
  Definition: Inexperienced individuals overestimate their abilities, while experts underestimate theirs.
  Impact Today: Misaligns perceived and actual competence, undermining professional growth, teamwork, and effective problem-solving.
- **Anchoring Bias:**
  Definition: Relying too heavily on initial information (the "anchor") when making judgments.
  Impact Today: Affects negotiations, pricing strategies, and forecasts, skewing market and consumer decisions.

- **Availability Heuristic:**
  Definition: Judging likelihood based on how easily examples come to mind.
  Impact Today: Media-driven exposure to dramatic events distorts perceptions of risk, from terrorism fears to health concerns.
- **Planning Fallacy:**
  Definition: Underestimating time, costs, and risks for future tasks.
  Impact Today: Leads to project overruns, policy delays, and missed deadlines in intricate global systems.
- **Optimism Bias:**
  Definition: Believing one is less likely to experience negative outcomes.
  Impact Today: Undermines rigorous contingency planning, insurance coverage, and preparedness in volatile environments.
- **Fundamental Attribution Error:**
  Definition: Attributing others' behavior to character rather than situational factors.
  Impact Today: Affects cross-cultural understanding, conflict resolution, and fairness in judgments about performance and accountability.
- **Halo Effect:**
  Definition: Allowing one positive trait or impression to overshadow other attributes, leading to overly favorable judgments.
  Impact Today: Influences hiring decisions, brand loyalty, and academic evaluations, potentially limiting diversity, fairness, and innovation.

Societal-Level Biases

- **Groupthink:**
  Definition: Valuing consensus over critical evaluation.
  Impact Today: Encourages flawed decisions in corporate boards, governmental committees, and global alliances, stunting adaptability.
- **In-Group Bias:**
  Definition: Favoring one's own group over others.
  Impact Today: Fuels discrimination, systemic inequalities, and polarized politics, undermining social cohesion and equitable resource distribution.
- **Filter Bubble Effect (Echo Chambers):**
  Definition: Algorithm-driven personalization reinforces existing views, limiting exposure to diverse perspectives.
  Impact Today: Polarizes public discourse, hinders democratic dialogue, and reduces collective problem-solving capacity.
- **Authority Bias:**
  Definition: Over-relying on experts or leaders without critical assessment.
  Impact Today: Concentrates decision-making power, risking systemic errors, unethical decisions, and unchecked influence in governance and business.
- **Information Cascade:**
  Definition: Individuals follow others' decisions based on the assumption that previous actors possess better information.
  Impact Today: Accelerates misinformation, stock market bubbles, and policy misdirections in a hyperconnected media ecosystem.
- **Status Quo Bias:**
  Definition: Preferring the current state, resisting change even when improvements are available.
  Impact Today: Hinders reform, innovation, and adaptation to emerging challenges like climate change or digital transformation.

- **Cultural Cognition:**
  Definition: Interpreting information through the lens of cultural values, reinforcing existing worldviews.
  Impact Today: Complicates consensus on issues like public health measures, environmental policies, and social justice reforms.
- **Misinformation Effect:**
  Definition: Post-event misinformation distorting the memory of the original occurrence.
  Impact Today: Shapes collective understanding of events, influencing public opinions, historical interpretations, and policy agendas.

Emerging and Technology-Related Biases

- **AI Confidence Bias:**
  Definition: Over-trusting AI outputs without critical evaluation.
  Impact Today: As AI informs everything from hiring to sentencing recommendations, this bias risks embedding systemic errors at scale.
- **Digital Identity Bias:**
  Definition: Distortions arising from discrepancies between online personas and offline selves.
  Impact Today: Affects mental health, brand perception, and trust in digital interactions, influencing consumer and social behaviors.
- **Information Overload Bias:**
  Definition: Difficulty making decisions due to overwhelming data, resorting to oversimplified heuristics.
  Impact Today: In data-rich industries, decision fatigue and superficial judgment lead to missed insights and strategic missteps.
- **Automation Bias:**
  Definition: Preferring automated system suggestions over human judgment.
  Impact Today: Reduces critical oversight in fields like transportation, healthcare, and finance, risking safety and equity.
- **Algorithmic Bias:**
  Definition: AI or machine-learning models perpetuating societal prejudices due to biased data or design.
  Impact Today: Reinforces discrimination in hiring, lending, law enforcement, and healthcare, questioning the fairness and ethics of automation.
- **Illusory Truth Effect:**
  Definition: Repeated statements are perceived as more truthful regardless of accuracy.
  Impact Today: Media repetition can legitimize misinformation and shape public opinion, even when data contradict these statements.
- **Bias Blind Spot:**
  Definition: Recognizing biases in others but not in oneself.
  Impact Today: Fosters overconfidence in personal judgment, reducing motivation for self-reflection, quality checks, and continuous improvement.

Medium-Impact and Contextual Biases

- **Herd Behavior:** (Closely related to Information Cascade)
  Definition: Individuals mimic the actions of a larger group.
  Impact Today: Affects consumer trends, stock market movements, and social campaigns – sometimes beneficial, often destabilizing.

- **Survivorship Bias:**
  Definition: Focusing on successes while ignoring failures, skewing perceptions of what leads to positive outcomes.
  Impact Today: Distorts entrepreneurial guidance, product development strategies, and educational methodologies by highlighting only "winners."
- **Ambiguity Aversion:**
  Definition: Preferring known risks over uncertain ones.
  Impact Today: May discourage innovative solutions or policy experiments in complex, uncertain global contexts, limiting progress and adaptability.

5.5.2 Why These Biases Matter Today

These biases achieve "high-impact" status not just due to their prevalence, but because of the conditions that magnify their effects:

- **Global Connectivity:** Decisions and information transcend national borders instantly, magnifying both positive and negative outcomes.
- **Technological Amplification:** Algorithms, AI, and automated systems can scale biases exponentially, affecting millions simultaneously.
- **Complex and Interdependent Challenges:** Issues like climate change, global health crises, and economic volatility demand nuanced, evidence-based solutions. Biases that oversimplify or misinform hinder constructive problem-solving.
- **Rapid Information Flows:** Instantaneous media cycles and social platforms fuel cascades and echo chambers, shaping public opinion and policy in days or hours.
- **Cultural and Ethical Imperatives:** In diverse, multicultural societies, biases that perpetuate discrimination or misinformation erode trust, fairness, and the collective capacity to address shared problems.

By reviewing this glossary, you better equip yourself to recognize, interpret, and address these biases – preparing for the subsequent chapter's practical approaches to continuous learning, reflective analysis, and more adaptive decision-making.

5.6 FROM THEORY TO PRACTICE: TOOLS FOR POST-DECISION ANALYSIS

Understanding cognitive biases is a critical step – but the true challenge lies in operationalizing this knowledge to shape better decisions moving forward. The strategies outlined here aim not just to correct individual errors, but to build capabilities and cultures that learn, adapt, and evolve. In a world defined by complexity, interdependence, and relentless change, these tools offer scaffolding to help leaders, teams, and organizations convert hindsight into foresight.

5.6.1 Reimagining Reflection as a Strategic Asset

Reflection is not merely an academic exercise. When applied thoughtfully, it becomes a competitive advantage, accelerating learning cycles and increasing resilience. By treating post-decision analysis as part of a living system – one that integrates technology, cultural context, human psychology, and ethical considerations – you can refine not only how you think, but how you collaborate, innovate, and lead.

5.6.1.1 Simple Frameworks for Reflection: Beyond Checklists

The 3-Step Decision Review Checklist (Enhanced):

1. **Identify Emotional Triggers and Initial Assumptions:**
   Go deeper than emotions – consider "cognitive states." Were you operating in a scarcity mindset, worried about budget cuts, or was there a subtle group dynamic that made voicing

dissent feel risky? Consider how organizational incentives, cultural norms, and technological constraints shaped your starting point. Recognizing these subtle influences helps uncover hidden structural biases.

2. **List Which Biases Influenced Reasoning:**
Move beyond labeling biases as errors to seeing them as "cognitive indicators." A strong confirmation bias in strategic planning might hint at an innovation gap – perhaps you lack systems for exploring contrarian research. Authority bias could reveal insufficient governance structures to challenge top-level directives. By interpreting biases as signals about the organization's learning mechanisms, you elevate this step from blame-assessment to capability-building.
3. **Consider an Alternative Decision with Different Assumptions:**
Instead of just imagining a different decision, think in scenarios: How would you decide if your team were more diverse in expertise or cultural background? What if you leveraged a rotating "contrarian role" to regularly test core assumptions? By weaving scenario planning into everyday reflection, you cultivate strategic agility – able to pivot swiftly as contexts evolve.

New Insights and Angles:

- **Cognitive Journaling:** Encourage individuals to keep a "cognitive journal" where they note recurring biases, personal triggers, and successful interventions. Over time, patterns emerge that can inform tailored training or structural changes.
- **External Sparring Partners:** Pair up with peers outside your industry or cultural context to review decisions. A global perspective can highlight biases normalized within your domain.

5.6.1.2 Learning from Mistakes: Institutionalizing Adaptive Intelligence

From Decision Logs to Collective Intelligence Platforms:

A simple decision log is helpful, but modern organizations can go further. Consider building an internal "Collective Intelligence Platform" (CIP) where project post-mortems, user feedback, data analyses, and lessons from past decisions converge. Incorporate visualization tools, sentiment analysis, and AI-driven pattern recognition to identify subtle correlations between certain biases and certain conditions (e.g., regulatory uncertainty, rapid market shifts).

- **Structured Post-Mortems and Cultural Narratives:** Transform post-mortems into storytelling sessions. Narratives help internalize lessons more deeply than bullet-point lists. For instance, share a story about how a marketing campaign failed due to confirmation bias in user research. Present it not as a cautionary tale but as a narrative of transformation – how the team adopted new user-testing protocols and saw improved outcomes next time.
- **Rotating Roles and Cross-Pollination:** To avoid institutional memory fading, periodically rotate team members who "own" the decision logs. Fresh eyes bring fresh interpretations. Similarly, cross-pollinate lessons between departments. A product team's experience with sunk cost fallacy could help a finance team recognize loss aversion in investment decisions, fostering a culture of interdisciplinary learning.

New Insights and Angles:

- **Internal "Bias Clinics":** Organize quarterly "bias clinics" where team members present real decisions and dissect them with a multidisciplinary audience. This creates a safe, educational forum that normalizes vulnerability and continuous improvement.
- **Global Benchmarking:** Partner with external organizations, universities, or think-tanks to benchmark your learning processes, ensuring you remain at the cutting edge of cognitive resilience practices.

5.6.1.3 Navigating Information Overload: Synthesizing Signal from Noise

Toward Curated Intelligence and Augmented Analysis:

Facing a deluge of data, the task is to move from raw information to curated intelligence. Consider integrating decision-support dashboards that highlight key metrics aligned with strategic goals, rather than drowning in every available KPI.

- **Advanced Prioritization Techniques:** Besides Pareto analysis or ABC prioritization, experiment with approaches like scenario-based filtering. Ask: "Which data points matter most if a new competitor enters the market? If a regulatory change looms?" Aligning data filtering with scenario triggers ensures that priorities shift smartly as conditions evolve.
- **Data Distillation and Cognitive Tagging:** Implement "cognitive tagging" where team members annotate data sources with notes like "high uncertainty," "requires diverse viewpoints," or "check cultural bias." Over time, this meta-layer of information streamlines distillation. Tools that allow voice annotations or quick polling can surface minority opinions quickly, ensuring contrarian insights aren't lost.

New Insights and Angles:

- **Cognitive Ergonomics:** Apply principles of design thinking to your information workflows. Just as ergonomic chairs improve physical comfort, ergonomically designed data interfaces reduce cognitive strain. Think about color-coding biases, creating intuitive navigation paths, and layering information so that the most critical insights are front-and-center.
- **Emotional Metadata:** Consider tagging decisions with emotional or contextual metadata – e.g., "high-stress decision" or "cross-cultural tension noted." Over time, AI-driven sentiment analysis might reveal that certain emotional states or team dynamics correlate with specific biases.

5.6.1.4 Counteracting Algorithmic Influence: Fostering Digital Integrity

Shaping Your Algorithmic Diet:

Just as individuals manage their nutritional intake for health, consider managing your "algorithmic diet" for cognitive well-being. This involves being mindful about which digital platforms you trust, how you configure feed preferences, and when you seek manual verification of automated suggestions.

- **Intentional Feed Diversification and Cognitive Curation:** Periodically "break the feed" by resetting preferences, following sources with opposing viewpoints, or using aggregation platforms that highlight global perspectives. Consider appointing a "diversity advocate" within teams who ensures exposure to unconventional insights – be it from emerging markets, niche academic fields, or grassroots organizations.
- **Augmenting Algorithms with Human Wisdom:** Pair AI tools with human advisory panels that include ethicists, cultural liaisons, and even citizen representatives. For major decisions – like launching a new product in a sensitive market or setting policy in a multicultural workplace – this hybrid approach ensures that automated data analysis doesn't overshadow ethical judgment, cultural nuance, and emotional intelligence.

New Insights and Angles:

- **Digital Sabbaticals and Cognitive Refresh:** Encourage periodic digital sabbaticals – short breaks from algorithmically curated environments. During these sabbaticals, rely on human networks, international magazines, or research from neutral think-tanks. Upon return, you'll notice biases or distortions you missed before.
- **Algorithmic Contradiction Playlists:** Just as music streaming services create playlists, consider creating an "Algorithmic Contradiction Playlist" of content that deliberately conflicts with your usual sources. Reviewing these curated contradictions can spark productive tension, challenging comfort zones and revealing blind spots.

5.6.1.5 Integrating These Practices into a Dynamic Culture

The ultimate objective is cultural transformation. Implementing these tools and insights is not about checking a box – it's about cultivating an environment where critical reflection, ethical reasoning, global sensitivity, and adaptive intelligence become second nature.

Think of It as Cognitive Fitness:
Just as physical fitness requires regular exercise and varied routines, cognitive fitness demands consistent engagement with challenging perspectives, disciplined reflection, and conscious recalibration of mental habits. Over time, your "cognitive muscles" strengthen, biases become more transparent, and decisions more robust.

Leadership and Accountability:
Leaders play a crucial role in normalizing these practices. By publicly acknowledging their own biases, celebrating teams that identify and correct flawed assumptions, and investing in tools that enhance reflective capabilities, they set the tone. Accountability mechanisms – like regular reviews by boards or ethics committees – ensure that improvements are not fleeting but embedded in the organization's DNA.

Future-Proofing with an Evolving Toolkit:
As emerging technologies, cultural shifts, and market complexities evolve, continuously update your approaches. Test new reflection exercises, adopt fresh AI-driven analytics tools responsibly, and remain open to international best practices. This dynamic approach guards against complacency, ensuring that your decision-making processes remain fit for the future.

5.6.2 Conclusion: Toward a Cognitive Ecosystem of Integrity and Innovation

By integrating these advanced insights and angles, you elevate post-decision analysis from a remedial step to a strategic engine of innovation. Embrace these practices as part of a "cognitive ecosystem" where individuals, teams, cultural contexts, and technologies interact to refine understanding, foster creativity, and anticipate emerging challenges.

In this ecosystem, biases become signals for where to improve, not just glitches to fix. Information overload becomes a source of selective advantage as you learn to distill signal from noise. Algorithmic influences become opportunities for curation and ethical engagement. Over time, this sustained practice transforms theoretical knowledge into tangible capabilities – enabling you to navigate complexity with integrity, adaptability, and a readiness to seize tomorrow's opportunities.

5.7 LONG-TERM STRATEGIES FOR INDIVIDUALS, ORGANIZATIONS, AND SOCIETIES

Cognitive biases are not static challenges to be "fixed" once and for all; they are dynamic phenomena influenced by evolving technologies, shifting cultural landscapes, and global interdependencies. Long-term resilience demands continuous adaptation, strategic foresight, and a willingness to invest in sustainable practices. By examining long-term strategies through three lenses – individual, organizational, and societal – we can align personal growth, institutional reforms, and broad policy frameworks to counter biases effectively and ethically.

5.7.1 A Tri-Level Ecosystem of Adaptation

Think of these three spheres as interconnected layers in a cognitive ecosystem:

- **Individuals:** The front line of awareness and action, where personal habits and mindsets influence everyday decisions.

- **Organizations:** The engines of collective endeavor, shaping cultural norms, decision-making protocols, and professional development.
- **Societies:** The overarching context in which laws, educational systems, media environments, and global collaboration frameworks define the baseline conditions for bias mitigation.

Enhancing bias resilience at each level, while recognizing their interconnections, sets the stage for holistic and enduring improvements.

5.7.1.1 Continuous Learning and Adaptation

At the Individual Level:

Personal growth lies at the heart of bias resilience. Individuals should commit to lifelong learning – seeking new perspectives, regularly revisiting their assumptions, and embracing discomfort.

- **Ongoing Education:** Take short courses in behavioral economics, social psychology, or cross-cultural communication. Periodically review your "cognitive journal" entries to identify emerging patterns.
- **Personal Bias Audits:** Every few months, pause to reflect: Have you relied heavily on familiar sources or avoided challenging viewpoints? Setting personal goals for bias reduction – such as reading one credible, contrarian analysis per week – ensures incremental progress.

At the Organizational Level:

Organizations act as incubators of collective intelligence. By institutionalizing reflection and adaptation, they create feedback loops that refine their strategic direction.

- **Periodic Re-education and Team Workshops:** Offer ongoing training modules and "bias refreshers," integrating new research and global best practices. Rotate trainers from different departments or regions to avoid intellectual inbreeding.
- **Structured "Bias Audits":** Twice a year, conduct audits assessing decision protocols, hiring outcomes, product launches, or negotiation results for bias indicators. These audits guide targeted interventions, policy changes, or revised incentive systems.

At the Societal Level:

Societies can foster a culture of critical thinking and intellectual humility at scale.

- **Curriculum Integration:** Incorporate bias awareness, media literacy, and critical thinking skills into national education standards. By teaching these skills early, societies lay the foundation for future generations of informed citizens.
- **Public Awareness Campaigns and Lifelong Learning Programs:** Governments, NGOs, and think-tanks can run campaigns encouraging citizens to question assumptions, diversify their media diets, and engage in respectful debate. National libraries or cultural centers might host discussion forums and workshops open to the public.

5.7.1.2 Cultural and Institutional Safeguards

At the Individual Level:

While individuals have limited power to overhaul institutions, they can practice ethical leadership and encourage integrity within their spheres of influence.

- **Championing Fairness and Inclusion:** Individuals who lead teams or projects can model bias-free decision-making, invite dissenting voices to meetings, and reward critical inquiry. Encourage a norm where pointing out potential biases is met with appreciation, not defensiveness.

At the Organizational Level:
Institutions hold significant sway over decision climates. By weaving bias awareness into corporate governance, policymaking, and organizational culture, they shape lasting norms.

- **Cultural Competency Modules and Structured Decision Protocols:** Require all employees to undergo cultural awareness training. Integrate pre-mortems, scenario planning, and devil's advocate roles into regular decision-making processes. Such protocols reduce the risk of groupthink and authority bias, fostering an environment where balanced judgments are standard practice.
- **Ethical and Diversity Committees:** Formalize bodies within organizations – comprising ethicists, data scientists, diverse community representatives, and junior staff – to review high-stakes decisions for potential biases. Transparent reporting of these reviews can build trust among stakeholders and reinforce a reputation for responsible stewardship.

At the Societal Level:
Societies can encourage ethical conduct and cultural sensitivity through policies, regulations, and international cooperation.

- **Policy Frameworks and Regulatory Oversight:** Governments can enact laws requiring transparency in algorithmic decision-making or mandate diversity on corporate boards. Regulatory bodies might audit sectors prone to biases – like lending, hiring, or law enforcement – to ensure fairness.
- **Global Standards and Multilateral Agreements:** International bodies (e.g., the UN, regional trade blocs) can establish guidelines and best practices, encouraging cross-border collaboration on bias mitigation. By harmonizing approaches, societies share insights, reduce cultural blind spots, and promote equitable global governance.

5.7.1.3 Documenting and Sharing Knowledge

At the Individual Level:
Personal reflection is a powerful tool for long-term improvement.

- **Decision Journals and Personal Knowledge Logs:** By consistently documenting decisions, outcomes, and identified biases, individuals build a personal repository of lessons learned. Over time, reviewing these logs reveals personal growth trajectories and helps maintain accountability.

At the Organizational Level:
Knowledge management transforms scattered lessons into collective intelligence.

- **Internal Case Studies and Best Practice Libraries:** Store success stories and failures in a centralized platform, tagged for easy retrieval. A marketing team can learn from an engineering team's approach to avoiding confirmation bias, fostering cross-functional insight exchanges.
- **Collaborative Platforms and Feedback Loops:** Implement internal forums, digital suggestion boxes, or "bias insight" Slack channels. Encourage employees to contribute real-time observations, articles, and tools. Over time, these platforms become dynamic knowledge ecosystems where lessons are refined and disseminated quickly.

At the Societal Level:
When societies invest in open knowledge-sharing, the benefits multiply.

- **Public Databases and Educational Repositories:** Universities, think-tanks, and libraries can maintain open-access databases of case studies from various sectors and cultures. Citizens, journalists, researchers, and policymakers can tap into these resources for evidence-based decision-making.

- **International Conferences and Cultural Exchanges:** Host global summits focusing on bias awareness and mitigation. Academics, practitioners, policymakers, and civil society leaders share experiences and innovations, bridging cultural divides and ensuring that best practices travel across borders.

5.7.1.4 Tools and Checklists: Building a Multi-Level Cognitive Toolkit

At the Individual Level:

Simple, easily accessible tools help maintain personal discipline and awareness.

- **Personal "Bias Check" Apps:** Use mobile tools that prompt reflection after key decisions – "Did you seek a contrarian viewpoint?" or "Have you considered cultural nuances?" Over time, these micro-interventions shape habits.

At the Organizational Level:

Organizations can provide structured toolkits that employees can rely on when facing complex decisions.

- **Appendix-Style Guides and Templates:**
  - *Decision Journal Template* for consistent recording
  - *Bias Awareness Prompts* for key meetings
  - *Feed Diversity Checklists* for balanced information consumption
- **Integrating Tools into Workflow:** Embed checklists into project management software, add bias-awareness steps into process maps, and use gamification (badges, recognition) to encourage utilization.

At the Societal Level:

Societies benefit from widely available tools that raise general standards of critical thinking.

- **Publicly Available Frameworks and Checklists:** NGOs or international organizations could publish open-source guides for media literacy, civic engagement, and balanced information diets. These resources empower citizens to navigate media landscapes responsibly.
- **Public–Private Partnerships:** Partner with tech companies to develop browser extensions or AI-driven alert systems that highlight potential bias patterns in public discourse. Such tools serve as cognitive supports for the entire citizenry.

5.7.2 Shaping a Bias-Resilient Future Together

In aligning strategies across individuals, organizations, and societies, we create synergy:

- **Individuals:** Develop personal resilience, maintain intellectual curiosity, and model ethical behavior.
- **Organizations:** Institutionalize reflection, create structural safeguards, share knowledge, and leverage diversity to achieve robust decision-making.
- **Societies:** Establish supportive legal, educational, and cultural frameworks that promote widespread bias awareness, ensuring policies and public discourse are grounded in critical thought and empathy.

5.7.3 Long-Term Vision and Ethical Leadership

Over time, these efforts cultivate a landscape where biases are no longer hidden liabilities but transparent signals guiding continuous improvement. By sustaining these efforts, we move closer to a world where innovation, fairness, and understanding flourish in tandem.

Ultimately, the distinction between personal habits, organizational protocols, and societal norms is not a separation but a tapestry. Each level enriches and informs the others, creating a resilient cognitive ecosystem. As individuals deepen their self-awareness, organizations institutionalize good practices, and societies set enlightened standards, we collectively evolve toward decisions that are wiser, more inclusive, and more ethically grounded.

5.8 EMBRACING MODERN CHALLENGES: TECHNOLOGY AND CULTURAL CHANGE – A CATALYST FOR RADICAL REIMAGINATION

The future of decision-making will not be defined by simply outsmarting existing biases; it will hinge on our ability to anticipate, co-create, and transform the cognitive environment itself. As technologies like AI, synthetic media, and immersive virtual worlds emerge, and as cultural identities and norms continually evolve, our challenge is no longer just keeping pace – it's leading the way. This demands a shift from reactive adaptation to proactive experimentation, from well-worn frameworks to agile mental architectures, and from insular perspectives to vibrant, cross-cultural coalitions.

Rethinking the Landscape:
The interplay of technology and culture isn't a distant horizon – it's the medium in which we now operate. Bias mitigation can become a driver of innovation rather than a brake on progress. By seeing biases as signals of where our cognitive systems strain under complexity, we gain clues about how to build more resilient, inclusive, and imaginative forms of sense-making. Through this lens, every emerging challenge is an invitation to re-engineer our thought processes and institutional designs.

5.8.1 Staying Ahead of Emerging Biases: Prototype, Hack, and Iterate

5.8.1.1 Individual Level: Explorers of Mental Frontiers

Individuals can become cognitive explorers – early adopters who prototype new modes of thinking as soon as novel technologies arise.

- **Mental Prototyping:** Treat your mind as a lab. When a new AI-driven assistant emerges, experiment with how it influences your choices. Document your cognitive "before and after," pinpointing fresh biases and brainstorming hacks to counter them.
- **Micro-Pilots of Behavior Change:** Regularly run personal experiments: Spend a week using a contrarian content feed, test decision-support apps that challenge your defaults, or adopt periodic "cognitive fasts" from certain digital environments. By iterating swiftly, you build a personal skill set to surf the waves of evolving biases.

5.8.1.2 Organizational Level: In-House Cognitive Innovation Hubs

Organizations can establish "Cognitive R&D units" that explore how shifting technologies and social values forge new bias patterns.

- **Bias Hackathons:** Host internal hackathons where teams compete to design quick interventions – like algorithmic filters, scenario simulations, or cultural exchange experiences – to tackle a newly identified bias. A single weekend sprint can yield transformative insights that keep your institution ahead of the curve.
- **Iterative Frameworks:** Instead of static policies, think in terms of rolling beta versions of decision protocols. Update them monthly or quarterly based on feedback from frontline employees, emerging AI tools, or social listening efforts capturing stakeholder sentiment.

5.8.1.3 Societal Level: Global Cognitive Sandboxes

Societies can create spaces – both physical and virtual – where the public, entrepreneurs, academics, and policymakers collaboratively experiment.

- **Civic Test Beds:** Cities or regions could volunteer as "cognitive innovation zones" where new policies or AI-driven systems are trialed with transparency and community input.

These controlled experiments yield data on how biases mutate under novel conditions, informing national or international guidelines.

- **Cross-Border Explorations:** Multilateral alliances can sponsor "bias foresight labs," bringing together diverse cultural perspectives to project future bias scenarios and preempt them with creative interventions. By pooling global talent, societies foster a collective resilience that transcends borders.

5.8.2 Global and Cross-Cultural Perspectives: Co-Creating a Cognitive Commons

5.8.2.1 Individual Level: Cultural Catalysts

Individuals can become cultural connectors, deliberately seeking out and integrating worldviews that challenge their native biases.

- **Cultural Prototyping:** Volunteer to mentor or learn from peers in different cultural contexts. For example, pair a startup founder in Berlin with a policy analyst in Nairobi to exchange decision-case stories. Such "cultural prototypes" expand your mental repertoire, making your reasoning more fluid and inventive.

5.8.2.2 Organizational Level: Knowledge Without Borders

Companies and institutions should treat global diversity as a strategic asset, a source of cognitive raw material to forge new insights.

- **Intercultural Decision Jams:** Prior to launching a global initiative, convene teams from multiple continents for a "decision jam," where they dissect assumptions and biases side-by-side. This raw intercultural collision often yields surprising solutions that no single team would have generated alone.
- **Cultural Intelligence Networks:** Create formal networks of cultural ambassadors within the organization – people whose role is to continuously scan the global environment, report on emerging trends, and recommend adjustments to decision protocols and training curricula.

5.8.2.3 Societal Level: A Global Cognitive Commons

Societies can commit to sharing not just data but also cognitive schemas, ethical guidelines, and educational resources openly.

- **Open-Source Cognitive Toolkits:** International consortia can develop free, adaptable bias-mitigation toolkits localized into multiple languages, ensuring communities worldwide have access to advanced methods.
- **Cultural Diplomacy in Cognitive Resilience:** Embassies and international NGOs might facilitate "cognitive diplomacy," running workshops that transfer successful bias mitigation practices from one region to another, effectively cross-pollinating solutions.

5.8.3 Scalability and Ethical Use: Designing for Trust and Fairness at Scale

5.8.3.1 Individual Level: Personal Codes of Cognitive Ethics

As individuals, codify your own cognitive ethics – your personal criteria for what technology, content, and decision aids you deem acceptable.

- **Personal Cognitive Charters:** Draft a mini-manifesto stating the principles you'll uphold – like always seeking a dissenting opinion before a major choice, supporting platforms that prioritize transparency, or committing to respectful debate. Living by such a charter empowers you to shape demand for ethical products and services.

5.8.3.2 Organizational Level: Radical Transparency and Co-Governance

Organizations can scale fairness and trust by embracing radical transparency and stakeholder co-governance.

- **Cognitive Commons within Corporations:** Invite customers, community members, and independent auditors to participate in decision-making reviews. This shared governance of cognitive ecosystems ensures ethical principles and accountability measures aren't top-down edicts but co-created commitments.
- **Ethical Product Roadmaps:** Publicly post your product development pipelines, highlighting how bias checks are embedded at each milestone. Encourage feedback loops with user communities, turning passive "consumers" into active partners who shape the trajectory of your offerings.

5.8.3.3 Societal Level: Planetary Governance of Cognitive Ecosystems

At scale, societies can reshape governance structures to uphold fairness and integrity in a hyper-connected world.

- **Global Cognitive Treaties:** Negotiate international agreements that set baselines for algorithmic explainability, data rights, and bias audits. Such treaties could treat cognitive fairness as a global public good, akin to environmental stewardship.
- **Civic Stakeholder Assemblies:** Host periodic global assemblies where citizens, experts, and innovators propose updates to ethical standards and accountability frameworks – an evolving "constitution" for collective cognition.

5.8.3.4 A Vision for the Future: Emerging as Cognitive Architects

This new perspective challenges us to see bias mitigation not as a defensive game but as a generative frontier. By embracing experimentation, cultural fluidity, and ethical design thinking, we evolve from mere problem-solvers into architects of cognitive ecosystems that foster trust, creativity, and shared growth.

From Resilience to Renaissance:
Instead of settling for bias resilience, aim for a cognitive renaissance. In this envisioned future:

- Individuals are cognitive explorers, continually refining their mental maps.
- Organizations are innovation crucibles, prototyping decision models that adapt at the speed of change.
- Societies are co-creative landscapes, aligning global ethics and local ingenuity to produce equitable outcomes.

By harnessing the interplay of technology and cultural change, we can transform biases into steppingstones toward a more enlightened, humane, and imaginative global dialogue. In this unfolding era, each new challenge becomes not a threat but an opportunity to reinvent how we think, collaborate, and decide.

5.9 CONCLUSION: TURNING INSIGHT INTO INCREMENTAL ACTION

As we close this chapter, it's vital to recognize that we stand not at an endpoint but at a threshold. The exploration of cognitive biases, their cultural and technological amplifiers, and the strategic frameworks for addressing them was never meant to produce a final blueprint. Instead, it offers an evolving toolkit – one that shifts perspectives and invites experimentation, continuous learning, and adaptive thinking.

From Insight to Evolutionary Practice:
We've moved beyond the simplistic notion that biases are merely "problems" to be fixed. Instead, we see them as complex signals embedded in our decision ecosystems. They highlight where mental shortcuts, cultural assumptions, and technological filters shape – and sometimes distort – our reasoning. By examining these patterns, we gain opportunities to refine our methods, challenge our assumptions, and embrace more nuanced, informed approaches.

5.9.1 Key Takeaways: Embracing Complexity and Opportunity

- **Neither Good nor Bad, but Informative:** Cognitive biases are not moral failings; they're cognitive indicators that reveal how our minds process information under uncertainty. Recognizing them doesn't mean shaming our past decisions but illuminating where we can grow. When we treat biases as feedback loops, we transform them into catalysts for reflective practice.
- **Systematic Analysis, Incremental Gain:** The essence of progress lies in discipline – consistently documenting decisions, exposing ourselves to diverse viewpoints, and revisiting outcomes with humility. Over time, these small steps accumulate. They don't guarantee perfect choices, but they create a persistent upward trend in decision quality, adaptability, and ethical consideration.

At the Individual Level:
You have the agency to evolve. By becoming aware of your biases, seeking contrarian perspectives, and maintaining a decision journal, you slowly but steadily enhance your cognitive toolkit. Over months and years, this dedicated effort yields sharper judgment, greater open-mindedness, and an enriched understanding of complex issues.

At the Organizational Level:
Institutions can embed bias-awareness into their cultures – through periodic audits, structured decision protocols, best-practice libraries, and intercultural dialogues. Such systematic approaches turn an organization into a learning entity, better equipped to adapt to market upheavals, technological shifts, and social demands. Over time, these accumulations of incremental refinements enhance trust, credibility, and innovation potential.

At the Societal Level:
Societies thrive when policies, educational curricula, media ecosystems, and legal frameworks reflect a shared commitment to fairness, transparency, and constructive dialogue. No single reform is a panacea; however, incremental improvements – from mandating AI explainability to fostering intercultural exchanges – gradually uplift public discourse and collective problem-solving. Each positive step nudges communities toward a more equitable, informed, and resilient future.

5.9.2 Incremental Progress: The Road to Lasting Impact

Perfection is not only unattainable; it's unnecessary. Our goal is not to eradicate biases – a futile endeavor – but to understand and work with them. By focusing on incremental progress, we acknowledge that decision-making contexts will always be fluid. Technologies will morph, cultural norms will shift, and new forms of complexity will arise. The virtue lies in readiness, curiosity, and the willingness to adapt.

- **Small Steps, Big Resonance:** A single bias audit at work might inspire a new policy that later influences market trends. Documenting one tough personal decision can, over time, reshape your approach to leadership, mentoring, or collaboration. A modest educational reform nudging students to question their sources can, over a generation, transform how a society engages with truth and consensus-building.
- **Enduring Mindsets, Expanding Horizons:** By approaching biases with curiosity and strategic action rather than frustration or complacency, we cultivate a mindset that's both humble and ambitious. This attitude enables us to pivot gracefully in response to new challenges and embrace complexity as a source of strength rather than confusion.

5.9.2.1 Looking Ahead: The Journey Continues

As we leave this chapter behind, consider it a launchpad rather than a final verdict. The insights gained here prepare you to enter dialogues about AI ethics, cultural evolutions, global governance, and organizational transformations with greater poise and discernment. Over time, the incremental actions you take – personally, professionally, and civically – coalesce into lasting, positive changes.

In a world that may seem ever more intricate, this approach offers a powerful reassurance: By turning insight into incremental action, we continuously refine our collective wisdom. We become more adept at navigating uncertainty, more skilled in bridging cultural divides, and more committed to ethical stewardship of technology and resources. Ultimately, each small improvement lays the groundwork for a future where constructive reasoning, empathy, and principled decision-making are not exceptions, but the ever-improving norm.

6 Enhanced Adaptive Social Cognition Theory (EASCT)

Core Premise:

1. **Non-Linear Adaptation** – Human behavior is dynamic, responding to complex and evolving global, digital, and social forces.
2. **Co-Creation of Environments** – Individuals actively shape cultural narratives and technological landscapes.
3. **Ethical Agency** – Adaptation must integrate moral reasoning, ensuring fairness and responsibility in digital and social interactions.
4. **Flourishing Over Survival** – The goal is not just resilience but growth, creativity, and ethical leadership.

Key Principles and Applications:

- **Cognitive Flexibility and Emotional Intelligence** – Training for better decision-making, digital literacy, and critical thinking.
- **Social Learning and Identity Formation** – Understanding how global interactions reshape values and group affiliations.
- **Technology and Human Agency** – Advocating for ethical Artificial Intelligence (AI), transparent algorithms, and responsible digital engagement.

EASCT **guides education, workplace innovation, and policymaking**, ensuring that individuals and institutions **proactively navigate complexity, embrace diversity, and drive ethical technological change**.

6.1 INTRODUCTION

In the 21st century, human cognition, emotion, and behavior unfold within a dynamic interplay of social, cultural, and technological forces that defy traditional boundaries. Individuals no longer simply adapt to fixed environments; instead, they co-create their ecosystems, forging identities, alliances, and solutions in real-time digital networks, multicultural workplaces, and fluid communities of interest. Against this backdrop, classical frameworks for understanding human behavior – rooted in stable roles, linear learning processes, or isolated cognitive modules – risk becoming obsolete. Emerging complexities demand a theory that not only integrates established insights but also embraces the accelerating pace of change.

The **Enhanced Adaptive Social Cognition Theory (EASCT)** responds to this need by presenting a unifying lens through which to understand how humans navigate their environments with increasing sophistication. Drawing upon and extending foundational models such as Social Cognitive Theory (Bandura, 1986), Social Identity Theory (Tajfel & Turner, 1986), the Theory of Planned Behavior (Ajzen, 1991), and self-regulatory frameworks (Carver & Scheier, 1998), EASCT emphasizes the radical impact of technology and cultural fluidity on human cognition and behavior. It posits that adaptation today is not just about surviving change; it's about leveraging complexity, engaging ethically with technology, and aligning personal values with global imperatives.

DOI: 10.1201/9781003640288-9

Where earlier theories often treated technology as a background factor or cultural variation as a peripheral concern, EASCT places them at the core of its explanatory power. It approaches technology not as a mere tool but as an active mediator that reshapes social norms, reconfigures interpersonal relationships, and redefines what it means to learn, achieve, and cooperate. Likewise, it treats cultural and individual differences not as static variables but as dynamic, evolving forces that enrich the tapestry of human adaptation.

In this chapter, we articulate the Core Premise of EASCT, detail its Key Principles, and demonstrate how its integrative perspective can inform research, practice, and policy. By viewing adaptation as an iterative, feedback-driven, and context-sensitive process, EASCT aspires to guide educators, organizational leaders, policymakers, and individuals toward strategies that foster resilience, creativity, and well-being in an era defined by ceaseless change and interconnection.

6.2 CORE PREMISE

At its heart, the EASCT posits that human behavior is a dynamic, co-creative process unfolding at the nexus of individual cognition, social structure, cultural variation, and technological mediation. Unlike earlier models that often treated adaptation as a mostly reactive or incremental adjustment to stable conditions, EASCT reframes adaptation as a proactive, context-sensitive, and ethically charged endeavor. In this view, individuals are not merely shaped by their environments; they are active participants in shaping, curating, and reinventing the ecosystems – both digital and physical – in which they operate.

6.2.1 Complexity and Non-Linear Adaptation

Classical frameworks of human behavior tended to assume relatively predictable paths of learning, motivation, and identity formation. They also often segmented influences into neat categories – cognition versus emotion, technology as a tool rather than a cultural force, social norms as fixed reference points. EASCT diverges by recognizing that contemporary environments are characterized by complexity, uncertainty, and rapid change. Adaptation here is not linear progress toward a known equilibrium; it is a continuous, iterative negotiation of multiple, interlocking variables. These include shifting social norms emerging in online communities, evolving cultural mores as global dialogues intensify, and ever-updating technological architectures that alter what is visible, who is connected, and how values are expressed.

6.2.2 Co-Creation of Environments and Narratives

EASCT introduces a new angle on human adaptation by emphasizing that individuals and environments are not separate entities but parts of an intricate co-creative cycle. Traditional models often portrayed individuals as learners who internalize norms, values, and technologies that others have produced. EASCT, however, underscores that individuals also generate content, shape narratives, design tools, and influence cultural currents. For instance, a single individual's social media post can spark global conversations, while a community's choice to develop ethical guidelines for AI usage can alter industry standards. Rather than viewing adaptation as aligning oneself with a pre-given environment, EASCT sees adaptation as participating in an ongoing cultural and technological innovation process – a form of adaptive "cultural engineering" or "cognitive stewardship."

6.2.3 Global Scale and Cultural Interpenetration

In contrast to frameworks developed in largely homogeneous or slow-changing contexts, EASCT addresses the global interpenetration of ideas, norms, and behaviors. Digital platforms allow memes, policy proposals, and moral frameworks to cross borders instantaneously, blending diverse cultural logics and ethical standpoints. EASCT's core premise holds that adaptation now involves dynamically negotiating and integrating multiple cultural reference points, often simultaneously. Individuals may blend a collectivist orientation from their familial heritage with entrepreneurial individualism encouraged by their online communities, all while navigating AI-driven content that nudges them toward certain worldviews. This global fluidity pushes adaptation beyond the familiar and into a realm where cultural dexterity, moral imagination, and intercultural empathy are essential adaptive capacities.

6.2.4 Ethical Agency and Reflective Engagement with Technology

A defining feature of EASCT's approach is the elevation of ethical agency in the adaptation process. Traditional theories often marginalized ethical considerations to a secondary status, focusing on efficiency, compliance, or stability. EASCT repositions ethics at the forefront: modern adaptation isn't just about adjusting one's behavior to "fit in" or "get by"; it's about actively considering how choices uphold fairness, transparency, well-being, and sustainability. In a world where technologies can amplify biases, influence elections, or commodify personal data, adaptation must involve critical literacy about digital affordances, algorithmic influences, and data ethics. Individuals who adapt successfully are not just those who learn to use technology efficiently, but those who understand and question its broader societal implications, carving out adaptive strategies that protect human dignity and democratic values.

6.2.5 Transcending Reactive Models – From Survival to Flourishing

Many traditional accounts framed adaptation as a process primarily concerned with coping or maintaining equilibrium under stress. EASCT, by contrast, envisions adaptation as growth-oriented and future-facing. Adaptive individuals do not merely survive turbulence; they learn to thrive in it. They leverage cognitive biases not as pitfalls but as heuristics that can be refined and strategically managed. They engage with cultural diversity not as an obstacle, but as an opportunity to expand their conceptual repertoires and empathy. They interact with technology not as passive users but as co-creators who can demand accountability, transparency, and human-centric design. Here, adaptation is not the endpoint of a learning curve but the continuous, creative redefinition of norms, relationships, and visions of the possible.

6.2.6 An Integrative, Cross-Disciplinary Foundation

Under EASCT's premise, adaptation is too complex to be explained by any single discipline. The theory weaves together insights from cognitive and social psychology, affective neuroscience, cultural anthropology, behavioral economics, information science, and critical media studies. This synthesis acknowledges that to understand how humans adapt today, one must consider how emotional states intersect with machine learning algorithms, how cultural narratives shape global discourse communities, how social identity formation mutates in digital spheres, and how self-regulatory capacities evolve when confronted with overwhelming data streams. The core premise is that only a richly interdisciplinary approach can capture the full spectrum of adaptive phenomena now at play.

EASCT's core premise redefines adaptation as a nuanced, ethically grounded, culturally and technologically intertwined process. Rather than seeing individuals as adapting to stable environments through incremental updates, EASCT sees them as active, morally responsible agents navigating fluid networks of information, power, and meaning. This new angle calls for rethinking how we measure adaptability, design interventions, and craft policies. It urges us to develop digital literacy curricula that include ethical reasoning, organizational models that embrace diversity and international collaboration, and public policies that champion transparency and fairness in AI-driven infrastructures.

Through this lens, adaptation emerges as a collaborative, future-building exercise – one in which human agency, cultural richness, emotional wisdom, and technological innovation converge to shape not just what is, but what could be.

> **IT IS EASIER TO FOOL PEOPLE**
>
> **THINKING BOX:**
>
> "How do we guard ourselves against misinformation in an era of endless content? The key, or some part of it, might be bound up with **42**."

6.3 KEY PRINCIPLES

Under EASCT, human adaptation emerges from a dynamic interplay of cognitive mechanisms, emotional states, cultural narratives, social identities, motivational drivers, and technological architectures. While each principle outlined here draws on established research traditions, EASCT reconfigures and extends these insights to address the complexity, global interconnectivity, and ethical agency central to contemporary life. These principles, taken together, form a holistic blueprint for understanding how people learn, collaborate, and transform themselves and their environments under conditions of rapid change.

6.3.1 Cognitive Processes and Emotional Regulation

EASCT acknowledges that cognition is never a purely rational or context-free process. Building on decades of scholarship on bounded rationality (Simon, 1957), heuristics and biases (Tversky & Kahneman, 1974), and affective influences on decision-making (Lerner et al., 2015), this principle positions emotions and cognition as intertwined forces guiding perception, attention, memory, and choice.

- **Interdisciplinary Integration:** By integrating affective neuroscience (Phelps, 2006) and emotional intelligence research (Salovey & Mayer, 1990), EASCT highlights that emotional regulation is not just a personal skill – it's a socially and technologically mediated competence. Individuals must learn to interpret emotionally charged content online, discern reliable signals amid digital noise, and remain empathetic even as they interact through mediated channels lacking traditional social cues.
- **New Angles and Complexity:** Unlike static models that view biases as errors, EASCT frames them as adaptive shortcuts that can be refined through education, technological design (e.g., decision-support systems that flag potential bias), and reflective exercises (e.g., mindfulness practices). Emotional regulation is similarly reframed: not simply as controlling emotions to maintain composure but as harnessing emotional insight to guide moral judgment, intercultural understanding, and innovative problem-solving.
- **Implications for Practice:** Training in emotional intelligence and bias awareness becomes vital in workplaces, educational curricula, and public workshops. Policymakers could incentivize digital platforms to incorporate features that support emotional well-being and cognitive reflection (e.g., prompting users to consider alternative perspectives before sharing content).

6.3.2 Social Learning and Cultural Transmission

Social learning theory (Bandura, 1986) and cross-cultural research have long established that humans acquire behaviors, norms, and values through observation and interaction. EASCT elevates this principle by foregrounding the unprecedented speed and scale of global information flows, and how cultural elements – from social norms to moral frameworks – are now disseminated and transformed across borders via digital infrastructures.

- **Global, Rapid, and Iterative:** Technologies such as social media, online learning platforms, and cross-border collaborations in science and activism enable cultural hybridization at a pace unimaginable in prior generations. EASCT stresses that adaptation now involves navigating rapidly changing cultural repertoires, often blending elements from multiple traditions, professional standards, or ideological communities into a personal tapestry of meaning.
- **Ethical and Reflective Dimensions:** This principle encourages considering how algorithms shape cultural transmission – amplifying some voices while silencing others – and how individuals can respond by seeking diverse sources and challenging homogeneous echo chambers. Ethical use of social learning capacity means not just absorbing trends but evaluating their integrity, fairness, and long-term implications.
- **Practical Applications:** Educators can design curricula that expose students to multicultural case studies, policymakers can support international exchange programs leveraging digital platforms, and organizations can encourage global teams to learn from each other's cultural strengths. Researchers can investigate how policy interventions (e.g., content moderation standards) influence cultural transmission online.

6.3.3 Motivation and Self-Regulation

Traditional motivation theories and self-regulation models (Carver & Scheier, 1998; Ryan & Deci, 2000) provide a foundation for understanding why people pursue certain goals and how they maintain effort over time. EASCT enriches this understanding by situating motivation within digital ecosystems that can either erode autonomy (through persuasive design or addictive loops) or enhance it (through empowering tools and personalized feedback).

- **Adaptive Motivational Landscapes:** In EASCT, motivation is not static; it is constantly renegotiated amid changing social expectations, technological affordances, and global uncertainties. Individuals must learn to sustain intrinsic motivation in contexts where extrinsic rewards proliferate (e.g., online ratings, social media "likes"), and where algorithmic nudges can distort perceived control.
- **New Angles on Grit and Perseverance:** Grit traditionally refers to persevering toward long-term goals (Duckworth & Quinn, 2009). Under EASCT, grit must also be culturally agile and digitally literate. For example, continuing professional development might involve iterating one's skillset in response to shifting market demands, learning new technologies, and balancing one's values against external pressures (like over-reliance on quick-fix digital rewards).
- **Guiding Interventions:** Organizations can foster autonomy-supportive environments, digital designers can incorporate features that respect user agency rather than exploiting attention, and policymakers can encourage consumer protection laws that support healthy digital habit formation.

6.3.4 Social Identity and Interpersonal Dynamics

Social Identity Theory (Tajfel & Turner, 1986) and attachment research inform how individuals affiliate with groups, form interpersonal bonds, and manage identities across contexts. EASCT expands these ideas to a global, digitally connected world where identities are often multifaceted and situationally fluid.

- **Multiple, Overlapping Identities:** Rather than assuming stable, singular group memberships, EASCT acknowledges that individuals may simultaneously identify with local communities, professional networks, global activist circles, and online fandoms. This

multiplicity of affiliation creates complex identity negotiations, demanding adaptability in impression management, empathy, and conflict resolution.
- **Ethical and Inclusive Dynamics:** EASCT encourages moving beyond simplistic in-group/out-group divides to foster bridging identities that transcend parochial boundaries. By cultivating inclusive, respectful online environments and designing digital platforms that promote cross-cutting affiliations, we can reduce polarization and improve collective problem-solving capacity.
- **Practical Approaches:** Leadership training can highlight intercultural competence and global citizenship, platform designers can introduce features encouraging dialogue between diverse interest groups, and public campaigns can celebrate hybrid identities and bicultural role models to inspire flexible identity negotiation.

6.3.5 Technology and Human Agency

While earlier models often cast technology as a background variable, EASCT centers technology as an active and evolving mediator. It moves beyond deterministic views (technology as destiny) or purely instrumentalist stances (technology as neutral) by positioning human users as ethically responsible co-creators with technological systems.

- **Algorithmic Environments and Agency:** Individuals now interact with recommendation engines, personalization algorithms, and AI-driven decision aids. EASCT insists that users can – through digital literacy, collective action, and policy engagement – shape how these systems operate. Instead of acquiescing to filter bubbles, individuals can seek out diverse content, question opaque algorithms, and demand more transparent digital governance.
- **Designing for Empowerment:** Technologists can incorporate user feedback loops, explainable AI (XAI) frameworks, and participatory design principles. Policymakers can enforce standards that protect user rights, ensure data privacy, and promote fair and inclusive algorithmic outcomes. By doing so, adaptation involves harnessing technology to broaden horizons, not narrowing them.

6.3.6 Cognitive Dissonance and Adaptation

Cognitive dissonance (Festinger, 1957) traditionally focused on how people restore coherence when faced with conflicting beliefs. EASCT reimagines dissonance as a potential driver of growth, creativity, and ethical reflection. In a rapidly changing world, individuals are constantly encountering contradictions – between their values and their actions, between their community's traditions and global challenges, between convenience and morality in technology use.

- **Leveraging Dissonance for Positive Change:** Rather than trivializing or repressing these conflicts, EASCT encourages conscious engagement with dissonance. For example, environmental activists may reflect on their own carbon footprint, social media influencers might re-evaluate their authenticity versus brand obligations, and consumers may confront the ethics of data exploitation by tech giants. Recognizing dissonance paves the way for adaptive realignment of habits, goals, and norms.
- **Institutional and Policy Dimensions:** Educational programs could incorporate exercises prompting students to identify and examine personal dissonances. Organizations can create reflective spaces or ethics committees that help employees confront moral ambiguities. Policymakers can incentivize transparency and accountability measures that surface and address societal-level dissonances (e.g., between declared values like equity and unequal algorithmic outcomes).

6.3.6.1 Integrative Force of the Principles

Together, these key principles establish EASCT as an innovative theoretical architecture that acknowledges complexity and seeks transformative adaptation. They show that cognition is not purely "in the head," but socially and technologically embedded; that motivation and identity are not static traits, but fluid, globally networked constructs; that emotion, bias, and dissonance can be harnessed for growth rather than suppressed; and that technology is not an external constraint but a domain for ethical co-creation.

By applying these principles in tandem, researchers can design rigorous studies that measure how interventions enhance emotional resilience and cognitive flexibility. Practitioners – educators, organizational leaders, designers – can craft strategies that foster inclusive dialogue, support moral courage in digital landscapes, and encourage proactive engagement with cultural diversity. Policymakers can build frameworks that protect digital rights, promote transparency, and incentivize platforms to support rather than undermine adaptive capacities.

In sum, EASCT's key principles articulate the foundational building blocks of a model that transcends conventional boundaries, enabling us to understand, study, and nurture human adaptation as a richly integrated and ethically meaningful process in an era of continuous change.

6.4 UNIFYING CONCEPTS

While EASCT's key principles detail the mechanisms and domains of adaptation, its unifying concepts provide the overarching lenses that bring these principles together into a coherent, future-facing theory. These concepts – Dynamic Adaptation, Feedback Loops and Systems Thinking, Role of Emotions and Emotional Intelligence, Technological Mediation with Human Agency, and Cultural and Individual Differences – serve as integrative anchors. They clarify how individuals navigate complexity, negotiate competing influences, and exercise ethical responsibility. Together, they form a blueprint for interpreting human behavior in contexts that are simultaneously global and localized, digital and embodied, stable and ever-evolving.

6.4.1 Dynamic Adaptation

Contemporary human behavior does not unfold along a steady linear trajectory; instead, it arises in an environment of continuous flux – shifting social norms, evolving technologies, and cultural encounters that transcend old boundaries. EASCT emphasizes that adaptation is not a final state one achieves; it is a perpetual process of recalibration. Just as an ecosystem sustains its balance through ongoing interactions among myriad species and forces, individuals sustain psychological health, social functionality, and moral purpose through iterative, context-sensitive adjustments.

- **Beyond Homeostasis:** Traditional models often implied that adaptation aims to reach a stable equilibrium. EASCT rejects this notion. Instead, it argues that thriving in the 21st-century demands comfort with uncertainty, openness to new inputs, and readiness to modify one's strategies as conditions evolve. Dynamic adaptation thus positions individuals as learners, creators, and ethical agents who can pivot and innovate.
- **Practical Implications:** This perspective encourages interventions that teach adaptive skills – such as learning how to learn, cultivating resilience, and developing scenario-planning competencies. It informs policies that anticipate change rather than merely respond to it, and organizational cultures that value experimentation and continuous professional development.

6.4.2 Feedback Loops and Systems Thinking

Adaptation in EASCT is not a linear cause-and-effect chain; it is a complex web of feedback loops. Individuals shape their environments through their behaviors, choices, and communications.

In turn, these environments – social circles, organizational structures, media ecosystems – reshape individuals' cognition, emotions, and values.

- **Holistic, Multi-Level Interactions:** Systems thinking (Meadows, 2008) reveals that no single factor can be understood in isolation. A change in organizational leadership style can alter team dynamics, which then influence innovation rates, which can inspire policy reforms, and so forth. EASCT positions every adaptive act as part of a reciprocal network of influence extending from personal habits to global cultural shifts.
- **Ethical Accountability in Systems:** Recognizing feedback loops also underscores the moral dimension of adaptation. Choices made by platform designers or policymakers reverberate through societal systems. Conversely, citizens who engage critically with technology can push for more equitable and transparent algorithms. This systems view encourages humility and responsibility, as no stakeholder is a detached observer.

6.4.3 Role of Emotions and Emotional Intelligence

Emotions are not distractions from rationality; they are indispensable guides that help individuals assign value, empathize with others, and find meaning in complexity. EASCT treats emotional intelligence as a crucial meta-skill for navigating globalized, technology-saturated contexts.

- **Emotions as Filters and Catalysts:** Emotional states color how information is perceived, prioritizing certain signals and discounting others. Well-regulated emotions can amplify creativity, moral discernment, and intercultural understanding. Emotional intelligence allows individuals to engage with cultural differences constructively, negotiate digital disagreements with grace, and sustain collective efforts in the face of adversity.
- **Designing Emotionally Sensitive Environments:** Educators, managers, and technologists can create emotionally intelligent environments – schools that teach empathy and perspective-taking, digital platforms that reduce incendiary content while promoting constructive dialogue, and workplaces that treat well-being as integral to performance. These efforts enhance adaptability by ensuring that emotional feedback loops contribute to growth rather than conflict.

6.4.4 Technological Mediation with Human Agency

Where older theories often viewed technology as a backdrop or a tool, EASCT insists that technology actively shapes and is shaped by human behavior. The theory argues that ethical agency and human creativity must guide how we engage with, design, and govern technologies, ensuring they empower rather than exploit users.

- **Beyond Technological Determinism:** EASCT replaces deterministic models – where technology either "inevitably" molds behavior – with a vision of negotiated co-creation. Users can challenge algorithmic biases, form online communities that support critical thinking, and call for open-source transparency. Meanwhile, developers and policymakers can embed fairness, explainability, and user agency in the architecture of digital platforms.
- **Opportunities for Co-Creation:** This approach encourages cooperative innovation: citizens, engineers, ethicists, and policymakers working together to craft standards that protect rights, celebrate diversity, and nurture long-term adaptability. It reframes adaptation

not as adjusting to whatever technology arrives, but as consciously shaping technological evolution to align with humanistic values.

6.4.5 Cultural and Individual Differences

EASCT posits that adaptation strategies vary widely due to cultural backgrounds, personality traits, life experiences, and social positions. While universal psychological mechanisms exist, they manifest differently across contexts. A method for managing digital overload that works in one cultural setting may need customization in another, and an intervention that boosts motivation for one personality profile might require adjustments for another.

- **Richness, Not Uniformity:** By honoring cultural nuance and individual variance, EASCT resists one-size-fits-all models. Adaptation gains complexity and depth when it embraces diverse lenses – feminist perspectives, indigenous knowledge systems, religious ethics, and neurodivergent cognition – all enrich our repertoire of adaptive tools.
- **Tailored Interventions and Global Synergies:** This concept translates into designing flexible programs, policies, and technologies that adapt to different linguistic, cultural, and experiential contexts. Organizations might localize training materials, educators might incorporate multicultural content, and policymakers might consult marginalized communities to ensure inclusive adaptation strategies.

6.4.5.1 Integrating the Unifying Concepts

When combined, these unifying concepts illustrate why EASCT stands as a transformative theoretical framework. Instead of treating global complexity as a problem to be tamed, EASCT embraces it as a generative resource. Dynamic adaptation acknowledges continuous change; feedback loops and systems thinking highlight interdependence and responsibility; emotional intelligence ensures that adaptation is human-centered; technological mediation with agency ensures that humans steer digital futures ethically; and cultural and individual differences guarantee richness and inclusivity in adaptive responses.

Collectively, these concepts chart a path forward: one where adaptation becomes not just a coping mechanism but an art of thriving amid uncertainty, a catalyst for moral imagination, and a medium for forging transnational solidarity. They encourage holistic research designs that measure interconnected variables, institutional cultures that reward critical reflection and experimentation, and digital ecosystems that celebrate curiosity, fairness, and humility. In doing so, they realize EASCT's vision of adaptation as a dynamic, participatory, and ethically guided project for an interconnected age.

6.5 APPLICATIONS OF EASCT

The true potential of EASCT lies not just in explaining how adaptation works, but in informing how educators, organizations, policymakers, and individuals can enhance adaptability, resilience, and ethical engagement amid global complexity. By integrating cognitive insights, emotional intelligence principles, motivational dynamics, cultural competence, and the constructive use of technology, EASCT serves as a design blueprint for interventions that cultivate both individual flourishing and collective well-being.

These applications underscore the theory's versatility: it can inform targeted training modules for educators, strategic HR policies for multinational corporations, data governance frameworks for governments, and personal development tools for individuals. Crucially, applying EASCT requires sensitivity to cultural contexts, continuous feedback loops for refinement, and a commitment to empirical validation – ensuring that interventions are not only theoretically sound but effective and equitable in real-world settings.

6.5.1 Education

EASCT-driven applications in education shift the focus from transmitting stable knowledge to fostering adaptive, critical, and ethically guided learners prepared to navigate a volatile, uncertain world.

- **Curriculum Integration and Design:** Curricula guided by EASCT principles would integrate emotional intelligence training (e.g., exercises in perspective-taking, emotion recognition, and conflict resolution) with cognitive skills that challenge biases (e.g., analyzing media sources, practicing reflective thinking, and learning decision-making heuristics). Additionally, incorporating global cultural content and ethics discussions on technology use can help students develop intercultural competence and moral reasoning abilities.
- **Pedagogical Innovations and Learning Environments:** Adaptive learning platforms – supported by AI algorithms – can personalize educational trajectories to learners' motivational profiles, linguistic backgrounds, and cultural values. For example, digital collaboration tools might connect classrooms from different countries, enabling students to practice social learning across cultural boundaries and engage in problem-based projects addressing global issues like climate action or public health.
- **Empirical Validation and Continuous Improvement:** Researchers can run longitudinal studies comparing cohorts who receive EASCT-informed instruction with those in traditional programs, measuring outcomes like critical thinking scores, emotional regulation under stress, intercultural empathy, and digital literacy. Policy reforms might tie accreditation or funding incentives to demonstrable improvements in these adaptive competencies.

6.5.2 Workplace

In a world of global supply chains, remote teams, and AI-enhanced workflows, organizations must cultivate adaptability in their workforce and leadership. EASCT empowers companies to build cultures and systems that support continuous learning, innovation, and ethical decision-making.

- **Employee Development and Training Programs:** EASCT-informed training could involve workshops on cognitive bias awareness, sessions on emotional intelligence, and simulations where employees navigate cultural misunderstandings or uncertain market conditions. Upskilling employees in digital literacy, data ethics, and scenario planning ensures they can respond flexibly to emerging technologies and dynamic client demands.
- **Organizational Culture and Structure:** Embedding EASCT principles into organizational norms means encouraging open dialogue, promoting diversity in teams, and leveraging technology ethically (e.g., transparent AI tools for talent management or augmented-reality interfaces for design collaboration). Mentorship programs can pair junior employees with cross-cultural mentors, fostering identity negotiation skills and broadening worldviews.
- **Metrics and Policy Alignment:** Organizations can measure changes in adaptation-related outcomes – innovation rates, team cohesion, employee well-being, cultural intelligence – using validated psychometric tools and performance indicators. They can align internal policies with EASCT's emphasis on fairness, emotional health, and global competence, potentially gaining a competitive advantage by nurturing an agile and ethically committed workforce.

6.5.3 Public Policy

Policymakers face unprecedented challenges: misinformation campaigns, AI-driven inequality, global health crises, and environmental emergencies. EASCT provides a conceptual framework to design interventions that enhance societal resilience, cultural integration, and trust in institutions.

- **Behavioral Interventions and Nudge Strategies:** Governments can apply EASCT by crafting policies that help citizens adapt to complex realities. For instance, they might use nudge interventions to promote sustainable behaviors (e.g., setting green defaults in public utilities), encourage pro-social digital habits (e.g., disclosure rules that label AI-generated content), or facilitate active citizenship through digital town halls that invite diverse perspectives.
- **Digital Governance and Ethical Technology Use:** EASCT-informed policymaking recognizes that technology shapes collective adaptation. Regulators can mandate algorithmic transparency, data privacy, and fairness audits of AI systems. Promoting digital inclusivity initiatives – like broadband access in marginalized communities – ensures that technological mediation does not entrench inequality, but rather expands participation, opportunity, and cultural exchange.
- **Cultural Sensitivity and Global Cooperation:** At the international level, EASCT inspires treaties and frameworks for ethical AI development, intercultural education exchanges, and collaborative crisis response protocols. Policymakers could measure success by tracking public trust, social cohesion, digital well-being, and the diversity of cultural contributions in global fora.

6.5.4 Personal Development

While systemic changes are vital, individuals also have the agency to implement EASCT's lessons in their personal lives. By cultivating self-awareness, emotional intelligence, moral discernment, and digital literacy, one can adapt proactively to shifting circumstances and evolving cultural landscapes.

- **Self-Regulation, Mindfulness, and Reflection Techniques:** Personal development strategies might include mindfulness practices that enhance emotional stability, journaling exercises that help individuals identify cognitive biases and cultural assumptions, or exposure to multicultural content that broadens moral perspectives. Users can adopt habit-tracking apps designed to improve digital hygiene, emotional well-being, and intercultural curiosity.
- **Adaptive Goal-Setting and Ethical Dimensions:** Setting adaptive goals – such as learning a new language to improve cultural competence or participating in a global online forum to expand social learning – aligns personal growth with EASCT principles. Ethical considerations come into play when individuals examine how their digital consumption patterns affect mental health or how their lifestyle choices impact global sustainability.
- **Evaluating Personal Growth:** Individuals can measure progress by self-assessing improvements in emotional regulation under stress, ability to empathize with differing cultural norms, or capacity to resist manipulative digital content. Specialized tools like personal adaptation inventories, resilience scales, or cultural intelligence assessments can provide tangible markers of growth.

6.5.4.1 Integrating Across Domains

The power of EASCT emerges most fully when these applications intersect and reinforce each other. An educator trained in EASCT principles who introduces global case studies and digital debate forums prepares students to become agile professionals who thrive in EASCT-informed workplaces. Policymakers who foster ethically governed digital ecosystems enable both organizations and individuals to adapt without sacrificing fairness or mental well-being. In turn, ethically minded organizations may lobby for policies that enhance digital inclusivity and support EASCT-inspired educational reforms.

6.5.4.2 Empirical Validation and Interdisciplinary Collaboration

To ensure that these applications deliver tangible benefits, stakeholders must engage in rigorous empirical validation. Researchers can conduct quasi-experimental studies comparing workplaces with and without EASCT-based interventions, analyzing productivity, job satisfaction, and adaptability metrics over time. Governments can pilot policies in controlled local contexts before scaling them up, measuring shifts in civic participation, trust indices, or cross-cultural understanding. NGOs might collaborate with academic institutions to design randomized controlled trials (RCTs) for educational programs, assessing changes in critical thinking, emotional resilience, and intercultural competence.

Moreover, these applications require sustained interdisciplinary dialogue. Cognitive scientists can help design better bias-mitigation tools, ethicists can guide the responsible development of AI-driven platforms, cultural anthropologists can advise on culturally sensitive curricula, and behavioral economists can propose incentive structures for positive behavioral shifts.

6.5.4.3 From Theory to Transformative Action

By demonstrating how EASCT can inform education, workplaces, policymaking, and personal development, this section illuminates the theory's broad relevance and transformative potential. EASCT's applications empower stakeholders to orchestrate complexity constructively, channeling cognitive and emotional capacities toward moral progress, and aligning cultural and technological trends with inclusive, humanistic values.

Ultimately, successful application of EASCT principles can foster societies where adaptation is not experienced as frantic accommodation or passive submission, but as a deliberate, ethically aware, and creatively empowering process. In doing so, it helps both individuals and communities navigate the present and shape the future with confidence, empathy, and wisdom.

6.6 LIMITATIONS AND FUTURE DIRECTIONS

While EASCT offers a rich, integrative framework for understanding and guiding human adaptation in complex, technologically mediated, and culturally diverse contexts, it does not claim to be a final or universal solution. Like any ambitious theoretical model, EASCT faces limitations and unresolved questions that must be confronted if it is to evolve into a robust, widely applicable tool for researchers, practitioners, and policymakers. Recognizing these challenges provides opportunities for constructive critique, empirical testing, and interdisciplinary dialogue, ensuring that EASCT remains a living framework capable of refinement and expansion.

6.6.1 Complexity, Operationalization, and Specificity

One fundamental challenge is the sheer breadth and complexity of EASCT. The theory weaves together multiple disciplines – cognitive and social psychology, affective neuroscience, cultural anthropology, behavioral economics, information science, and ethics – to capture the full tapestry of adaptive behavior in modern life. While this holistic approach is a strength, it can make operationalizing EASCT's constructs unwieldy. Researchers may ask: How to measure "adaptive emotional intelligence" or "ethical agency in technology use" consistently across cultures and contexts?

- **Paths Forward:** To address complexity, future work should focus on developing and validating specialized instruments. For example, psychometric tools that quantify intercultural adaptability, digital literacy scales linked to ethical reasoning, or indices of organizational cultures aligned with EASCT principles. Experimental and longitudinal designs, combined with mixed-methods approaches (including qualitative interviews, ethnographic fieldwork, and big-data analytics), can help translate conceptual richness into testable hypotheses and actionable guidelines.

6.6.2 Empirical Validation and Cross-Cultural Comparison

EASCT's hypotheses and proposed interventions need empirical backing. While existing research underpins many of its principles, the integrated model has not yet been subjected to comprehensive testing. Moreover, cultural variation requires that we ask how EASCT's applicability differs across societies with distinct value systems, economic infrastructures, and historical trajectories.

- **Research Opportunities:** Researchers can design cross-cultural, multi-site studies to examine whether EASCT-inspired educational reforms improve critical thinking, emotional regulation, and intercultural empathy in diverse classrooms worldwide. Longitudinal studies tracking how organizational policies informed by EASCT affect innovation, employee well-being, and retention over time are also needed. By comparing outcomes in different cultural and technological ecosystems, scholars can identify context-specific adaptations of EASCT's principles, enhancing the model's global relevance and fairness.

6.6.3 Individual, Cultural, and Structural Nuances

EASCT does not assume a "universal subject" responding uniformly to interventions. It acknowledges that personal history, personality traits, socioeconomic status, and cultural identity shape how individuals navigate complexity. Similarly, structural inequalities – such as limited access to education, broadband gaps, or language barriers – can constrain how well some populations can engage with digital tools or intercultural dialogues.

- **Tailoring Interventions and Policies:** Future research should develop frameworks for customizing EASCT-based interventions to particular demographics, professions, or localities. For instance, a program for enhancing emotional intelligence and media literacy might require adapting content for rural communities with limited internet bandwidth, or translating resources into indigenous languages to ensure cultural resonance. Policy analyses must consider how institutional barriers – like censorship, discrimination, or rigid bureaucracies – hamper the realization of EASCT's ideals and propose solutions that address these root causes.

6.6.4 Ethical Considerations and Responsible Implementation

EASCT's emphasis on ethical agency and moral reflection is both a strength and a challenge. Promoting responsibility in a digital age raises controversial questions: How do we balance freedom of expression with harm reduction online? What are the trade-offs between personalization and privacy? How should we govern AI systems that shape public opinion or guide medical decisions?

- **Ethical and Regulatory Frameworks:** Future directions include fostering interdisciplinary consortia where technologists, ethicists, policymakers, citizen groups, and psychologists collaboratively draft guidelines for "EASCT-aligned" tech design and governance. Regulatory bodies may consider EASCT principles when shaping data protection laws or algorithmic auditing standards. Educational institutions can incorporate ethics case studies into digital literacy and leadership programs, ensuring that tomorrow's decision-makers internalize these values from the outset.

6.6.5 Balancing Theory, Practice, and Continuous Feedback

As EASCT moves from conceptual architecture to practical implementation, it must remain open to critique and revision. Early interventions might not always yield expected results, revealing oversights, unintended consequences, or contextual factors previously underestimated. Ensuring that EASCT remains dynamic and iterative is vital: it should inspire not only well-intentioned reforms but also ongoing evaluation, critical self-reflection, and responsive adaptation.

- **Feedback Loops for Theory-Building:** Scholars and practitioners can create online repositories or open-access platforms to share case studies, best practices, and lessons learned from EASCT-inspired initiatives. Conferences or workshops can facilitate dialogues between researchers who develop measurement tools, educators who apply them in classrooms, corporate trainers who integrate them into workshops, and policymakers who scale them into national strategies. This continuous feedback loop ensures that EASCT evolves through evidence-based refinements, improved theoretical precision, and enhanced ecological validity.

6.6.6 Long-Term Horizons and Global Crises

The complexity of global challenges – climate emergencies, pandemics, geopolitical conflicts, and mass migrations – means that adaptive capabilities must be sustainable and robust. While EASCT provides a framework for adaptability, we must also consider long-term horizons. Will the interventions inspired by EASCT help societies become more resilient in the face of disruptions that may occur decades into the future?

- **Future-Orientation and Scenario Planning:** Future research might pair EASCT principles with scenario planning methods, designing simulations that test how different adaptation strategies fare under plausible future conditions. Could EASCT approaches make societies less vulnerable to misinformation during crises, more agile in restructuring economies after automation shocks, or more compassionate in integrating climate refugees? Answering such questions can guide policymakers, organizations, and individuals in preparing for tomorrow's uncertainties.

6.6.6.1 In Summary: Charting the Path Ahead

The limitations and future directions outlined here underscore that EASCT is an evolving project, not a completed doctrine. Acknowledging complexity, seeking empirical validation, accommodating cultural and individual differences, addressing ethical dilemmas, and committing to continuous refinement are all essential steps on the path toward fully realizing EASCT's promise.

Rather than diminishing EASCT's significance, these challenges enhance its credibility and usefulness. They demonstrate that EASCT embraces complexity and invites critical scrutiny – traits befitting a theory that aims to guide adaptation in a world where change is constant and stakes are high. By collaborating across disciplines, incorporating diverse perspectives, and committing to long-term thinking, we can ensure that EASCT serves as a generative force for scholarship, governance, innovation, and human flourishing in the decades to come.

YOU ARE NOT YOUR THOUGHTS

THINKING BOX:

"Who or what silently observes your mental chatter? This might be worth mulling over – perhaps even **42** times."

7 The Meta-Layer
Integrating Knowledge Systems and Mastering Complexity

Key Tools for Integration:

1. **Integration Canvas** – Maps frameworks, stakeholders, and conflicts to visualize complexity.
2. **Contradiction Navigator** – Resolves competing priorities (e.g., fairness vs. efficiency) through structured trade-offs.
3. **Adaptive Integration Matrix** – Prioritizes decisions based on urgency and complexity.
4. **Wardley Maps** – Identifies evolving technological and strategic landscapes for long-term decision-making.

Practical Applications:

- **AI Ethics Implementation** – Balancing privacy, fairness, and speed while maintaining compliance.
- **Digital Transformation** – Managing employee resistance and operational stability in innovation adoption.

By mastering **meta-level thinking**, organizations and individuals can **foresee complexity, harmonize contradictions, unlock innovation, and future-proof decision-making**. This **shifts problem-solving from reactive firefighting to strategic clarity**, empowering leaders to **thrive in uncertainty**.

7.1 INTRODUCTION: THE CHALLENGE OF INTEGRATION

7.1.1 Opening Scenario: The AI Ethics Dilemma

Picture yourself as the Chief Technology Officer (CTO) at a booming tech company on the brink of deploying a groundbreaking AI system for customer service. It's a potential game-changer, promising faster responses, personalized interactions, and improved customer satisfaction. Yet, this monumental innovation comes wrapped in a web of challenges that demand immediate attention:

1. **Privacy Concerns**: The AI's performance improves exponentially with access to detailed personal data. But every additional data point requested feels like walking a tightrope. Customers are vocal about their concerns over data security. Regulators, with GDPR and AI Act mandates, loom large. A misstep here could not only derail your project but also spark public outrage.
2. **Fairness Issues**: Early tests reveal implicit bias within the system – certain demographics receive less accurate or respectful responses. While algorithms can be tuned, addressing fairness involves more than code; it touches on societal values and trust. Balancing these sensitivities within a competitive timeline feels like steering through a minefield.
3. **Economic Pressure**: Competitors are already in the market, flaunting similar systems. The CEO and board demand speed to market, worried about losing a critical

DOI: 10.1201/9781003640288-10

edge. The pressure to launch quickly threatens to overshadow concerns about ethical and technical readiness.

4. **Cultural Impact**: Internally, the workforce is divided. Employees express fears about job displacement. Managers question the system's reliability, worried it could disrupt workflows or alienate loyal customers. The stakes are high – not just for your customers but also for your organization's culture and morale.

7.1.2 The Storm of Competing Priorities

This situation isn't just complex; it's a crucible of competing priorities:

- **Ethics** insists on prioritizing fairness and transparency but offers no guidance on balancing these ideals with corporate deadlines.
- **Technical optimization** demands access to more data, even as privacy concerns mount.
- **Business imperatives** pressure you to move fast, leaving little room to deliberate or refine.

Every choice seems to come with a price:

- Ensuring **privacy** might throttle AI performance, reducing its appeal to customers.
- Rushing to fix **bias** could jeopardize the system's market debut, ceding ground to competitors.
- Prioritizing **speed** risks alienating employees and eroding public trust.

Each framework – whether ethical, technical, or business-driven – shines a spotlight on part of the problem. But none offer a holistic solution. Attempting to resolve one issue often ignites new conflicts. Without an integrated approach, you risk paralysis – or worse, decisions that escalate the problem instead of solving it.

7.1.3 Why Single Frameworks Fail

This scenario exposes the **inherent limitations of siloed thinking**:

- Ethical frameworks often focus on abstract principles, offering little on how to navigate real-world trade-offs.
- Technical guidelines prioritize efficiency and functionality but ignore the human elements that drive adoption and trust.
- Business strategies, laser-focused on speed and profitability, may leave ethical and cultural risks unaddressed.

Relying on any one framework means viewing the problem through a narrow lens – unable to see how different elements interact and exacerbate one another. It's like trying to solve a Rubik's Cube while only looking at one face.

The result? Unintended consequences, from a loss of public trust to a diminished employee morale, to regulatory scrutiny.

7.1.4 The Need for a Meta-Level Approach

What's needed is a **meta-level framework** – an integrative, dynamic system that:

1. **Synthesizes** insights across disciplines, creating a cohesive strategy.
2. **Navigates contradictions**, resolving tensions between competing goals like privacy and performance.
3. **Prioritizes clarity** in complexity, ensuring timely, effective decisions without sacrificing long-term objectives.

Such an approach doesn't just patch over conflicts – it transforms them into opportunities for innovation and resilience. By integrating ethical foresight, technical precision, and business acumen, a meta-level framework provides a roadmap for navigating complexity.

7.1.5 A Preview of Solutions

This chapter introduces tools to operationalize this integration, turning chaos into clarity:

1. **The Integration Canvas**: A visual tool for mapping stakeholders, conflicts, and opportunities across frameworks.
2. **The Contradiction Navigator**: A structured method for identifying and resolving opposing priorities.
3. **The Adaptive Integration Matrix**: A guide to selecting and combining frameworks based on urgency and complexity.

7.1.6 The Competitive Advantage of Integration

Integration is not just a survival skill – it's a competitive advantage. Leaders who can master the art of synthesis are better equipped to:

- Build trust with stakeholders by addressing their concerns holistically.
- Deliver solutions that balance speed with substance, ethics with performance.
- Thrive in a world where complexity is the norm, not the exception.

By the end of this chapter, you'll be equipped with practical, actionable tools to bring harmony to competing priorities, turning challenges into triumphs. Together, we'll transform the way you approach complexity – empowering you to lead with clarity, confidence, and impact.

7.2 QUICK START GUIDE: YOUR INTEGRATION TOOLKIT

In the face of multi-dimensional challenges like the AI ethics dilemma, success demands more than just problem identification – it requires dynamic, actionable tools that cut through complexity and empower decisive action. This chapter introduces a **three-part integration toolkit** designed to help you navigate conflicting priorities, align stakeholder needs, and balance urgency with long-term impact.

Each tool complements the others, forming a cohesive system that transitions you from analysis paralysis to empowered decision-making.

7.2.1 Integration Canvas: A Bird's-Eye View of Complexity

The **Integration Canvas** is your **map for complexity**, allowing you to visualize how frameworks, stakeholders, and impacts intersect. It reveals potential conflicts and opportunities for alignment, ensuring no critical variable is overlooked.

7.2.1.1 What It Does

The Integration Canvas:

- Organizes diverse frameworks (e.g., ethics, technical, and business) into a unified view.
- Highlights stakeholder priorities and their intersections.
- Surfaces hidden conflicts and uncovers synergies.

7.2.1.2 How It Works

1. Identify and list the frameworks relevant to the decision.
2. Map stakeholders and their concerns (e.g., customers, employees, regulators).
3. Visualize conflicts and overlaps, such as competing priorities or alignment opportunities.
4. Pinpoint integration points to guide balanced solutions.

7.2.1.3 Example: Mapping AI System Impacts

Imagine using the Integration Canvas to assess changes to your AI system:

- **Frameworks**: Privacy laws, fairness principles, performance metrics.
- **Stakeholders**:
  - **Customers**: Prioritize data privacy and unbiased responses.
  - **Employees**: Concerned about job security and workload shifts.
  - **Management**: Driven by speed-to-market and ROI.
  - **Regulators**: Monitor compliance with data protection laws.
- **Potential Conflicts**: Balancing privacy safeguards with data-driven performance.
- **Integration Points**: Privacy enhancements that also build customer trust while maintaining operational efficiency.

The Integration Canvas becomes your **decision dashboard**, visually connecting all elements and enabling data-driven alignment.

7.2.2 Contradiction Navigator: Cutting Through Conflict

When priorities clash, progress halts. The **Contradiction Navigator** is a **conflict resolution powerhouse**, helping you untangle opposing priorities and chart actionable paths forward. It's particularly effective for situations requiring trade-offs or innovative synthesis.

7.2.2.1 What It Does

The Contradiction Navigator:

- Identifies key conflicts (e.g., fairness vs. efficiency, privacy vs. functionality).
- Quantifies the impacts and risks of opposing priorities.
- Guides you toward trade-offs, creative solutions, or hybrid strategies.

7.2.2.2 How It Works

1. Identify conflicting priorities and their respective impacts.
2. Use structured prompts to evaluate trade-offs:
   - What's gained or lost by prioritizing each side?
   - Where are synergies or win-win scenarios possible?
3. Map the outcomes and develop actionable strategies for resolution.

7.2.2.3 Example: Resolving Fairness vs. Efficiency

Suppose your AI system faces a trade-off between fairness and efficiency:

- **Fairness Focus**: Reduces bias but delays time-to-market.
- **Efficiency Focus**: Speeds up deployment but risks reputational backlash.

Using the Contradiction Navigator, you:

- Quantify the risks of bias (e.g., regulatory fines, lost trust).

- Identify hybrid strategies, such as:
  - Deploying with baseline fairness safeguards while scheduling iterative updates.
  - Communicating the roadmap transparently to regulators and customers.

This tool turns "either-or" dilemmas into **strategic opportunities** for innovation and balance.

7.2.3 Adaptive Integration Matrix: Precision Under Pressure

Not all decisions require equal urgency or complexity. The **Adaptive Integration Matrix** allows you to **triage challenges**, ensuring that you apply the right tools to the right problems at the right time. It bridges the gap between strategic thinking and on-the-ground execution.

7.2.3.1 What It Does

The Adaptive Integration Matrix:

- Assesses urgency and complexity to prioritize decisions.
- Matches tools and frameworks to the problem's characteristics.
- Guides action based on the time and resources available.

7.2.3.2 How It Works

1. Evaluate the **urgency** (How soon must a decision be made?) and **complexity** (How many variables are involved?).
2. Place the challenge in one of four quadrants:
   - **High Urgency, Low Complexity**: Quick, straightforward decisions.
   - **High Urgency, High Complexity**: Focused prioritization and phased action.
   - **Low Urgency, High Complexity**: Strategic planning with deep analysis.
   - **Low Urgency, Low Complexity**: Routine decisions requiring minimal attention.
3. Tailor your approach based on quadrant placement.

7.2.3.3 Example: Prioritizing During a High-Stakes Launch

Your AI system launch falls into the **High Urgency, High Complexity** quadrant. The Adaptive Integration Matrix helps you:

- Focus immediate efforts on critical, high-impact issues (e.g., regulatory compliance).
- Defer less urgent, complex challenges (e.g., full-scale bias audits) to subsequent phases.
- Use iterative progress tracking to adapt quickly as conditions evolve.

This framework ensures you **act decisively without sacrificing long-term goals**.

7.2.3.4 Bringing It All Together

These tools work in concert, providing a comprehensive approach to complexity management:

1. **Integration Canvas**: Maps the landscape, clarifying relationships and potential conflicts.
2. **Contradiction Navigator**: Resolves competing priorities with actionable strategies.
3. **Adaptive Integration Matrix**: Prioritizes efforts, ensuring the right focus for each scenario.

Together, these tools don't just help you manage complexity – they empower you to **thrive within it**, turning even the most daunting challenges into structured, solvable opportunities.

COMPLIANCE AND RISK ASSESSMENT

THINKING BOX:

"Are we actively preventing hazards or just dodging responsibility? Like the ultimate question, the real answer might hover near **42**."

7.3 MAP OF CONNECTIONS: BUILDING THE META-LAYER

This chapter serves as the **critical bridge** that connects the foundational concepts explored in earlier parts of the book with their **practical application** to real-world challenges. By integrating insights from ethics, cognition, and resilience, this chapter introduces a **meta-layer framework** – a unifying approach to address complex, multidimensional problems effectively.

7.3.1 Ethical Foundations (Chapters 1–3)

The ethical underpinnings from Part 1 shape the tools and strategies introduced here, grounding decision-making in **moral clarity** and **systemic awareness**.

- **Complexity and the Model Boundary Paradox:** The paradox illustrates the **limits of any single model** in capturing the complexity of real-world scenarios, reinforcing the need for a multi-framework approach.
  - *Connection:* The Integration Canvas builds on this insight by mapping where models fall short and identifying points of synergy across frameworks.
  - *Example:* Designing an AI system where data privacy conflicts with operational efficiency highlights the need for cross-framework integration.
- **Causal Relevance Ethics:** This framework provides a **structured lens** for evaluating the ethical ripple effects of decisions, helping prioritize actions based on their causal impact.
  - *Connection:* The Adaptive Integration Matrix leverages causal analysis to determine which frameworks to prioritize in time-sensitive decisions.
  - *Example:* Balancing ethical considerations and speed-to-market in AI deployment by evaluating which impacts carry the greatest moral weight.
- **Byung-Chul Han's Philosophical Critiques:** Han's exploration of societal erosion and digital overload offers a **human-centric perspective** on the unintended consequences of technological decisions.
  - *Connection:* The Contradiction Navigator aligns with Han's insights by helping resolve tensions between human and technical priorities.
  - *Example:* Mitigating employee fears of job displacement by integrating cultural and ethical considerations into decision-making.

7.3.2 Cognitive Insights (Chapters 4–6)

The tools in this chapter extend the cognitive frameworks discussed in Part 2, offering **strategic clarity** and **mental adaptability** for navigating competing priorities.

- **Bias Awareness:** Cognitive biases often cloud judgment in complex situations. The strategies in this chapter are designed to identify and mitigate these distortions, enhancing **decision accuracy**.

  - *Connection:* The Contradiction Navigator incorporates bias mitigation techniques to counteract distortions like confirmation bias or groupthink.
  - *Example:* Avoiding sunk cost bias when reassessing investments in AI systems that show ethical flaws.
- **Enhanced Adaptive Social Cognition Theory (EASCT):** EASCT equips readers with tools to **adapt to shifting social and technological landscapes**, making it indispensable for stakeholder management.
  - *Connection:* The Integration Canvas incorporates EASCT principles to map and address stakeholder behaviors and dynamics.
  - *Example:* Predicting and managing resistance to change in a team affected by automation technologies.

7.3.3 Resilience and Well-Being (Chapters 8–10)

The resilience-building practices from Part 3 underpin the integration tools in this chapter, providing a foundation for **thriving in uncertainty** and leveraging complexity as an opportunity.

- **Practical Resilience Strategies:** The mental toolkits and daily practices introduced earlier are extended here, equipping readers with the stamina to handle high-pressure, multifaceted problems.
  - *Connection:* The Adaptive Integration Matrix incorporates resilience strategies to prioritize actions under stress.
  - *Example:* Using resilience techniques to manage competing demands during a high-stakes product launch.
- **Anti-fragility Frameworks:** Viewing challenges as opportunities to grow stronger becomes central to this chapter's approach to managing contradictions and complexity.
  - *Connection:* The Contradiction Navigator builds on anti-fragility by transforming conflicting inputs into pathways for innovation.
  - *Example:* Turning a privacy vs. efficiency conflict into a unique selling proposition through creative compromise.

7.3.3.1 Unifying Insights: The Meta-Layer

The meta-layer introduced in this chapter consolidates the diverse approaches of ethics, cognition, and resilience into a **flexible, integrative framework**.

1. **Why the Meta-Layer Is Crucial?**
   - Single frameworks often create **blind spots** and **unintended consequences**.
   - Complex problems require **diverse perspectives** to achieve balanced, effective solutions.
   - The meta-layer reduces cognitive overload by offering structured tools that **simplify complexity** while maintaining depth.
2. **What the Meta-Layer Achieves?**
   - **Unified Decision-Making:** Integrates insights from earlier chapters into a cohesive system.
   - **Conflict Resolution:** Provides tools to harmonize opposing priorities.
   - **Strategic Flexibility:** Adapts to the evolving demands of dynamic challenges.
3. **How the Meta-Layer Works?**
   - Tools like the Integration Canvas and Adaptive Integration Matrix help **synthesize frameworks**, enabling clear, actionable pathways.
   - The Contradiction Navigator transforms **conflicting inputs into complementary outcomes**, ensuring no perspective is ignored.

7.3.3.2 Visual Roadmap: Connecting the Parts

This chapter weaves together the previous sections of the book into a clear visual roadmap:

Part 1: Ethical Foundations

↓

Ethical complexity (Model Boundary Paradox)

Causal decision-making (Causal Relevance Ethics)

Societal impact (Han's critiques)

↓

Part 2: Cognitive Insights

↓

Bias management

Adaptive social cognition (EASCT)

↓

Part 3: Resilience and Well-Being

↓

Resilience strategies

Anti-fragility

↓

Part 4: The Meta-Layer

↓

Unified tools for integration and adaptability

7.3.3.3 Empowering Readers Through Integration

This chapter doesn't just teach readers to **combine insights** from earlier sections; it equips them with the tools to **master integration as a skill**. By applying the meta-layer, readers can:

- Balance ethical, technical, and human considerations in **real-world scenarios**.
- Navigate the **uncertainties of the modern world** with confidence and clarity.
- Harness the full potential of the frameworks presented in this book to thrive in complexity.

The result? Readers move from understanding individual frameworks to **operating at a higher, integrative level**, ready to face the most demanding challenges of our interconnected age.

7.4 THE ARCHITECTURE OF INTEGRATION

7.4.1 Knowledge System Mapping

To create a cohesive and actionable strategy for complex challenges, it's essential to recognize the different forms of knowledge that shape decision-making. Each type of knowledge plays a distinct role, and understanding their strengths, weaknesses, and interactions allows us to harness their collective power effectively.

7.4.1.1 Theoretical Knowledge: Foundational Principles

Definition:

Theoretical knowledge consists of established principles, frameworks, and models developed through systematic inquiry. It provides a structured lens for understanding complex problems and offers consistent rules or guidelines to inform decisions.

Purpose:

Theoretical knowledge functions as a **compass**, offering direction in decision-making. By anchoring our decisions in well-established frameworks, we can evaluate priorities and maintain a principled approach.

Examples in Practice:

- **Utilitarian Ethics:** Applying the principle of maximizing overall benefits to prioritize AI features that help the majority of users while minimizing harm.
- **Game Theory in Business Strategy:** Anticipating competitor moves in a rapidly evolving AI market by simulating possible scenarios.
- **Causal Relevance Ethics:** Identifying actions with the highest long-term impact, such as designing AI algorithms that balance privacy with fairness.

Application Context:

Theoretical knowledge is most valuable in:

- **Strategic Decision-Making:** Defining long-term goals or addressing systemic issues.
- **Policy Development:** Creating ethical guidelines or regulatory compliance frameworks.

Key Insight:

While theoretical knowledge provides structure and predictability, it often lacks the **nuance needed for practical application**, such as accounting for real-world constraints or dynamic stakeholder needs.

7.4.1.2 Empirical Knowledge: Data-Driven Insights

Definition:

Empirical knowledge stems from observations, measurements, and research. It's rooted in evidence and provides **quantifiable clarity** to validate or challenge assumptions.

Purpose:

Empirical knowledge acts as a **magnifying glass**, focusing on specific areas of interest and offering measurable insights. It ensures that decisions are informed by the realities of the situation rather than assumptions or generalizations.

Examples in Practice:

- **User Behavior Analytics:** Evaluating how customers interact with an AI chatbot to refine its language and reduce bias.
- **Performance Metrics:** Using error rates or response times to improve the efficiency of an AI system.
- **Market Research:** Identifying customer pain points to guide product feature development.

Application Context:

Empirical knowledge is essential for:

- **Validation:** Testing whether theoretical frameworks align with real-world outcomes.
- **Refinement:** Iteratively improving systems based on feedback and performance data.

Key Insight:

Empirical knowledge provides actionable precision but often lacks the **broad vision** or interpretative flexibility of theoretical or intuitive knowledge.

7.4.1.3 Practical Knowledge: Hands-On Expertise

Definition:

Practical knowledge arises from experience in implementing strategies, managing teams, and solving problems. It reflects the lessons learned from real-world successes and failures.

Purpose:

Practical knowledge serves as the **toolbox** of integration, offering adaptive strategies for specific situations. It helps bridge the gap between abstract concepts and tangible actions.

Examples in Practice:

- **Leadership Strategies:** Drawing from prior experience to address team concerns about job displacement during AI deployment.
- **Crisis Management Protocols:** Leveraging tested methods to mitigate risks when a product faces public backlash.
- **Project Rollout Expertise:** Learning from past implementation challenges to streamline future product launches.

Application Context:

Practical knowledge is critical for:

- **Operational Efficiency:** Ensuring that strategies are feasible and align with organizational capabilities.
- **Stakeholder Engagement:** Navigating interpersonal and organizational dynamics to secure buy-in.

Key Insight:

Practical knowledge excels in **adaptability and immediacy**, but its reliance on personal experience can limit its scalability or applicability in unfamiliar contexts.

7.4.1.4 Intuitive Knowledge: Tacit Understanding

Definition:

Intuitive knowledge is rooted in instinct, emotional intelligence, and situational awareness. It represents a **subconscious synthesis** of experience and context.

Purpose:

Intuitive knowledge is the **compass of ambiguity**, guiding decisions in high-pressure situations where data or clear frameworks are unavailable. It helps leaders act decisively when faced with uncertainty.

Examples in Practice:

- **Team Morale Assessment:** Recognizing unspoken signs of burnout during a high-stakes project.

- **Stakeholder Management:** Navigating conflicting priorities with empathy to foster collaboration.
- **Risk Sensing:** Detecting subtle signals of potential failure and preemptively addressing them.

Application Context:

Intuitive knowledge is invaluable for:

- **Leadership:** Building trust and cohesion in teams.
- **Ambiguous Scenarios:** Making decisions when evidence is incomplete or conflicting.

Key Insight:

While powerful, intuitive knowledge can be **subject to bias** and requires validation through empirical or theoretical perspectives.

Integrating Knowledge Systems: A Unified Map

Knowledge Type	Interaction	Examples	Challenges
Theoretical ↔ Empirical	Validation flows: Empirical data confirms theoretical models.	Using user analytics to validate ethical frameworks.	Data may not always align with theoretical predictions.
Practical ↔ Intuitive	Integration points: Practical actions informed by intuitive and emotional insights.	Using emotional intelligence in leadership decisions.	Intuition may conflict with operational realities.
Theoretical ↔ Practical	Synergy opportunities: Combining frameworks with hands-on expertise.	Applying decision theory in project management.	Theories may oversimplify real-world complexities.
Empirical ↔ Intuitive	Conflict zones: Resolving mismatches between intuition and data.	Using market feedback to refine instinct-driven designs.	Data may challenge established intuitive practices.

Key Interactions

- **Validation Flows**: Empirical data confirms or challenges theoretical models.
- **Integration Points**: Practical actions are often shaped by both theoretical knowledge and intuitive understanding.
- **Synergy Opportunities**: Leveraging complementary strengths between theoretical, empirical, practical, and intuitive knowledge.
- **Conflict Zones**: Addressing tensions when different knowledge types provide conflicting insights.

Key Insights for Knowledge System Mapping

1. **Complementary Knowledge:** Integration thrives when different knowledge systems reinforce each other. For example:
   - Empirical data validates a theoretical model.
   - Practical expertise aligns with intuitive decision-making to address team concerns.
2. **Conflicting Knowledge:** Misalignments arise when knowledge systems pull in opposite directions. For instance:
   - **Theory vs. Practice:** A theoretical model prioritizing fairness may conflict with the practical need for speed.
   - **Intuition vs. Data:** A leader's instincts might contradict customer feedback, requiring careful reconciliation.

3. **Unified Decision-Making:** Effective integration involves leveraging each system's strengths to mitigate its weaknesses. For example:
   - Using theoretical frameworks to set a strategic direction.
   - Validating with empirical data to ensure feasibility.
   - Drawing on practical and intuitive knowledge to execute plans effectively.

System Interactions in Practice

1. **Example: AI Ethics Dilemma**
   - **Theoretical Knowledge:** Apply ethical principles like fairness to evaluate potential bias in the AI system.
   - **Empirical Knowledge:** Use customer feedback to identify areas of concern.
   - **Practical Knowledge:** Address rollout challenges by applying past project management lessons.
   - **Intuitive Knowledge:** Gauge employee sentiment to preemptively address resistance to change.
2. **Conflict Resolution Example**
   - Privacy concerns (theory) conflict with performance needs (practice).
   - Resolution: Develop a hybrid solution, such as anonymized data to balance privacy and functionality.

By mapping and integrating these diverse forms of knowledge, we create a robust foundation for tackling complexity. This understanding paves the way for applying the tools and frameworks introduced in subsequent sections.

7.4.1.5 Introducing Wardley Maps: Bridging the Meta-Layer and Strategic Detail

While the Integration Canvas, Contradiction Navigator, and Adaptive Integration Matrix help clarify complexity and align decision-making, there's still a need for a strategic, big-picture lens that can connect meta-level understanding to concrete actions. Wardley Maps fill this niche perfectly, acting as a layer between your conceptual meta-framework and the finer details of implementation.

7.4.1.6 What Are Wardley Maps?

Wardley Maps are a visual strategy tool developed by Simon Wardley. They help you represent the value chain of a product, service, or initiative, placing each component according to its stage of evolution and value to the end-user. By mapping activities, practices, and technologies, you gain situational awareness – understanding where you stand, what's changing, and where potential moves exist.

Why Wardley Maps Fit the Meta-Level Approach:

1. **From Abstract to Concrete:** EASCT and other meta-level concepts provide a high-level understanding of adaptation, ethics, and cultural intelligence. Wardley Maps translate these insights into strategic positioning. Instead of just knowing you must balance ethics and efficiency, a Wardley Map shows where ethical safeguards fit into the value chain – are they commodity services, custom builds, or evolving components that require strategic investment?
2. **Identifying Strategic Plays:** At the meta-layer, you acknowledge complexity and contradictions. With a Wardley Map, you can locate where these tensions arise in your value chain. For instance, privacy protections may appear as evolving capabilities in a certain part of the map – seeing them in context might reveal opportunities for standardization, outsourcing, or innovation.
3. **Aligning Frameworks to Evolving Landscapes:** Your existing tools guide decision-making (who to involve, which conflicts to resolve first), but Wardley Maps help you see

which components are shifting from novel to commodity. This can inform when to apply the Integration Canvas (e.g., mapping new stakeholders as novel activities evolve), or when to use the Contradiction Navigator (e.g., if a rapidly commoditizing technology introduces ethical or cultural tensions).

Practical Steps to Use Wardley Maps:

1. **Select a Target Scenario:** Pick a complex challenge – like introducing a new AI tool – and outline its value chain: data input, algorithm development, compliance checks, user interface, customer feedback loops.
2. **Place Each Component:** Plot each component on the map according to its visibility to the end-user and its stage of evolution (from genesis to commodity). For example, general computing infrastructure might be highly commoditized, while custom fairness algorithms may be novel and evolving rapidly.
3. **Overlay Frameworks and Concerns:** Once the value chain is visualized, integrate your meta-layer insights:
   - Use the Integration Canvas to understand which stakeholders influence each component.
   - Apply the Contradiction Navigator to components that are particularly contentious (e.g., novel privacy features that conflict with performance goals).
   - Refer to the Adaptive Integration Matrix to decide where urgent attention is needed – perhaps on those components that are evolving fast and have high stakeholder visibility.
4. **Identify Strategic Moves:** Wardley Maps help you see where to invest, where to outsource, and where to standardize. Aligning these moves with your meta-layer decisions ensures that each action is not just theoretically sound but strategically positioned.

Example: AI Ethics Integration using Wardley Maps:

- **Scenario:** Balancing privacy and performance in a customer service AI system.
- **Mapping Components:**
  - User Data Gathering (evolving from bespoke solutions to more standardized anonymization tools)
  - Fairness Algorithms (currently novel and experimental, high complexity)
  - Compliance Reporting (a more established practice, heading toward standardization)
  - End-User Interface (highly visible, currently a custom design)

Overlaying Meta-Layer Tools:

- **Integration Canvas:** Highlights that anonymization (a novel practice) requires collaboration between legal, IT, and user experience teams.
- **Contradiction Navigator:** Focuses on the fairness algorithms (novel and volatile) to resolve tensions between ethics and performance optimization.
- **Adaptive Integration Matrix:** Guides prioritization – addressing the most novel and visible components first, given their strategic importance.

Benefits of Adding Wardley Maps:

- **Enhanced Situational Awareness:** Move beyond abstract reasoning to understand the strategic landscape at a glance.
- **Contextualized Decision-Making:** Each meta-layer tool is applied where it matters most, turning broad insights into targeted interventions.
- **Evolving with the Market:** As components commoditize or new technologies appear, your map updates, ensuring ongoing alignment with dynamic environments.

7.4.2 Complexity Reduction Strategies

Navigating complex problems requires a careful balance between simplification and maintaining an understanding of the broader context. This section provides practical strategies for reducing complexity while retaining the depth and nuance necessary for effective decision-making. It explores tools to streamline problem-solving, prioritize efforts, and select the most suitable approaches based on situational demands.

7.4.2.1 Simplification Without Oversimplification

Why This Matters:

Complex challenges often involve layers of interconnected issues, from ethical dilemmas to technical constraints. Simplifying these challenges into manageable elements is crucial to avoid decision fatigue, but oversimplification risks ignoring critical factors. The key is **focusing on actionable insights** while ensuring broader systemic implications are considered.

Using Decision Trees to Break Down Complexity

Decision trees are a structured approach to untangling multifaceted problems by breaking them into smaller, sequential choices. These trees not only provide clarity but also help visualize potential trade-offs and outcomes.

- **How to Construct a Decision Tree:**
  1. **Define the Problem Statement:** Start with a clear articulation of the challenge. For instance, "How can we optimize AI system performance while safeguarding user privacy?"
  2. **Identify Key Decisions:** Break the challenge into sub-decisions (e.g., data collection, user consent mechanisms, and security protocols).
  3. **Map Possible Outcomes:** For each decision, visualize the potential impacts to evaluate trade-offs and implications.
- **Example Decision Tree:**

Option	Description	Pros	Cons
Option A	Collect anonymized data	Low privacy risk Moderate performance	Limited personalization
Option B	Gather detailed user data	High performance Full personalization	High privacy concerns Regulatory risks
Option C	Hybrid approach	- Balanced trade-offs	- Increased complexity - Higher costs

Balancing Actionable Insights with Context Awareness

Simplification should not come at the cost of understanding the broader context. While decision trees provide clarity for immediate action, maintaining an overarching perspective ensures solutions align with **long-term goals and systemic impacts**.

- **Key Techniques:**
  1. Highlight essential data points (e.g., privacy laws, stakeholder preferences).
  2. Create checkpoints for alignment with broader objectives (e.g., ethical commitments or market positioning).
  3. Document the rationale for each decision to ensure transparency and adaptability for future adjustments.

7.4.2.2 Framework Triage

Why This Matters:

Not all problems require the same level of attention or resources. By categorizing challenges based on their **scope, complexity, and impact**, you can allocate efforts strategically and avoid spreading resources too thin. This ensures focus on initiatives that deliver the most value while balancing short-term wins with long-term transformations.

Framework Triage Categories:

1. **Quick Wins**
   - **Definition:** Straightforward problems that can be resolved quickly but offer significant immediate benefits.
   - **Examples:**
     - Adding clear disclaimers to the AI system to address transparency concerns.
     - Training employees on basic user privacy protocols.
   - **Approach:** Focus on fast, high-impact actions that build momentum.
2. **Strategic Projects**
   - **Definition:** Medium- to long-term initiatives that require planning and collaboration to achieve meaningful results.
   - **Examples:**
     - Developing comprehensive AI ethics guidelines.
     - Creating workflows to monitor and mitigate biases in AI outputs.
   - **Approach:** Structure these projects with defined milestones, dedicated teams, and measurable outcomes.
3. **Transformative Initiatives**
   - **Definition:** Large-scale, complex efforts aimed at achieving groundbreaking changes in strategy or operations.
   - **Examples:**
     - Redesigning the AI system architecture to ensure both fairness and high performance.
     - Restructuring organizational priorities to integrate ethical considerations across all functions.
   - **Approach:** Emphasize cross-functional collaboration, iterative development, and robust risk management.

Prioritization Template:

Scope	Complexity	Category	Recommended Action
Local	Low	Quick Wins	Implement immediately
Global	Medium	Strategic Projects	Develop and plan
Global	High	Transformative Initiatives	Execute long-term

7.4.2.3 Analog vs. Digital Tools

Why This Matters:

The tools you use can shape how effectively you manage complexity. Analog tools are often more intuitive and tactile, ideal for brainstorming and conceptualization, while digital tools excel in scalability, collaboration, and dynamic updates. Selecting the right tool depends on the nature of the challenge and the team's needs.

Analog Tools for Simplicity

1. **Whiteboards:**
   - **Best For:** Brainstorming and visualizing complex problems in a collaborative setting.
   - **Example Use Case:** Mapping how changes to the AI system affect stakeholders, such as customers, employees, and regulators.
2. **Sticky Notes and Clustering:**
   - **Best For:** Identifying patterns or prioritizing issues during team workshops.
   - **Example Use Case:** Grouping ethical, technical, and business priorities to highlight interdependencies.
3. **Paper Templates:**
   - **Best For:** Quickly iterating through ideas or testing new frameworks.
   - **Example Use Case:** Drafting an early version of the Integration Canvas to visualize stakeholder impacts and conflicts.

Digital Tools for Scalability

1. **Interactive Dashboards:**
   - **Best For:** Visualizing live data and monitoring project progress.
   - **Example Use Case:** Tracking AI system performance metrics alongside user feedback.
2. **Collaboration Software (e.g., Miro, Notion):**
   - **Best For:** Distributed teams working on shared problems.
   - **Example Use Case:** Using a digital Integration Canvas to coordinate input from various departments.
3. **Data Analysis Platforms:**
   - **Best For:** Analyzing large datasets or evaluating system performance.
   - **Example Use Case:** Identifying patterns in AI behavior across diverse demographic groups to address bias.

Tool Selection Criteria:

Context	Analog Tools	Digital Tools
Small Teams	Whiteboards, Stickies	Collaboration Platforms
Limited Time	Templates, Post-Its	Interactive Dashboards
High Complexity	Not Ideal	Data Analysis Software
Dynamic Updates	Not Ideal	Real-Time Tracking Tools

Key Takeaways for Complexity Reduction

1. **Streamline Without Losing Depth:** Decision trees and triage frameworks allow for simplified decision-making while maintaining the awareness of broader systemic factors.
2. **Adapt Tools to Context:** Analog tools are ideal for initial problem framing, while digital tools excel in execution and scalability.
3. **Iterate and Refine Continuously:** Complexity reduction is an ongoing process. Reassess and adjust your strategies as new information and challenges emerge.

By employing these strategies, you can navigate even the most intricate challenges, aligning actionable insights with long-term organizational and ethical goals.

7.5 INTEGRATION IN PRACTICE

7.5.1 Standardized Case Study Format

When tackling complex, multidimensional challenges, a structured approach to case studies ensures clarity, consistency, and the ability to extract actionable insights. Below is an enhanced framework for presenting case studies that emphasizes practical application and long-term learning.

7.5.1.1 Initial Situation: Define the Challenge and Its Scope

Paint a vivid picture of the problem, setting the stage for its complexity and stakes:

- **Core Challenge**: Articulate the primary issue with precision.
  Example: An AI-powered customer service tool struggles to balance user privacy with the need for high performance, creating tension among stakeholders.
- **Scope and Scale**: Specify the boundaries and scale of the issue – organizational, regional, or global.
  Example: A mid-sized company navigating privacy laws while expanding globally.
- **Stakeholders**: Identify the key players and their roles.
  Example: Customers demanding transparency, regulators enforcing compliance, and employees concerned about job security.

7.5.1.2 Framework Selection: Detail Which Frameworks Were Chosen and Why

Break down the reasoning behind the chosen frameworks and their alignment with the challenge:

- **Primary Framework**: The leading tool or approach that addresses the most critical aspects of the issue.
  Example: Causal Relevance Ethics to navigate privacy vs. fairness dilemmas.
- **Supporting Frameworks**: Secondary tools to handle complementary or niche aspects.
  Example: EASCT to manage employee concerns; DMN insights to analyze decision-making biases.
- **Selection Process**: Provide a clear justification for prioritizing certain frameworks.
  Example: Regulatory pressure made ethical considerations paramount, supported by empirical insights to validate decisions.

7.5.1.3 Integration Challenges: Identify Conflicts and Roadblocks

Illuminate the real-world complexity by highlighting contradictions and obstacles:

- **Framework Conflicts**: Detail instances where frameworks offered conflicting advice.
  Example: Ethical frameworks favor minimal data use, while technical needs demand detailed analytics.
- **Technical Barriers**: Limitations in available technology, tools, or expertise.
  Example: Existing infrastructure lacked capabilities for a hybrid data-collection approach.
- **Human Factors**: Emotional, cultural, or organizational resistance to proposed solutions.
  Example: Employee fear of automation was compounded by a lack of clear communication.

7.5.1.4 Resolution Approach: Explain How Tools Resolved Conflicts

Provide a step-by-step account of how the tools were used to integrate frameworks and resolve challenges:

- **Integration Canvas**: Mapped out stakeholders, impacts, and conflicts, providing a comprehensive view of the problem landscape.
  Example: Demonstrated how balancing privacy enhanced customer trust but required internal culture shifts.

- **Contradiction Navigator**: Pinpointed trade-offs and developed balanced solutions.
  Example: Hybrid data-collection strategy that met privacy concerns without sacrificing performance.
- **Adaptive Integration Matrix**: Allowed prioritization of tasks by urgency and complexity.
  Example: Focused initial efforts on regulatory compliance to avoid fines, followed by long-term AI training improvements.

7.5.1.5 Lessons Learned: Summarize Insights Gained

Capture and distill the key takeaways to inform future challenges:

- **Framework Effectiveness**: Identify what worked well and why.
  Example: The Integration Canvas clarified stakeholder priorities early, avoiding costly misalignments.
- **Unforeseen Challenges**: Highlight unexpected issues and their resolution.
  Example: Employees required emotional support to overcome fears of job automation.
- **Scalable Solutions**: Point out approaches that could be reused or adapted.
  Example: The hybrid strategy became a template for other departments.

7.5.1.6 Follow-Up Results: Measure and Analyze Outcomes

Showcase measurable successes and lessons from the process:

- **Quantitative Metrics**: Data that illustrates the tangible impact.
  Example: Customer satisfaction scores increased by 25% after privacy improvements.
- **Qualitative Insights**: Stakeholder feedback and broader perceptions.
  Example: Employees expressed greater trust in management after being involved in the process.
- **Next Steps**: Actions planned to capitalize on gains or address lingering issues.
  Example: Expansion of AI tools to new regions while maintaining ethical safeguards.

Enhanced Framework for Reporting Case Studies

Case Study Element	Key Details
Initial Situation	Define the challenge, its scope, and stakeholder dynamics.
Framework Selection	Justify the chosen frameworks, distinguishing between primary and supporting tools, and explain their alignment with the challenge.
Integration Challenges	Highlight contradictions between frameworks, technical obstacles, and human resistance to change.
Resolution Approach	Detail how tools like the Integration Canvas, Contradiction Navigator, and Adaptive Integration Matrix were applied to integrate frameworks and resolve conflicts.
Lessons Learned	Summarize actionable insights, including framework effectiveness, unforeseen challenges, and scalable solutions for future application.
Follow-Up Results	Present quantitative and qualitative outcomes and outline next steps for building on success or addressing unresolved issues.

Why This Structure Elevates Integration

This case study format transforms anecdotal problem-solving into a replicable model, enabling leaders to:

- Learn from practical applications of integration tools.
- Anticipate potential challenges in similar scenarios.
- Refine their use of meta-frameworks for increasingly complex problems.

The added structure and precision ensure the case studies are not only informative but also immediately applicable for readers facing real-world integration dilemmas.

7.5.2 Expanded Case Studies

7.5.2.1 AI Ethics Implementation: Striking the Balance Between Privacy, Fairness, and Speed

7.5.2.1.1 Challenge Overview

In the rush to deploy a game-changing AI customer service tool, competing demands created a high-stakes balancing act:

- Privacy Concerns: The AI relied on user data for personalization, sparking fears of overreach and regulatory backlash.
- Fairness Issues: Initial testing revealed algorithmic bias that could erode user trust and reinforce harmful stereotypes.
- Economic Pressures: Market competition demanded rapid deployment to maintain a competitive edge.
- Cultural Considerations: Employees were apprehensive about job automation and ethical missteps tarnishing the company's reputation.

Tools in Action

1. **Integration Canvas: Clarifying Stakeholder Concerns and Aligning Priorities**
   The Integration Canvas visually mapped the complex web of stakeholder interests:
   - Customers: Demanded data security and fairness in interactions.
   - Regulators: Focused on strict compliance with privacy laws.
   - Executives: Pressed for speed-to-market without sacrificing brand reputation.
   - AI Development Teams: Sought resources to improve algorithm transparency and reduce bias.

   The canvas identified critical intersections:
   - Conflict Zones: Privacy regulations clashed with the need for rich user data.
   - Synergy Opportunities: Aligning fairness improvements with regulatory compliance.
   - Shared Priorities: Building user trust while maintaining competitive functionality.
2. **Contradiction Navigator: Resolving Privacy vs. Performance Dilemmas**
   The Contradiction Navigator tackled opposing priorities:
   - Option A: Privacy-First Approach
     - Pros: High compliance, user trust.
     - Cons: Reduced personalization and performance.
   - **Option B: Performance-First Approach**
     - Pros: Superior functionality, market leadership.
     - Cons: Privacy concerns, potential reputational harm.
   - **Option C: Hybrid Strategy (Selected Solution)**
     - Balanced trade-offs by anonymizing sensitive data while enhancing algorithmic efficiency.
     - Introduced user consent options to empower individuals while meeting regulatory expectations.
3. **Adaptive Integration Matrix: Strategizing a Phased Rollout**
   Urgency and complexity shaped the implementation timeline:
   - Immediate Actions: Addressed compliance with privacy laws and mitigated critical biases in the AI.
   - Mid-Term Goals: Enhanced data anonymization methods and developed fairness benchmarks.
   - Long-Term Vision: Established ongoing bias audits and expanded AI capabilities through iterative updates.

Outcome

- Short-Term Success: A phased rollout restored user trust and satisfied regulators, avoiding costly delays.
- Mid-Term Wins: Transparent bias reduction measures became a competitive advantage.
- Long-Term Gains: The company positioned itself as an ethical AI leader in a competitive market.

4. **Digital Transformation: Breaking Resistance to Foster Innovation**

7.5.2.1.2 Challenge Overview

A legacy enterprise sought to modernize its digital infrastructure but faced resistance:

- Cultural Barriers: Employees feared losing roles to automation and resisted change.
- Innovation vs. Stability: Management wrestled with balancing innovation against operational continuity.
- Timeline Pressures: The company was lagging behind competitors in adopting modern systems.

Tools in Action

1. Integration Canvas: Identifying Cultural and Operational Barriers
   Using the canvas, the leadership team uncovered critical barriers:
   - Employee Resistance: A lack of communication about the benefits of digital tools fueled mistrust.
   - Operational Concerns: Managers feared that rapid adoption could destabilize critical workflows.
   - Leadership Hesitation: A cautious approach stifled necessary risk-taking.

   Insights Gained:

   - Communication Gaps: Employees didn't understand how digital transformation aligned with company values.
   - Training Needs: A skills gap prevented effective adoption of new tools.
2. **Contradiction Navigator: Balancing Innovation and Stability**
   The navigator unraveled competing goals:
   - Innovation-First Approach
     - Pros: Accelerated rollout maintained competitiveness.
     - Cons: Risked operational disruptions.
   - Stability-First Approach
     - Pros: Protected workflow reliability.
     - Cons: Slowed modernization efforts.
   - Balanced Strategy (Selected Solution)
     - Introduced new systems incrementally, focusing first on low-risk areas.
     - Combined hands-on training sessions with leadership-driven change initiatives to build trust and capability.
3. **Adaptive Integration Matrix: Timing Milestones for Adoption**
   The matrix-clarified implementation phases:
   - Urgent/Low Complexity: Deploying basic collaboration tools to boost short-term efficiency.
   - Urgent/High Complexity: Migrating core systems to cloud platforms, ensuring business continuity.
   - Non-Urgent/High Complexity: Implementing advanced analytics post-adoption to unlock transformative insights.

Outcome

- Cultural Alignment: Employees embraced the changes after seeing tangible benefits from the early successes.
- Operational Stability: The phased rollout maintained workflow reliability while minimizing disruption.
- Strategic Leadership: The company leapfrogged competitors by combining modern tools with a motivated workforce.

Key Takeaways from Expanded Case Studies

Aspect	AI Ethics Implementation	Digital Transformation
Core Challenge	Balancing privacy, fairness, and speed	Overcoming cultural resistance to system changes
Tools Applied	Integration Canvas, Contradiction Navigator, Adaptive Matrix	Integration Canvas, Contradiction Navigator, Adaptive Matrix
Actionable Strategy	Phased rollout with a hybrid privacy-performance model	Incremental adoption with tailored training programs
Outcome	Restored trust, regulatory compliance, ethical leadership	Operational continuity, cultural alignment, competitive edge

These expanded case studies highlight the power of the Integration Toolkit in transforming theoretical concepts into actionable, real-world solutions. By combining clarity, stakeholder alignment, and practical prioritization, the tools deliver sustainable outcomes even in the most challenging environments.

7.6 ADVANCED META-LEVEL THINKING

7.6.1 Meta-Thinking Principles

7.6.1.1 Adaptive Thinking: The Symphony of Dynamic Framework Switching

In a world of relentless complexity, where problems are multi-dimensional and evolving, adaptive thinking is the maestro that orchestrates harmony among diverse frameworks. It's not about rigidly applying a single tool –, it's about knowing when to pivot, blend, or invent approaches as new challenges arise.

- **Core Idea**: Adaptive thinking is the bridge between strategy and agility. It empowers you to embrace uncertainty, process real-time data, and recalibrate your approach without losing sight of your goals.
- **Key Steps**:
  i. **Situation Sensing**: Develop an acute awareness of environmental and stakeholder shifts.
     - *Example*: During a cybersecurity breach, begin with containment frameworks; later, pivot to communication strategies to restore customer trust.
  ii. **Framework Fit Check**: Continuously evaluate whether the current framework still aligns with the problem's evolving context.
     - *Question*: Is this tool still effective, or has the terrain shifted beneath us?
  iii. **Hybrid Framework Development**: Where no single framework suffices, craft hybrid solutions by combining complementary approaches.
     - *Example*: Pair risk mitigation frameworks with behavioral change models to address both technical and human factors.

Enhanced Tool: Adaptation Velocity Matrix

Change Trigger	Framework Response	Timeline for Adjustment	Outcomes Expected
Stakeholder demands shift	Incorporate participatory tools	Immediate	Improved stakeholder alignment
Emerging risks or threats	Integrate risk analysis	Short-term	Enhanced decision readiness
Game-changing innovation	Add innovation frameworks	Medium-term	Strategic foresight and agility

7.6.1.1.1 Real-World Power Play

Consider how Tesla rapidly shifted its battery technology frameworks when cobalt shortages emerged. Instead of waiting for a crisis, they adopted adaptive strategies – testing new chemistries and partnering with suppliers to diversify resources.

7.6.1.2 Abstraction and Prioritization: Mastering the Art of Strategic Focus

Meta-level thinkers are both the eagle surveying the landscape and the architect laying the foundation. They can zoom out to discern global patterns and zoom in to ensure every bolt is tightened. This dual capability is the essence of abstraction and prioritization.

- **Seeing the Patterns (Abstraction)**: Recognize systemic trends and recurring dynamics to shape long-term strategy.
  - *Example*: Observing a consistent surge in data privacy demands across geographies can guide enterprise-wide product policies.
- **Focusing on Impact (Prioritization)**: Identify which details or tasks matter most in the immediate term to align with broader goals.
  - *Example*: During a new product launch, prioritizing compliance in key markets while setting secondary timelines for regions with less stringent regulations.

Practical Framework: The Abstraction Ladder

Level	Focus Area	Key Question	Action Example
Level 5	Philosophy and Purpose	What legacy are we building?	Align all decisions with the organization's mission.
Level 4	Value Principles	Are we honoring our core values?	Evaluate proposals for ethical alignment.
Level 3	Strategic Themes	What overarching goals must we achieve?	Set quarterly objectives around these themes.
Level 2	Tactical Frameworks	What frameworks will guide execution?	Select agile or waterfall approaches based on scope.
Level 1	Immediate Actions	What must be done now to keep momentum?	Assign tasks with immediate deliverables.

The Balance of High and Low:

- Use **abstraction** to find your "North Star," ensuring all efforts align with ultimate goals.
- Use **prioritization** to map the most efficient pathway to reach that star, step by step.

Example of Strategic Precision:
A retail company planning global expansion uses:

- **Abstraction**: A high-level vision of sustainability and customer-centric growth.
- **Prioritization**: Initiating in regions with minimal regulatory hurdles while building internal capacity for longer-term challenges.

Refined Summary of Meta-Thinking Principles

Principle	Key Insight	Power Move
Adaptive Thinking	Flexibly adjust and combine frameworks in real-time	Integrate new tools to respond to changing data or risks
Abstraction and Prioritization	Balance long-term vision with short-term execution	Align immediate actions with overarching strategic goals

Meta-level thinking isn't just a skill; it's a strategic advantage. By mastering adaptability and focus, leaders transform complexity into clarity, ensuring decisions not only solve today's challenges but also pave the way for tomorrow's breakthroughs.

7.6.2 Expanded Complexity Management

7.6.2.1 Decision Triage Trees: Simplifying Decision Pathways

Decision-making in complex scenarios often feels overwhelming, with competing demands, uncertain outcomes, and high stakes. **Decision Triage Trees** serve as a dynamic tool to cut through the noise, guiding leaders toward clarity and action without losing sight of nuance.

7.6.2.1.1 Core Idea

A **Decision Triage Tree** is not just about breaking decisions into steps; it's about tailoring each pathway to the urgency, impact, and resources available, ensuring that even complex challenges are addressed systematically and effectively.

Structure of a Decision Triage Tree

Step	Key Question	Outcome
Define the Problem	What is the root cause or conflict?	Establish the scope, identify key players, and align goals.
Assess Urgency	How soon must this be resolved?	Classify as Immediate, Short-Term, or Long-Term.
Measure Impact	How critical is the decision's outcome?	Categorize as High (strategic), Medium (tactical), or Low (operational).
Evaluate Resources	What's available to support this decision?	Align available time, talent, and tools with the scale of the challenge.
Select the Pathway	Which frameworks and tools are best suited here?	Create a streamlined plan with specific frameworks and clear next steps.

7.6.2.1.2 Real-World Application: Privacy vs. Performance Decision Tree

In the AI ethics example, balancing privacy with performance requires careful navigation through conflicting goals.

1. **Start**: What is the core challenge?
   - Privacy impacts performance – how to balance them?
2. **Urgency Assessment**: Is the decision time-sensitive?
   - Immediate: Customer trust and compliance issues.

3. **Impact Evaluation**: What's at stake?
   - High: Trust erosion, user churn, and regulatory penalties.
4. **Resource Analysis**: Do you have what's needed?
   - Sufficient: Skilled teams and budget for a phased rollout.
5. **Pathway Selection**: Which framework is optimal?
   - Use the **Contradiction Navigator** to identify trade-offs, followed by a **Hybrid Solution** that prioritizes user trust.

Printable, User-Friendly Triage Flowchart

Question	Immediate	Short-Term	Long-Term
Is this decision urgent?	Resolve immediately.	Plan for near-future action.	Schedule strategic discussions.
Does it have a high impact?	Leadership involvement.	Delegate to skilled teams.	Plan cross-departmental effort.
Are resources readily available?	Proceed with execution.	Adjust scope and timeline.	Build capacity incrementally.

Outcome: Clear guidance from ambiguity to action, fostering rapid alignment without oversimplification.

7.6.2.2 Contradiction Resolution: Synthesis Techniques for Competing Goals

Complex decisions often force leaders into a false binary: prioritize one goal at the expense of another. **Contradiction Resolution** provides the tools to transform these conflicts into opportunities for innovation and synergy.

7.6.2.2.1 Core Idea

Rather than compromising, contradiction resolution aims to **synthesize** opposing priorities, leveraging advanced tools to find new pathways that serve multiple objectives.

Advanced Techniques for Contradiction Resolution

Technique	How It Works	Example
Trade-Off Analysis	Systematically evaluate the benefits and costs of each option to determine acceptable concessions.	Deciding how much personalization is worth sacrificing to meet stricter privacy standards.
Hybrid Solutions	Combine strengths of conflicting priorities to create a balanced approach.	Using anonymized data for most users and detailed data for opt-ins to balance privacy and performance.
Reframing the Problem	Shift the perspective to find a unifying question or goal.	Moving from "privacy vs. performance" to "what builds long-term user trust and engagement?"
Conflict Mapping	Visually diagram tensions and intersections to identify potential alignments.	Charting the overlap between customer trust, regulatory compliance, and competitive differentiation.
Collaborative Synthesis	Engage diverse stakeholders to co-create innovative solutions.	Bringing engineers, legal experts, and customers together to design transparent, user-friendly policies.

Real-World Application: Privacy vs. Performance Synthesis

1. **Challenge**: User data improves AI accuracy but creates privacy risks.
2. **Tools Used**:
   - **Conflict Mapping**: Identifies overlapping priorities like transparency and control.
   - **Hybrid Solution**: Opt-in personalization for trust-sensitive users, default anonymization for others.

3. **Outcome**: A phased implementation strategy balancing performance and trust.

Printable Contradiction Mapping Table

Goal A: Privacy	Goal B: Performance	Overlapping Opportunities	Resolution Pathway
Anonymized data for trust	Detailed data for accuracy	User transparency fosters trust and personalization opt-ins.	Hybrid approach with user-controlled settings.
Compliance with regulations	Feature diversity	Align transparency policies with innovative data tools.	Co-design privacy protocols with stakeholders.
Empowerment through choice	Efficiency for scaling	Empowered users can opt for tailored services.	Offer a simple opt-in process.

7.6.2.2.2 The Edge of Meta-Level Mastery

Both **Decision Triage Trees** and **Contradiction Resolution** elevate decision-making from tactical firefighting to strategic innovation. Together, they create a robust system for mastering complexity, enabling leaders to navigate competing priorities without losing sight of long-term goals. This is **integration in action** – where clarity meets creativity, and innovation thrives.

7.7 TOOLS FOR INTEGRATION: ENHANCED APPROACH

The ability to navigate complexity stems from using the right tools in the right ways – and often, at the right times. This section introduces and then expands upon a suite of methods designed to clarify conflicts, negotiate contradictions, prioritize actions, and adapt over time.

7.7.1 The Integration Canvas

Purpose:

- Visualize and organize complex relationships among frameworks, stakeholders, and conflicts.
- Provide clarity, identify synergies, and create a roadmap for resolution.

Template: Advanced Integration Canvas

Component	Description	Example
Frameworks	Ethical, technical, and business frameworks influencing the decision.	Privacy laws (GDPR), AI fairness protocols, ROI models.
Stakeholders	Key players and their interests.	Customers: Privacy concerns; Developers: Performance needs.
Conflicts	Areas where priorities or frameworks clash.	Privacy vs. Functionality; Fairness vs. Speed.
Integration Points	Opportunities where frameworks align or complement each other.	Privacy safeguards that enhance customer trust while supporting functionality.
Proposed Actions	Steps to address conflicts and exploit synergies.	Implement anonymization for data while maintaining algorithm accuracy.
Metrics	Measurable outcomes to track success.	Bias reduction, user satisfaction, compliance rate.

Example: AI Ethics Case Application

1. **Frameworks:**
   - **Privacy:** GDPR compliance
   - **Fairness:** Causal Relevance Ethics to assess long-term societal impact
   - **Performance:** Technical KPIs for speed and accuracy

2. **Stakeholders:**
   - **Customers:** Concerned about data misuse
   - **Engineers:** Seeking more data for improved AI accuracy
   - **Regulators:** Focusing on legal compliance
3. **Conflicts:**
   - Tension between customer expectations of privacy and the system's need for data
4. **Integration Points:**
   - Employ anonymized data processing to balance both needs
5. **Proposed Actions:**
   - Conduct bias testing and establish transparency in communication with users
6. **Metrics:**
   - Monitor the number of bias-free outcomes and track regulatory approval timelines

Enhanced Visualization for Printing

Component	Example
Frameworks	GDPR, Causal Ethics, KPIs
Stakeholders	Customers, Engineers, Regulators
Conflicts	Privacy vs. Functionality
Integration Points	Anonymized Data, Transparency
Proposed Actions	Bias Testing, User Communication
Metrics	Bias-Free Outcomes, Compliance

7.7.2 The Contradiction Navigator

Purpose:

- Resolve conflicting priorities by framing trade-offs and creating win-win strategies.

Template: Contradiction Navigator Framework

Priority A	Priority B	Trade-Offs	Resolution Strategy
Speed	Ethical Fairness	Sacrificing safeguards for a quicker rollout.	Phased rollout with bias testing.
Innovation	Cultural Resistance	Employees fear redundancy; innovation slows.	Reskilling programs for employees.
Profitability	Sustainability	Immediate ROI vs. long-term environmental impact.	Hybrid models for balanced growth.

Example: Innovation vs. Cultural Resistance

1. **Conflict:** Rapid AI adoption is resisted by employees fearing job losses.
2. **Trade-Offs:**
   - **A:** Faster adoption builds market advantage but alienates employees.
   - **B:** Slower adoption retains trust but risks competitive edge.
3. **Resolution:**
   - Roll out AI tools as *aids* rather than replacements.
   - Provide reskilling programs alongside technology adoption.

7.7.3 Adaptive Integration Matrix

7.7.3.1 Purpose:

- Match decisions and challenges to their appropriate level of urgency and complexity, ensuring resources and attention are allocated effectively.

Matrix for Decision Prioritization

Urgency	Low Complexity	High Complexity
High	Quick-win solutions	Comprehensive crisis frameworks
Low	Incremental optimizations	Strategic, long-term planning

7.7.3.2 Example: Digital Transformation Case

1. **Challenge:** Integrating a cloud-based system while addressing employee concerns.
2. **Approach:**
   - **High Urgency + Low Complexity:** Pilot the system in low-risk departments.
   - **High Urgency + High Complexity:** Provide extensive training for IT teams while phasing out legacy systems.
   - **Low Urgency + High Complexity:** Develop long-term plans for integrating cross-departmental workflows.
3. **Outcome:**
   - Initial quick wins create confidence, paving the way for smooth, organization-wide adoption.

Enhanced Decision Support Table

Context	Recommended Action
High Urgency, Low Complexity	Rapid response framework.
High Urgency, High Complexity	Multi-team task force.
Low Urgency, Low Complexity	Incremental improvements.
Low Urgency, High Complexity	Long-term strategic alignment.

7.7.3.2.1 Elevating Your Toolkit

By combining the **Integration Canvas**, **Contradiction Navigator**, and **Adaptive Integration Matrix**, leaders can:

- **Deconstruct Complexity:** Map relationships and conflicts clearly.
- **Balance Trade-Offs:** Resolve contradictions thoughtfully.
- **Prioritize Effectively:** Align actions with urgency and resources.

7.7.4 Integration in Action: The Four-Tool Cycle

While each tool provides unique value, their true power emerges when used **together** as an integrated system for navigating complexity:

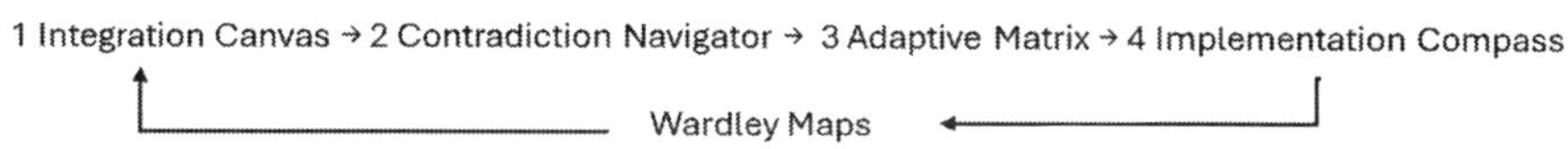

7.7.4.1 The Cycle in Motion

a. **Integration Canvas**
   - **Maps the current landscape.**
   - Identifies stakeholders, frameworks, and conflicts.
   - Creates visibility of the whole system.

b. **Contradiction Navigator**
    - **Takes conflicts from the Canvas.**
    - Develops balanced solutions or hybrid approaches.
    - Identifies potential synergies.
c. **Adaptive Matrix**
    - **Prioritizes solutions** identified by the Navigator.
    - Matches resources to urgency and complexity.
    - Creates a clear action timeline.
d. **Implementation Compass**
    - **Guides execution** of prioritized actions.
    - Gathers feedback and measures outcomes.
    - Generates learning loops back to the Canvas.

Key Components of the Implementation Compass:

1. **Execution Tracking** – ensuring planned solutions are carried out.
2. **Feedback Collection** – capturing user data, stakeholder input, and resilience evidences.
3. **Learning Integration** – refining or revisiting the earlier steps as new insights emerge.
4. **Adaptation Triggers** – deciding when significantly new information warrants major changes.

7.7.4.2 Wardley Maps as Strategic Overlay

Wardley Maps provide crucial strategic context at each step in this cycle:

- **Canvas**: Shows where each component stands in its evolution, ensuring you grasp both the big picture and the immediate system.
- **Navigator**: Informs trade-off decisions by highlighting which elements are or will soon become commoditized vs. those offering unique advantage.
- **Matrix**: Helps time your strategic moves – knowing when to invest in emerging capabilities or pivot from mature ones.
- **Compass**: Tracks evolution progress, linking it back to broader strategic direction.

Example: AI Ethics Case Through the Cycle

1. **Integration Canvas**
    - Maps stakeholders (users, developers, regulators).
    - Identifies conflicts (privacy vs. performance).
    - Shows current component positioning (e.g., novel data-handling methods).
2. **Contradiction Navigator**
    - Resolves the privacy-performance tension.
    - Creates a hybrid solution approach that addresses user trust while maintaining a competitive speed.
    - Considers evolution trajectories (some tech might quickly become standard or commodity).
3. **Adaptive Matrix**
    - Prioritizes immediate compliance with regulators (high urgency, high complexity).
    - Schedules longer-term improvements for system transparency (low urgency, high complexity).
    - Aligns these moves with overall strategic timing.
4. **Implementation Compass**
    - Tracks real-world solution deployment (pilot tests, user acceptance).
    - Gathers feedback on fairness and performance.

- Monitors when significant new data triggers a revisit of the Canvas or Navigator.
- Ensures continuous alignment with evolving conditions.

Enhanced Decision Support Reference

Tool	Primary Function	Feeds Into	Strategic Overlay
Canvas	System mapping	Conflict resolution	Component positioning
Navigator	Trade-off balance	Action planning	Evolution-aware choices
Matrix	Resource allocation	Implementation	Strategic timing
Compass	Execution and learning	System remapping	Evolution tracking

The power of this integrated approach lies in its cyclical nature. Each tool's outputs enhance the others, creating a continuous improvement loop that grows more refined with every pass. You gain not only solutions but a living philosophy of adaptation, ensuring each new challenge becomes an opportunity for ethical innovation and resilience.

7.8 COMMON PITFALLS AND SOLUTIONS

Mastering integration at the meta-level brings unparalleled clarity and effectiveness, but the journey is riddled with challenges. Anticipating and overcoming these common pitfalls ensures sustained momentum and impactful outcomes.

7.8.1 Integration Traps

7.8.1.1 Framework Overload

Description: When too many frameworks are applied simultaneously, the sheer volume creates confusion, dilutes focus, and leads to stagnation.

Example: A team attempting to integrate ethical principles, technical requirements, customer needs, and regulatory mandates into every decision point, causing endless deliberation.

Consequences:

- Sluggish decision-making.
- Stakeholder frustration.
- Overlap and duplication of effort.

Solution:

1. **Streamline with the Integration Canvas**: Use it to visualize overlaps and identify which frameworks are essential.
2. **Perform Regular Framework Audits**: Ensure every framework actively contributes to outcomes. Remove or deprioritize redundant ones.
3. **Set Boundaries**: Limit the number of frameworks for any single decision to three.

7.8.1.2 False Equivalence

Description: Treating all priorities or frameworks as equally important, ignoring their varied relevance or impact.

Example: Equating short-term financial gains with long-term ethical responsibilities during an AI product launch.

Consequences:

- Misaligned decisions.
- Inefficient resource allocation.
- Strategic missteps.

Solution:

1. **Apply the Contradiction Navigator**: Assess the weight of competing priorities and resolve conflicts through evidence-backed trade-offs.
2. **Use a Weighting Matrix**: Score each factor based on its impact and urgency to guide decision-making.
3. **Adjust for Time Horizons**: Recognize differences between immediate needs and long-term objectives.

7.8.1.3 Integration Paralysis

Description: Analysis becomes an endless loop, and action is postponed as teams fixate on every possible conflict and contingency.

Example: A digital transformation project delayed due to prolonged debates about which system features to prioritize.

Consequences:

- Missed opportunities.
- Diminished stakeholder trust.
- Erosion of competitive advantage.

Solution:

1. **Leverage the Adaptive Integration Matrix**: Match frameworks and actions to urgency, enabling a phased approach.
2. **Establish Decision Deadlines**: Set fixed timelines for deliberations to ensure timely progress.
3. **Adopt an Iterative Process**: Start with small, testable actions and refine based on results.

Common Integration Traps and Solutions

Trap	Description	Key Solutions
Framework Overload	Too many frameworks lead to confusion and stagnation.	Use the Integration Canvas; streamline with audits.
False Equivalence	Treating all factors as equally important.	Weight priorities with a matrix; adjust for timelines.
Integration Paralysis	Over-analyzing without moving to action.	Set decision deadlines; use phased frameworks.

7.8.2 Simplification Strategies

7.8.2.1 Prioritization Matrices

Purpose: Focus efforts on the highest-value actions while avoiding unnecessary distractions.

Example: Impact/Urgency Matrix

Impact	High	Low
Urgency High	Immediate action required	Monitor for escalation.
Urgency Low	Strategic projects	Low-priority tasks.

Steps to Implement:

1. **Map Tasks**: List all decisions and categorize them by impact and urgency.
2. **Focus Resources**: Allocate attention to high-impact, high-urgency items first.
3. **Defer Lower Priorities**: Save less urgent tasks for later stages.

Application: In a digital transformation, prioritize addressing cultural resistance (high impact, high urgency) before optimizing the backend performance (low urgency).

7.8.2.2 Framework Audits

Purpose: Prevent inefficiency by ensuring that every applied framework directly contributes to decision-making.

Steps to Implement:

1. **Inventory Current Frameworks**: List all tools, methods, and models in use.
2. **Evaluate Utility**: Assess each framework's relevance to the problem at hand.
3. **Streamline**: Retire or pause any framework that offers minimal or redundant value.

Example: A company focusing on user satisfaction during an AI rollout may pause frameworks unrelated to customer experience to streamline decision-making.

7.8.2.3 Simplification Without Oversimplification

Purpose: Balance the need for clarity with the complexity inherent in nuanced challenges.

Techniques:

1. **Use Decision Trees**: Map decision pathways to isolate critical elements without neglecting the context.
2. **Apply the Integration Canvas**: Represent complex relationships visually for clearer understanding.
3. **Emphasize Actionable Insights**: Focus on steps that lead to tangible progress while retaining broader awareness.

Example: For an AI ethics rollout, a decision tree helps evaluate trade-offs between user privacy and performance metrics, preserving ethical considerations while moving forward.

Simplification Strategies

Strategy	Purpose	Example Application
Prioritization Matrices	Focus on high-impact, urgent tasks.	Address cultural resistance first.
Framework Audits	Eliminate frameworks with low utility.	Retire tools unrelated to customer focus.
Simplification Tools	Streamline decisions without losing nuance.	Use decision trees for privacy trade-offs.

By recognizing and mitigating these pitfalls, decision-makers can confidently navigate complexity, harnessing the full power of integration while maintaining focus, clarity, and forward momentum.

7.9 FUTURE DIRECTIONS: THRIVING IN COMPLEXITY

As technological and societal landscapes evolve, the need for integration frameworks that adapt, innovate, and remain resilient becomes critical. This section challenges readers to anticipate emerging complexities and embrace continuous learning as a core competency. The goal? To stay ahead of disruption while fostering ethical, adaptive, and forward-thinking decision-making.

7.9.1 Emerging Challenges: The Expanding Role of AI in Decision-Making

7.9.1.1 The Challenge

AI is no longer a mere tool; it is a partner, advisor, and, in some cases, an autonomous decision-maker. This shift introduces profound complexities:

- **Opportunities**: Speed, scalability, and insights beyond human capacity.
- **Risks**: Bias, ethical ambiguity, and unintended consequences.

Three Critical Questions for AI Integration:

1. **Bias Management**: How do we ensure AI aligns with societal fairness while learning from flawed historical data?
2. **Oversight vs. Autonomy**: When should humans intervene, and when should AI be allowed to operate independently?
3. **Evolving Ethics**: How do we govern AI as its capabilities – and ethical dilemmas – outpace existing frameworks?

7.9.1.2 Solutions and Strategies

1. **Evolving Ethical Frameworks**
   - **Modular Ethics Layers**: Introduce ethical modules into AI systems that adapt to context and real-time data.
     Example: Layered protocols for privacy, fairness, and safety that dynamically adjust based on user interaction.
2. **Human-AI Collaboration Models**
   - **Human-in-the-Loop Systems**: Balance AI autonomy with human oversight.
     Example: AI suggests options for medical diagnostics, but doctors make final decisions.
3. **Responsible Deployment Protocols**
   - Develop and enforce pre-launch checks:
     - **Ethical Viability**: Does the AI meet ethical benchmarks?
     - **Crisis Preparedness**: What's the fallback if AI fails or behaves unexpectedly?
     - **Impact Testing**: Simulate long-term societal impacts in controlled environments.

7.9.1.3 Adaptive Approaches for Emerging AI Scenarios

Scenario	Potential Impact	Response Strategy
AGI Emergence	Ethical and existential risks	Iterative global consensus on governance
Quantum Computing AI	Disruption of encryption and privacy	Develop quantum-resilient frameworks
Autonomous AI Systems	Loss of control in critical industries	Mandatory kill-switch protocols and fail-safes

7.9.2 Continuous Learning: Evolving Frameworks for an Adaptive Future

7.9.2.1 Why Continuous Learning is Essential

Static frameworks are relics in a world defined by exponential change. Integration frameworks must evolve through ongoing adaptation to remain relevant.

Key Catalysts for Change:

1. **Technological Revolutions**: Quantum AI, blockchain-driven governance, and neural interface technologies.
2. **Cultural Shifts**: The growing emphasis on sustainability, inclusion, and shared societal responsibility.
3. **Global Crises**: Pandemics, climate change, and geopolitical instability reshaping priorities.

7.9.2.2 Strategies for Future-Proof Integration

1. **Framework Evolution Protocols**
   - **Monitor Change**: Establish trend-watching teams to identify shifts in technology, ethics, and behavior.
   - **Test Iteratively**: Pilot changes in low-risk environments before full-scale implementation.
   - **Embed Feedback Loops**: Incorporate stakeholder insights to refine frameworks dynamically.

2. **Lifelong Learning for Individuals and Teams**
   - **Dynamic Training Modules**: Update team skills in real-time, focusing on adaptability and foresight.
     Example: Simulations that stress-test ethical and operational decision-making in complex AI scenarios.
   - **Incentivized Learning Cultures**: Reward organizations and individuals for experimentation and growth.

7.9.2.3 Visionary Tools for Continuous Adaptation

1. **Scenario-Based Simulations**: Test integration frameworks against speculative futures, from utopian to catastrophic.
2. **Ethical Sandbox Environments**: Simulate the ethical consequences of decisions before real-world deployment.
3. **Real-Time Dashboarding**: Live tracking of framework effectiveness and readiness against emerging global trends.

7.9.3 Building Future Resilience Through Anticipation

7.9.3.1 Strategic Scenario Planning

Scenario Type	Key Questions	Anticipated Response
Best-Case Scenario	How do we scale success ethically?	Expand tested frameworks into new domains
Worst-Case Scenario	What safeguards prevent catastrophic failure?	Strengthen oversight, redundancy, and contingency
Wild Card Disruption	How do we prepare for unforeseeable game-changers?	Cultivate agility and innovation in all systems

7.9.3.2 Readiness Assessment Metrics

Dimension	Metric	Target	Gap Analysis
Integration Depth	% of frameworks seamlessly linked	90%	Identified integration lag
Adaptation Speed	Time to adjust to new challenges	<3 months	Bottlenecks in feedback
Ethical Alignment	Compliance with evolving norms	100%	Areas of ethical ambiguity

7.9.4 Vision for the Future

7.9.4.1 Closing Reflection

The ability to thrive in an increasingly complex world depends on the capacity to integrate not only knowledge systems but also the wisdom to anticipate change. Meta-level thinking equips leaders to navigate the chaotic intersection of technology, ethics, and human dynamics with precision and foresight.

7.9.4.2 Call to Action

Embrace the evolving meta-layer not as a challenge but as an opportunity to lead. Equip yourself with tools, strategies, and an adaptive mindset to transform uncertainties into possibilities. Only through intentional integration can we turn today's challenges into tomorrow's breakthroughs.

7.10 PRACTICAL APPLICATIONS: TURNING INTEGRATION INTO ACTION

This section bridges theory and practice, empowering you to internalize meta-level thinking in your personal and professional life. Through structured challenges, reflection exercises, and measurable tracking, these practical tools help you master complexity and integration.

7.10.1 Personal Development

Meta-thinking isn't just a skill – it's a practice. By embedding integration exercises into your daily routine, you can cultivate sharper decision-making, greater adaptability, and a holistic view of complex systems. Below are three structured pathways to guide your journey.

7.10.1.1 Weekly Integration Challenges: Building Muscle Memory

Weekly challenges are tailored to build foundational skills, from recognizing integration points to resolving conflicts. Each week adds a layer of complexity, ensuring steady growth and confidence.

Week	Focus Area	Challenge	Goal
Week 1	Awareness	Identify conflicting goals in a personal or work scenario.	Develop skill in spotting integration points.
Week 2	Simplification	Use a decision tree to break down a complex problem.	Hone focus on essential, actionable elements.
Week 3	Visualization	Map out stakeholders and impacts using the Integration Canvas.	Build clarity in understanding relationships.
Week 4	Contradiction Resolution	Resolve a competing priority using the Contradiction Navigator.	Balance opposing objectives without compromise.
Week 5	Adaptive Thinking	Match a framework to a scenario using the Adaptive Integration Matrix.	Increase agility in selecting and applying tools.

7.10.1.1.1 Example Challenge

Scenario: A team project conflicts with personal time commitments.

1. **Week 1**: Identify conflicting goals (e.g., project completion vs. personal well-being).
2. **Week 2**: Use a decision tree to isolate non-negotiables and flexibilities.
3. **Week 4**: Apply the Contradiction Navigator to prioritize both team outcomes and personal recovery.

7.10.1.2 Reflection Prompts: Cultivating Self-Awareness

Reflection is the anchor for growth. Structured prompts help you evaluate successes, learn from setbacks, and refine your approach.

Dimension	Reflection Questions
Framework Selection	Were the frameworks I chose appropriate? How well did they address the situation?
Challenges	What roadblocks emerged? How could I navigate them better in the future?
Personal Growth	What has practicing integration taught me about my cognitive and emotional patterns?
Outcomes	Did my approach deliver the desired results? If not, why?

7.10.1.2.1 Deep Dive Example

After a workplace negotiation:

- "Did I balance long-term stakeholder needs with short-term business pressures?"
- "What trade-offs worked, and which created unnecessary friction?"

7.10.1.3 Progress Tracking: Seeing Growth in Action

Consistent tracking transforms abstract growth into tangible progress. Use analog and digital tools to monitor challenges, frameworks used, and results achieved.

Tracking Method	What to Record	Example Output
Journal	Daily integration attempts, outcomes, and lessons	"Applied Contradiction Navigator: Resolved privacy vs. performance issue."
App	Digital logs of frameworks used, resolved conflicts, trends.	"Resolved 4 conflicts this month, success rate: 85%."
Dashboard	Visual representation of metrics like frameworks used and resolutions achieved.	"Integration Canvas applied 3 times; 2 conflicts resolved successfully."

7.10.1.3.1 Visualization Idea

A **progress wheel** where each segment tracks:

- Frameworks mastered.
- Conflicts resolved.
- New insights gained.

Call to Action: Commit to Meta-Level Mastery

Start Today: Choose one tool – be it the Integration Canvas or Contradiction Navigator – and use it this week.
Reflect: Note the process, challenges, and outcomes.
Build Momentum: With each challenge, refine your skills and expand your toolkit.

The journey to meta-level mastery begins with action. Each step strengthens your ability to navigate complexity, balance competing demands, and lead with clarity.

7.10.2 Organizational Implementation

Achieving organizational mastery in integration requires both deliberate training and methodical implementation. Below are two structured pathways to infuse meta-thinking into your organization.

7.10.2.1 Team Training: Empowering Collective Mastery

Workshops offer an interactive way to introduce integration tools, build team cohesion, and align everyone on strategic priorities. A well-designed training program ensures that every team member—from executives to practitioners – becomes proficient in applying meta-level thinking.

Workshop Module	Content Focus	Exercises	Outcome
Introduction	Overview of meta-level thinking and its organizational benefits.	Group discussion: Reflect on recent challenges.	Build awareness of integration challenges.
Tool Deep-Dive	Detailed training on the Integration Canvas, Contradiction Navigator, and Adaptive Integration Matrix.	Case study: Apply each tool to a mock scenario.	Develop confidence in tool application.
Conflict Resolution	Strategies for resolving competing priorities.	Role-play: Resolve cultural vs. technical conflicts.	Build agility in navigating organizational tensions.
Advanced Integration	Using multiple tools in tandem to solve complex problems.	Team project: Integrate tools into a hypothetical business decision.	Master holistic, team-based problem-solving.
Implementation Planning	Strategies for embedding tools into daily workflows.	Collaborative planning: Map current processes and add integration steps.	Ensure long-term adoption of meta-level thinking.

7.10.2.1.1 Example Workshop Flow

1. **Morning Session**: Introduction and tool training.
2. **Afternoon Session**: Case studies and team exercises.
3. **Outcome**: Teams leave with practical experience and tailored action plans.

7.10.2.2 Implementation Plans: Embedding Integration into Workflows

Workflows are the backbone of organizational operations. By embedding integration tools into these processes, organizations can ensure that meta-thinking becomes a habitual practice.

Phase	Action Steps	Tools	Metrics
Preparation	Assess current workflows and identify integration opportunities.	Process mapping software, team interviews.	Map of existing workflows with gaps noted.
Pilot	Test tools with a small team or department.	Integration Canvas, Contradiction Navigator.	Success rate of resolved conflicts.
Deployment	Roll out tools across the organization with structured training and resources.	Training materials, dashboards.	% of teams trained, adoption rate of tools.
Optimization	Monitor usage, gather feedback, and refine processes based on outcomes.	Progress tracking tools, feedback forms.	Tool usage frequency, conflict resolution metrics.

7.10.2.2.1 Key Implementation Steps

1. **Integrate Tools into Existing Processes**:
   - Example: Add the **Contradiction Navigator** to project kickoff meetings to resolve early conflicts.
   - Example: Use the **Adaptive Integration Matrix** during quarterly reviews to prioritize initiatives.
2. **Create Role-Specific Guidelines**:
   - Executives: Use tools for strategic decision-making.
   - Managers: Facilitate team alignment with the Integration Canvas.
   - Practitioners: Apply tools to daily problem-solving.

7.10.2.3 Measuring Success: Organizational Impact

Embedding tools isn't the final step – measuring their impact ensures continuous improvement. Develop key metrics to track effectiveness and refine implementation.

Metric	Target	Example
Tool Adoption Rate	% of teams using tools regularly.	85% adoption across all departments.
Conflict Resolution Time	Time taken to resolve competing priorities.	30% reduction in resolution time.
Employee Engagement	% increase in employee satisfaction.	20% improvement in annual engagement scores.
Decision Quality	% of decisions yielding desired outcomes.	90% success rate in project outcomes.

Call to Action: Build Organizational Resilience

Start Today:
- Schedule a pilot workshop for your leadership team.
- Identify one workflow to embed integration tools this month.

Measure and Adjust:
- Track tool adoption and resolve roadblocks swiftly.

Scale for Impact:
- Expand training programs and update workflows as tools prove their value.

By fostering a culture of meta-thinking, your organization can thrive in complexity, resolve competing priorities with precision, and navigate change with resilience.

7.11 CONCLUSION: THRIVING THROUGH INTEGRATION

This chapter has unveiled a dynamic, actionable roadmap for mastering complexity through the integration of diverse knowledge systems. Let's distill the essence of this approach, equip you with tools for self-assessment, and reflect on the transformative power of meta-thinking as a cornerstone for navigating multidimensional challenges.

7.11.1 Implementation Roadmap: Steps for Applying the Meta-Framework

The meta-framework is not just a theoretical construct; it's a practical guide to turning overwhelming complexity into structured clarity.

Phase	Key Actions	Outcome
1. Context Analysis	Define the problem's scope, stakeholders, and urgency.	A clear map of challenges, priorities, and gaps.
2. Framework Selection	Select and align the most appropriate primary and secondary frameworks.	Tools and strategies tailored to the problem's nature.
3. Stakeholder Mapping	Use the Integration Canvas to identify impacts and potential conflicts.	Stakeholder alignment and an inclusive decision-making process.
4. Conflict Navigation	Apply the Contradiction Navigator to reconcile competing priorities and values.	Solutions that balance ethical, technical, and practical needs.
5. Strategic Execution	Prioritize actions with the Adaptive Integration Matrix, leveraging urgency and complexity scales.	Seamless, timely, and high-impact implementation.
6. Continuous Feedback	Monitor progress, assess outcomes, and refine the approach with feedback loops.	Evolving solutions that adapt to new data and insights.

7.11.2 Self-Assessment Tools: Gauging Your Integration Success

Evaluate your ability to integrate frameworks and navigate complexity with these practical tools.

7.11.2.1 Integration Success Checklist

Ask yourself or your team these critical questions:

Dimension	Self-Assessment Questions	Indicators of Mastery
Clarity	Are goals clearly articulated and shared among stakeholders?	Unified understanding of goals and alignment of actions.
Framework Synergy	Were frameworks complementary, and how well did they integrate?	Minimal friction and synergy across frameworks.
Conflict Resolution	Were competing priorities balanced effectively?	Trade-offs resolved with fairness and innovation.
Stakeholder Inclusion	Was everyone's voice heard and valued?	High engagement and collective commitment.
Outcome Impact	Did the implementation yield tangible and measurable improvements?	Positive shifts in defined metrics and stakeholder satisfaction.

7.11.2.2 Reflective Prompts for Continuous Growth

1. Where did integration flow smoothly, and where did friction arise?
2. What frameworks proved most versatile across scenarios?
3. How effectively were stakeholder conflicts mediated?
4. How can these insights inform future decision-making processes?

7.11.3 Final Reflection: The Transformative Power of Meta-Thinking

Meta-thinking isn't just a methodology – it's a mindset. It redefines how we perceive, dissect, and solve problems. By operating at the meta-level, we gain the ability to:

- **Foresee Complexity**: Address challenges before they escalate.
- **Harmonize Contradictions**: Fuse ethics, logic, and practicality seamlessly.
- **Unlock Innovation**: Generate creative, sustainable solutions through cross-disciplinary collaboration.
- **Foster Adaptability**: Build frameworks that flex and evolve alongside real-world shifts.

In a world increasingly shaped by rapid technological change, systemic challenges, and unpredictable crises, meta-thinking equips us with the tools to thrive rather than merely survive. It is both a skillset and a mindset that future-proofs our ability to navigate uncertainty.

Call to Action: Embrace Your Meta-Level Potential

1. **Start Small**: Identify a current challenge and experiment with one integration tool this week.
2. **Cultivate the Habit**: Incorporate regular reflection and assessment into your decision-making.
3. **Expand the Circle**: Share these tools with your team or community to amplify impact.

Meta-thinking is not just about solving problems – it's about reimagining how we engage with complexity, creating clarity where others see chaos, and turning obstacles into opportunities. Let this mindset guide you as you shape not only your path but also the future.

THE BLUE DOT EFFECT

THINKING BOX:

"Have we honed in on every minor flaw so much that we miss the bigger picture? Shifting your perspective might involve gazing toward **42**."

Part 3

Well-Being and Resilience in Modern Society

TEASER: CULTIVATING STRENGTH AND MEANING IN A CHAOTIC AGE

Armed with **ethical** perspectives (Part 1) and **cognitive** strategies (Part 2), **Part 3** centers on **personal and communal well-being**. Complexity might be the norm, but that doesn't mean it has to undermine your emotional health or relationships. Here, we fuse advanced theories with **practical** approaches, revealing how personal resilience blossoms through synergy with the broader community.

WHY READ ON?

- **Turn Theory into Everyday Triumphs**: Complexity and bias management are more than abstract ideas; they become daily routines, mental exercises, and emotional check-ins that **uplift** your quality of life.
- **Embrace the Power of Community**: True resilience isn't a lone pursuit. This part highlights how communal connections – friends, family, support groups – form a **vital** bulwark against the storms of modern existence.
- **Integrate Love, Habit-Tracking, and Anti-fragility**: From the science and philosophy of love to baseline monitoring of mental states, you'll learn how a multi-layered approach fosters **growth**, even under stress or upheaval.

WHAT AWAITS YOU?

- **In-Depth Well-Being Methods**: Concrete techniques – like journaling, baseline monitoring, and stress reappraisal – help you handle daily chaos without losing yourself.
- **Emotional and Social Anchors**: Explore how empathy and emotional intelligence anchor your mental health, forging stronger bonds in an era that often promotes **digital isolation**.
- **Roadmap to Anti-fragility**: Move beyond just surviving adversity – acquire the mindset that uses setbacks as catalysts for a **stronger, wiser** you. By the end of Part 3, you'll not only feel more resilient individually, but also more equipped to contribute to collective well-being.

DOI: 10.1201/9781003640288-11

8 The Science and Philosophy of Love

Scientific Foundations:

- **Neurology of Love:** The **Triune Brain Model** explains love's instinctual (Reptilian Brain), emotional (Limbic System), and rational (Neocortex) dimensions. **Neuroplasticity** enables long-term bonding and adaptation.
- **Neurotransmitters and Hormones: Dopamine (pleasure), oxytocin (bonding), serotonin (obsession), and endorphins (contentment)** shape different stages of love.
- **Evolutionary Role:** Love fosters **mate selection, parental investment, and social cohesion** for survival.

Philosophical and Cultural Perspectives:

- **Historical shifts:** From **courtly love to modern romantic ideals**, love has evolved with societal norms.
- **Cultural variations:** Western cultures prioritize **romantic choice**, while Eastern traditions emphasize **familial approval**.
- **Greek Love Typology: Eros (passion), Philia (friendship), Storge (family), Agape (selfless love), Pragma (long-term love), Ludus (playful love), Philautia (self-love).**

Practical Applications:

- **Narrative Therapy:** Reshaping love stories fosters relationship resilience.
- **Exercises for Couples: Value Wheel, Maslow's Hierarchy of Needs, Journaling, Communication Tools** to align values and deepen connection.
- **Shadow Work:** Addressing personal insecurities enhances emotional intimacy.

By understanding **love's scientific and narrative dimensions**, couples can **co-author meaningful, enduring relationships based on shared values and mutual growth**.

8.1 INTRODUCTION

8.1.1 Purpose of the Chapter

Love is a timeless and multifaceted emotion that has captivated human beings for centuries. This text delves into the depths of love, exploring its neurological, biological, philosophical, and cultural dimensions. By understanding the scientific foundations and the narratives that shape our experiences of love, we aim to provide practical strategies for sustaining and nurturing enduring relationships.

8.1.2 Overview of Key Themes

In this comprehensive exploration, we will embark on a journey that combines the latest insights from neuroscience with profound philosophical perspectives, cultural narratives, and personal stories. We will also provide practical tools and exercises designed to help individuals and couples cultivate and maintain lasting love.

DOI: 10.1201/9781003640288-12

8.1.3 TL;DR Summary

This chapter offers a comprehensive journey through the complexities of love, covering:

1. **The Science of Love**: Understanding the neurological and biological foundations of love.
2. **Philosophical and Cultural Dimensions**: Exploring how historical and cultural narratives shape our perceptions of love.
3. **Personal Narratives and Psychological Insights**: Examining the impact of personal stories and psychological development on love.
4. **Practical Applications and Tools**: Providing actionable strategies and exercises for sustaining and nurturing love.

What to Expect

1. **The Science of Love**:
   - **Neurological Underpinnings**: Learn about the triune brain model and how different brain regions contribute to the experience of love. Understand neuroplasticity and how our relationships can shape our neural pathways.
   - **Biological Aspects**: Discover the roles of key neurotransmitters and hormones in the various stages of love, from initial attraction to deep emotional bonds. Explore genetic and evolutionary perspectives on love and its role in human development.
2. **Philosophical and Cultural Dimensions**:
   - **Historical and Cultural Narratives**: Trace the evolution of love through different cultures and historical periods, from medieval courtly love to modern romantic ideals.
   - **Impact of Media and Literature**: Analyze how media and literature influence our perceptions of love, setting benchmarks and shaping societal expectations.
   - **Sociocultural Influences and Variations**: Appreciate the diversity of love's expression across different cultures, understanding how cultural background shapes our experiences and expressions of love.
   - **The Greek Words for Love**: Explore the different Greek words for love and how they offer nuanced perspectives on various types of love.
3. **Personal Narratives and Psychological Insights**:
   - **Psychological Impact of Personal Narratives**: Reflect on how early life experiences and psychological development shape our approach to love and relationships.
   - **Narrative Therapy**: Discover how narrative therapy can help individuals and couples navigate and reshape their love stories, fostering healthier relationships.
   - **Diverse Personal Stories**: Highlight the diverse narratives of individuals from different backgrounds, illustrating how varied experiences influence our understanding of love.
4. **Practical Applications and Tools**:
   - **Strategies for Sustaining Love**: Emphasize the importance of communication, shared goals, and mutual understanding in building lasting relationships.
   - **Practical Exercises and Tools for Couples**: Offer guided journaling prompts, communication exercises, and activities to help couples explore and align their values, aspirations, and needs.

COUNTERING THE BLUE DOT EFFECT

THINKING BOX:

"How do we train ourselves to notice positives amidst the noise? Maybe we need a different vantage – somewhere near **42**."

8.2 THE SCIENCE OF LOVE

8.2.1 Neurological Underpinnings

8.2.1.1 The Triune Brain Model

Understanding love from a neurological perspective requires exploring the triune brain model (Figure 8.1), which divides the brain into three regions, each contributing uniquely to our experiences and actions.

1. **Reptilian Brain**:
   - **Role**: Governs instinctual behaviors and biological drives, such as attraction and sexual response.
   - **Function**: Ensures survival by managing basic functions like heart rate, breathing, and reproduction.
   - **Impact on Love**: This part of the brain is activated during the initial stages of attraction, driving the primal urge to seek out a mate and ensuring the continuation of the species.
2. **Limbic System**:
   - **Role**: The emotional center responsible for forming attachments and emotional memories.
   - **Function**: Regulates emotions like love, joy, and attachment, which are crucial for bonding with others.
   - **Impact on Love**: The Limbic System plays a significant role in the emotional bonding process, helping to create the feelings of love and attachment that keep couples together over time.
3. **Neocortex**:
   - **Role**: Oversees higher cognitive functions, including planning, language, and conscious thought.

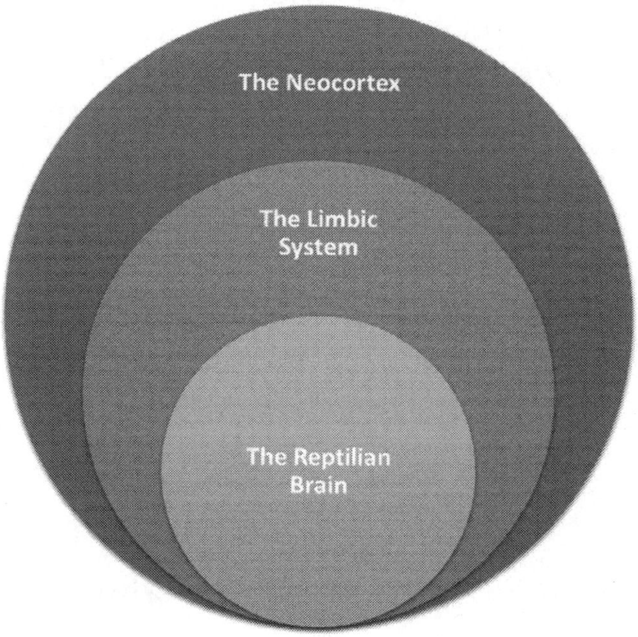

FIGURE 8.1 The Triune Brain model

- **Function**: Enables rational thinking, decision-making, and long-term commitments, playing a vital role in sustaining relationships.
- **Impact on Love**: The Neocortex is involved in the more thoughtful and strategic aspects of love, such as planning a future together, making compromises, and maintaining a long-term commitment.

8.2.1.2 Diagram: The Triune Brain Model

Disclaimer: The Triune Brain model, while a useful visualization of different cognitive functions, does not reflect the actual structure or evolutionary development of the brain; it is used here purely as an illustrative framework rather than a literal representation of neuroscience.

8.2.1.3 Neuroplasticity and Love

Neuroplasticity refers to the brain's ability to change and adapt in response to experiences. This capacity is essential for forming and sustaining loving relationships.

- **Positive Interactions**: Regular positive interactions with a loved one can strengthen neural pathways associated with empathy, understanding, and emotional connection.
- **Adaptation and Growth**: Our brains adapt and grow through shared experiences, challenges, and conflicts, highlighting the dynamic nature of love.
- **Resilience and Recovery**: Neuroplasticity also plays a role in how we recover from relationship challenges and heartbreaks, enabling us to heal and build new connections.

8.2.1.4 Example of Neuroplasticity in Relationships

Imagine a couple who regularly practice open communication and empathy. Over time, their brains become more adept at understanding each other's emotions and needs, leading to a stronger, more resilient relationship.

8.2.2 Biological Aspects

8.2.2.1 Role of Neurotransmitters and Hormones

Neurochemicals play a pivotal role in the various stages of love, from initial attraction to deep emotional bonds. Here's how some key neurotransmitters and hormones influence our experiences of love:

1. **Dopamine**:
   - **Role**: Fuels feelings of pleasure and euphoria.
   - **Function**: Activates the brain's reward circuits, making us feel happy and motivated when we are with a loved one.
   - **Impact on Love**: High levels of dopamine are associated with the intense pleasure and excitement of new love, often described as the "honeymoon phase."
2. **Serotonin**:
   - **Role**: Regulates mood and focus.
   - **Function**: Often decreases during early romantic attraction, contributing to obsessive thoughts about a partner.
   - **Impact on Love**: Lower levels of serotonin in the early stages of love can lead to preoccupation and idealization of the loved one, similar to symptoms of obsessive-compulsive disorder.
3. **Noradrenaline**:
   - **Role**: Heightens alertness and arousal.
   - **Function**: Increases heart rate and excitement, often experienced during the initial stages of romance.
   - **Impact on Love**: The rush of noradrenaline can cause physical symptoms like sweating and a racing heart, which are commonly associated with falling in love.

4. **Endorphins**:
   - **Role**: Contribute to feelings of comfort and contentment.
   - **Function**: Act as natural painkillers, promoting a sense of well-being and security in long-term relationships.
   - **Impact on Love**: Endorphins help to sustain long-term relationships by providing feelings of peace and stability, reducing stress, and enhancing emotional bonding.
5. **Oxytocin**:
   - **Role**: Promotes bonding and trust.
   - **Function**: Released during physical touch and intimacy, fostering deep emotional connections.
   - **Impact on Love**: Often referred to as the "cuddle hormone," oxytocin strengthens the bond between partners, especially during physical closeness like hugging, kissing, and sexual intimacy.
6. **Phenethylamine (PEA)**:
   - **Role**: Creates excitement and attraction.
   - **Function**: Induces a sense of exhilaration and passion, often present in the early stages of love.
   - **Impact on Love**: PEA is responsible for the euphoric feelings often experienced when falling in love, contributing to the "butterflies in the stomach" sensation.

8.2.2.2 Genetic and Evolutionary Perspectives

Love has evolved as a survival mechanism, ensuring the continuation of our species. Here are some genetic and evolutionary aspects of love:

1. **Mate Selection**:
   - **Genetic Diversity**: Choosing a genetically diverse mate increases the chances of healthy offspring.
   - **Symmetry and Attraction**: Evolutionary psychology suggests that certain physical traits, such as facial symmetry, are universally attractive because they signal genetic fitness.
   - **Parental Investment Theory**: This theory posits that the sex making the larger investment in offspring (typically females) will be more selective in mate choice, leading to more intense competition among males.
2. **Parental Investment**:
   - **Ensuring Offspring Survival**: Love and attachment encourage parents to invest time and resources in raising their children, increasing their chances of survival.
   - **Attachment Bonds**: Strong emotional bonds between parents and children promote social stability and cohesion, ensuring the transfer of knowledge and skills across generations.
3. **Social Cohesion**:
   - **Group Survival**: Love fosters social bonds and cooperation, which are crucial for group survival and success.
   - **Alloparenting**: The practice of individuals other than the biological parents providing care to offspring (such as grandparents or other relatives) enhances the survival of the young and strengthens social bonds within the group.

8.2.2.3 Example of Evolutionary Influence on Love

Consider the bond between a mother and her child. This strong attachment ensures that the mother will protect and nurture her child, increasing the child's chances of survival and continuing the genetic line.

8.3 THE EXPERIMENT

8.3.1 Introduction to the Experiments

In exploring the multifaceted nature of love and its sustainability, we introduce two distinct but complementary experiments. These are designed to delve into the dynamics of human connection – starting from the spark of initial interaction to the complexities of maintaining these connections over time.

8.3.2 Experiment 1a: Eye-Gazing

8.3.2.1 Framework

This experiment involves two participants, ideally strangers or non-romantic acquaintances, who engage in sustained eye contact for two minutes. The setting is controlled to minimize external distractions, focusing solely on the non-verbal communication facilitated through eye-gazing.

8.3.2.2 Observations

- **Initial Discomfort**: A common initial reaction is discomfort due to the intimate nature of sustained eye contact, challenging social norms and personal boundaries.
- **Shift to Connection**: Over time, we hypothesize a transformation in the participants' emotional state – from discomfort to a sense of understanding, empathy, or connection, illustrating the potential for initial awkwardness to evolve into meaningful interaction.

8.3.2.3 Hypothesis

The eye-gazing experiment symbolizes the journey of building and deepening connections that form the bedrock of love. It demonstrates how enduring love can grow from the willingness to engage with and transcend initial discomfort, fostering a deeper bond.

8.3.3 Experiment 1b: Prolonged Handshake

8.3.3.1 Framework

In this scenario, two participants engage in a handshake that extends well beyond the typical few seconds, lasting up to a minute or more. This experiment contrasts with Experiment 1a by focusing on prolonged physical contact rather than non-verbal visual communication.

8.3.3.2 Observations

- **Initial Comfort**: Unlike eye-gazing, a handshake is a socially accepted form of greeting, which makes participants initially comfortable.
- **Increasing Discomfort**: Over time, as the handshake continues beyond societal norms, participants begin to experience discomfort and awkwardness.
- **Social Norms and Intimacy**: This heightened discomfort underscores the role of social norms in governing our perceptions of intimacy and personal space, contrasting with the eye-gazing experiment where emotional barriers can be broken down with time.

8.3.3.3 Hypothesis

The prolonged handshake experiment highlights the boundaries of physical interaction in forming and maintaining connections. It serves to illustrate how certain forms of physical contact, when extended beyond societal norms, do not necessarily lead to a deeper sense of connection, offering a counterpoint to the eye-gazing experiment.

8.3.4 Recap

Together, Experiments 1a and 1b offer a nuanced exploration of the initial steps toward forming connections that could underpin love. By understanding these dynamics – how we navigate the discomfort of vulnerability and the boundaries of physical interaction – we gain insights into the complexities of sustaining love. These experiments encourage readers to reflect on their interactions and the conscious efforts required to nurture and maintain the bonds of love.

8.3.5 Additional Explanations

Understanding the intricacies of sustaining love requires delving into the neurological orchestration within our brains. The processes of forming, experiencing, and nurturing love involve a sophisticated interplay across three critical areas of the brain, each contributing uniquely to the dance of love:

1. **The Neocortex**: The forefront of logical thinking, this region oversees higher cognitive functions, including planning, language, and conscious thought. It's instrumental in the rational aspects of love, such as shared values, mutual understanding, and long-term commitments. The Neocortex allows us to reflect on our relationships, make decisions about our partners, and plan a future together, thereby playing a vital role in sustaining love.
2. **The Limbic System**: This emotional center is crucial for forming attachments and emotional memories. It is where feelings of love, joy, and emotional bonding flourish. The Limbic System's ability to regulate emotions is essential in navigating the ups and downs of relationships, ensuring that love can endure through challenges and conflicts.
3. **The Reptilian Brain**: Governing instinctive behaviors and biological drives, including attraction and sexual response, this region lays the foundation for the initial sparks of love. However, for love to last, the initial instincts and attractions must evolve into deeper emotional and intellectual connections facilitated by the other brain regions.

8.3.6 The Interplay and Communication Between Brain Regions

The journey of love begins with the instinctual and emotional reactions driven by the Reptilian Brain and the Limbic System. These reactions are vital for the initial attraction and emotional bonding. However, the development and maintenance of deep, enduring love engage the Neocortex, where thoughtful consideration, mutual respect, and shared life goals come into play.

- **Neuroplasticity's Role in Love**: The brain's capacity for change, or neuroplasticity, underscores the potential for relationships to shape our neural pathways. Positive interactions and deep connections can strengthen the circuits associated with empathy, understanding, and love, illustrating how sustained emotional and cognitive engagement contributes to the flourishing of love.
- **Conscious Effort in Sustaining Love**: Transitioning from initial attraction to lasting love requires more than just emotional engagement; it demands a cognitive and psychological investment. Through the Neocortex's involvement, individuals can navigate challenges, manage conflicts thoughtfully, and cultivate a deep understanding and appreciation for their partner. This conscious effort to regulate emotions, resolve conflicts, and share positive experiences is pivotal in sustaining love.
- **Learning and Growing Together**: The dynamic nature of love necessitates continual learning and adaptation. The Neocortex facilitates this aspect of love, enabling couples to incorporate new experiences, overcome obstacles, and evolve together. This capacity for adaptation and growth is essential for maintaining a vibrant, enduring relationship.

In essence, sustaining love is a complex process that encompasses our entire neurological being – from the instinctual desires rooted in our Reptilian Brain, through the emotional connections housed in our Limbic System, to the thoughtful considerations and decisions made in our Neocortex. This holistic engagement across all layers of our brain highlights the multifaceted nature of love, underscoring the importance of nurturing every aspect of our connections to keep love alive.

8.3.7 Brief Excursion: The Gut-Brain Axis

An intriguing aspect that merits attention in our exploration of love and connection is the gut-brain axis, a complex communication network linking the gastrointestinal system with the brain. This axis plays a significant role in influencing neurotransmitters, which are further discussed later in this text. The pathways between the brain and gut, primarily facilitated by the vagus nerve, represent a fascinating area of neurogastroenterology. This direct link means that over 90% of the information transmitted to the brain originates from the gut.

The gut-brain axis's impact on our emotional state and decision-making processes is profound. It suggests that our "gut feelings," so to speak, are not just metaphorical but have a physiological basis that can affect our mood, choices, and by extension, our interpersonal relationships. The neurotransmitters involved in this communication, such as serotonin and dopamine, play crucial roles in regulating mood and behavior, underscoring the interconnectedness of our digestive health with our emotional well-being.

However, delving deeply into the gut-brain axis's intricacies would complicate the simplicity of our experiments. Therefore, in the context of this text, it is regarded as an important "source of information" or a factor influencing decision-making rather than a primary focus. The acknowledgment of the gut-brain axis enriches our understanding of the biological foundations of love and connection, highlighting how our physical state can influence our emotional experiences and interpersonal dynamics.

This excursion into the gut-brain axis underlines the holistic nature of love, suggesting that sustaining deep, meaningful connections goes beyond mere emotional and cognitive engagement. It involves a complex interplay of physiological processes, including those governed by the often-overlooked communication between our gut and brain. Recognizing this connection invites us to consider how our overall health and well-being can impact our capacity to love and maintain lasting relationships.

8.4 PHILOSOPHICAL AND CULTURAL DIMENSIONS

8.4.1 Historical and Cultural Narratives

Love is not a static concept; it has evolved significantly across different cultures and historical periods. By examining the changing narratives of love, we gain insight into how societal transformations, technological advancements, and shifting cultural norms have influenced our understanding and expression of love.

8.4.2 Medieval Courtly Love

In medieval Europe, courtly love was characterized by chivalric romance, often involving knights and noblewomen. This form of love was idealized and unattainable, emphasizing virtue, honor, and a profound emotional connection.

8.4.3 Romanticism of the 18th Century

The Romantic era brought a shift toward personal emotions and individualism. Love was celebrated as a passionate, almost mystical experience, often accompanied by deep emotional suffering and artistic expression.

8.4.4 Modern Romantic Ideals

Contemporary notions of love are influenced by a blend of historical romanticism and modern values of equality and partnership. Today's love narratives often emphasize mutual respect, personal fulfillment, and the balance between independence and togetherness.

8.4.5 Cultural Variations

- **Western Cultures**: Typically emphasize romantic love, individual choice, and emotional connection.
- **Eastern Cultures**: Often value arranged marriages, familial approval, and the collective good over individual desires.
- **Indigenous Cultures**: May prioritize community bonds, spiritual connections, and the integration of love within broader social structures.

8.5 IMPACT OF MEDIA AND LITERATURE

Media and literature profoundly influence our perceptions of love, often setting benchmarks for what love is or should be. Analyzing these representations can help us critically assess and adopt more realistic and diverse love narratives.

8.5.1 The Role of Movies and Television

Films and TV shows often depict idealized versions of love, with dramatic conflicts and fairy-tale resolutions. While entertaining, these portrayals can create unrealistic expectations and pressures.

8.5.2 Literature's Influence

From Shakespearean sonnets to modern romance novels, literature has shaped our understanding of love by exploring its complexities, joys, and tragedies. Books provide a deeper, more nuanced exploration of love than visual media.

8.5.3 Social Media's Impact

Platforms like Instagram and Facebook present curated images of love, often showcasing only the highlights of relationships. This can lead to comparisons and feelings of inadequacy in real-life relationships.

8.5.4 Example: Comparing Media Influence

Consider how different media portray a first kiss. A Hollywood movie might depict it as a perfect, passionate moment, while a novel might explore the awkwardness and emotion behind it. Social media, on the other hand, might show a filtered, idealized version that omits the context.

8.6 SOCIOCULTURAL INFLUENCES AND VARIATIONS

Cultural background significantly impacts how we experience and express love. By exploring diverse cultural perspectives, we can appreciate the variability and richness of love's expression worldwide.

8.6.1 Family and Community Roles

In many cultures, love is not just an individual affair but a collective one involving families and communities. Understanding these dynamics can help us appreciate the different ways love is nurtured and sustained.

8.6.2 Rituals and Traditions

Various cultures have unique rituals and traditions surrounding love and marriage. These practices often reflect deeper values and beliefs about the nature of love and relationships.

8.6.3 Example: Cross-Cultural Love Practices

- **A Traditional Eastern European Couple**: In many Eastern European cultures, strong family ties and community involvement are pivotal. This couple may navigate their relationship with a significant emphasis on familial approval and support, which influences their love narrative.
- **A Rural and Urban Couple**: A couple where one partner grew up in a rural setting and the other in an urban environment may face different challenges and bring unique perspectives to their relationship. Understanding and integrating these varied experiences can strengthen their bond.

8.6.4 Impact on Modern Relationships

Globalization and cultural exchange have led to a blending of love practices and narratives. Understanding these influences can help us navigate modern relationships with greater empathy and awareness.

8.7 THE GREEK WORDS FOR LOVE

The ancient Greeks had several words to describe different types of love, offering a nuanced understanding of this complex emotion. By exploring these terms, we can better appreciate the diverse manifestations of love and their implications for relationships.

1. **Eros (ἔρως)**:
   - **Description**: Eros represents passionate, romantic love, often associated with physical attraction and desire.
   - **Implications**: While eros can be intense and exhilarating, it is often fleeting and primarily driven by hormonal changes and physical attraction.
2. **Philia (φιλία)**:
   - **Description**: Philia denotes deep friendship and affectionate love, characterized by mutual respect, shared values, and common interests.
   - **Implications**: Philia forms the foundation of lasting friendships and strong emotional bonds, crucial for long-term relationships.
3. **Storge (στοργή)**:
   - **Description**: Storge refers to familial love, the natural affection between parents and children, and other family members.
   - **Implications**: Storge emphasizes the importance of nurturing, protective love within families, creating a stable environment for growth and support.
4. **Agape (ἀγάπη)**:
   - **Description**: Agape is selfless, unconditional love, often associated with spiritual and altruistic feelings.

- **Implications**: Agape transcends personal desires, focusing on the well-being and happiness of others. It is considered the highest form of love, fostering compassion and empathy.

5. **Ludus (λῆδον)**:
   - **Description**: Ludus represents playful, flirtatious love, characterized by teasing, laughter, and fun.
   - **Implications**: Ludus can add excitement and spontaneity to relationships, but it often lacks depth and long-term commitment.
6. **Pragma (πράγμα)**:
   - **Description**: Pragma is pragmatic, enduring love that develops over time, rooted in commitment, understanding, and mutual goals.
   - **Implications**: Pragma is essential for long-lasting partnerships, emphasizing the importance of patience, compromise, and practical considerations.
7. **Philautia (φιλαυτία)**:
   - **Description**: Philautia refers to self-love, which can be healthy (self-respect) or unhealthy (narcissism).
   - **Implications**: Healthy philautia is crucial for self-esteem and well-being, enabling individuals to form balanced, loving relationships with others.

8.7.2 Conclusion on Greek Words for Love

Understanding these different types of love helps us recognize the multifaceted nature of our relationships. Each type plays a unique role in our lives, contributing to the rich tapestry of human connections. By acknowledging and nurturing these diverse forms of love, we can build more fulfilling and resilient relationships.

8.8 HOMO NARRANS: THE STORYTELLING HUMAN

Humans are inherently storytelling beings, often referred to as "Homo narrans." This concept underscores the importance of narratives in shaping our experiences and understanding of love. Unlike the fleeting influence of hormones, narratives are constructed and sustained in the Neocortex, which plays a crucial role in overriding the more transient hormonal effects to create enduring connections.

8.8.1 The Role of Narratives in Love

8.8.1.1 Constructing Personal Narratives

Our personal narratives about love are influenced by cultural, familial, and personal experiences. These stories shape how we perceive love, how we seek it out, and how we maintain it.

8.8.1.2 Example: Personal Love Narratives

Consider an individual who grew up witnessing a loving, supportive partnership between their parents. This narrative likely influences their expectations and behavior in their own romantic relationships, emphasizing values such as mutual respect, support, and long-term commitment.

8.8.1.3 Overcoming Hormonal Influences

While hormones like dopamine and oxytocin play a significant role in the initial stages of love, it is our narratives that sustain long-term relationships. The Neocortex allows us to reflect on our relationships, make conscious decisions, and navigate challenges, ensuring that love endures beyond the initial hormonal rush.

8.8.1.4 Experiment Insights: Eye-Gazing and Prolonged Handshake

The experiments conducted (eye-gazing and prolonged handshake) vividly illustrate how social norms and personal narratives impact our experiences of connection.

- **Eye-Gazing**: Initially violates personal and social boundaries, creating discomfort. Over time, as participants engage, they may begin to construct a new narrative of connection and empathy, overriding initial discomfort and building a deeper bond.
- **Prolonged Handshake**: Starts within the comfort zone of social norms but becomes uncomfortable as it continues. This discomfort challenges the narrative of personal space and social conduct, highlighting the boundaries within which our narratives operate.

8.8.1.5 Sustaining Love Through Narratives

By engaging with and consciously shaping our love narratives, we can sustain deep, meaningful relationships that go beyond the initial hormonal attraction. The Neocortex's role in reflection, adaptation, and growth is crucial in this process, allowing us to build enduring love stories that are rich, complex, and resilient.

8.8.1.6 Example: Narrative Therapy

Narrative therapy helps individuals and couples rewrite their love stories, fostering healthier and more fulfilling relationships. By changing the narrative, people can alter their emotional experiences and relationship dynamics, demonstrating the power of the Neocortex in sustaining love.

8.8.2 Conclusion on Narratives and Love

To experience long-lasting love, it is essential to seek a partner who shares similar or complementary narratives. These shared narratives, especially those that touch on inner values, play a crucial role in sustaining a relationship. Unlike daily behaviors, which can change and adapt, core narratives and values form the foundation of a strong, enduring partnership. By understanding and aligning these narratives, couples can navigate the complexities of love with greater clarity and resilience.

AI SINGULARITY

THINKING BOX:

"What if machines exceed our intellect faster than we anticipate? The implications – like **42** – could be both absurd and profound."

8.9 PERSONAL NARRATIVES AND PSYCHOLOGICAL INSIGHTS

8.9.1 Psychological Impact of Personal Narratives

Our personal narratives about love are shaped by our early life experiences, psychological development, and cultural contexts. These narratives significantly impact how we approach relationships, how we perceive love, and how we deal with challenges in our romantic lives.

8.9.2 Early Life Experiences

Our childhood and formative years play a crucial role in shaping our understanding of love. Witnessing the relationships of parents, caregivers, and other significant adults can influence our expectations and behaviors in our own relationships.

8.9.3 Example: Influence of Parental Relationships

If an individual grows up in a household where love is expressed through constant support and open communication, they are likely to seek similar traits in their romantic relationships. Conversely, those who witness conflict or emotional neglect might struggle with trust and intimacy.

8.9.4 Psychological Development

As we grow, our psychological development continues to influence our love narratives. Factors such as attachment styles, self-esteem, and personal coping mechanisms play a crucial role.

- **Attachment Styles**: Secure, anxious, and avoidant attachment styles can dictate how we form and maintain relationships. Understanding one's attachment style can provide insights into relationship behaviors and challenges.
- **Self-Esteem**: High self-esteem often correlates with healthier relationship dynamics, while low self-esteem can lead to dependency, jealousy, or conflict.
- **Coping Mechanisms**: How we deal with stress and emotional pain affects our relationships. Healthy coping strategies can enhance resilience, while unhealthy ones can create barriers.

8.10 NARRATIVE THERAPY AND ITS ROLE IN SHAPING LOVE

Narrative therapy is a therapeutic approach that helps individuals and couples understand and reshape their personal narratives. By rewriting these narratives, people can alter their emotional experiences and improve their relationship dynamics.

8.10.1 Principles of Narrative Therapy

- **Externalizing Problems**: Separating the person from their problems allows for a more objective view and reduces self-blame.
- **Re-authoring Stories**: Encouraging individuals to create new, empowering narratives that foster positive change and growth.
- **Highlighting Strengths**: Focusing on strengths and achievements helps build resilience and self-efficacy.

8.10.2 Example: Narrative Therapy in Action

A couple struggling with communication issues might benefit from narrative therapy by identifying and altering the stories they tell themselves about their relationship. By re-authoring these stories to emphasize understanding and collaboration, they can develop more effective communication strategies.

8.10.3 Benefits for Relationships

Narrative therapy can help couples:

- **Identify and Address Core Issues**: By understanding the underlying narratives, couples can address root causes rather than just symptoms.
- **Build Empathy and Understanding**: Sharing and understanding each other's narratives fosters empathy and deeper emotional connections.
- **Enhance Problem-Solving Skills**: Re-authoring stories can lead to more constructive approaches to resolving conflicts and challenges.

8.11 DIVERSE PERSONAL STORIES AND THEIR IMPLICATIONS

Highlighting diverse personal narratives from different backgrounds demonstrates how varied experiences shape our perceptions of love. These stories emphasize the importance of understanding and respecting different perspectives in building strong relationships.

8.11.1 Example: Cross-Cultural Narratives

- **A Traditional Eastern European Couple**: In many Eastern European cultures, strong family ties and community involvement are pivotal. This couple may navigate their relationship with a significant emphasis on familial approval and support, which influences their love narrative.
- **A Rural and Urban Couple**: A couple where one partner grew up in a rural setting and the other in an urban environment may face different challenges and bring unique perspectives to their relationship. Understanding and integrating these varied experiences can strengthen their bond.

8.11.2 Implications for Relationships

Understanding diverse personal stories can help couples:

- **Foster Inclusivity**: Embracing diverse narratives promotes inclusivity and empathy in relationships.
- **Adapt to Differences**: Recognizing and respecting differences in personal narratives can enhance adaptability and mutual support.
- **Strengthen Bonds**: Sharing and understanding each other's stories can deepen emotional connections and build stronger bonds.

8.12 PRACTICAL APPLICATIONS AND TOOLS

To make the text more actionable and impactful, this section provides practical tools and exercises based on the research and insights presented. These tools are designed to help individuals and couples identify and align their inner narratives, fostering deeper connections and sustaining long-lasting love.

8.13 STRATEGIES FOR SUSTAINING LOVE

Effective communication, shared goals, and mutual understanding are foundational for a lasting relationship. Here are some strategies focused on aligning inner narratives:

8.13.1 Identifying Core Narratives

- **Narrative Sharing Sessions**: Set aside time to share personal stories that shaped your understanding of love. Discuss how these stories influence your current relationship.
- **Value Exploration**: Identify and discuss your core values. Reflect on how these values shape your personal narratives about love.

8.13.1.1 Example: Life Story Exercise

Each partner takes turns sharing significant life events that influenced their view on love and relationships. This helps both partners understand each other's backgrounds and inner narratives.

8.13.2 Aligning Inner Narratives

- **Shared Narrative Creation**: Work together to create a shared narrative that reflects both partners' values and goals. This narrative should incorporate elements from each partner's individual story.
- **Vision Alignment**: Discuss and align your visions for the future, ensuring that both partners' aspirations and values are represented.

8.13.2.1 Example: Creating a Relationship Vision Board

Together, create a vision board that represents your collective dreams, goals, and values. Use images, words, and symbols that resonate with both partners' narratives.

8.13.3 Verifying Compatibility of Inner Values

- **Deep Values Discussion**: Engage in conversations about your deepest values and beliefs. Assess how these align and where there might be differences.
- **Scenario Planning**: Discuss hypothetical scenarios and how you would each respond based on your inner narratives and values. This helps reveal underlying beliefs and compatibilities.

8.13.3.1 Example: Values Alignment Exercise

Create a list of core values and discuss each one, sharing why it is important to you and how it shapes your narrative about love. Identify common values and explore any differences.

8.13.4 The Value Wheel

The value wheel is a visual tool that helps partners compare their inner narratives on various levels of their relationship (Figure 8.2). It categorizes values into different layers, from deep inner values to everyday behaviors.

- **Inner Values**: The core beliefs and principles that are fundamental to each partner.
- **Norms**: The social and cultural expectations that shape behavior.
- **Rules**: The agreed-upon guidelines for behavior within the relationship.
- **Daily Behaviors**: The routine actions and habits in everyday life.

8.13.4.1 Diagram: Value Wheel

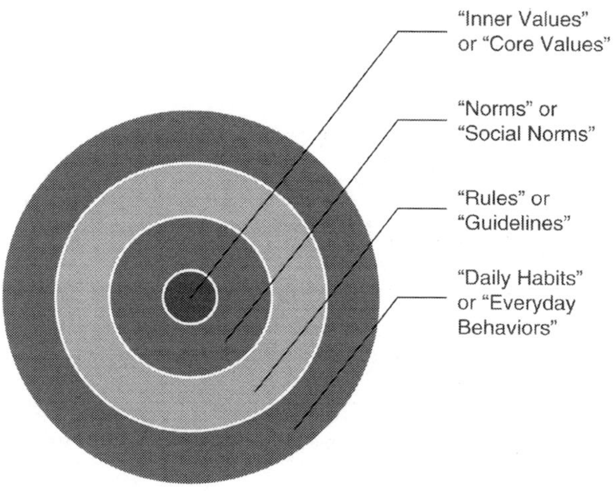

FIGURE 8.2 Value wheel

8.13.4.2 Example: Using the Value Wheel

Compare your values on each level of the wheel. Discuss any significant differences, especially in the core inner values, and find ways to align your narratives.

8.13.5 Maslow's Hierarchy of Needs

Maslow's hierarchy of needs is a psychological theory that prioritizes human needs from basic survival to self-actualization (Figure 8.3). Understanding where each partner is on this hierarchy can help align goals and expectations.

- **Physiological Needs**: Basic survival needs such as food, water, and shelter.
- **Safety Needs**: Security and stability.
- **Love and Belonging**: Emotional relationships and connections.
- **Esteem**: Respect, self-esteem, and recognition.
- **Self-Actualization**: Achieving one's full potential and creative fulfillment.

8.13.5.1 Diagram: Maslow's Hierarchy of Needs

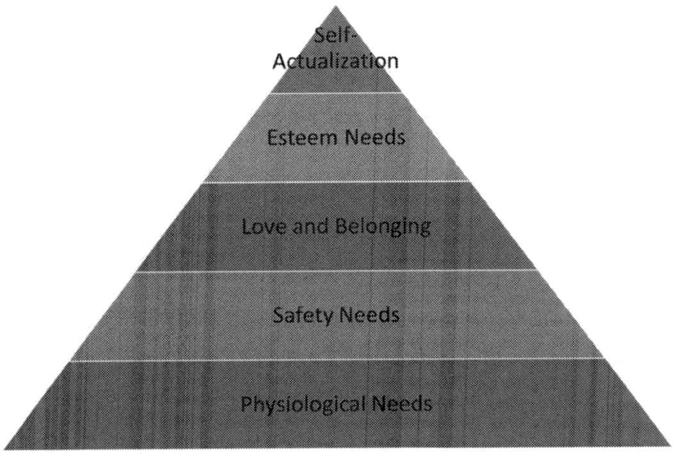

FIGURE 8.3 Maslow's hierarchy of needs

8.13.5.2 Example: Applying Maslow's Hierarchy

Identify where each partner is on the hierarchy. Discuss how to support each other in meeting these needs and how to work together toward higher levels of self-actualization.

8.14 PRACTICAL EXERCISES AND TOOLS FOR COUPLES

To further enhance the relationship, here are some practical exercises and tools focused on identifying and aligning inner narratives:

8.14.1 Guided Journaling Prompts

Encourage self-reflection on personal love narratives with these journaling prompts:

- **Reflect on Early Influences**: Write about your earliest memories of love and how they have shaped your current understanding of relationships.

- **Explore Core Values**: List your core values and reflect on how they influence your approach to love.
- **Envision the Future Together**: Describe your ideal relationship in the future. What qualities and values do you want to prioritize?

8.14.1.1 Example: Journaling Session

Set aside time each week for a journaling session where you both write and then share your reflections. This can enhance understanding and empathy.

8.14.2 Communication Exercises

Foster open and honest dialogue with these exercises:

- **Narrative Sharing Circles**: Take turns sharing stories from your past that illustrate your inner narratives about love. Listen actively and ask questions to deepen your understanding.
- **Mirroring Technique**: Repeat back what your partner says to ensure understanding. This technique helps prevent misunderstandings and demonstrates active listening.

8.14.2.1 Example: Weekly Communication Practice

Dedicate one evening each week to practice these communication exercises. Use this time to discuss any concerns or simply to connect on a deeper level.

8.14.3 Activities for Identifying and Refining Love Narratives

Engage in activities that help identify and refine your love narratives:

- **Relationship Timeline Creation**: Map out significant events in your relationship, noting both positive and challenging moments. Reflect on what you learned from each experience.
- **Storytelling Night**: Take turns telling stories about your relationship. Focus on moments that highlight your growth and connection.

8.14.3.1 Example: Relationship Review

Once a month, review your relationship timeline and discuss what you've learned from past experiences. Use these insights to strengthen your bond and plan for the future.

8.15 SHADOW WORK IN LOVE

Shadow work involves confronting and integrating the darker, often hidden aspects of our personality. By addressing these "shadow" parts, individuals can achieve greater self-awareness and authenticity, which are crucial for deep, meaningful relationships.

8.15.1 Principles of Shadow Work

- **Identifying Shadows**: Recognize and acknowledge the parts of yourself that you tend to hide or deny, such as fears, insecurities, or past traumas.
- **Acceptance and Integration**: Accept these shadow aspects as part of your whole self and work to integrate them into your conscious identity.
- **Healing and Growth**: Use the insights gained from shadow work to heal past wounds and foster personal growth.

8.15.2 Example: Shadow Work in Relationships

A couple may engage in shadow work by sharing their insecurities and past traumas with each other. This process can foster deeper understanding and empathy, as well as strengthen their emotional bond.

8.15.3 Benefits for Relationships

- **Deepened Intimacy**: By sharing and accepting each other's shadows, couples can build a stronger, more intimate connection.
- **Improved Communication**: Understanding your own and your partner's shadows can enhance communication, reducing misunderstandings and conflicts.
- **Resilience and Growth**: Integrating shadow work into your relationship can foster resilience and continuous growth, helping both partners to support each other through challenges.

8.15.4 Example: Shadow Work Exercise

Each partner writes a letter to their "shadow" self, acknowledging its presence and expressing acceptance. They then share these letters with each other, discussing how these shadow aspects have impacted their relationship and how they can support each other in integrating them.

8.16 CONCLUSION ON PRACTICAL APPLICATIONS AND TOOLS

By incorporating these practical tools and exercises into your relationship, you can foster a deeper connection, mutual understanding, and long-lasting love. These strategies emphasize the importance of identifying and aligning inner narratives and values, effective communication, and ongoing personal and relational growth.

8.17 FINAL THOUGHTS

Enduring love is a narrative co-authored by partners, continuously rewritten through shared experiences and growth. It requires conscious effort, mutual understanding, and adaptability. By engaging with love as a complex interplay of biology, narrative, and culture, we can actively participate in crafting our love stories, fostering connections that are not only enduring but also enriching and fulfilling.

8.17.1 Combining Key Insights

To achieve a fulfilling and long-lasting love, consider these important strategies derived from deeper self-reflection and understanding:

1. **Self-Discovery of Love Narratives**:
   - Reflect on the movies, books, and stories that resonate with you. What common themes and narratives about love do they share?
   - Analyze the relationships you have observed or experienced. Identify the narratives that shaped these relationships and how they influence your own understanding of love.
2. **Strategic Partnership Selection**:
   - Seek a partner whose love narrative aligns well with your own. While identical narratives are not necessary, a significant overlap in core values and long-term goals is crucial.
   - Use tools like the Value Wheel and Maslow's Hierarchy of Needs to assess compatibility and alignment on fundamental levels.
3. **Understanding and Aligning Narratives**:
   - Engage in open discussions with your partner about your individual and shared narratives. Identify areas of alignment and divergence.

- Develop a shared narrative that incorporates elements from both partners' stories, ensuring that it supports mutual growth and fulfillment.

4. **Avoiding Negative Narratives**:
   - Be mindful of the impact of negative narratives on your relationship. Just as positive stories can strengthen bonds, negative ones can undermine them.
   - Focus on creating and reinforcing positive narratives that emphasize growth, resilience, and mutual support.
5. **Practical Exercises and Ongoing Growth**:
   - Regularly revisit and refine your shared narrative through activities like relationship reviews, storytelling nights, and guided journaling.
   - Adapt and evolve your narrative as you grow together, ensuring that it continues to reflect your shared values and goals.
6. **Library of Wisdom (Bibliothek der Weisheiten)**:
   - **Analogy of Love and Knowledge**: In a library, you will find a few books on linear algebra, indicating a topic with generally agreed-upon principles. In contrast, the section on self-help and love contains numerous books with diverse perspectives, suggesting there is no single definitive answer to love. This diversity highlights the complexity and individuality of love narratives.
   - **Personalized Approach**: Just as in a library, where different books cater to various needs and interests, your love narrative should be tailored to your unique experiences and values. Embrace the diversity of perspectives and find what resonates with you and your partner.

8.17.2 Conclusion: The Essence of Lasting Love

To experience a fulfilling and long-lasting love, it is essential to move beyond biological explanations and delve into the narratives that shape our understanding of relationships. By consciously crafting and aligning these narratives with a partner who shares similar values and goals, we can build a strong foundation for enduring love.

As you embark on this journey, remember that love is not a static state but a dynamic process of continuous growth and adaptation. Embrace the power of storytelling, the wisdom of shared experiences, and the strength of aligned values to create a love that is not only lasting but also deeply fulfilling.

8.17.3 Additional Thoughts

- **Neutrality and Reality**: Understand that reality is often a subjective interpretation influenced by our narratives. Strive to maintain a neutral perspective when evaluating your relationship dynamics.
- **Impact of Cultural and Social Norms**: Recognize the influence of cultural and social norms on your narratives about love. While these norms can provide a framework, your personal and shared narratives should ultimately guide your relationship.
- **Continuous Learning and Adaptation**: Relationships require continuous learning and adaptation. Embrace changes and challenges as opportunities to deepen your connection and enhance your shared narrative.

By integrating these insights and strategies, you can navigate the complexities of love with greater clarity and resilience, fostering a partnership that is both enduring and deeply rewarding.

9 Baseline Monitoring for Human Resilience

Key Steps to Establish a Baseline:

1. **Weekend Recharge:** Observe natural sleep, energy, and mood patterns during a low-stress weekend to define your baseline.
2. **Tracking Key Metrics:** Monitor **sleep quality, energy levels, stress, emotional well-being, diet, and physical activity** using low-tech journaling or data-driven tools (apps, spreadsheets, wearables).
3. **Identifying Deviations:** Recognize **gradual declines, stress spikes, and behavioral shifts** as early warning signs requiring intervention.
4. **Making Targeted Adjustments:** Small tweaks – **adjusting bedtime, improving diet, or reducing screen time** – can restore balance before issues escalate.
5. **Long-Term Adaptation:** Life events, seasons, and environmental factors influence well-being; periodic baseline re-evaluation ensures sustained resilience.

By **embracing complexity as a guide rather than an obstacle**, baseline monitoring provides a **proactive, personalized approach to well-being**, fostering **adaptability, long-term resilience, and self-trust**.

9.1 INTRODUCTION: THE FOUNDATION OF PERSONAL WELL-BEING

Buzzing notifications, ever-changing demands, and subtle stressors lurking in your daily routine, it's easy to lose touch with what "feeling good" naturally looks like for you. Much like cybersecurity experts establish a baseline for normal network activity to detect intrusions, you can establish a baseline for your mind and body – a reference point of balanced well-being. This baseline is your personal measure of what it means to be rested, nourished, calm, focused, and emotionally steady under relatively stable conditions.

Why Baseline Monitoring Matters:

- **Early Warning System:** Just as unusual network traffic signals a potential hack, shifts from your health baseline – such as lingering fatigue, reduced focus, or mood swings – indicate that something is off. By noticing these signs early, you can make timely adjustments, preventing small issues from escalating into major burnout or illness.
- **Empowerment Through Awareness:** Instead of waiting passively for stress to overwhelm you, baseline monitoring puts you in control. You become the "admin" of your personal health network, actively seeking out and mitigating vulnerabilities before they cause harm.
- **Long-Term Resilience:** Over time, tracking your baseline builds self-understanding. You'll know how much sleep you truly need, which foods support steady energy, which activities restore calm, and how your environment affects your mood. This insight equips you with strategies to stay balanced, even when life throws curveballs.

  Real-Life Example (Sarah's Story): Consider Sarah, a 35-year-old project manager often feeling drained. By setting aside a weekend for quiet observation (as we'll detail in later sections), she discovered her baseline included feeling energized after a morning walk. Skipping it consistently led to lower energy and irritability by midday – an

DOI: 10.1201/9781003640288-13

"anomaly" compared to her balanced state. Armed with this knowledge, Sarah made her morning walk non-negotiable. The result? Improved mood, sustained productivity, and a daily routine that supports her natural equilibrium.

What This Chapter Will Offer: This chapter takes a practical, low-tech approach to personal baseline monitoring. You don't need fancy gadgets (though they can help if you're inclined), nor do you have to become a health fanatic. Instead, think of it as a guided process to understand your "normal" patterns so you can spot and address changes effectively. We'll show you how to:

i. **Identify Your Baseline Condition:** We'll propose a "Weekend Recharge" method – a calm, screen-light, restorative period to observe how you sleep, eat, move, socialize, and feel when free from usual pressures. By capturing these observations in a simple notebook, you create a reference point for what "feeling good" looks like, untainted by day-to-day stress.

ii. **Track Key Metrics in a Manageable Way:** We'll explore which aspects of life to monitor – sleep quality, energy levels, stress markers, emotional well-being, dietary habits, and more. You'll learn simple, approachable methods like journaling or basic habit trackers to gather useful data without turning this into an overwhelming chore.

iii. **Use Tools That Suit Your Style:**
   - **For Simplicity Seekers:** A pen and notepad are enough. Each morning and evening, jot down how you slept, how your body feels, and any noticeable shifts in mood.
   - **For the Nerds and Data Enthusiasts:** If you love numbers, consider wearable trackers for sleep and steps, simple spreadsheets for logging meals and moods, or even open-source software to analyze trends. You might set up a personal dashboard that visualizes your energy levels or mood scores over time, helping you pinpoint correlations between late-night screen time and morning sluggishness.

iv. **Interpreting Changes and Taking Action:** Once you know your baseline, the real power emerges: spotting deviations early. If your baseline suggests you normally feel refreshed after seven hours of sleep but suddenly seven hours leave you groggy, you'll treat this as a "signal" that something – maybe stress at work or an inadequate bedtime routine – is interfering. Instead of ignoring it, you can experiment with small adjustments: an earlier bedtime, a cup of herbal tea instead of a caffeinated drink late at night, or five minutes of stretching before bed.

The Cybersecurity Analogy in More Detail: In cybersecurity, continuous baseline monitoring catches unusual login attempts, traffic spikes, or suspicious file transfers. Similarly, continuous self-monitoring helps you catch unusual "behavior" in your health – disrupted sleep, persistent worries, or dwindling enthusiasm for activities you once enjoyed. Each deviation is like a red flag raised by your personal "security system," prompting you to investigate, understand root causes, and apply remedies.

Long-Term Vision: Building a Strong Resilience Framework: This isn't about perfection or turning your life into a spreadsheet. It's about cultivating mindful awareness. Over months or years, you'll refine your baseline, deepen your understanding of what supports balance, and become adept at swift course-corrections when life shifts. This resilience isn't just about dodging burnout; it's about growing stronger, more adaptable, and more aligned with your true needs.

How This Chapter Fits Into the Bigger Picture: We've discussed complexity in organizations, technologies, and societies. But that same complexity affects you personally – your health and happiness face constant influences from work, family, technology use, and cultural contexts. By applying baseline monitoring, you bring the power of complexity-based thinking down to the personal level, using it to maintain stability and thrive amidst change.

In essence, baseline monitoring personalizes the complexity management strategies we've explored. As you proceed through this chapter, you'll gain tools and approaches to understand your natural "operating state," catch anomalies early, and make targeted improvements that sustain your well-being over the long haul.

9.2 ESTABLISHING YOUR BASELINE: THE WEEKEND RECHARGE

Just as cybersecurity analysts pick a quiet period to map out normal network activity – when minimal updates or stressors occur – your best chance to establish a personal health baseline is during a relatively calm weekend. This is your opportunity to observe yourself in a low-pressure environment, free from the usual onslaught of notifications, deadlines, and high-stimulation media. By dedicating one or two days to gentle activities, minimal screen time, and mindfulness, you effectively "scan" your system when it's unencumbered, revealing what balanced well-being truly feels like.

Why a Weekend Works: Most people can carve out a Saturday or Sunday with fewer obligations. Think of these days as a maintenance window where IT staff would monitor a system with fewer users online. In your life, fewer external demands mean you can tune in to your natural rhythms – when you wake, how you move, what you crave to eat, and how your mood fluctuates without work pressures or social overload.

9.2.1 Planning Your Weekend

1. **Choose Restorative Activities:** Your goal is to identify what "healthy traffic" in your personal health network looks like. Pick activities that help you unwind and reconnect with yourself. Avoid high-adrenaline sports or binge-watching series; instead, choose pursuits that let you notice subtle changes in your body and mind.
   - **Nature Immersion:** A leisurely walk in a park, a brief hike, or even tending to a small garden. Like observing a network under normal load, time in nature offers stable, calming "data."
   - **Creative Pursuits:** Painting, sketching, writing in a journal, playing an instrument, or crafts. Creativity often puts you in a state of "flow," providing insight into how relaxed and focused you can be when not bombarded by tasks.
   - **Mindfulness Practices:** Yoga, meditation, or simple breathing exercises. These practices slow down your internal "traffic" enough to notice small variations – like tension in your shoulders or a restless mind – helping define what "no stress" or "normal tension" feels like.

   Avoiding Screen Time: Consider limiting social media, news, and work emails. If a cybersecurity expert wants to detect unusual data patterns, they first need a baseline without extra noise. Similarly, avoiding digital distractions reduces mental "white noise," making it easier to perceive your authentic baseline state.

   Highlighting Nature's Benefits (Example): Studies show that spending time outdoors can lower cortisol (stress hormone) levels and improve mood and concentration. Think of it as a quieting background that makes any emotional or physical "anomalies" stand out more clearly.

   Reflection Question: What activities make you genuinely feel more "like yourself"? Which pursuits help you sense subtle shifts in energy or mood?
2. **Allow Flexibility:** While having a loose plan for your weekend is helpful, give yourself permission to adjust based on how you feel. If you wake up and crave a short nap in the

afternoon – something you never do on weekdays – let it happen. These spontaneous preferences help you understand your true baseline. Are you someone who naturally needs a midday rest when unstressed? That's valuable information.

3. **Minimize External Stressors:** To get a clear "signal" of your baseline, reduce "network noise." Set boundaries with family or friends if needed ("I'm taking this weekend to recharge and won't be responding to work calls"). Create a comfortable environment – tidy your space, adjust lighting, maybe play soft instrumental music. Think of this as ensuring no heavy downloads or suspicious spikes occur during your observation period.

9.2.2 Incorporating Mindfulness and Stress-Reduction Techniques

To understand how calm and balanced you can feel at your baseline, integrate a few mindfulness exercises. They help you spot subtle mental fluctuations, much like running a diagnostic tool that reveals hidden processes in your system.

- **Mindfulness Exercise:** On your morning walk, focus on your senses: the sound of birds, the texture of the ground beneath your feet, the scent of fresh air. By honing in on these details, you notice how your mind and body respond to simple stimuli without external stress.
- **Breathing Technique (4-7-8 Method):**
  - i. Inhale quietly through your nose for four seconds.
  - ii. Hold your breath for seven seconds.
  - iii. Exhale forcefully through your mouth for eight seconds.

    Repeat this cycle three times. This practice can calm your nervous system and offer a reference point: "This is what feeling relaxed and stable feels like."

    Reflection Question: After a day of mindful walking, creative play, or gentle yoga, how does your mental "network" feel compared to a normal workday? Are you more at ease, less hurried, or more attentive to your body's cues?

9.2.3 Monitoring During the Weekend

Throughout the weekend, record simple observations about your well-being. Keep it low-pressure – a bullet journal, a note on your phone, or a single sheet of paper is enough. You're not logging every detail obsessively, just sketching a rough map of your normal physical, mental, and emotional states.

Actionable Worksheet (Enhanced Example):

Time	Activity/ Context	Physical State	Emotional/Mental State	Notes
8:00 AM	Woke naturally	Rested (8/10 energy)	Calm, optimistic	Woke before alarm, no grogginess
10:00 AM	Light yoga	Muscles relaxed	Focused, steady mood	Felt good stretch, no tension
2:00 PM	Reading novel	Slightly sleepy (6/10)	Content, engaged	Lost track of time, peaceful
6:00 PM	Short walk	Energized (8/10)	Happy, curious	Enjoyed nature sounds, birds chirping
9:00 PM	Relaxing music	Comfortable (7/10)	Tranquil, thoughtful	Thought about long-term goals calmly

If you're a data enthusiast, you could assign scores (like energy or mood from 1 to 10) for easy reference. If you prefer a holistic approach, descriptive words are enough. The key is consistency.

Low-tech vs. Nerdy Approaches:

- **Low-Tech:** A pen and notebook, one short note per morning, afternoon, and evening, rating energy and mood simply.
- **Nerdy/Data-Driven:** If you have a smartwatch or fitness tracker, log sleep hours, resting heart rate, or steps taken. For mental states, you might try mood-tracking apps that let you tag activities and see patterns over time. For the truly ambitious, a spreadsheet with daily entries and conditional formatting can highlight trends – e.g., cells turn red if energy dips below a certain threshold three days in a row.

Tip for Staying Motivated: Remember, this is not about perfection. If you miss logging one afternoon, that's fine. The purpose is to get a rough sense of your baseline. Treat any data you gather as helpful clues, not strict measurements that must be flawless.

9.2.3.1 Bringing It All Together

By the end of this weekend, you'll have a snapshot of "healthy normal." You'll know about what time your body naturally awakens, which meals leave you feeling best, whether afternoon breaks improve mood, and how certain activities (like reading quietly vs. attending a noisy event) influence your emotions.

Just as IT professionals feel more confident defending their networks once they know normal patterns of activity, you'll feel more empowered to handle daily challenges after understanding your normal well-being baseline. This knowledge sets the stage for detecting anomalies early – just like spotting suspicious network traffic – allowing timely adjustments that keep your mental and physical systems resilient.

In essence, the "Weekend Recharge" is your personal baseline mapping session: you're observing, not judging; learning, not forcing change. With this baseline, you enter the following weeks better equipped to notice when life's stresses push you off track – and better prepared to gently steer yourself back toward balance.

BALANCING AI REGULATION

THINKING BOX:

"How can we fuel innovation while guarding against harm? Striking that harmony might be as puzzling as **42** itself."

9.3 COMPREHENSIVE MONITORING FRAMEWORK: CHARTING YOUR PERSONAL "TRAFFIC PATTERNS"

Having spent a weekend establishing your baseline, the next step is to formalize a simple, steady monitoring practice. Think of this as setting up your personal "health dashboard," much like how IT professionals configure dashboards to track network performance metrics. This framework isn't about rigid rules or perfection – it's a flexible guide you can tailor to your preferences, schedule, and personal goals.

Why a Framework Matters: Complexity in our daily lives – shifting work priorities, evolving personal goals, and changing seasons – makes consistent self-awareness challenging. A monitoring framework ensures you're not just reacting emotionally to a bad day or guessing about what's wrong. Instead, you rely on a structured, regularly updated perspective that highlights deviations from your baseline. Over time, you'll see patterns emerge, just as network admins learn to distinguish normal fluctuations (e.g., a slight midday energy dip) from actual anomalies (e.g., persistent afternoon crashes in mood or motivation).

Key Principle: Track just enough information to notice patterns, without turning self-care into a stressful data-collection exercise. Start small and build up only if you find it helpful.

9.3.1 Physical Metrics: Understanding Your "Hardware"

Just as network hardware monitoring checks for unusual CPU loads or memory spikes, physical metrics help you spot bodily "stress signals" early.

1. **Sleep Quality:**
   What to Note:
   - Time to fall asleep (how long it takes you to drift off)
   - Nighttime awakenings
   - Morning restfulness (rate from 1 = not rested to 5 = fully rested)
   - Environmental factors (room temperature, noise, light)

   Real-Life Example: John logs his sleep for a week and notices that on nights he goes to bed at 10:30 PM, in a quiet, cool room, he rates morning restfulness as a 4 or 5. When he stays up past midnight or leaves the TV on, it drops to a 2 or 3. This pattern guides John to tweak his bedtime and room conditions.
2. **Energy Levels:**
   What to Note:
   - Daily fluctuations: morning, midday, evening
   - 1–10 scale for easy scoring (1 = exhausted, 10 = highly energetic)

   Nerdy Approach: Use a spreadsheet or a simple note-taking app and plot these numbers over time. After a few weeks, you might see that energy dips align with heavy carb lunches or late-night social media use. If you're into data, try a simple line graph or even a basic statistical correlation (does late bedtime correlate with lower morning energy?).
3. **Physical Discomfort or Pain:**
   What to Note:
   - Intensity and location of discomfort (e.g., mild lower-back ache after long hours at the desk)
   - Type of pain (dull, sharp, throbbing) and any triggers (e.g., certain workouts, poor posture)

   Adjustments Over Time: If a mild headache emerges every Wednesday afternoon, you might discover it's triggered by a recurring work meeting that runs long without breaks. With that insight, you can insert a short stretch or hydration break to prevent it.
4. **Nutrition and Hydration:**
   What to Note:
   - Meals and snacks (just a quick note: "oatmeal breakfast, chicken salad lunch")
   - Water intake (approximate glasses per day)

   For the Data Enthusiast: You can break down macronutrients (proteins, fats, carbs) using a simple food tracking app. Over time, see if certain dietary changes affect mood, sleep, or energy. Perhaps a protein-rich breakfast correlates with steadier morning energy.

   Day-to-Day Tool: A small notebook or a phone note where each morning you rate last night's sleep, midday you note energy, and evening you quickly log meals and overall physical comfort. Just two to three minutes total per day.

9.3.2 Mental Metrics: Mapping Your "Software State"

Where physical metrics reflect "hardware performance," mental metrics show how stable your "mental software" is. If you're suddenly experiencing mental "glitches" – prolonged stress or lack of focus – this could signal external "attacks" on your well-being, or internal factors that need attention.

1. **Stress Levels:**
   What to Note:
   - Rate perceived stress daily (1 = no stress, 10 = extremely stressed)
   - Note physical stress signals (tight jaw, headaches, racing thoughts)
   - Identify triggers (e.g., confrontational emails, long commutes)

   Adaptive Practice: Over time, if you notice that stress spikes correlate with certain tasks or times of day, try micro-interventions: a five-minute midday walk, a quick breathing exercise before an afternoon meeting, or decluttering your workspace.
2. **Mood and Positive Emotions:**
   What to Note:
   - Moments of joy, calm, or inspiration (what sparked it?)
   - Rate mood overall (1–10) or choose a few descriptive words (e.g., "content, hopeful")

   In-Depth Analysis for the Curious: If you track mood daily, you can later review and see patterns. Maybe Fridays are always sunnier after a certain hobby class. Knowing this, you might try introducing hobby-like activities earlier in the week to balance out stress.
3. **Focus and Cognitive Sharpness:**
   What to Note:
   - Concentration level during work or hobbies (rate from 1 = easily distracted to 10 = laser-focused)
   - Activities that boosted or drained focus (e.g., quiet reading increased focus, multitasking reduced it)

   Integration with Other Metrics: Cross-reference focus levels with sleep data or nutritional habits. Maybe you'll discover that on days you have a protein-rich breakfast and no late-night Netflix, your focus scores 8–9 out of 10 by midday.

9.3.3 Behavioral Metrics: Observing Your "Network Usage Patterns"

Behavioral patterns represent how you "move data" in your life – social interactions, screen time, and routines that affect mental traffic flow and resource allocation.

1. **Social Interactions:**
   What to Note:
   - Frequency, duration, and quality of interactions (positive, neutral, draining)

   Real-Life Example: If you notice you're irritable after too much socializing, you might schedule more solo downtime. Conversely, if isolation correlates with low mood, plan short coffee breaks with a friend midweek.
2. **Screen Time and Digital Overload:**
   What to Note:
   - Total daily screen time, with rough categories (work, social media, entertainment)

   Simple Adjustments: If too much late-night scrolling disrupts your baseline sleep pattern, try a screen curfew. Over a couple of weeks, see if returning to baseline restfulness becomes easier.

9.3.4 Environmental Metrics: Context Matters

In IT, network performance can depend on external factors – peak usage times, seasonal sales, global events. Similarly, your well-being baseline is influenced by environment – light, noise, air quality, and seasons.

1. **Light Exposure:** Note time spent outside or near windows during daylight. If you notice better mornings after even a brief sunlit walk, you'll appreciate how environment shifts affect mood and energy.
2. **Noise Levels:** Track whether you're in quiet or noisy settings. If noise correlates with irritability or headaches, you might adopt noise-canceling headphones or take breaks in quieter spaces.
3. **Air Quality and Comfort:** Even small adjustments, like opening a window for fresh air or using a humidifier in winter, can stabilize your baseline by preventing physical discomfort.

9.3.4.1 Integrating and Iterating: A Living Framework

Your framework isn't static – it evolves. As life changes (a new job, a newborn, a move to a different climate), your baseline may shift. Just as cybersecurity teams periodically reassess normal network traffic patterns in response to new devices or user behaviors, you should revisit and update your understanding of normal health markers.

Regular Check-Ins: Maybe once a month, review your notes. Have you noticed that shorter daylight hours in winter slightly lower your baseline energy? Or that a recent shift to remote work altered your social interaction patterns? These insights guide new strategies – like investing in a light therapy lamp or scheduling virtual catch-ups with colleagues.

Adjusting the Frequency and Detail of Tracking: If daily logging feels cumbersome, try weekly summaries. If you crave more detailed data for a particular issue (e.g., ongoing low-grade headaches), you can intensify monitoring for a short period. Remember, this is your tool – adapt it to fit your comfort level and goals.

9.3.4.2 Practical Tools and Approaches

For the Minimalist:

- A single-sheet paper on the fridge with morning/evening quick notes (e.g., "Today's sleep: Good (4/5), Mood: Steady, Focus: Decent").

For the Nerdy Data Fan:

- A Google Sheet or Excel file with columns for date, sleep hours, energy scores, mood words, steps count, and simple formulas to generate averages over the week.
- A note-taking app with tags, e.g., #energy_high, #mood_calm, #noisy_evening, allows you to search patterns later.

For the Experimenter:

- Try small "intervention weeks" – for instance, on one week, add a 10-minute meditation practice daily and compare results to baseline. Another week, replace your nighttime scrolling with reading a physical book and see if it improves morning energy.

9.3.4.3 Embrace Flexibility and Curiosity

This comprehensive monitoring framework is not about burdening yourself with chores; it's a map guiding you through the complexity of daily life. Think of yourself as a detective: you're collecting clues, not writing a rigid manual. Over time, patterns become clearer, correlations stand out, and you gain the power to make informed changes.

Motivational Note: As you gather data and reflect, remember that noticing anomalies early is a strength. It means you're attentive and proactive, turning complexity and change into allies rather than obstacles. With this framework, you're not just surviving life's twists – you're using them to grow, adapt, and refine your well-being strategy.

In short, the Comprehensive Monitoring Framework transforms raw observations into actionable insights. It helps you understand where you stand, detect subtle shifts in your personal "system," and maintain resilience no matter what uncertainties arise.

9.4 TRACKING AND ANALYZING DATA: TURNING OBSERVATIONS INTO INSIGHTS

You've established a baseline and begun monitoring various metrics – sleep, energy, mood, dietary habits, and more. Now it's time to make sense of that information. Just as cybersecurity experts don't just log data but also analyze it to detect anomalies and respond effectively, your notes, journals, or app-logged metrics need interpretation. Through regular review and pattern recognition, you can transform raw observations into actionable insights, guiding you to maintain or restore your well-being equilibrium.

Key Idea: Think of data tracking as continuous feedback loops. By periodically reviewing what you've recorded, you spot trends, confirm what's working, and identify emerging challenges before they become bigger issues.

9.4.1 Daily Tracking Tools: Simple, Consistent Logging

For most people, the simplest approach is best. Just as network admins glance daily at system dashboards, you can skim your notes or a quick app entry to keep your finger on the pulse of your well-being. The goal: small, easy habits that feel supportive, not burdensome.

Low-Tech Methods:

- **Mini Mood Journal:** Each morning and evening, jot down how you feel physically (rested, tense, energetic?), mentally (calm, anxious, focused?), and note any standout factors. A few sentences or bullet points suffice.
- **Checklists or Calendars:** If you prefer more structure, use a habit-tracking calendar. For example, mark a green dot on days you exercised, a blue dot on days you felt well-rested, a yellow dot if stress was moderate, and a red dot if stress was high. Over a few weeks, you can literally see patterns in color-coded form.

Nerdy/Data-Driven Methods:

- **Spreadsheets:** Create a table with columns for date, sleep hours, morning energy level, midday mood rating, step count, and any notes on significant events. Over time, you can sort or filter data, or use simple formulas to find averages or detect trends.
- **Apps and Devices:** If you love data, wearable fitness trackers or mood-tracking apps can automatically generate graphs. For example, a sleep tracker can show that on nights you get seven to eight hours of quality sleep, your next-day mood scores 8/10 on average. When sleep dips below six hours, mood drops to 5/10. Recognizing such correlations turns vague hunches into concrete insights.

Tip for Accessibility: If writing every day feels tedious, consider weekly summaries. Maybe on Sunday evenings, review your week's notes and highlight key observations: "Noticed feeling sluggish on Wed and Thu – coincided with late-night screen time." This approach still yields valuable insights without daily data entry.

9.4.2 Analyzing Trends: Identifying Patterns and Correlations

After a few weeks of logging data, patterns emerge. Like spotting persistent suspicious IP addresses in network logs, you might notice that tension headaches frequently appear after certain tasks, or that slightly earlier bedtimes boost morning productivity. The magic lies in connecting the dots and asking "why" certain variables move together.

What to Look For:

1. **Consistent Deviations from Baseline:** If your baseline showed steady afternoon energy, but three weeks in a row you've felt drained by 2 PM, investigate. Is it a heavier workload lately? A new dietary habit? Late-night streaming sessions?
2. **Correlations Between Variables:** Check if changes in one metric align with changes in another. Do stressful mornings correlate with reduced lunch appetite or poor evening sleep? Finding these links helps you understand cause and effect in your personal "system."
3. **Impact of External Events:** Just as a sudden surge in suspicious network traffic might coincide with a holiday sale or global event, consider if personal anomalies correlate with external factors. Did a recent project deadline coincide with mood dips or insomnia? Did family visits lift your spirits consistently?

Practical Examples:

- **Scenario 1 (Physical-Focused):** Reviewing a month of notes, Sam notices he consistently rates morning energy lower after heavy late dinners. By adjusting his evening meal composition – maybe lighter, earlier dinners – he tests whether energy improves. When it does, he's found a key lever: dietary timing affects energy next day.
- **Scenario 2 (Mental-Focused):** Over several weeks, Priya observes that on days she spends at least 20 minutes outdoors before starting work, her stress ratings at noon are lower. This suggests a direct correlation: morning nature exposure might "buffer" her against midday stress.
- **Scenario 3 (Behavioral-Environmental Link):** Tim tracks mood (1–10 scale), daily screen time, and bedtime. He realizes that nights he doom-scrolls social media late (high screen time), he logs poorer morning mood. Cutting back screen time after 9 PM might restore baseline mood levels.

For the Data Enthusiast: If you have enough data, you could try basic statistical tools:

- **Correlation Coefficients:** Check if higher sleep hours correlate with higher daily mood scores.
- **Rolling Averages:** Smooth out daily ups and downs by looking at weekly averages of energy or stress to identify gradual trends.
- **Conditional Formats/Filters:** Highlight days where mood drops below a certain threshold, then see what else occurred those days.

Remember: The goal is not to become a data scientist, unless you enjoy that. Even simple pattern recognition – "I notice I'm always feeling better on days after a proper breakfast" – is powerful.

9.4.3 Interpreting Changes as Early Warning Signals

A key principle borrowed from cybersecurity is that any steady deviation from known norms signals a potential "breach" in your health system. It's not always a crisis; it might be an early sign that you need minor adjustments.

- **Gradual Declines:** Maybe your baseline involved feeling content most mornings. Now, you observe a slow drift toward feeling neutral or even grumpy over two weeks. This slow shift can be a subtle alarm: perhaps new work demands are creeping in, or seasonal

changes are affecting your sleep. Acting early – maybe adding a short yoga session or rearranging your schedule – can prevent deeper burnout.
- **Sudden Spikes in Stress or Fatigue:** If energy or mood plunges abruptly over a few days with no obvious cause, consider this a high-priority alert. It might mean a new stressor you overlooked (like poor indoor air quality due to closed windows in winter) or a nutritional gap (maybe you skipped fruits and veggies lately). Respond by experimenting with quick interventions: opening a window for fresh air, adding a nutrient-rich snack, or trying a brief midday walk to restore normal levels.

Adaptive Response: Your data review sessions – weekly or monthly – become opportunities to refine your well-being "defense strategy." If a certain tweak consistently improves your scores, integrate it permanently. If a supposed improvement doesn't help, discard it and try something else. This continuous iteration mirrors how IT admins refine firewall rules or load-balancing strategies over time.

9.4.4 Making Decisions Based on Observations

Just as IT staff might add a new security protocol after noticing certain network vulnerabilities, you apply targeted interventions based on your observations. These decisions can range from minor lifestyle changes to deeper personal or professional shifts.

Examples of Decision-Making:

- **If Afternoon Energy Is Always Low:** Consider adjusting lunchtime meals (lighter, balanced options), scheduling a quick outdoor break, or experimenting with a short power nap – then review if the change restores your baseline energy pattern.
- **If Stress Levels Spike Mid-Week:** Maybe front-loading certain tasks earlier in the week can reduce mid-week crunch. Or integrate a Wednesday yoga class to break up the routine. Reassess if stress metrics improve, and if they do, maintain the new strategy.
- **If You're Consistently Happier After Time with Friends:** Plan more social engagements strategically placed to maintain stable mental well-being. This is like adding a beneficial redundancy to your system – ensuring a regular source of positive emotional "bandwidth."

The "Try-Observe-Adjust" Cycle: Think of each intervention as a mini-experiment. You try something new, observe its effect for a week or two, then adjust based on results. Over time, this cycle hones your routines, making your life more resilient and aligned with your baseline of well-being.

9.4.4.1 Embracing Complexity and Continuous Adaptation

It's natural to feel overwhelmed at first – so many metrics, so many potential correlations. But remember, complexity is an asset. As you track and analyze data, patterns emerge that guide more strategic decisions.

Staying Flexible: Your baseline and environment evolve. Perhaps after a year, you realize new job responsibilities alter your sleep needs or parenting duties shift your daily schedule. Revisit your baseline periodically and note if the "new normal" has changed. Adjust your benchmarks accordingly – just like a system admin updates baseline metrics as the network infrastructure grows.

Fostering Creativity and Self-Discovery: Beyond crisis prevention, analyzing your well-being data can inspire creativity. Maybe you discover an unusual link: playing the piano for ten minutes lifts your evening mood more than a standard TV break. Understanding these relationships lets you incorporate more enriching activities into your life, enriching your "health architecture."

In essence, tracking and analyzing your data is how you turn raw numbers and notes into meaningful insights. By regularly reviewing patterns, asking questions, and testing changes, you shift from merely reacting to problems into proactively shaping a healthier, more balanced future.

9.5 TROUBLESHOOTING CHALLENGES: OVERCOMING OBSTACLES AND MAINTAINING MOMENTUM

Even the most well-designed systems face hiccups. In cybersecurity, a network admin might run into unexpected bugs, changing threat landscapes, or resource constraints. Similarly, when monitoring your personal baseline, you'll occasionally encounter obstacles – time crunches, motivation slumps, confusing data, or life events that temporarily derail your routines. The key is to approach these challenges with flexibility and a problem-solving mindset rather than giving up or letting your progress slip away.

Core Perspective: Just as IT teams debug scripts, update firewall rules, or reconfigure servers in response to evolving conditions, you can adjust your baseline tracking approach. Think of "troubleshooting" as fine-tuning your personal monitoring system so it remains effective and supportive, not stressful.

9.5.1 Addressing Common Misconceptions

Myth 1: "I need to track everything perfectly."

Reality: Perfection isn't the goal. In cybersecurity, not every packet is analyzed in exhaustive detail; instead, admins sample data and focus on key indicators. The same applies to your well-being. If you miss a day of logging or forget to rate your energy once in a while, it's not a failure. Partial data still provides insights. Over time, patterns emerge even from imperfect logs.

Practical Tip: If daily tracking feels too rigid, scale back. Maybe log only in the morning and evening, or summarize your week on Sundays. Your health "firewall" still works fine with slightly less data.

Myth 2: "If I miss a day, I've failed."

Reality: Missing a day is normal. Just as a network admin might skip a routine check occasionally due to other priorities, you can afford small lapses. The real strength lies in consistency over the long term, not flawless execution every single day.

Practical Tip: Forgive occasional lapses immediately. Next day, just resume. Think of it like patching a minor vulnerability after a brief oversight – no drama needed.

9.5.2 Staying Motivated When Tracking Feels Tedious

Challenge: Over time, you might lose the initial excitement. Maybe at first tracking felt empowering, but now it's a chore.

Strategies:

1. **Set Micro-Goals:** In cybersecurity, admin teams celebrate small wins (like blocking a new malware variant). For you, celebrate small personal wins – e.g., "I tracked my mood consistently for five days" or "I identified that skipping breakfast affects my focus." Recognizing these achievements keeps you motivated.
2. **Vary Your Format:** If you're bored with text journaling, try a voice note or a quick video log. If the same routine feels stale, switch to a simpler scoring system. The goal is to keep the process engaging.

3. **Pair Tracking With a Reward:** After you log your day's metrics, treat yourself to something enjoyable – a piece of dark chocolate, a quick stretch, or a favorite song. Associating tracking with a positive experience can maintain momentum.

9.5.3 Handling Overload: When Data Feels Overwhelming

Symptom: You find yourself with too many notes, graphs, or metrics, unsure how to interpret them. It's like a cybersecurity analyst flooded with too many logs, struggling to isolate meaningful signals.

Solutions:

1. **Focus on Key Indicators:** If you've tried tracking ten metrics and feel swamped, scale back to the two to three most crucial ones. For instance, prioritize sleep, daily mood rating, and energy level. These core metrics often reflect changes in other areas indirectly.
2. **Periodic Summaries:** Instead of analyzing data daily, set a weekly or bi-weekly "review session." During these sessions, look for general trends rather than dissecting every minor fluctuation.
3. **Ask Guiding Questions:** Instead of staring at data blankly, ask questions to structure analysis:
   - "Have I felt more or less stressed this week than last?"
   - "Did the change in my bedtime improve my morning energy?"
   - "Which interventions seemed to restore my baseline faster?"

Nerdy Approach: If you love data analysis, consider simple charts or use conditional formatting in a spreadsheet to highlight values above or below certain thresholds. This visual cue helps you identify anomalies without getting lost in raw numbers.

9.5.4 Adapting When Life Changes Drastically

Your baseline and routines aren't static. Just as networks evolve – new servers, different traffic patterns – your life events (new job, moving to another city, starting a family) alter your context. The baseline you established might need recalibration.

Approach:

1. **Re-Establish a New Baseline:** After a major life change, repeat the "Weekend Recharge" process. A new job might mean different sleep/wake times or new stress triggers. A couple of low-pressure days to observe your new normal can reset your references.
2. **Accept Temporary Deviations:** If you know a short-term project will cause sleep deprivation or stress, don't panic over temporary anomalies. Just note them and plan to recover after the project ends. This is like adjusting firewall rules during a known event (like a big software update) and then returning to normal settings afterward.
3. **Incremental Changes:** Start small. If a new schedule forces you to wake up earlier, see if shifting your bedtime by 15 minutes helps restore morning alertness. Over time, find a stable pattern in the new context.

9.5.5 Seeking Support and Outsourcing Certain Tasks

If you repeatedly face roadblocks – maybe interpreting results is too confusing, or maintaining motivation is tough – consider external help:

- **Mentors or Accountability Partners:** Share insights with a friend or family member who supports your well-being journey. They might provide outside perspective on recurring issues, similar to a second IT specialist reviewing suspicious network logs for confirmation.

- **Professional Guidance:** If persistent sleep problems or mood swings occur, consider consulting a therapist, nutritionist, or doctor. Your baseline data can help them understand your patterns quickly. Just as cybersecurity experts might call in a specialist for complicated threats, medical professionals can offer expert solutions where your personal interventions fall short.

9.5.6 Ensuring Long-Term Sustainability

Challenge: You fear baseline tracking and adjustments might feel like a never-ending project.

Solution:

1. **Seasonal Reviews:** Every few months, reassess your metrics. If you've achieved stable energy and mood levels consistently, maybe you can track less frequently. If new challenges arise, intensify tracking temporarily. Think of it like periodic network audits – only ramping up monitoring when required.
2. **Rotate Focus Areas:** If you've stabilized sleep but still struggle with stress levels, shift your main focus to stress metrics for a while. This dynamic allocation of attention keeps the process fresh and aligned with current priorities.
3. **Emphasize Results, Not Process:** The end goal is a healthier, happier life, not a perfect tracking record. Keep reminding yourself why you started: to catch anomalies early, prevent burnout, and maintain balance. If you're seeing improvements, your method is working, even if not every log entry is perfect.

9.5.6.1 Putting It All Together: Troubleshooting as a Growth Opportunity

Troubleshooting challenges in baseline monitoring isn't just about fixing problems – it's about learning and adapting. Each obstacle you overcome refines your approach, making your personal well-being "system" more robust, flexible, and responsive. Whether it's scaling down the metrics you track, changing how often you review data, or seeking a friend's perspective, these adjustments ensure your baseline monitoring remains a valuable, long-term ally in managing life's complexities.

In essence, just as network admins continuously refine their systems to handle new threats and conditions, you'll keep fine-tuning your well-being monitoring process. This resilience keeps you agile, ensuring that when challenges arise, you know how to pivot, re-strategize, and keep moving toward sustained well-being.

ALAN WATTS' KEY INSIGHTS

THINKING BOX:

"Could 'letting go' be the true secret to thriving in a digital age? This notion resonates with the timeless mystery of **42**."

9.6 LONG-TERM IMPLEMENTATION AND ADAPTATION + INCORPORATING NATURE AND ENVIRONMENT

Establishing a baseline and refining your personal monitoring system is not a one-time project – it's an evolving practice. Just as cybersecurity teams periodically reassess their defensive strategies as new threats emerge and technologies evolve, you'll need to revisit and tweak your well-being approach as life circumstances shift. Over months or years, certain routines may lose their effectiveness, new stressors may appear, and your preferences for monitoring might change.

Additionally, one crucial but often overlooked element is your environment. Just as network performance can depend on external conditions – peak usage hours, seasonal sales, global events – your mental and physical health also interacts continuously with the natural world, your immediate surroundings, and cultural factors. By mindfully integrating nature and considering environmental influences, you can ensure that your baseline practices remain not only stable but also enriched by the world around you.

9.6.1 Long-Term Implementation and Adaptation

1. **Recognizing Evolving Needs:** Your life may not look the same next year as it does today. Maybe you move to a new city, change careers, become a parent, or pick up a new hobby. Each shift can alter your baseline, prompting a recalibration of what "normal" health patterns mean.

Adaptive Strategy:

- **Periodic Resets:** Every three to six months, consider doing a mini "weekend baseline scan" again. This is akin to updating firewall rules or running new diagnostics in cybersecurity. Ask: Have my sleep patterns changed? Are new stresses emerging at work or home? By re-checking your baseline occasionally, you stay aligned with current conditions.
- **Modular Tracking:** Over time, you might focus on different metrics. If you've mastered sleep regulation, you may shift attention to managing social energy or refining your exercise routines. Treat your metrics like modular components you can swap in and out depending on what matters most at the moment.

Nerdy Approach for Data Evolution: If you've been logging data in a spreadsheet for several months, consider plotting longer-term trends. Are you gradually improving in certain areas but plateauing in others? Using simple trend lines or rolling averages helps you see if interventions you introduced months ago still yield benefits or if it's time to try something new.

Practical Example:

- Suppose you initially focused heavily on identifying and resolving afternoon energy slumps. After six months, you've stabilized that issue by adjusting your lunch meal composition and adding a brief midday walk. You might now find your main challenge is emotional resilience during busy work seasons. Based on this, you stop micro-tracking energy levels (they're steady now) and start monitoring stress triggers more closely, introducing a short meditation break on stressful days to maintain balance.

Maintaining Motivation over the Long Haul:

- **Set Evolving Goals:** As conditions change, update your objectives. Initially, your goal might be to prevent afternoon fatigue. Later, it might be to improve creative focus or manage seasonal mood variations. Evolving goals keep the process fresh and purposeful.
- **Celebrate Milestones:** Just like IT teams celebrate successfully thwarted cyberattacks, reward yourself when you identify and fix a recurring wellness problem. Maybe treat yourself to a massage after maintaining steady mood for a month or celebrate by trying a new outdoor activity.

9.6.2 Incorporating Nature and Environment: Contextualizing Your Baseline

Your baseline and monitoring efforts don't exist in a vacuum. Environmental factors – light levels, noise pollution, greenery, weather patterns – shape your daily life. Engaging thoughtfully with your environment transforms complexity from a source of random influence into a supportive ally.

1. **Nature as a Stabilizer:** Research shows that spending time outdoors can reduce stress, boost concentration, and improve mood. Consider integrating nature into your routine so it becomes part of your baseline "normal." For example:
   - **Morning Sunlight for Better Sleep:** A quick morning walk in the sun helps regulate your circadian rhythm, improving sleep quality. Over time, your baseline might shift toward more stable energy and improved mood.
   - **Green Spaces for Emotional Reset:** If you notice daily mood dips, experiment with a short break outdoors. If the data shows a positive impact, you've found a leverage point in your environment.

Example: Carlos, who works from a home office, noticed persistent low-level irritability in his baseline measurements. Introducing a ten-minute break in a nearby park after lunch improved afternoon mood scores, revealing that environmental exposure to greenery acted like a calming "patch" in his system.

2. **Adjusting Indoor Spaces:** Just as cooling servers and optimizing infrastructure improves network performance, enhancing your indoor environment can help maintain your baseline:
   - **Lighting:** Replacing harsh overhead lighting with softer, more natural-toned bulbs, or setting up a workspace near a window, might improve focus and reduce eye strain. Over time, track if better lighting correlates with reduced afternoon headaches.
   - **Soundscapes:** Noise-canceling headphones, soft background music, or simply rearranging your workspace away from high-traffic areas can stabilize your environment. If your baseline analysis shows frequent tension when working near a noisy hallway, try soundproofing solutions. Re-check metrics to see if the tension dissipates.
3. **Seasonal and Cultural Considerations:** The environment also includes seasonal shifts and cultural contexts:
   - **Seasonal Shifts:** In winter, reduced daylight might lower your baseline energy or mood. Plan interventions – like a light therapy lamp or a short midday outdoor break – to counteract this. Over time, confirm whether these measures restore your baseline as seasons cycle.
   - **Cultural Rhythms and Events:** Holidays, festivals, and local traditions affect sleep patterns, social habits, and dietary changes. If you notice that certain cultural celebrations lead to a temporary spike in stress or fatigue, adapt your approach for that period – maybe schedule rest days after major events.

Tip for Aligning Environment with Baseline Goals: Experiment with small adjustments in your surroundings and log any observed changes. If rearranging your desk to face a window improves your morning mood scores, it's as if you've optimized network conditions to reduce latency – simple but effective.

9.6.2.1 Synthesis: Integrating Long-Term Adaptation with Environmental Awareness

By combining long-term adaptation with environmental context, you ensure that baseline monitoring remains both relevant and effective as life evolves. Each major life change, seasonal shift, or environmental tweak is an opportunity to refine your system. This responsive approach lets you navigate complexity gracefully.

Holistic View:

- Long-term adaptation ensures your baseline monitoring isn't static but grows with you as your needs, goals, and challenges change.
- Incorporating environmental factors acknowledges that your well-being is intertwined with the world around you – sunlight, fresh air, quiet spaces, cultural rhythms – and leverages these resources to maintain or enhance your equilibrium.

Practical Example: After a year of steady baseline tracking, Priya moves to a colder, darker climate. She notices a downturn in her baseline mood each winter morning. By adding a morning light therapy session and a plant-filled corner in her living room, she partially restores her previous mood levels. Over weeks, the data shows steady improvement, confirming her environmental adjustments worked.

9.6.2.2 Embrace Complexity, Evolve Continuously

The journey of baseline monitoring is never "done." Like a cybersecurity system that must continually adapt to new threats and technologies, your personal well-being strategy thrives on ongoing refinement. Each new season, job role, family addition, or environmental shift is not a setback but a prompt for curiosity, creativity, and growth.

Remember:

- **Flexibility is Strength:** Don't be discouraged by changes in your baseline; treat them as natural invitations to adjust and innovate.
- **Environment as Ally:** Leverage nature and ambient conditions to enrich your baseline stability rather than viewing them as external stressors.
- **Steady Evolution:** Over the long term, this approach keeps you resilient, able to handle complexity not as an overwhelming puzzle but as a landscape where you navigate skillfully with the right tools, strategies, and awareness.

In essence, long-term success in baseline monitoring is about staying responsive. By embracing ongoing adaptation and integrating the environmental context, you craft a dynamic, evolving system that supports your well-being journey through all seasons and shifts in life's complex terrain.

9.7 SUMMARIZE KEY TAKEAWAYS

1. **Establish a Calm Baseline:** Just as IT teams first observe normal network traffic to detect anomalies, your initial step is to understand your natural state of health and mood. Choose a low-stress weekend or a peaceful holiday window. By noting how you sleep, eat, move, and feel when relatively relaxed and unplugged, you create a reference point for what "good health" looks like.

   Practical Tip: Spend one quiet weekend without heavy digital use or major obligations. Log simple observations – sleep patterns, energy fluctuations, emotional notes – two or three times a day. This snapshot becomes your personal "normal traffic" record.
2. **Track Changes Over Time:** Once your baseline is set, consistent tracking helps you spot gradual changes. A simple notebook or a mobile note-taking app is enough for daily logs. If you're more data-inclined, a spreadsheet or wearable device can add detail. Over weeks and months, patterns emerge that guide adjustments.

   Nerdy Approach: Use conditional formatting in a spreadsheet to highlight days when mood or energy drop below a certain threshold, making deviations pop visually. If you prefer coding, a simple Python script could analyze correlations (e.g., using pandas library) between sleep hours and mood ratings.
3. **Detect Early Deviations:** When something feels off – persistently lower energy, creeping anxiety, or unusual irritability – treat it like a network anomaly. Investigate. Ask what might have changed: diet? Sleep? Social habits? Work stress? Small, strategic tweaks – adjusting bedtime, adding a short midday walk, limiting screen time at night – can restore your baseline before problems become severe.

Everyday Tools:

- A weekly reflection session to review logs: "On Wednesday, mood dipped after late-night work sessions. Try stopping work by 9 PM next week."
- Introducing a simple habit: a warm bath or herbal tea before bed, to see if it improves next-morning energy.

4. **Evolve Continuously:** Your baseline isn't static. As life transitions – new job, different seasons, altered family routines – your "normal" may shift. Periodically re-run your "weekend baseline check" or adapt your metrics if certain issues resolve while new concerns arise. Flexibility ensures this practice remains helpful, not a stagnant chore.

Example: If originally you focused on stabilizing sleep but now sleep is steady, you can scale back nightly logging and pay more attention to stress management techniques. If six months later you move to a colder climate, revisit your baseline to note how shorter daylight affects mood or energy and adjust accordingly.

9.7.1 Conclusion: A Lifelong Practice for Resilience

By treating your mind and body as a complex system and yourself as its diligent monitor – akin to a cybersecurity analyst maintaining a secure network – you gain control over your well-being. Instead of waiting passively for burnout or health issues to escalate, you identify subtle shifts early. This proactive stance transforms complexity from a source of overwhelm into a strategic ally.

Long-Term Vision:

- **Anti-fragility Over Fragility:** Anti-fragility means growing stronger under stress, not merely resisting it. Baseline monitoring fosters anti-fragility by continually teaching you about your personal "infrastructure." When a stressor appears – like a new project at work or seasonal affective dips – your established baseline and tracking methods help you adapt swiftly. Over time, these adaptations accumulate, making you more skilled at navigating changes.
- **Building Trust in Yourself:** The knowledge gained from monitoring builds self-confidence. You know that if energy dips, you have tools and insights to restore it. If mood wavers, you're not helpless; you have past data showing what worked before – whether it was a slight dietary tweak, a walk in nature, or a mindfulness session. This trust in your capacity to respond reduces anxiety about uncertainty.
- **Continuous Learning and Innovation:** Each anomaly you address and each environmental change you adapt to enriches your understanding. You become more creative in your solutions, more open to trying new strategies, and more aware that life's complexity is not an enemy. Just as networks constantly evolve, so can your methods of maintaining health. Perhaps down the road, you'll integrate biofeedback tools, or you'll collaborate with friends to share tips and experiences, similar to how IT experts exchange best practices.

In Daily Life:

- If a stressful month approaches, you pre-emptively introduce calming routines rather than waiting for stress to spike.
- If winter clouds dim your mood, you add light therapy or more midday outdoor breaks before the slump deepens.
- If a dietary experiment improves afternoon energy levels, you incorporate that permanently into your routine.

Over the years, these iterative improvements yield a lifestyle that's responsive, balanced, and robust against the complexities of the modern world.

Final Encouragement: Complexity in your health and surroundings can seem daunting, but with baseline monitoring, regular review, and adaptive changes, you transform unpredictability into an advantage. Like a secure, well-monitored network that thrives amid shifting traffic and emerging threats, you become vigilant, flexible, and prepared. This is the essence of resilience – engaging complexity not as a foe to be tamed but as a dynamic resource that, when understood and embraced, guides you toward lasting well-being and strength.

In essence, by adopting a baseline-based, data-informed, and environment-sensitive approach, you embark on a lifelong journey of resilience. Through steady refinement, humility, and creative adaptation, you ensure that complexity becomes your ally in achieving enduring physical and mental vitality.

10 Mental Toolkits for Thriving

Key Toolkits:

1. **Foundations of Well-Being – Stabilizing Your Internal State**
   - **Mindfulness Toolkit:** Micro-meditations, sensory awareness, and nature immersion for focus and calm.
   - **Emotional Regulation Toolkit:** Labeling emotions, progressive muscle relaxation, and journaling to process feelings.
   - **Stress Management Toolkit:** Prioritization, time-blocking, and setting boundaries to prevent burnout.
2. **Relationships and Communication – Building Stronger Connections**
   - **Communication Toolkit:** Active listening, "I" statements, and confirming understanding to avoid conflict.
   - **Healthy Relationships Toolkit:** Gratitude, boundary-setting, and trust-building rituals for deeper bonds.
3. **Growth and Achievement – Pushing Forward with Purpose**
   - **Goal Setting Toolkit:** SMART goals, progress tracking, and breaking tasks into manageable steps.
   - **Creativity Toolkit:** Brainstorm sprints, mind mapping, and lateral thinking for innovation.
   - **Adaptive Learning Toolkit:** Post-action reviews, feedback loops, and incremental experimentation.
4. **Resilience and Adaptation – Turning Challenges into Growth**
   - **Resilience Toolkit:** Positive reframing, resilience journaling, and small stress exposures.
   - **Adaptation Toolkit:** Regular strategy audits, scenario testing, and continuous learning.

Integration and Long-Term Growth:

By **mixing and adapting toolkits**, you **transform challenges into learning opportunities**, fostering **anti-fragility, self-efficacy, and lasting resilience**.

10.1 INTRODUCTION: TURNING INSIGHT INTO ACTION

Modern life can feel like navigating a shifting maze – just when you think you have a handle on your routine, a new challenge emerges. Perhaps deadlines pile up unexpectedly, relationships grow strained, or your motivation plummets without warning. It's easy to feel that complexity is winning, leaving you reactive, drained, and unsure of how to move forward.

But imagine if, instead of being caught off guard, you had a set of versatile strategies ready to deploy. When stress rises, you apply a technique to restore calm. When negative thoughts intrude, you know how to reframe them. When relational tensions spike, you have communication methods at hand to foster understanding. Instead of merely coping, you adapt and thrive, turning each challenge into an opportunity for growth.

From Awareness to Action: Previous chapters introduced the idea of establishing a personal baseline for well-being (so you can detect subtle changes early) and explored the concept of resilience evidences – tangible proof that you can handle adversity and grow stronger from it. This chapter picks up where awareness leaves off, guiding you from "I notice something's off" to "I know exactly what to do next." It provides an arsenal of mental toolkits – sets of practical skills and approaches – that help you transform complexity into a proving ground for resilience, creativity, and meaningful relationships.

DOI: 10.1201/9781003640288-14

10.1.1 Why Toolkits?

Think of complexity as a landscape of ever-changing conditions. A single static solution won't cut it. Instead, you need a repertoire of adaptable methods, much like having multiple apps on your phone. Some days you might need a stress management technique to handle a busy afternoon; on others, emotional regulation might be key after a tense family call. On yet another day, a communication strategy could smooth a team conflict. Over time, these tools help you build resilience evidences – small, repeated confirmations that you are capable, resourceful, and strong enough to navigate whatever comes your way.

How to Use This Chapter: Each section introduces a specific toolkit focused on a core area of well-being and growth, such as mindfulness, emotional regulation, stress management, communication, goal setting, creativity, resilience, and cognitive restructuring. You'll find:

- **Practical Techniques:** Simple, everyday approaches you can try immediately, from one-minute breath exercises to five-minute brainstorm sessions.
- **Structured Methods for Deeper Engagement:** For those who enjoy more data-driven or systematic approaches, we offer suggestions like rating your mood before and after a technique, or setting small, trackable goals to measure effectiveness.
- **Real-Life Examples and Step-by-Step Instructions:** Each toolkit includes scenarios that show how others have applied these methods to real challenges, demonstrating how you can adapt them to your own life.
- **Adaptability and Cultural Sensitivity:** We acknowledge that everyone's context is unique. If writing feels hard, consider voice notes; if cultural norms shape how you express emotions, adapt techniques to feel more natural. The goal is not rigid compliance but flexible customization.
- **Encouragement to Experiment and Integrate:** You don't need to master every toolkit at once. Start with one that resonates, test it for a week, and note any changes. Over time, mix and match toolkits to address complex challenges from multiple angles. This is how you build long-term resilience and anti-fragility – layer by layer, skill by skill.

Aligning with Life's Complexity: Our world is dynamic and uncertain – new technologies emerge, cultural shifts occur, personal circumstances evolve. These toolkits equip you to respond gracefully rather than be overwhelmed. By gradually applying these methods, you train your mind to handle complexity as a source of innovation and personal development. Each time you successfully navigate a hurdle – perhaps using mindfulness to calm pre-meeting nerves, or applying creativity to solve a tricky problem – you add another resilience evidence to your internal library of strengths.

Long-Term Benefits: Over time, these toolkits become less about crisis management and more about continuous improvement. You don't just return to baseline after a setback; you learn from it, refining your approach and updating your strategies. This process – observe, act, reflect, adapt – turns challenges into catalysts for growth, strengthening your self-efficacy and reinforcing the notion that you can thrive in complexity, not just endure it.

A Journey That Builds On Itself: While this chapter stands on its own as a guide to mental toolkits, it naturally complements previous insights. If you've monitored your baseline (Chapter 9) and grasped the concept of resilience evidences (Chapter 11), you'll recognize that these toolkits are how you translate awareness and philosophy into tangible skill development. But even if you haven't read those chapters, you'll find plenty of value here: these strategies work independently as well, helping you cultivate focus, manage stress, enhance communication, set goals, spark creativity, nurture resilience, and transform your mindset.

Final Encouragement: Embrace these toolkits as living resources. The world isn't static, and neither is your growth. As you try these methods, some will click immediately, while others may need tweaking or timing adjustments. Keep experimenting. Record your experiences. Celebrate small victories as resilience evidences that prove your capacity for growth. With practice, you'll find that life's complexities no longer feel like insurmountable burdens, but rather rich opportunities to build strength, adaptability, and lasting fulfillment.

Now, let's dive in and explore each toolkit in detail, discovering how to turn insight into action and complexity into a wellspring of personal thriving.

DOPAMINE INEQUALITY

THINKING BOX:

"Is the online world rewiring our brains for instant highs? The consequences – like the legend of **42** – may be more profound than we think."

10.2 FOUNDATIONS OF WELL-BEING

Overview: Before tackling complex interpersonal challenges or ambitious personal goals, it's wise to start with the basics – your internal equilibrium. This section focuses on three core toolkits that form the bedrock of mental well-being: **Mindfulness, Emotional Regulation**, and **Stress Management**. Think of them as the essential first-line responses that help you stay centered, respond thoughtfully rather than reflexively, and handle everyday pressures with greater ease.

By mastering these foundational skills, you give yourself a stable platform from which to engage more advanced toolkits later. Whether you're experiencing sudden mood dips, tension at work, or a pile-up of tasks, these foundational approaches help restore balance, guiding you back toward a steady mental "baseline" and laying the groundwork for future growth and resilience.

10.2.1 The Mindfulness Toolkit

Purpose: Mindfulness helps you stay present, turning down the volume on mental chatter and anxiety. When life's complexity feels overwhelming, mindfulness techniques allow you to pause, observe, and respond with clarity rather than reacting impulsively.

Everyday Techniques:

1. **Micro-Meditations (one to three minutes):**
   - Sit comfortably, close your eyes, and focus solely on your breathing. Each time your mind wanders, gently bring your attention back to the breath.
   - Practical Application: If you notice from daily reflections that your stress spikes before afternoon meetings, try a one-minute mindful breathing break just beforehand. Over a week, see if post-meeting irritability decreases.
2. **Mindful Pauses During Transitions:**
   - Before switching tasks – e.g., from answering emails to starting a presentation – take 30 seconds to breathe and notice your surroundings.
   - Practical Application: If your energy logs show afternoon slumps, introduce a brief sensory check (What do I see/hear/feel?) between tasks to maintain alertness.
3. **Nature Integration:**
   - If available, spend a couple of minutes each day noticing a tree outside your window or walking in a quiet park. The sensory immersion in nature often boosts relaxation and focus.
   - Practical Application: If nighttime restlessness appears in your logs, try a short evening walk to calm your mind before sleep.

For the Nerdy/Data-Driven:

- **Calmness Index:** Rate calmness (1–10) before and after a short mindfulness practice. Log these pairs for two weeks. Check if average calmness post-practice is higher, confirming that mindfulness sessions yield measurable benefits.
- **Frequency Tracking:** Record how many mindfulness breaks you take daily and correlate frequency with overall mood or stress scores. More frequent breaks might correlate with steadier mood lines in your personal data.

Cultural/Accessibility Considerations:

- If formal meditation feels foreign, adapt to culturally familiar practices, like focusing on a comforting prayer, mantra, or a well-loved piece of music.
- If sitting still is challenging, try mindful walking or lying down scanning body sensations.

Practical Tip: Set a daily alarm labeled "Just Breathe." Even a single daily reminder helps build the habit. Observe if weeks with consistent mindfulness reminders show fewer mood fluctuations.

10.2.2 The Emotional Regulation Toolkit

Purpose: Emotions drive our reactions. Emotional regulation doesn't suppress feelings; it helps you process them constructively. By learning to acknowledge and direct your emotions, you reduce impulsive outbursts and cultivate emotional resilience.

Everyday Techniques:

1. **Name and Acknowledge the Emotion:**
   - When feeling anger or sadness rising, silently say "I feel angry" or "I feel sad." Research shows naming emotions can reduce their intensity.
   - Practical Application: If you've logged recurring post-work irritability, try labeling the emotion next time it surfaces. Note if irritability duration shortens.
2. **Progressive Muscle Relaxation (PMR):**
   - Tense and relax muscle groups from toes upward. This calms the nervous system, making intense emotions more manageable.
   - Practical Application: If you note a spike in anxiety before virtual meetings, do PMR for two minutes beforehand. Track if anxiety ratings drop over successive meetings.
3. **Emotional Journaling:**
   - Write down a brief account of what triggered your emotions and how you responded. Consider alternative responses for next time.
   - Practical Application: If sadness often emerges after certain social interactions, journaling can help identify triggers and brainstorm healthier coping methods.

For the Nerdy/Data-Driven:

- **Emotion Intensity Charts:** Assign numeric intensities to emotions before and after applying a regulation technique. Over weeks, graph these changes. If the average intensity decreases, you have evidence that these methods work for you.
- **Trigger Maps:** Use a spreadsheet to log emotions and potential triggers. Highlight recurring patterns and see if emotional regulation techniques reduce intensity in similar future situations.

Cultural/Accessibility Considerations:

- Emotional expression norms vary widely. If direct confrontation feels uncomfortable, consider gentle expressions (like drawing emotions or using a feelings wheel to pick a label).
- If writing is tough, voice-record your emotional reflections briefly each evening.

Practical Tip: When a tough emotion arises next, pick one technique (like PMR) and note if the intensity rating is lower the next time that trigger occurs. This confirmation motivates continued practice.

10.2.3 The Stress Management Toolkit

Purpose: Stress is inevitable, but unmanaged stress saps your energy and resilience. Stress management techniques help you prioritize tasks, set boundaries, and introduce restorative breaks, ensuring stress doesn't accumulate unchecked.

Everyday Techniques:

1. **Morning Priority-Setting (The Top-3 List):**
   - Each morning, list three must-do items. Complete them first. This creates a sense of accomplishment and reduces the "I'm behind!" feeling later.
   - Practical Application: If afternoon stress spikes in your logs, see if completing key tasks early reduces afternoon stress readings.
2. **Time Blocking and Scheduled Breaks:**
   - Allocate focused work periods (e.g., 45 minutes) followed by a 5-minute break.
   - Practical Application: If your mood crashes after long uninterrupted work sessions, time blocking might prevent that slump. Check if shorter, more structured sessions correlate with steadier mood levels.
3. **Boundary Setting (Screen Curfews or Work Cutoffs):**
   - If you sleep poorly after late-night emails, set a "no-email-after-9-PM" rule. Note if morning irritability lessens over time.
   - Practical Application: Adjust boundaries as needed. If 9 PM doesn't help, try 8 PM next week, monitoring changes until you find a sweet spot.

For the Nerdy/Data-Driven:

- **Task and Stress Correlation:** Record daily tasks and stress ratings. After two weeks, introduce one stress management technique (like boundaries) and see if the average daily stress rating drops.
- **Work-Rest Ratios:** Track work minutes vs. break minutes. Experiment with ratios (e.g., 45-minute work:5-minute break vs. 25-minute work:5-minute break) to find what yields the best stress reduction scores.

Cultural/Accessibility Considerations:

- Cultural expectations about work hours vary. Start with a small boundary (like one email-free evening per week) and scale up.
- If physical or cognitive challenges make time management tough, consider using simple timers or visual schedules to externalize planning tasks.

Practical Tip: Implement one small boundary this week – such as no phone during lunch – and see if stress ratings improve on those days. If yes, you've found a simple, effective stress management tactic.

10.2.4 Integrating the Foundations

Each of these three foundational toolkits – Mindfulness, Emotional Regulation, and Stress Management – lays the groundwork for more advanced skills. By first stabilizing your internal environment, you'll find the relationship-focused and growth-oriented toolkits easier to implement. Over time, mixing these foundational techniques will feel natural. For example, if midday stress leads to emotional strain, you might first apply a mindfulness break to regain composure, then use emotional regulation journaling to process feelings, and finally adjust your afternoon schedule for less overload.

Adapting Over Time:

- Start with one technique from one toolkit and track its effect. Once comfortable, add another technique or introduce a different toolkit's approach.
- Periodically review what's working. If a certain breathing exercise consistently improves mood, keep it. If journaling shows minimal impact, try a different emotional regulation method like creative expression.

Long-Term Benefits: Over weeks and months, these foundational toolkits not only enhance day-to-day well-being but also generate "resilience evidences" – concrete instances where you handled stress, managed emotions, or maintained focus under pressure. These evidences build confidence in your capacity to adapt, forming a stable base from which you can expand into more advanced communication, relationship, growth, and resilience toolkits later in the chapter.

In summary, Section 10.1 focuses on stabilizing your internal equilibrium. By practicing mindfulness, improving emotional regulation, and managing stress effectively, you create a solid platform for facing complexity head-on. These foundational skills ensure that when life's challenges arise, you're equipped not just to cope, but to adapt, learn, and thrive.

10.3 RELATIONSHIPS AND COMMUNICATION

Overview: Human connections profoundly shape our resilience and capacity to thrive. Supportive relationships act as a buffer against stress, while healthy communication prevents misunderstandings from escalating into larger conflicts. This section introduces two toolkits **– The Communication Toolkit** and **The Healthy Relationships Toolkit –** designed to help you refine how you interact with others, cultivate trust, and build vibrant, supportive networks.

By learning to communicate openly, set boundaries, and show genuine appreciation, you transform relationships from sources of tension into pillars of stability. Over time, positive interactions become resilience evidences, proving that you can navigate complexity not only on your own but together with the people who matter to you.

10.3.1 The Communication Toolkit

Purpose: Good communication is the oil that keeps the gears of human interaction running smoothly. Whether at work, among friends, or within the family, effective communication reduces conflict, clarifies expectations, and strengthens mutual respect. When stress patterns or mood dips correlate with certain conversations (as noted from your baseline logs), applying these tools can shift the dynamic, turning once-draining interactions into constructive exchanges.

Everyday Techniques:

1. **Active Listening Drills:**
   - **Method:** During your next conversation, focus entirely on the speaker. Maintain eye contact, nod occasionally, and summarize their main points before responding.

- **Practical Application:** If your logs show that work meetings leave you frustrated, try active listening in the next one. Later, note if you feel less misunderstood or annoyed. Even a small improvement is a resilience evidence – proof that a communication tweak can enhance emotional outcomes.

2. **Assertive "I" Statements:**
   - **Method:** Replace accusatory language ("You never do X ….") with "I" statements ("I feel unsupported when ….").
   - **Practical Application:** If tension arises after household chores are discussed, say, "I feel stressed when we don't divide tasks clearly," instead of "You never help!" Track if fewer conflicts appear over the following weeks.
3. **Check for Understanding:**
   - **Method:** After someone speaks, ask, "Do I understand correctly that you mean …?" This clarifies points and prevents minor misunderstandings from snowballing.
   - **Practical Application:** If logs show lingering stress after certain friend calls, confirm their meaning when something seems ambiguous. See if stress ratings drop as clarity improves.

For the Nerdy/Data-Driven Approaches:

- **Positive vs. Negative Interaction Tally:**
  Assign a simple "+" or "–" to each significant conversation (e.g., after a meeting or a call). Introduce one communication technique (like active listening) for a week and compare the ratio of positive to negative interactions before and after. If positives rise, that's quantifiable evidence of success.
- **Feedback Surveys:**
  For work teams, consider a brief anonymous survey asking colleagues if they felt heard or valued during interactions. Check if scores improve as you implement assertive "I" statements or empathy-building methods.

Cultural/Accessibility Considerations:

- Different cultures have unique norms around directness, eye contact, or emotional expressiveness. Adjust techniques accordingly. If direct confrontation feels unnatural, focus on active listening or subtle affirmations.
- If verbal communication is challenging, consider written summaries or pictorial cues. Some apps transcribe speech to text or help translate multiple languages, ensuring inclusivity.

Practical Tip: Start small. Choose one communication skill – say, active listening – and apply it in a low-stakes conversation. If that interaction improves, try it in higher-stakes conversations. Document each success as a resilience evidence, showing that better communication reduces emotional tension and confusion.

10.3.2 The Healthy Relationships Toolkit

Purpose: Resilient individuals rarely stand alone. Strong, supportive relationships multiply your resilience and adaptive capacity. This toolkit helps you establish trust, maintain healthy boundaries, and regularly reinforce bonds with appreciation. Over time, you'll find that nurturing these connections can stabilize your well-being baseline, making life's ups and downs easier to handle.

Everyday Techniques:

1. **Regular Gratitude Expressions:**
   - **Method:** Once a week, send a short message or say something appreciative to a friend, partner, or colleague.

   - **Practical Application:** If logs show feelings of isolation or low mood, see if weeks with a gratitude practice correlate with improved emotional ratings. Each positive social exchange you facilitate becomes a resilience evidence that relationships can lift your mood.
2. **Setting and Respecting Boundaries:**
   - **Method:** If certain interactions leave you feeling drained (e.g., a friend who calls late at night), communicate a gentle boundary: "I'm not available after 9 PM, but I'd love to chat earlier in the evening."
   - **Practical Application:** Track your mood after enforcing a boundary. If emotional exhaustion decreases over time, you've proven that boundary-setting enhances relationship quality and personal well-being.
3. **Trust-Building Rituals:**
   - **Method:** Schedule periodic check-ins with key people (e.g., a weekly coffee break with a close friend or a monthly video call with a distant relative). Consistent, positive interactions reinforce trust and reliability.
   - **Practical Application:** If relational stress often spikes before big life events (e.g., holidays), regular check-ins can preempt tension. Note if smoother holiday interactions appear in your records after consistent small gestures of connection leading up to the event.

For the Nerdy/Data-Driven Approaches:

- **Relationship Satisfaction Ratings:** Every two weeks, rate your satisfaction with key relationships (1–10). Introduce one relationship-building method (like gratitude notes or monthly check-ins) and see if average satisfaction ratings improve over a month or two.
- **Interaction Quality Logs:** Keep brief notes on how you feel post-interaction. If applying boundaries or trust-building measures leads to more neutral or positive post-interaction emotions, you've found a successful tweak.

Cultural/Accessibility Considerations:

- Some cultures prioritize indirect communication. If expressing gratitude openly feels unusual, try subtle ways – small helpful acts or sharing a meaningful article.
- If mobility or time constraints limit face-to-face interactions, consider virtual alternatives – video calls, voice messages, or online interest groups can cultivate closeness without physical presence.

Practical Tip: Test a simple relational enhancement – like a weekly thank-you message to a friend – and note any shift in your emotional baseline or stress readings. If results are positive, layer more techniques. Over time, you establish a network of supportive bonds that make life's challenges more manageable and even collaborative.

10.3.3 Integrating the Relationship and Communication Toolkits

These two toolkits naturally reinforce each other: clear communication paves the way for healthier relationships, and supportive relationships make communication easier and more fulfilling. For example, if you struggle with a misunderstanding at work:

1. Use the Communication Toolkit: Practice active listening in the next meeting.
2. Apply Relationship-Building Strategies: Follow up with a gratitude note to a helpful colleague who clarified a point.

Track if subsequent meetings feel less tense or if you feel more supported at work. Improved relational dynamics become resilience evidences – tangible proof that you can influence and improve your social environment.

Adapting Over Time:

- If one technique feels off (e.g., direct gratitude messages are awkward), tweak it. Maybe show appreciation through small helpful acts or a supportive text instead.
- If you find communication improvements stall, try combining emotional regulation (from Section 10.1) before tough conversations, or add creativity exercises (from later sections) to brainstorm new ways to connect.

Long-Term Benefits: Over months, you'll likely notice more consistent emotional support, fewer draining conflicts, and a stronger sense of belonging. Each positive shift in relational quality is not just a fleeting good moment – it's a resilience evidence reflecting your ability to cultivate healthier networks. As these toolkits help you navigate relationships skillfully, you build a social safety net that buoys you during life's storms, reinforcing confidence in your capacity to handle complex interpersonal and emotional landscapes.

In essence, Section 10.2 transforms your social world from a source of unpredictable tension into a landscape where trust, empathy, and clear communication guide you toward stability and growth. By applying these communication and relationship-building tools, you make your connections more resilient – and by extension, you become more resilient yourself.

10.4 GROWTH AND ACHIEVEMENT

Overview: While the first two sections focused on staying centered and building strong, supportive relationships, this section shifts attention to moving forward – intentionally shaping your future through purposeful goals, innovation, and continuous learning. When life's complexity threatens to stunt your growth or undermine ambition, these toolkits provide the means not only to cope, but to thrive – harnessing complexity as a driving force for personal evolution.

The three toolkits – **Goal Setting**, **Creativity**, and **Adaptive Learning** – empower you to set meaningful directions, generate fresh solutions, and refine your approach over time. Together, they help transform inertia or confusion into forward momentum, ensuring that success doesn't remain a distant dream but becomes a structured, measurable, and ever-improving reality.

10.4.1 The Goal Setting Toolkit

Purpose: Goals convert vague aspirations into tangible targets, providing a roadmap through complexity. Instead of drifting or feeling overwhelmed, you gain direction and a sense of agency. Achieving small goals repeatedly creates resilience evidences, reinforcing your belief in your capacity to adapt and excel.

Everyday Techniques:

1. **SMART Goals (Specific, Measurable, Achievable, Relevant, Time-bound):**
   - **Method:** Define your objective clearly (e.g., "Write 500 words daily" instead of "Improve writing").
   - **Practical Application:** If baseline data shows feelings of aimlessness or low motivation, set a single SMART goal this week. Track if your mood or energy improves after hitting these small milestones.
2. **Short-Term Action Steps:**
   - **Method:** Break large goals into weekly or even daily steps.
   - **Practical Application:** If you aim to learn a new language, start with ten minutes of vocab study each morning. Note if consistent small steps yield a steady sense of progress, uplifting your daily stress or mood scores.

3. **Progress Visualization:**
    - **Method:** Use a simple chart or a habit-tracking app to mark completed steps.
    - **Practical Application:** After a month, review your chart. If hitting goals correlates with improved well-being metrics (like less stress), you have concrete evidence that structured ambition strengthens resilience.

For the Nerdy/Data-Driven Approaches:

- **Goal Completion Rate:** Document how often you meet daily or weekly targets. Compare these rates with emotional or productivity ratings in your personal logs. Are higher completion weeks associated with better mood or less evening fatigue?

Cultural/Accessibility Considerations:

- If certain goals clash with cultural norms (e.g., seeking assertiveness in a culture that values modesty), adapt the goal's expression. Maybe focus on polite assertiveness or seeking mentors within those norms.
- If reading or writing is challenging, use simple voice notes or pictorial checklists to plan and track your steps.

Practical Tip: Start ultra-small – like doing five push-ups daily – and measure if completing this micro-goal improves your self-confidence. Success in micro-goals often cascades into tackling bigger dreams.

10.4.2 The Creativity Toolkit

Purpose: When routines feel stale or traditional solutions fail, creativity breaks you free. It encourages viewing obstacles as puzzles, inviting experimentation and innovation. Each creative solution you devise – and confirm as effective – becomes a resilience evidence proving you can navigate uncertainty with ingenuity.

Everyday Techniques:

1. **Brainstorm Sprints (five minutes):**
    - **Method:** Set a timer and generate as many ideas as possible for a current challenge without judging them.
    - **Practical Application:** If logs indicate recurring blocks in problem-solving at work, a five-minute idea sprint each morning could freshen your perspective. Check if problems get resolved faster or more smoothly afterward.
2. **Mind Mapping for Clarity:**
    - **Method:** Draw a central issue in the center of a page, branching out related concepts, potential solutions, and resources.
    - **Practical Application:** If emotional strain arises in your logs when facing complex decisions (like a career shift), mind mapping may reveal hidden paths, reducing decision-related stress over time.
3. **Lateral Thinking Exercises:**
    - **Method:** Challenge assumptions. Ask: "What if we did the opposite?" or "How would a child solve this?"
    - **Practical Application:** If stress spikes before project deadlines, try lateral thinking to discover shortcuts or delegate tasks differently. Note if deadline stress lessens after applying these methods.

For the Nerdy/Data-Driven Approaches:

- **Idea Count vs. Problem Resolution Time:** Track how many ideas you generate weekly and how long it takes to solve a recurring problem. If generating more ideas correlates with shorter resolution times, creativity proves its value quantitatively.

Cultural/Accessibility Considerations:

- Different cultures have varied norms about "out-of-the-box" ideas. If bold creative efforts feel uncomfortable, start with mild adaptations – slightly altering an existing method rather than a complete overhaul.
- If writing or drawing is challenging, consider voice-based brainstorming apps or digital mind mapping tools with built-in accessibility features.

Practical Tip: In moments of stuckness, say "I'll find three alternative approaches in the next 3 minutes." Tracking these creativity bursts over months may show that creativity sessions consistently lower frustration ratings, reinforcing their role as resilience evidences.

10.4.3 The Adaptive Learning Toolkit (New Tool Idea)

Purpose: Adaptive learning emphasizes continuous improvement – reflecting on outcomes, seeking feedback, and adjusting strategies. It ensures that you don't just set goals or solve problems once, but learn from each attempt, evolving to handle increasingly complex situations with greater ease.

Everyday Techniques:

1. **Post-Action Reviews:**
   - **Method:** After completing a challenging task, pause and ask:
     - What worked well?
     - What didn't go as planned?
     - What will I do differently next time?
   - **Practical Application:** If certain tasks consistently raise stress, apply a short post-action review. Note if stress gradually decreases as you refine your approach.
2. **Feedback Loops:**
   - **Method:** Seek input from mentors, peers, or even family members about your performance or approach.
   - **Practical Application:** If baseline observations show recurring difficulties in time management, ask a colleague for suggestions. Implement one and track if the next week's stress ratings improve.
3. **Incremental Experiments:**
   - **Method:** Try small tweaks – like shifting your focus blocks from 60 to 45 minutes – and note the impact on productivity or mood. Keep iterating until you find the best fit.
   - **Practical Application:** If morning energy is low, experiment with a different breakfast and measure subsequent energy levels. Once you find a combination that yields improvement, you've validated a resilience evidence through adaptation.

For the Nerdy/Data-Driven Approaches:

- **Before/After Comparison:** For any given method, log your baseline metrics (stress, mood, efficiency) for a week, then apply a new strategy (like changing break intervals) for another week. Compare the two data sets to confirm if adaptation improved outcomes.

- **Scenario Testing:** Create hypothetical "what-if" scenarios (e.g., "What if I take a break at 10 AM instead of noon?") and try them out, using logs to evaluate changes systematically.

Cultural/Accessibility Considerations:

- Adaptive learning embraces diversity. Tailor methods to cultural or personal constraints. If seeking external feedback feels uncomfortable, start with self-reflection or online communities where anonymity is easier.
- If data tracking is hard, simplify: just note "Better," "Same," or "Worse" after each tweak, looking for patterns over time.

Practical Tip: Think of each improvement as a tiny software update for your mental operating system. Document each adjustment and its outcome, confirming that you can refine your strategies and become more efficient and less stressed over time.

10.4.4 Integrating Growth and Achievement Toolkits

These three toolkits – Goal Setting, Creativity, and Adaptive Learning – work synergistically:

- **Set Goals** to provide direction, ensuring you know what you're working toward.
- **Use Creativity** when standard approaches fail or become stale, discovering new routes to success.
- **Embrace Adaptive Learning** to refine and improve your methods continuously, turning each success and failure into wisdom.

For example, if a complex project leaves you overwhelmed, you might:

1. Use Goal Setting to break it down into manageable weekly targets.
2. Deploy Creativity when stuck on a particular step, brainstorming alternative solutions.
3. Apply Adaptive Learning by reviewing what worked and what didn't afterward, refining your approach for the next project phase.

By recording outcomes and noting improvements, you accumulate resilience evidences – proof that you can navigate complexity more skillfully each time. Over months, these achievements reinforce self-efficacy and breed anti-fragility, making challenges feel less daunting and more like opportunities to sharpen your skills.

Adapting Over Time:

- If a certain goal-setting method feels too rigid, experiment with more flexible milestones.
- If creativity sessions yield many ideas but no practical results, focus on idea evaluation steps.
- If adaptive learning reviews become repetitive, rotate the questions you ask or seek feedback from different people.

Long-Term Benefits: Over time, as you repeatedly set and achieve goals, innovate solutions, and refine strategies, you build a track record of competence and adaptability. Each success is a resilience evidence that assures you: no matter how complex life becomes, you have the mindset and skill set to progress, improve, and thrive.

In summary, Section 10.3 moves you from maintaining equilibrium to actively forging your path forward. By applying goal setting, creativity, and adaptive learning, you don't just endure complexity – you leverage it, transforming obstacles into catalysts for personal growth and lasting achievement.

10.5 RESILIENCE AND ADAPTATION

Overview: Up to now, we've covered foundational stability, improving relationships and communication, and fueling personal growth and achievement. This final section focuses on resilience and adaptation – capabilities that ensure you don't just weather life's storms, but emerge from them stronger and more confident.

Resilience isn't simply returning to where you were before a setback; it's learning from adversity and using that knowledge to improve. Adaptation ensures you can adjust your strategies as conditions evolve, maintaining momentum rather than getting stuck. Together, these toolkits help you transform hardship into fuel for growth, proving that each challenge can serve as a catalyst for personal evolution.

10.5.1 The Resilience Toolkit

Purpose: Resilience empowers you to rebound after setbacks – an essential skill in a complex, unpredictable world. Instead of feeling defeated by failures or stressors, resilience techniques help you absorb the shock, learn from the experience, and carry forward new strengths. Over time, repeatedly overcoming obstacles forms a library of resilience evidences – proof that you can handle adversity and thrive.

Everyday Techniques:

1. **Positive Reframing:**
    - **Method:** When a setback occurs (like a job rejection), identify at least one lesson or silver lining ("I learned how to improve my portfolio" or "Now I know I should apply earlier next time").
    - **Practical Application:** If your logs show that negative events drag down your mood for days, try reframing the next setback. Check if recovery time shortens, indicating you're building resilience.
2. **Resilience Reflection Journaling:**
    - **Method:** After facing a challenge – a tough conversation, a failed project, a missed opportunity – write down:
        - What went wrong?
        - What did I do that worked?
        - What can I do differently next time?
    - **Practical Application:** Over weeks, note if repeated journaling leads to quicker emotional rebounds and improved coping strategies. Each quicker recovery is resilience evidence.
3. **Small Exposure to Manageable Discomfort:**
    - **Method:** Introduce controlled "stress tests" in low-stakes environments (e.g., speaking up briefly in a small meeting if public speaking is a fear).
    - **Practical Application:** Over time, see if tackling manageable discomfort correlates with reduced anxiety in bigger, more challenging scenarios. This confirms that incremental challenges build resilience.

For the Nerdy/Data-Driven Approaches:

- **Recovery Time Metrics:** Track how many days (or hours) it takes to return to baseline mood or energy after a setback. After introducing resilience strategies, see if this time decreases. If previously it took you 3 days to bounce back and now it takes one to two days, that's quantifiable improvement.
- **Frequency of Successful Coping Instances:** Log each instance where you applied a resilience technique and rate how effective it felt. Over months, look for a pattern of increasing effectiveness.

Cultural/Accessibility Considerations:

- Cultural views on failure differ widely. If openly discussing failures feels uncomfortable, try private reflection methods, like silently acknowledging lessons.
- If writing is tough, consider voice memos or short video reflections to record resilience evidences.

Practical Tip: Start small. After the next minor setback – maybe a delayed train that spoils your evening plans – apply positive reframing. If you notice you bounce back more swiftly than before, you've confirmed that resilience techniques make a tangible difference.

10.5.2 The Adaptation Toolkit

Purpose: Adaptation ensures that as life changes – new job demands, shifting relationships, cultural transitions – you can adjust your strategies fluidly. Where resilience focuses on rebounding, adaptation is about evolving. You don't just return to baseline; you learn to anticipate shifts, refine tactics, and maintain a growth trajectory amid complexity.

Everyday Techniques:

1. **Regular Strategy Audits:**
   - **Method:** Every few weeks, review your routines, habits, and approaches:
     – What's still working?
     – What feels outdated or ineffective?
     – What new conditions have emerged that you need to address?
   - **Practical Application:** If data shows your morning routine no longer stabilizes mood (maybe due to seasonal changes or new responsibilities), adjust it. Over time, confirm if the updated routine restores balance, showing you can adapt effectively.
2. **Experimentation and Iteration:**
   - **Method:** Treat your life strategies like prototypes. Try small tweaks – e.g., shifting your workout to the morning rather than evening – and measure changes in energy or mood.
   - **Practical Application:** If logs show afternoon slumps remain unresolved, try changing lunch composition or adding a short mindful walk. If subsequent logs show improvement, adaptation was successful.
3. **Embrace Continuous Learning:**
   - **Method:** When faced with a recurring issue, seek new knowledge – read an article, watch a tutorial, ask a mentor. Integrate what you learn into your approach.
   - **Practical Application:** If certain relationship tensions persist, study conflict resolution tips. Try them next time and note if conflicts resolve faster.

For the Nerdy/Data-Driven Approaches:

- **Before/After Comparisons:** For any tactic – like changing your sleep schedule – log a baseline week, then implement the change for another week and compare. If mood or productivity improves, you have data-backed adaptation success.
- **Scenario Testing:** Create "if-then" plans: "If I feel stressed at 3 PM, then I'll do a 2-minute stretch." Document if this reduces stress frequency.

Cultural/Accessibility Considerations:

- Adaptation might mean respecting cultural norms while still evolving your personal habits. For instance, if direct feedback is uncommon in your culture, adapt conflict resolution methods to more subtle communication.
- If certain adaptive steps are physically demanding, consider assistive technologies or gentler adaptations that still achieve the desired outcome.

Practical Tip: Start by adapting one small aspect of your daily routine – such as changing your midday break activity. Observe if this minor shift yields improved afternoon focus or mood. Each successful adaptation is proof you can navigate changing circumstances effectively.

10.5.3 Integrating Resilience and Adaptation Toolkits

Resilience and adaptation naturally complement each other:

- **Resilience** gives you the ability to recover from setbacks.
- **Adaptation** ensures you learn from each challenge and refine your approaches, making future setbacks less impactful and more growth-oriented.

For example, if you face a challenging project at work:

1. Resilience: Use positive reframing and journaling to recover quickly after a tough deadline.
2. Adaptation: Adjust your planning strategy next time to prevent similar stress – maybe start tasks earlier or delegate differently.

Track these interventions over time. If you see that subsequent projects cause less stress or are completed more efficiently, you have hard evidence of building anti-fragility – growing stronger through challenges rather than just surviving them.

Adapting Over Time:

- If a particular resilience technique (like reframing) feels less effective over time, rotate in another method (like seeking social support or controlled exposure to mild stressors) to keep evolving.
- Regularly experiment with small adaptations. If some yield no improvement, discard them and try others. This iterative approach keeps your resilience and adaptability fresh and responsive.

Long-Term Benefits: Over months and years, applying resilience and adaptation strategies repeatedly creates a library of personal success stories. Each overcome setback and effective adaptation is a resilience evidence that strengthens self-efficacy and bolsters your confidence in facing future complexities. Eventually, your approach to adversity shifts from dread to curious challenge, knowing you have the resilience to bounce back and the adaptability to improve continuously.

In summary, Section 10.4 elevates you from maintaining equilibrium and achieving personal goals to consistently learning, evolving, and growing stronger through life's uncertainties. By weaving resilience and adaptation into your daily life, you ensure that every challenge enriches your skillset, deepens your confidence, and cements your anti-fragile mindset – preparing you for whatever complexity the future may hold.

THE PARADOX OF CHOICE

THINKING BOX:

"Does an abundance of options bring freedom or paralysis? The real resolution might lie somewhere in the domain of **42**."

10.6 INTEGRATION AND CONCLUSION

10.6.1 Integration and Conclusion

Overview: You've explored a range of mental toolkits across four sections – foundational well-being, relationships and communication, growth and achievement, and resilience and adaptation. Each toolkit is valuable on its own, yet their true power emerges when you integrate them, customizing your approach to fit specific challenges and evolving needs. By mixing and matching these tools, you can transform complexity from a daunting puzzle into a dynamic environment rich in opportunities for personal development.

Why Integration Matters:

- **Holistic Problem-Solving:** A single technique might help you manage stress, but combining it with emotional regulation or communication skills can address the root causes of that stress more thoroughly.
- **Adaptive Mastery:** Life isn't static; as new conditions arise – changing jobs, relocating to a new city, shifts in personal values – you'll need to adapt. Having multiple toolkits lets you pivot smoothly, ensuring continuous growth.
- **Rich Resilience Evidences:** Each successful combination of techniques forms a stronger evidence base for your resilience and anti-fragility. Over time, you reinforce the belief that you can handle complexity in various forms.

Practical Approaches to Integration:

1. **Start with One Core Area:**
    - If you've identified from your personal logs or reflections that midday stress is your primary challenge, begin with something foundational (like a brief mindfulness break each noon).
    - Once you see some improvement, add a second toolkit: maybe emotional regulation to handle lingering frustration, or stress management to structure your afternoon tasks better.
2. **Mix Toolkits for Specific Challenges:**
    - **Example: Handling a Tough Work Project**
        - **Foundational Well-Being:** Start your day with a micro-meditation (Mindfulness Toolkit) to center yourself.
        - **Relationships and Communication:** If the project involves a tense team, use assertive "I" statements (Communication Toolkit) to clarify roles and reduce conflict.
        - **Growth and Achievement:** Set a small daily goal related to the project (Goal Setting Toolkit) to maintain motivation.
        - **Resilience and Adaptation:** After each project milestone, reflect on what worked or didn't (Resilience Toolkit) and consider a slight adjustment to your approach (Adaptive Learning Toolkit).

          By applying multiple toolkits, you tackle emotional well-being, teamwork dynamics, personal motivation, and long-term adaptability all at once, turning a stressful project into a structured growth opportunity.
3. **Leverage Data and Feedback:**
    - If you enjoy tracking data, use simple metrics for each toolkit applied and note outcomes (mood changes, stress ratings, relationship satisfaction). Over weeks, identify which combos yield the best results.
    - Ask for feedback from trusted friends or mentors. Did your communication tweaks improve group morale? Did focusing on small daily goals visibly boost your productivity?

4. **Periodic Integration Reviews:**
   - Every month or quarter, review your toolkit usage. Which techniques still resonate? Which ones feel stale or less effective?
   - If mindful breathing sessions no longer ease your pre-meeting tension, try pairing mindfulness with creative brainstorming beforehand. Adjust and recheck the results.

Cultural, Accessibility, and Contextual Adaptations:

- Integration doesn't mean rigid uniformity. Adapt techniques to your cultural norms, personal preferences, and physical or cognitive abilities.
- For instance, if assertive communication feels aggressive in your cultural context, blend the Communication Toolkit's empathic listening with the Relationship Toolkit's trust-building activities to create a supportive dialogue style that feels comfortable and authentic.

Integration Plan Worksheet (Sample):

Challenge or Goal	Toolkits to Combine	Plan to Integrate Practices	Checkpoints and Adjustments
Midday Stress at Work	Mindfulness (one-minute breathing) + Stress Management (short breaks) + Emotional Regulation (quick journaling)	Before lunch: one-minute breathing. After lunch: five-minute break and if emotions spike, two-minute journal.	Check logs after two weeks. If stress persists, try adding goal setting (finishing a small task early).
Improve Communication in Team	Communication (active listening) + Emotional Regulation (label emotions) + Adaptive Learning (post-meeting reviews)	During meetings: summarize others' points, if tension arises label your emotion silently, after meeting review what worked/didn't	After one month, check if meeting tension scores drop. If not, add Creativity (brainstorm solutions to team conflicts).
Feeling Stagnant in Personal Growth	Goal Setting (one small weekly goal) + Creativity (five-minute idea sprint) + Resilience (reframing failure)	Each Sunday set a small skill-building goal, daily five-minute ideas to tackle obstacles, reframe setbacks immediately when they occur	After four weeks, see if you feel more motivated. If low progress, try another adaptation strategy.

This worksheet illustrates how to outline a challenge, pick toolkits, plan application, and establish checkpoints to refine your approach.

Long-Term Vision:

- As you repeatedly integrate these toolkits, you build a personalized mental "operating system" for thriving. Complexity no longer feels like a threat but a landscape you know how to navigate.
- Over time, you accumulate resilience evidences – tangible proof of your ability to handle stress, communicate effectively, achieve goals, and adapt to new realities. These evidences reinforce a growth mindset and confidence in your anti-fragility.

Sustaining Momentum:

- Celebrate every improvement. Even small positive shifts – fewer conflicts after using communication techniques, slightly improved stress ratings after integrating mindfulness – are wins worth acknowledging.
- If progress stalls, tweak your toolkit combination or try a new technique. The aim is continuous learning and iterative refinement, not chasing perfection.

Connecting to Your Life Beyond This Chapter:

- While this chapter provides a robust set of tools, remember that life is fluid. You'll face new challenges that may prompt you to revisit old techniques or learn new ones.
- Consider these toolkits not as a finished product but as evolving skills. With each stage of life, integrate differently: maybe you lean more on resilience and adaptation when starting a new job, or focus on relationships and communication when building a family.

10.6.2 Conclusion: From Complexity to Empowerment

By now, you've assembled an array of mental toolkits – mindfulness, emotional regulation, stress management, communication, relationship-building, goal setting, creativity, resilience, and adaptation. Each toolkit addresses a facet of well-being and growth. When integrated, they offer a flexible, evolving system that transforms adversity into a catalyst for improvement.

Key Takeaways:

- **Start Where You Are:** Pick one area to improve – be it managing stress or enhancing communication – and apply a relevant toolkit.
- **Mix and Match:** Don't hesitate to combine methods. Stress plus relational tension? Mindfulness plus communication adjustments might help. Feeling stuck professionally? Goal setting plus creativity and adaptation can break through plateaus.
- **Measure and Reflect:** Use simple logs, ratings, or reflections to confirm what works. Each success is a resilience evidence strengthening your self-efficacy.
- **Evolve Continuously:** Life changes – your strategies should too. Regularly reassess and refine your approaches, ensuring they remain aligned with your evolving goals and environments.

Final Encouragement: These toolkits don't just help you survive complexity; they enable you to thrive. With each proactive step, each small improvement, and each successful adaptation, you build a mental repertoire that makes you more agile, confident, and equipped to face an ever-changing world. Embrace the journey of experimentation and growth. By turning insight into action, complexity becomes an ally, guiding you toward a life of enduring resilience, meaningful achievement, and genuine fulfillment.

11 Resilience Evidence

A Comprehensive Framework for Developing Resilience and Anti-fragility

Key Concepts:

1. **Beyond the Illusion of Knowledge:** Knowing about resilience isn't enough – real growth comes from **practical experience** and **systematic skill-building**.
2. **Overcoming Helplessness: Learned helplessness** (failure-induced passivity) and **self-imposed helplessness** (negative self-beliefs) prevent action. **Small wins** break these patterns and build confidence.
3. **Building Resilience Evidences:** Every **incremental success** – handling stress, solving problems, or facing fears – becomes a **record of competence** that **reinforces self-efficacy and anti-fragility**.
4. **Bridging Big Dreams with Practical Steps:** Ambition must be **supported by smaller, trackable achievements**, avoiding **psychological barriers** to success.
5. **Dr. Steve Peters' Chimp-Human-Computer Model:** The **Chimp (emotional), Human (rational), and Computer (habit storage)** work together – controlled exposure, repetition, and adaptation **program resilience into your mental operating system**.
6. **Resilience in a Web of Interdependence:** Cultural, technological, and societal systems **influence resilience** – leveraging **collective resources** enhances adaptability.

By **transforming adversity into measurable proof of growth**, resilience evidences **turn uncertainty into opportunity**, making complexity an **engine for lifelong learning, confidence, and anti-fragility**.

11.1 INTRODUCTION: TURNING EXPERIENCE INTO RESILIENCE EVIDENCES

What if instead of viewing challenges as setbacks, you saw them as stepping stones to greater strength? This is the essence of resilience evidences – tangible proof of your ability not only to handle adversity but to emerge from it more capable, confident, and adaptable than before. These evidences come in many forms: the first time you stood your ground in a difficult conversation, the project you rescued under pressure, the fear you conquered by taking small, steady steps. Each experience shapes your belief in what you can do next.

From Concepts to Concrete Proof: Resilience isn't just a comforting idea; it's a skill set forged through action and reflection. It takes root when you convert abstract knowledge into lived experience – when you move from knowing that personal growth is possible to accumulating actual instances of having grown. These instances are your resilience evidences: reminders that you've navigated complexity before and can do so again, more effectively and creatively each time.

Building Self-Efficacy and Anti-fragility: Every resilience evidence you gather reinforces self-efficacy – the belief that you can influence outcomes. With higher self-efficacy, challenges don't

DOI: 10.1201/9781003640288-15

feel insurmountable; they become manageable tasks you can break down and tackle. Over time, this steady accumulation of proof also fuels anti-fragility, a state where stressors and uncertainties make you stronger rather than wearing you down.

A Roadmap for Growth: This chapter provides a structured approach to identifying, cultivating, and leveraging resilience evidences. You'll learn to:

- **Move Beyond Illusions of Knowledge (Section 11.1):** Understand why theoretical awareness isn't enough, and how hands-on effort and experience produce genuine growth.
- **Overcome Helplessness (Section 11.2):** Confront patterns that erode confidence and discover simple steps that rebuild your sense of agency, one small win at a time.
- **Systematically Build Resilience Evidences (Section 11.3):** Learn to treat discomfort as a signal to grow, turn failures into lessons, and accumulate a track record of steady improvement.
- **Marry Big Dreams with Practical Foundations (Section 11.4):** Translate lofty aspirations into achievable milestones, generating resilience evidences at every stage of your journey.
- **Apply Dr. Steve Peters' Framework (Section 11.5):** Use controlled exposure, embracing discomfort, repetition, and flexibility to create mental "routines" that embed resilience into your daily life.
- **Brief Mentions of Stress Factors, Mental Patterns, and Societal Context (Integrated Briefly):** Recognize stress triggers and mental tendencies not as dead ends, but as cues for applying the strategies you've learned. Acknowledge cultural and social dimensions as additional layers that shape how resilience evidences are formed and recognized.

A Life Rich in Proof of Your Strength: As you work through these concepts, you'll find that each challenge overcome leaves a tangible imprint – an evidence of resilience. Over time, your library of evidences grows, affirming your capacity to adapt and excel amid complexity. By repeatedly converting adversity into opportunity, you foster a mindset where uncertainty spurs creativity, stress prompts mastery, and ambition aligns with practical action.

Moving Forward: This chapter doesn't promise a life free of obstacles. Instead, it offers the tools to ensure each obstacle strengthens your foundation. Embrace the process, record your successes, and learn from setbacks. In doing so, you transform from someone who hopes to be resilient into someone who has the evidence – over and over again – that you already are.

11.2 BEYOND THE ILLUSION OF KNOWLEDGE – CONFRONTING THE ILLUSION OF EXPLANATORY DEPTH

Core Idea: Have you ever felt certain you understood something – like how a zipper works, or how to navigate a challenging conversation – only to discover, when pressed, that your grasp was more superficial than you realized? This discrepancy between perceived and actual understanding is known as the **Illusion of Explanatory Depth**. It can give us false confidence, making us believe we're ready for life's hurdles until we face them and stumble.

For building resilience, this illusion is a hidden trap. Intellectual familiarity with concepts of adaptability, stress management, or problem-solving isn't the same as being able to apply them under real pressure. If we rest on a theoretical sense of mastery without testing and refining our skills, we miss opportunities to produce resilience evidences – the tangible proof of growth that emerges only when we engage in real actions, learn from failures, and confirm what truly works.

Why the Illusion of Explanatory Depth Matters:

1. **Fragile Confidence Under Pressure:** The illusion makes you think, "I can handle a crisis," based on having watched others do it or read about strategies. But in the actual crisis,

if your skills aren't tested beforehand, that confidence can collapse, leaving you unsure and stressed.

2. **Barriers to Growth:** Overestimating your competence reduces your incentive to practice, get feedback, or experiment in low-stakes environments. Without these trials, you never accumulate resilience evidences. Your capabilities remain unproven, making true resilience fragile and untested.
3. **A Missed Opportunity for Mastery:** Recognizing the illusion doesn't condemn you to ignorance; it invites you to transform superficial knowledge into embodied skill. Every time you replace theoretical understanding with hands-on attempts, you generate resilience evidences – confirmations of your real-world ability to adapt and thrive.

Deeper Insight into the Illusion of Explanatory Depth: Cognitive scientists Rozenblit and Keil illustrated that people often believe they understand everyday mechanisms (like how a toilet flushes or a zipper closes) deeply. But when asked to explain these mechanisms step-by-step, their confidence wanes. Similarly, you might think you're ready to handle a heated negotiation after watching tutorials or observing a mentor. Yet, when you attempt it yourself, you realize you lacked practical nuance – how to respond to unexpected objections, how to stay composed under personal criticism, or how to time your pauses.

Implications for Resilience:

- **Without Action, No Proof:** If resilience were a language, the illusion of explanatory depth is like claiming fluency after reading a grammar book but never conversing. Reading about negotiation techniques isn't the same as successfully negotiating a pay raise. You only gain the "fluency" of resilience – its practical, ready-to-use skill – by diving into real interactions.
- **From Theoretical Safety to Hands-On Evidence:** A single successful attempt at calmly resolving a conflict under stress is more valuable than a dozen chapters read about conflict resolution. That one real incident forms a resilience evidence that permanently alters your self-perception, reinforcing self-efficacy and anti-fragility.

Bridging the Gap – Transforming Assumptions into Resilience Evidences:

1. **Reality Checks and Simulation:**
   - **Method:** Start by explaining a skill or concept to yourself or someone else. If you stumble, you've found a knowledge gap. Then simulate a related scenario in a low-stakes setting.
   - **Example:** If you think you can handle giving tough feedback at work, first try outlining exactly how you'd do it. If you can't detail it clearly, practice with a friend role-playing as a colleague. If during the role-play you manage to stay calm and assertive, you've earned a resilience evidence affirming your newfound readiness.
2. **Start with Manageable Challenges:**
   - **Method:** Don't jump straight into high-risk situations. Begin with mild challenges that partially replicate the conditions you'll face later.
   - **Example:** If public speaking terrifies you, first explain a simple topic to a family member. Next, try a small internal presentation at work. Each successful step provides incremental resilience evidences, confirming that you can operate under increasing levels of pressure.
3. **Iterative Learning and Adjustment:**
   - **Method:** After each attempt, reflect on what worked and what didn't. Adjust your approach and try again. Over time, repeated practice embeds resilient responses into your habits.
   - **Example:** After a trial negotiation with a friend, you realize you interrupt too often. Next time, focus on listening more. When this yields a more constructive outcome, you've converted theory into practice and speculation into solid evidence of resilience.

4. **From Intellectual Comfort to Adaptive Expertise:**
   - **Method:** Continue challenging assumptions. If you think you know how to handle personal stress – perhaps you've read about breathing techniques – test them the next time you sense anxiety rising. If the breathing stabilizes your nerves effectively, that's an immediate resilience evidence. If not, try another method (like progressive muscle relaxation or a brief walk) until you find what genuinely works for you.

For the Data-Oriented:

- **Pre/Post Confidence Ratings:** Assign a confidence score (1–10) about a skill before testing it. After the attempt, assign another score. Compare results. If your post-attempt confidence is more realistic and eventually improves as you refine your approach, you're witnessing your journey from illusion to real competence.
- **Tracking Successful Trials:** Keep a log of each skill tested, the scenario, and the outcome. Each success is a resilience evidence. Each failure guides you to refine, leading to eventual success and thus another evidence.

Cultural/Accessibility Considerations:

- Adapt the complexity of challenges to your comfort and abilities. If certain cultural norms discourage direct confrontation, practice more subtle, culturally fitting communication approaches. If a physical condition limits certain stress tests, design cognitive or verbal simulations.

Practical Tip: This week, pick one skill you believe you've "got down" (like staying calm under mild stress). Create a low-risk test environment – maybe an intentionally rushed errand. Observe your reactions. If you handle it calmly, great – you have a resilience evidence. If not, adjust your strategy and try again. This process ensures you move beyond illusion, incrementally building a portfolio of proven capabilities.

In essence, Section 11.1 highlights the imperative of action and experience. By challenging the Illusion of Explanatory Depth, you transform superficial confidence into solid, evidence-backed resilience. Each successful real-world application adds another resilience evidence to your mental archive, reinforcing your self-efficacy and paving the way for adaptive growth in more complex scenarios.

11.3 FROM HELPLESSNESS TO AGENCY – OVERCOMING LEARNED AND SELF-IMPOSED CONSTRAINTS

Core Idea: Resilience flourishes when you believe in your ability to influence outcomes. Yet two pervasive mental traps – **learned helplessness** and **self-imposed helplessness** – erode that belief. Learned helplessness arises from repeated exposures to seemingly uncontrollable adversity. Over time, you conclude that no matter what you do, nothing changes. Self-imposed helplessness, by contrast, emerges from the inside out: you set mental limits and adopt negative narratives, underestimating your capacity or refusing to recognize available opportunities.

Both forms of helplessness undermine your capacity to accumulate resilience evidences. Why? Because they deter you from taking the very actions that could provide proof of your adaptability and strength. Instead of testing your skills and seeking incremental wins, you remain stuck in inaction and doubt.

Bridging to the Nocebo Effect: In a previous chapter, we explored the nocebo effect – the phenomenon where negative expectations lead to worse outcomes. Helplessness, especially the self-imposed variety, is a psychological nocebo: you tell yourself change is impossible, and this expectation makes it so. This self-sabotaging mindset prevents you from even attempting solutions, thus blocking the generation of resilience evidences that prove you're more capable than you think.

11.3.1 Learned Helplessness: When Past Failures Become Self-Fulfilling

How It Forms: Learned helplessness often follows genuine setbacks. Perhaps you tried to secure a job multiple times and failed, or struggled academically despite studying hard. After enough unsuccessful tries, you may stop trying altogether, convinced that effort is pointless.

Why It's Harmful:

- **Defeatist Feedback Loop:** Each time you avoid trying, you forfeit the chance to earn a resilience evidence. Your inaction confirms your belief in powerlessness, reinforcing the cycle.
- **Undermining Self-Efficacy:** Without fresh attempts, you never learn whether new strategies or different approaches could help. Your potential remains locked behind the perception of futility.

Breaking Free:

1. **Awareness and Reality Checks:**
   - Question fatalistic assumptions: "Is it truly impossible, or have I just not found the right method yet?"
   - Example: If you failed several job interviews, try smaller steps: update your resume, practice with a friend, apply to a less intimidating role. Just one improved outcome, even a better feeling after an interview, serves as a resilience evidence that effort can influence results.
2. **Micro-Goals and Small Wins:**
   - Start with easily attainable targets. Prove to yourself that one small step (like sending a refined application) leads to a more positive response or at least a more confident feeling.
   - Each small success rewrites your internal narrative, proving that actions matter and that you're not doomed to fail.

11.3.2 Self-Imposed Helplessness: The Limits You Set for Yourself

A New Concept: Self-imposed helplessness differs from learned helplessness in origin: it's not formed primarily by external failures, but by internal stories you tell yourself before even trying. It's rooted in negative expectations, akin to a psychological nocebo effect, where you predict failure or inadequacy so strongly that you never test your capabilities.

Examples:

- You might say, "I'm just not good at dealing with people," so you never attempt that networking event where you could learn and adapt. Without testing your social approach, you never gain evidence that you can improve.
- Or "I can't handle stress without freaking out," becomes a self-made barrier. You avoid slightly challenging tasks, never collecting resilience evidences that prove otherwise.

Why It's Harmful:

- **Preemptive Surrender:** You concede defeat before facing the challenge, ensuring no resilience evidence ever forms. Your skill set remains untested and underdeveloped.

- **The Nocebo of Mindset:** Like ingesting a harmless pill while believing it'll harm you, these negative narratives create the very difficulty you fear. By expecting failure, you never allow yourself the chance to build competence.

11.3.3 Overcoming Both Learned and Self-Imposed Helplessness

To break free, you must take conscious steps to restore agency, test assumptions, and gather concrete proof of your capabilities.

1. **Cultivate Awareness – Spot the Internal Dialogues:** Identify recurring phrases like "I can't" or "There's no point." Write them down. Ask: "Is there evidence this is true, or have I accepted it without testing?" Challenging these beliefs is the first step toward reclaiming your power.
2. **Reframe Failures and Inaction:** Instead of labeling a past failure as personal doom, treat it as an informative experiment. If job rejections battered your confidence, see each one as data: What can I improve in my approach? Once you apply a new tactic – better research before interviews, seeking a mentor's advice – you increase the chance of a slightly better outcome next time. That outcome, even a more positive interviewer reaction, is a resilience evidence that effort changes results.
3. **Start Extremely Small to Build Momentum:** If facing a big challenge directly feels impossible, start microscopically. Overcome self-imposed helplessness by choosing a trivial action in the same domain. Fear social events? Practice saying "hello" to a store clerk. If this interaction feels easier next time, you've earned a resilience evidence showing incremental improvement is possible.
4. **Leverage Reflection and Feedback:** After each attempt, reflect on what worked. Document it. Did sending just one more application result in a friendly response, even if not a job offer? That's progress, proof you're not helpless. Over time, these small improvements weave together into a powerful narrative of growth.

For the Data-Oriented:

- **Before/After Comparisons:** Assign helplessness a rating – how powerless do you feel (1–10)? After taking a small action and noting a slightly improved result or feeling, rate again. If the feeling of helplessness decreases over repeated attempts, that's measurable evidence of rising resilience.
- **Success Log:** Keep a simple log of each attempt. Even minor positives (like getting constructive feedback instead of a blunt rejection) count. Over weeks, these records show a trend away from helplessness toward meaningful engagement.

Cultural/Accessibility Considerations:

- Cultural norms may affect how you perceive control or the willingness to try new approaches. Adapt small goals that respect these norms, focusing on gradual change.
- If certain actions are physically challenging, break them down into accessible steps or seek assistive tools. The key is to prove to yourself that you can influence outcomes in a manner that feels achievable and authentic.

Practical Tip: This week, pick one area where you feel "stuck" (maybe you haven't applied for a promotion, or you keep avoiding a personal conversation). Take one small action in that domain – send an inquiry email, schedule a brief chat. If the response is even slightly positive, record it as a resilience evidence. If not, tweak your approach and try again. This process transforms abstract helplessness into tangible, improvable scenarios.

In essence, Section 11.2 emphasizes that helplessness – learned or self-imposed – is a psychological barrier rather than a permanent state. By challenging your assumptions, seeking small wins, and embracing trial-and-error, you break the cycle of inaction. Each successful action, however minor, serves as a resilience evidence, proving you're not condemned to stagnation. Instead, you're capable of influencing your life's trajectory, step by step.

THE LAW OF UNINTENDED CONSEQUENCES

THINKING BOX:

"Can we ever fully predict what our innovations will unleash? Navigating the unpredictable may require a dash of **42**."

11.4 BUILDING RESILIENCE EVIDENCES – FROM ADVERSITY TO ANTI-FRAGILITY

Core Idea: Resilience evidences are the backbone of a growth-oriented, adaptable life. Until now, we've seen how facing illusions of mastery spurs action (Section 11.1) and how dismantling helplessness frees you to attempt challenging tasks (Section 11.2). Now we focus on systematically accumulating these evidences – transforming adversity from a threat into a workshop where skills sharpen and confidence soars.

These evidences aren't confined to a particular phase of life. From a child mastering the courage to ride a bike without training wheels, to an adult navigating a career shift, to a senior adapting gracefully to changing health conditions – every stage offers opportunities to earn resilience evidences. Over time, they become the scaffolding that supports anti-fragility, enabling you to draw strength from complexity rather than shrinking from it.

11.4.1 The Lifelong Journey of Resilience Evidences

Childhood:

- **Example:** A kid scared of the dark sleeps with a nightlight. One night, they manage to fall asleep with it turned off. That small triumph – facing a fear – becomes their first resilience evidence. They learn that a bit of bravery can yield comfort and confidence.

Adulthood:

- **Example:** A new professional facing tough projects learns to break them into smaller tasks, steadily improving efficiency. Each successfully managed deadline confirms that stress can be managed, not just endured. This repeated success is resilience evidence that they can handle higher responsibilities.

Senior Years:

- **Example:** Adjusting to retirement or health changes might be daunting. Yet discovering new hobbies, mastering digital tools to stay connected, or finding gentle exercise routines can each serve as evidence that adaptation never stops. Even late in life, each new skill proves you can continue to grow stronger through change.

11.4.2 Why Emphasize the Whole Life Cycle?

Resilience isn't a skill you acquire once; it's an evolving capacity. By viewing every age and circumstance as fertile ground for building resilience evidences, you maintain a growth mindset and remain open to transforming challenges into incremental achievements.

11.4.3 Emotions as Signals, Discomfort as a Teacher

As established earlier, emotions like fear or frustration aren't enemies. They signal areas where your skills can expand. Discomfort points to the edges of your comfort zone – a place where new resilience evidences await.

Practical Steps:

1. **Identify the Growth Zone:** When you feel uneasy or intimidated by a task, ask: "What skill could I strengthen here?" If it's public speaking, for example, start by speaking up in a small meeting. A calm, successful contribution is immediate evidence of your capacity to handle public discourse.
2. **Reframe Setbacks:** A botched presentation or a tense family discussion isn't evidence of failure; it's raw material for improvement. Reflect on what happened, note what to tweak, and try again. The first time you handle a similar situation more effectively, that improvement is a resilience evidence.

11.4.4 Incremental Achievements: The Compound Effect of Small Wins

Large achievements often rest on foundations of small victories. Each mini-success builds confidence, making bigger steps less daunting. Over time, a series of small wins compounds, demonstrating not just one-time ability but sustained adaptability.

For Example:

- To handle anxiety in social settings, start by greeting a cashier, then join a small interest group, and later attend a larger networking event. Every step up – no matter how slight – confirms that with gradual exposure, your comfort and competence expand.

Benefit: Small achievements quickly convert theory into practice, preventing the nocebo effect (where negative expectations sabotage performance) and reinforcing that effort leads to progress. Every small win is a micro resilience evidence, proving you can influence outcomes and encouraging you to attempt more ambitious tasks.

11.4.5 Turning Experiences Into Resilience Evidences: A Practical Framework

1. **Deliberate Challenges:**
   - **Method:** Intentionally pick tasks slightly above your current comfort level – just difficult enough to feel like growth, not so daunting that you won't attempt them.
   - **Outcome:** Each successful try, even partially successful, is evidence that effort matters. By recording these increments, you create a personal narrative of growing mastery.
2. **Reflection and Documentation:**
   - **Method:** After attempting something challenging, write down what happened. Did you perform better than last time? Even a small improvement – less anxiety, a clearer mind, a quicker solution – counts as a resilience evidence.
   - **Outcome:** This record not only cements the memory but also provides a tangible reference. On tough days, revisiting these documented improvements reminds you of your adaptive potential.
3. **Feedback and Iteration:**
   - **Method:** Seek feedback from peers, mentors, or friends. Listen openly to suggestions and try revised approaches.
   - **Outcome:** Each adaptation leading to better results confirms that flexibility and learning from others also generates resilience evidences. By refining strategies over time, you link complexity with creativity and adaptability.

For the Data-Oriented:

- **Before/After Metrics:** Rate stress or confidence (1–10) before and after attempting a challenging task. Over multiple attempts, watch ratings improve, confirming tangible growth.
- **Frequency of Successful Attempts:** Track how often you manage to handle something that once felt tough. Rising success rates are quantitative resilience evidences.

11.4.6 Integrating Past Lessons

Recall from Section 11.1: turning illusion into action gave you initial successes. From Section 11.2: breaking helplessness allowed you to try small steps. Now, as you accumulate these small steps and reflect on them, you're weaving all previous insights into a cohesive process. You're not just occasionally succeeding; you're building a repertoire of proven strategies and capabilities.

As these resilience evidences pile up – adapting to stress factors, shifting from rumination to focused action, and, as will be explored later, using frameworks like Dr. Steve Peters' methods – you transition from mere resilience to anti-fragility. Anti-fragility means adversity doesn't just fail to break you; it becomes a catalyst for even greater resilience evidences in the future.

11.4.7 Life-Long Adaptation and Evolution

Remember, this isn't a one-time achievement. Your environment, goals, and challenges evolve. As a child, resilience might mean conquering fear of the dark. In mid-life, it might mean mastering interpersonal skills in a volatile job market. In later years, it might mean staying socially engaged or learning new technologies. Each phase brings fresh opportunities to generate resilience evidences, ensuring you remain mentally agile and hopeful, regardless of age.

Practical Tip: Identify one area you've improved in over time – even if slightly. Perhaps you've become more patient with family members, or better at problem-solving at work. Note this improvement as a resilience evidence: proof that you've adapted and grown. This realization encourages you to seek out further, slightly bigger challenges, secure in the knowledge that you can keep adding to your personal archive of strengths.

In essence, Section 11.3 shows that resilience evidences form through intentional engagement with challenges, guided reflection, and incremental advancement across the entire life span. By consistently converting adversity into achievements – no matter how small – you accumulate a robust inner database of competence and adaptability. Over time, these evidences align you with anti-fragility, empowering you to greet complexity not with apprehension, but with constructive confidence.

11.5 BIG DREAMS, PSYCHOLOGICAL BARRIERS, AND THE SWEET SPOT OF EVIDENCE

Core Idea: Ambitious goals can inspire, driving us toward innovation, personal transformation, or uncharted achievements. But these big dreams come with a risk: if you never gather personal evidence that you can achieve incremental steps along the way, the fall from high expectations can be crushing. Without resilience evidences to substantiate your path, a grand vision may become a source of stress, self-doubt, and disillusionment.

The Danger of Unsupported Ambition: Dreaming big isn't inherently bad. The problem arises when your vision soars far above your current abilities and you refuse to acknowledge the gap. If you pin all your hopes on a distant pinnacle without building small successes, you're essentially betting everything on a huge leap you're not prepared for. If that leap fails, there's no safety net of intermediate resilience evidences to cushion the blow. The resulting disappointment can reinforce helplessness, nocebo-like negativity, and retreat from future challenges.

The Roger Bannister Effect (Four-Minute Mile Phenomenon): To understand why evidence matters, consider the story of Roger Bannister. For years, athletes believed running a mile under four minutes was impossible – a psychological barrier as much as a physical one. Then Bannister broke it in 1954. Almost immediately afterward, many others achieved the same. What changed? The barrier wasn't just physical – it was mental. Seeing someone else do it proved the limit was not absolute. This "Bannister Effect" teaches us that so-called ceilings often crumble once evidence surfaces that the goal can be reached.

- **Application to Resilience Evidences:** While Bannister's success served as a global resilience evidence for other runners, you need personal resilience evidences to break your own "4-minute mile" in whatever domain you choose. Merely knowing others have achieved something may reduce disbelief, but until you notch small personal wins (a slightly longer run, a slightly better performance), your confidence remains abstract. Each increment you achieve provides a personal data point – evidence that you can chip away at psychological barriers.

11.5.1 The Sweet Spot: Big Dreams with Incremental Proofs

The key is not to abandon big dreams, but to ensure they're supported by incremental steps and resilience evidences:

1. **Set a Progressive Path:** Don't try to leap directly to the equivalent of a "sub-4-minute mile" in your field, whether it's launching a tech startup or becoming an acclaimed artist. Start by improving smaller metrics. For instance, if you dream of being a tech entrepreneur, begin by successfully developing a small application or service and note how each version release, bug fix, and user feedback loop acts as a resilience evidence. These proofs gradually erode the psychological barrier that "only super-geniuses can succeed in tech."
2. **Use Technological Tools to Document Growth:** In modern society, apps and online platforms can track your progress:
   - Fitness apps log your running times, proving you're getting faster.
   - Coding platforms track how many problems you've solved, confirming you're building skills to meet your startup idea head-on.
   - Online courses record your completed modules, providing tangible evidence of skill acquisition.

   This digital trail of incremental achievements stands as resilience evidences you can revisit, especially when doubts arise.
3. **Adapt as Conditions Evolve:** In a fast-moving, tech-driven world, goals and the paths to achieve them may shift rapidly. The adaptability learned through incremental gains ensures that even if the environment changes – market trends, new technologies, shifting social norms – you have the confidence to pivot. Each adaptation and successful reorientation becomes another resilience evidence affirming your anti-fragility.
4. **Balance Visionary Thinking with Grounded Action:** By all means, dream of being the next breakthrough innovator or top performer in your field. Let that vision motivate you. But don't rely on the dream alone. Turn every small breakthrough – completing a challenging project, learning a new coding language, presenting a minor workshop – as evidence that you're progressing from theory to practice.

Tackling Psychological Barriers in a Modern Context: Modern life bombards us with stories of overnight successes and highlight reels on social media, often reinforcing the illusion that giant leaps occur without foundational steps. Such narratives can set unrealistic benchmarks, acting like psychological barriers that dissuade you from starting small. Counter this by documenting each small improvement – like improving a prototype after user feedback or gaining 100 more subscribers on your channel – and recognize these as resilience evidences. Over time, these data points reaffirm that success is cumulative, not a magical event.

Avoiding the Nocebo of Unattainable Visions: If you continually tell yourself that your ultimate goal is too hard and gather no evidence to the contrary, you may create a nocebo-like scenario, where your negative expectations hinder progress. Instead, each incremental success is proof that the challenge is surmountable, chipping away at mental blocks just as Bannister's success chipped away at the four-minute mile myth. It moves you from "I can't" to "I can, step by step," and eventually "I did."

In essence, Section 11.4 emphasizes that big dreams are not the enemy. They are powerful motivators, but only if balanced by practical evidence-building actions that show you that you, personally, can advance toward them. By dismantling psychological barriers through incremental successes – whether running faster miles, improving a prototype, or learning complex skills gradually – you accumulate resilience evidences that fortify your journey. Thus, you navigate the tension between ambition and reality, transforming lofty visions into achievable futures supported by a growing archive of proof points.

11.6 DR. STEVE PETERS' FRAMEWORK – HARNESSING THE CHIMP, HUMAN, AND COMPUTER WITHIN

Core Idea: Dr. Steve Peters, in *The Chimp Paradox*, offers a powerful analogy for understanding the mind's inner workings: we each have a **Chimp**, a **Human**, and a **Computer** coexisting within our brains. The **Chimp** is your emotional, impulsive side – the part reacting with fear, anger, or anxiety when challenged. The **Human** is your rational, logical mind that can plan, reflect, and learn from experience. The **Computer** stores habits, automatic responses, and learned strategies, acting as a "reference library" or autopilot system.

Resilience building is about ensuring that the **Human** orchestrates growth opportunities, the **Chimp** is managed and not left to run wild, and the **Computer** is continuously updated with positive, adaptive "programs." By using controlled exposure to challenges, embracing discomfort, practicing repeatedly, and staying flexible, you accumulate resilience evidences – tangible proof that you can master adversity. These evidences don't just reassure you rationally; they become new "code" written into the Computer, helping you respond more effectively and calmly in future complexities.

11.6.1 The Three Mind Components and Resilience Evidences

1. **The Chimp (Emotional, Impulsive):**
   - **Nature:** The Chimp reacts first, often governed by survival instincts, insecurity, or instant gratification. When faced with adversity, the Chimp might panic, avoid the situation, or become defensive.
   - **Challenge:** If you rely solely on the Chimp's impulses, you struggle to gain resilience evidences because the Chimp prefers comfort and avoids stress. This leads to stagnation.
   - **Strategy:** By applying Dr. Peters' principles, you don't try to eliminate the Chimp, but to manage it. Controlled exposure to mild stressors shows the Chimp that not all challenging situations are threats. Over time, each successful encounter ("Hey, that presentation wasn't so bad!") is a resilience evidence that reassures the Chimp. The Chimp learns discomfort isn't catastrophic, reducing emotional hijacks.
2. **The Human (Rational, Reflective):**
   - **Nature:** The Human plans, sets goals, reflects on outcomes, and learns from experience. The Human's role is to guide the process of incremental challenges, ensuring you don't jump from zero to an overwhelming test of skill.

- **Challenge:** Without the Human's deliberate approach, attempts at building resilience might be haphazard – too easy or too hard – failing to produce meaningful resilience evidences.
- **Strategy:** The Human selects appropriate challenges, slightly above comfort level, and after each attempt, the Human analyzes what happened. This reflection cements the resilience evidences in memory. The Human also decides when to push further, when to adjust tactics, and when to celebrate small wins. This rational orchestration ensures that every success, however small, is recorded and appreciated.

3. **The Computer (Autopilot, Habit Storage):**
   - **Nature:** The Computer stores learned behaviors, responses, and beliefs. Initially, it may contain unhelpful "programs" (like avoidance patterns or negative narratives). Over time, you rewrite these programs through repeated positive experiences – resilience evidences.
   - **Challenge:** If the Computer is filled with old scripts – "I can't handle stress," "I always fail under pressure" – you'll revert to them automatically in tough times.
   - **Strategy:** Each time you handle a challenge slightly better than before, you enter a new "line of code" into the Computer: "In this situation, I can remain calm and adapt." After multiple iterations, your response to adversity becomes more automatic – resilience transforms from a conscious effort into a default routine. The Computer's updated programming is essentially your collection of resilience evidences made functional.

11.6.2 Applying Peters' Principles to Create Resilience Routines

1. **Controlled Exposure:**
   - **Concept:** Expose yourself to manageable difficulties rather than throwing your Chimp into panic mode.
   - **Example:** If public speaking triggers fear, start by speaking briefly to a supportive group or recording a short video for yourself. Handle it successfully, and log this as a resilience evidence – a stored program in the Computer that says, "You survived and learned something here."
2. **Embracing Discomfort for Growth:**
   - **Concept:** Discomfort signals the Chimp's alarm. The Human must interpret it as a growth opportunity.
   - **Example:** Confront a mild conflict at work rather than avoiding it. When you emerge calmer and more effective than expected, you produce a resilience evidence that tells the Computer, "We can manage this emotional terrain." Repeated exposures gradually desensitize the Chimp, showing it that discomfort leads to skill acquisition, not danger.
3. **Repetition as Reprogramming:**
   - **Concept:** Doing something once may not stick. Repetition ensures new responses become ingrained.
   - **Example:** If time management is your challenge, repeatedly apply a chosen technique (like time-blocking or short breaks). Each improvement in productivity or reduced stress is a resilience evidence. Over weeks, the Computer internalizes these patterns. Now, under pressure, your autopilot offers constructive routines instead of panic or procrastination.
4. **Flexibility and Adaptation:**
   - **Concept:** The Human, guided by data and reflection, periodically updates strategies. If one approach stops working (maybe a certain stress technique no longer calms the Chimp), try another.
   - **Example:** If journaling stopped reducing nighttime anxiety, switch to a brief relaxation exercise or try a mindfulness app. When this yields improved sleep, that's another resilience evidence confirming adaptability. This ensures the Computer doesn't get stuck on outdated code – it evolves, maintaining anti-fragility as technology, environment, and personal circumstances shift.

11.6.3 Tying It All Together with Technology and Modern Contexts

In a tech-driven era, you can reinforce these principles using digital tools:

- **Simulations and Virtual Reality:** Practice public speaking or negotiation in VR scenarios. Each smoother virtual attempt is stored in the Computer as a workable blueprint for real situations.
- **Habit-Tracking Apps:** Track improvements in stress response times or productivity metrics. Seeing data-backed progress convinces the Chimp and the Human that positive routines are forming.
- **Adaptive E-Learning and Online Feedback Loops:** Engage in platforms that adjust challenge levels as you improve, ensuring you're always at the sweet spot. Each incremental success logged and displayed visually is a resilience evidence that you can surpass perceived limitations.

Benefit: This synergy of Peters' framework with modern tools accelerates the transition from deliberate effort to ingrained resilience routines. Over time, you won't just believe you can handle complexity; your Chimp, Human, and Computer will have collectively experienced it, recorded it, and rely on that proven capability as second nature.

In summary, Section 11.5 shows that Dr. Steve Peters' chimp-human-computer model offers a clear roadmap for building resilience evidences into your mental architecture. By controlling exposure to challenges, embracing discomfort as a signpost to growth, repeating new strategies until they become habits, and adapting flexibly to changing conditions, you transform theoretical resilience into practical, code-like routines. Each incremental success – be it a calmer negotiation, a more focused work sprint, or a steadier emotional response to unexpected news – is written into your Computer's library of solutions, forming a resilient mental operating system ready to navigate modern complexities with confidence.

11.7 RESILIENCE IN A WEB OF INTERDEPENDENCE – SOCIETAL, CULTURAL, AND TECHNOLOGICAL LAYERS

Core Idea: Resilience does not exist in isolation. While we've focused on personal action, self-efficacy, and incremental achievements, it's essential to see that individuals operate within a vast, interwoven web of relationships, infrastructures, cultures, and historical legacies. Each attempt you make to build resilience evidences – be it tackling a new skill, adapting to stress, or aiming for a big dream – unfolds against a backdrop of physical, cognitive, social, economic, and historical contexts that both support and shape your journey.

Understanding these dimensions of human interdependence not only helps you access resources for building resilience but also broadens your perspective on what "adversity" truly means. Complex challenges often arise from or interact with larger systems. By acknowledging and engaging with these systems, you enhance your ability to produce resilience evidences at multiple levels – personal, communal, and even generational.

11.7.1 Interdependent Dimensions of Resilience

1. **Physical and Material Dependencies:**
    - **Infrastructures and Technology:** Roads, electricity grids, internet connections, and global supply chains form the material scaffold supporting daily life. A single resilience evidence – like handling a stressful remote work deadline – might rely on stable internet (provided by engineers and technicians) and a functioning device (created by global teams).

Implication for Resilience: When these systems falter (a blackout, a server crash), your ability to adapt comes into play. If you manage to complete your work differently – perhaps using offline tools or shifting your schedule – and succeed, you create a resilience evidence that you can handle disruptions in the material environment. This success is not just individual; it's anchored in understanding and navigating the physical systems upon which you rely.

- **Food and Resource Chains:** Every meal involves countless contributors: farmers, distributors, market dynamics, environmental conditions. If a supply chain hiccup forces you to adapt your diet or cooking habits, successfully adjusting confirms you can handle changes in fundamental necessities. Another resilience evidence earned, rooted in your interaction with physical resource networks.

2. **Cultural and Cognitive Dependencies:**
   - **Language and Knowledge Systems:** Language, inherited and evolved over generations, enables you to learn new skills, communicate solutions, and understand the principles of resilience. Every time you apply a learned concept – perhaps a coping strategy discovered in a cultural proverb or a piece of wisdom from historical figures – and it works, you accumulate resilience evidence.

     Implication for Resilience: By drawing on cultural narratives, collective problem-solving methods, or historical success stories (like breaking the "4-minute mile"), you realize limits are often psychological. Integrating these insights into your personal growth journey transforms second-hand inspiration into first-hand resilience evidence.
   - **Cultural Frameworks and Values:** Some societies emphasize communal support, others celebrate individual ingenuity. Understanding your cultural environment lets you select strategies aligned with prevalent norms and resources. If, for instance, your culture values patience and collective effort, cooperating with neighbors during a local crisis not only solves a problem but provides a resilience evidence that communal action is effective.
3. **Social and Economic Dependencies:**
   - **Direct Relationships (Family, Friends, Colleagues):** Emotional support, mentorship, and resource sharing from your network often make the difference between a challenge feeling overwhelming or manageable. When a colleague's tip helps you handle a stressful project more smoothly, that success story (a resilience evidence) is partly a product of these relational ties.
   - **Economic Systems and Public Services:** Jobs, markets, financial institutions, education, and healthcare services all influence the landscape of your challenges. Navigating a sudden job loss by quickly retraining (perhaps using online platforms and grants) and securing a new role transforms social and economic connections into catalysts for producing resilience evidences.
4. **Historical and Evolutionary Dimensions:**
   - **Inherited Technologies and Wisdom:** Modern conveniences, knowledge, and cultural traditions are cumulative outcomes of countless past efforts. The smartphone you use to attend a stress-management webinar or the online forum where you gain peer support are products of historical progress. Successfully applying a tip gleaned from such a webinar – perhaps learning a breathing technique that calms you before a big meeting – creates a resilience evidence that merges personal agency with historical, collective knowledge.
   - **Cultural Evolution and Adaptation Over Time:** Humanity's resilience is intertwined with centuries of adaptation to changing climates, crises, and innovations. Recognizing that your personal steps reflect a continuum of human adaptability can reinforce your confidence. Each resilience evidence you accumulate adds a new thread to the tapestry of human coping strategies, potentially inspiring others in the future.

11.7.2 Leveraging Interdependence for Greater Resilience

From Individual to Collective Resources: When you recognize these interdependencies, you see that your resilience evidences aren't formed solely by your internal processes. You can draw strength from diverse cultural practices (like using an Ubuntu-inspired communal approach), rely on digital tools, or learn from historical success stories. Each time you adapt to a system-level disruption – like shifting to remote work, redesigning routines due to environmental changes, or learning new digital skills – you confirm that complexity can be navigated with the help of these broader networks.

Adaptive Learning and Emotional Intelligence in Context:

- **Emotional Intelligence (EI):** EI helps you collaborate effectively in a team or community. If you handle a group conflict productively and learn from it, that's a resilience evidence tied to social and cultural dimensions. EI transforms group tensions into opportunities for collective growth.
- **Adaptive Learning:** Digital platforms and global knowledge networks adjust to your skill level, continuously challenging you at the right difficulty. Progressing through these adaptive challenges, verifying each improvement with actual performance metrics, creates resilience evidences that confirm you can learn in sync with evolving conditions.

Reinforcing Anti-fragility Through Collective Contexts: Over time, these layers of interdependence mean your resilience isn't just personal insurance – it contributes to a more robust community. Your success in adapting to local disruptions might encourage neighbors to try similar approaches, spreading resilience strategies. Likewise, societal resources (e.g., emergency services, policy frameworks) can reduce the fallout from major challenges, allowing you to focus on strategic adaptation rather than mere survival. The synergy between personal action and collective support enhances the steady accumulation of resilience evidences, making anti-fragility not just an individual trait, but a communal property.

In summary, Section 11.6 highlights that your resilience evidences are woven into a broader tapestry of human interdependence. Physical infrastructures, cultural knowledge, social relationships, economic systems, and historical legacies all shape the arena where you test and refine your abilities. Recognizing these layers helps you leverage collective strengths and adapt more swiftly, ensuring that resilience isn't merely an individual aim but a shared, evolving resource that thrives in a complex, interconnected world.

11.8 TRANSFORMING POTENTIAL INTO PROVEN STRENGTH

Core Message: Throughout this chapter, we've explored the journey from theoretical understanding and wishful thinking to actionable, proven resilience. You began by recognizing how knowing about resilience differs from living it. You learned to dismantle helplessness – both learned and self-imposed – so you could attempt small steps that generate tangible improvements. You discovered how discomfort signals opportunities for growth and how each incremental triumph, no matter how modest, becomes a resilience evidence: a piece of living proof that you can handle more than you once believed.

You saw that big dreams are not the problem; aiming high without building a trail of smaller achievements sets you up for a harder fall. By breaking ambitions into manageable increments, every success along the way chips away at mental barriers, increasing self-efficacy and reducing fear. You learned to operate in that "sweet spot" of challenge, where growth is steady and supported by incremental evidence, rather than hinging entirely on a single, risky leap.

The Dr. Steve Peters Framework: Peters' chimp-human-computer model, combined with principles like controlled exposure, embracing discomfort, repetition, and flexibility, provides a blueprint for systematically embedding these resilience evidences into your mental "operating system."

Every deliberate action you take – practicing a skill under mild stress, iterating until calm responses become routine – programs your "Computer" with adaptive strategies. Over time, resilience isn't just something you aspire to; it becomes integrated into your default reactions.

Life-Long and Multi-Dimensional: Resilience evidences accumulate over a lifetime, from childhood trials to late-in-life adaptations. They are not restricted to personal milestones but emerge within a rich tapestry of cultural narratives, social connections, technological supports, and historical legacies. Understanding these layers helps you see that resilience is not only about personal grit but also about how you navigate and leverage the interconnected world around you. Drawing on societal resources, cultural wisdom, and collective problem-solving enhances your capacity to build and rely on resilience evidences.

From Self-Efficacy to Anti-fragility: With each evidence you earn, you reinforce self-efficacy – the confidence that your actions influence outcomes. Collectively, these evidences also nurture anti-fragility: the principle that stress and uncertainty can make you stronger, not weaker. Instead of fearing complexity, you learn to engage with it strategically, using each challenge to refine your capabilities, deepen your toolkit, and confirm your adaptive potential.

A New Narrative of Strength: As you look back on the methods and insights in this chapter, you can rewrite your internal story. You no longer rely solely on hope or abstract knowledge. You have a growing archive of resilience evidences – concrete instances where you overcame adversity, learned a new skill under pressure, built stepping stones toward a big goal, or handled complexity with composure. These evidences aren't just motivational anecdotes; they are structural supports in your mind, turning uncertainty from a threat into a stage for ongoing development.

Continuous Improvement and Future Challenges: The world will keep changing, and new obstacles will emerge. But with the approaches discussed – from facing illusions, breaking helplessness, building incremental achievements, handling big dreams realistically, systematically programming resilience routines, and recognizing societal layers – you are equipped not just to survive changes, but to thrive because of them. Each future challenge is a chance to add another resilience evidence, further solidifying your anti-fragile mindset.

Final Encouragement: Remember that resilience is a dynamic skill, not a static trait. By consistently applying these principles – testing your skills, celebrating small wins, leveraging communal and cultural strengths, and updating your strategies – you ensure that your capacity for adaptation grows over time. In this way, resilience evidences become the currency of personal growth, buying you confidence, clarity, and a calm, constructive approach to whatever complexity life offers next.

Your journey continues beyond this chapter. As you move forward, keep accumulating these evidences of resilience. With each new proof point, you move closer to living in a world where complexity is not something to fear, but a constant invitation to evolve, innovate, and flourish.

THE CONCEPT OF ANTI-FRAGILITY

THINKING BOX:

"What if we could *gain* from chaos instead of merely surviving it? That shift in mindset might hold echoes of **42**."

Part 4

Building Tomorrow

Human Resilience in a Tech-Driven World

TEASER: SCALING PERSONAL AGENCY INTO SYSTEMIC TRANSFORMATION

Having established ethical grounding (Part 1), cognitive mastery (Part 2), and resilience practices (Part 3), **Part 4** propels you into the **macro** realm – where personal insights catalyze **systemic** evolution. Technology's rapid rise, climate challenges, and social upheavals demand that your enhanced grasp of complexity be harnessed for **far-reaching** change. Now, your **personal agency** becomes a powerful tool for shaping how societies adapt and thrive in a rapidly changing world.

WHY READ ON?

- **Bridge Individual and Global**: See how the same frameworks that bolster personal well-being can **restructure organizations**, guide AI development, and inform holistic governance.
- **Co-Evolve with Technology**: Understand how to direct AI, big data, and digital platforms ethically, so human values drive progress rather than get overshadowed by purely algorithmic imperatives.
- **Confront Urgent Challenges**: We live in an era of climate crises, economic flux, and cultural divides. By learning large-scale strategies (model evaluation, collaborative policymaking, new economic narratives), you'll be ready to **steer** society toward inclusive, sustainable solutions.

WHAT AWAITS YOU?

- **Blueprints for Societal Resilience**: Discover how resilience-based thinking, once applied personally, can scale to transform entire communities, businesses, or governments in a **rapidly changing** environment.
- **Techno-Philosophical Governance**: Explore new governance frameworks (like a "Modern Society OS") that integrate moral reflection, agility, and advanced tech.

DOI: 10.1201/9781003640288-16

- **Foresight and Policy Innovation**: Learn how to channel **competition** into collaborative benefit, forge ethical AI policies, rethink capitalism, and tackle emergent crises with bold ingenuity.

Armed with deeper **personal** capacities for adaptation and an **expanded** lens on ethics and complexity, you'll close Part 4 ready to **participate** in – and even **lead** – the design of a future that respects human dignity, environmental balance, and purposeful technological advancement.

12 The Fundamental Limitations of Models

Seeing Beyond the Map

Key Concepts:

1. **The Model Boundary Problem**: Models exclude variables, leading to **overconfidence** in their predictive power. The **Boeing 737 MAX crashes** exemplify the dangers of **unrecognized model limitations**.
2. **Model Ouroboros**: A model's **outputs can shape reality**, reinforcing assumptions (e.g., **credit-scoring AI** altering borrower behavior).
3. **False Certainty and Overconfidence**: Precise-looking numbers create an **illusion of control**, even when key variables remain **unmodeled**.
4. **Proxy Metrics and Scenario Planning**: **External signals** (like social media sentiment or supply chain backups) **catch blind spots** early, while **stress tests** expose hidden failures.
5. **AI Amplification and Feedback Loops**: **AI-driven decisions** scale errors rapidly (e.g., **biased hiring algorithms**), requiring **human oversight** and multimodel validation.
6. **Over-Optimization Paradox**: **Efficiency kills resilience** – systems fine-tuned for stability collapse under **unexpected stress** (e.g., **lean supply chains during COVID-19**).
7. **Resilient Modeling**: **Diverse inputs, redundancy, and interdisciplinary validation** strengthen models against systemic failures.

Key Takeaway:

No model is reality – only a map. Understanding **what's excluded, monitoring for weak signals, and designing for adaptability** ensures models **guide rather than dictate** decisions.

12.1 INTRODUCTION: FROM PERFECT MIRRORS TO DISTORTED MAPS

12.1.1 Introduction

In 2018 and 2019, two Boeing 737 MAX aircraft crashed under puzzling circumstances, killing 346 people. The cause was traced back to a new flight-control feature known as the Maneuvering Characteristics Augmentation System (MCAS). On paper, MCAS was designed to prevent the aircraft from stalling by automatically pushing the nose down if an excessively high angle of attack is detected. In flight simulators and internal tests, the system appeared sound. Its engineers had modeled specific flight conditions, input ranges, and pilot response times – everything seemed to check out. Yet, in two real-world emergencies, those same assumptions led to catastrophic failures.

The core issue: **MCAS relied on data from a single sensor** and made certain assumptions about *how quickly* and *in what manner* pilots would respond to unexpected downward pitch corrections. This left no room for the possibility of sensor malfunction, human confusion, or even mild divergences from the "standard pilot" model used in simulations. Put simply, MCAS was optimized within the boundaries of its own logic and data, yet it could not account for factors lying outside that tight bubble. Once those real-world factors came into play – like a faulty sensor reading and a cockpit crew that didn't immediately identify the source of the erroneous nose-down commands – MCAS repeatedly overrode pilot control, culminating in the tragic crashes.

DOI: 10.1201/9781003640288-17

This stark example highlights a critical truth about models of all kinds, whether they govern software, business decisions, or safety systems: **no matter how intricate the model, no matter how polished the simulations, it always reflects only a fraction of reality.** By their very design, models simplify complexity; they're forced to neglect some variables, assumptions, or edge conditions. These omissions might seem minor at first, but in high-stakes scenarios – like ensuring an airplane's stability – small blind spots can quickly spiral into lethal flaws.

The Boeing 737 MAX crashes demonstrate the danger of trusting a seemingly "perfect" model without scrutinizing what it leaves out. They also illustrate an equally important theme: **the greatest risks often lurk in the model's blind spots**, not in the well-understood territory that the model covers. If managers, regulators, or even pilots had monitored the *potential stress on a single sensor system*, listened more closely to *pilot feedback* in simulator training, or tracked *prior near-incidents* as a proxy metric of risk, the MCAS flaws might have been identified – and corrected – far earlier.

By examining how a "perfect mirror" in simulation turned into a "distorted map" in reality, we confront the core challenge of this chapter: **to reveal why models, for all their utility, can never fully capture the messy richness of the real world.** It's a reminder that, while models help us predict and control, they can only guide us within the boundaries we set for them. Outside those boundaries, even small oversights or narrow assumptions can have an outsized impact. As the rest of this chapter will explore, this lesson applies to everything from financial forecasting and social media algorithms to supply chain optimizations and climate projections.

12.1.2 Models vs. Reality

The tragic Boeing 737 MAX story reveals an uncomfortable but vital truth: **all models, by their very nature, simplify reality.** They must. Any attempt to replicate every nuance of the real world – every possible input, interaction, or fringe condition – would be too vast and unwieldy to be practical. In exchange for being computationally or conceptually manageable, models inevitably come with **blind spots**. These blind spots might be unmodeled variables (like human stress reactions or sensor failures), overlooked feedback loops, or implicit assumptions that everyone tacitly accepts until something goes wrong.

When we **overestimate** how closely a model mirrors reality, we risk believing that "if the model says so, it must be true." This **overconfidence** can have profound implications, especially in high-stakes environments – finance, aviation, healthcare, or critical infrastructure. A model that seems precise (offering neat numerical outputs or alluringly crisp predictions) can lull decision-makers into complacency. Once we treat the model's outputs as infallible fact, systemic consequences become more likely. Misaligned economic forecasts can lead to market crashes; flawed autopilot systems can result in tragic accidents; biased AI-driven hiring tools can silently exclude entire segments of the workforce.

12.1.2.1 The Nine Key Aspects of Model Limitations

In this chapter, we'll explore **nine core aspects** of why models fail, fall short, or cause unforeseen damage:

1. **Models Are Never Perfect**: They approximate reality but cannot reproduce it entirely.
2. **The Model Boundary Problem**: Each model has explicit or implicit limits on what it includes – and excludes.
3. **Blind Spots from Within**: It's often impossible to see that a model is fundamentally wrong if you rely exclusively on the model itself.
4. **Proxy Metrics for Early Warnings**: Indirect signals can expose hidden flaws when the primary model does not.
5. **The Validation Challenge**: Traditional metrics like accuracy or precision can be misleading in complex systems.

6. **AI Amplification**: As AI usage grows, small model errors can multiply swiftly across large systems.
7. **Tools for Testing Boundaries and Assumptions**: Practical methods exist but must be applied rigorously.
8. **Optimization Dangers**: Refining a model within a flawed framework may make problems worse, not better.
9. **Over-Optimization Kills Resilience**: Striving for maximum efficiency often eliminates the buffers needed to handle surprises.

12.1.2.3 Seven Perspectives

In addition to these nine aspects, we'll introduce **seven perspectives** that further illuminate the limitations of models and how they interact with the world:

1. **Model Ouroboros**: Explores how a model's outputs can loop back into and reshape the reality they model.
2. **Invisible Hand of Models**: Reveals how algorithmic decisions subtly direct collective behavior – often without our awareness.
3. **Grandmother Paradox**: Highlights scenarios where simple, experience-based wisdom can outperform sophisticated, data-heavy models.
4. **Conservation of Complexity**: Shows how complexity can't be eliminated – only moved around, sometimes creating hidden vulnerabilities.
5. **Model Ecosystems**: Treats different models as interconnected entities that evolve, compete, and influence one another.
6. **Model Archaeology**: Emphasizes the need to document and preserve a model's developmental history for future understanding.
7. **Language of Models**: Frames each model as its own "language" for **simplifying and** capturing reality, with unique strengths and blind spots.

By blending these **nine key aspects** and **seven new perspectives**, we'll see **why** models can fail and **how** to spot their weaknesses before they do harm. The next sections will show that the world is invariably richer and messier than our most elegant simulations. Acknowledging that fact – and building models that admit their own imperfection – gives us the flexibility to adapt, detect early warnings, and minimize the fallout when (not if) our simplifications prove insufficient.

12.1.3 Why It Matters?

Having seen how models unavoidably simplify reality and risk creating blind spots, the natural question arises: **So what?** Why should we worry about these limitations, especially when models often help us make sense of complexity? The answer: **recognizing and addressing these limitations is crucial to building more resilient, ethically sound, and trustworthy systems** – in business, technology, governance, or anywhere models guide decision-making.

1. **The Core Aim**
   - **A Practical Compass**: This chapter doesn't seek to discourage modeling or diminish its importance. Instead, it aims to **equip you with practical tools** and **fresh perspectives** for detecting where models might go wrong and how to mitigate potential harm.
   - **Building Resilience**: When models fail dramatically – such as in high-profile financial crises, flawed safety systems, or biased AI algorithms – the ripple effects can be enormous. By understanding core pitfalls, you can design models (and the organizations around them) that withstand shocks and adapt to rapidly changing conditions.

2. **Actionable Insights**: By the end of this chapter, you'll have a **concrete roadmap** for responsibly using and refining models, even in the face of uncertainty. In particular, you'll learn how to:
    - **Spot Blind Spots**: Discover warning signs that your model is omitting vital factors or over-relying on narrow assumptions.
    - **Use Proxy Metrics**: Identify indirect signals (e.g., user feedback, sensor redundancy) that can uncover hidden risks your model doesn't directly measure.
    - **Avoid Over-Optimization**: Understand how chasing maximum efficiency can sometimes sap a system's ability to handle unexpected stressors – leading to brittle outcomes.
    - **Maintain Humility**: Accept that every model has boundaries and blind spots, no matter how "precise" it appears, and foster the organizational culture to address them.
3. **Beyond Theory**
    - **Real-World Impact**: These ideas are more than academic; they apply to everyday decisions about finance, product design, cybersecurity, environmental planning, and much more.
    - **Future-Proofing**: Models and data analytics are only growing in influence – especially as AI systems increasingly automate choices that once required human deliberation. Recognizing fundamental limitations isn't optional; it's essential for **long-term sustainability** in any model-driven environment.

In short, **why this matters** is simple: **models shape our world**, and how well we navigate their limitations can be the difference between beneficial outcomes and disasters. With that in mind, let's delve into the core concepts and frameworks that will help you use models as powerful tools – rather than unassailable oracles.

THE ILLUSION OF CONTROL

THINKING BOX

"Is the quest for total control itself an illusion? Maybe the cosmic joke, much like **42**, is that certainty is overrated."

12.2 THE MODEL BOUNDARY PROBLEM

Key Aspect Addressed: (#2 "Knowing the limits of our knowledge")

12.2.1 Defining the "Model Boundary"

Imagine drawing a circle that represents everything your model aims to include – its **assumptions**, **variables**, and **intended scope**. Everything inside this circle is what the model can measure, predict, or process. Everything outside it is effectively invisible to the model. This demarcation is the **model boundary**. It's where the model's internal logic stops and the unmodeled rest of reality begins.

12.2.1.1 What Is a Model Boundary?

- **Core Definition**: A model boundary is the **limit of what a model represents** – all those variables, assumptions, and contexts that a model was explicitly (or implicitly) designed to handle.
- **Purpose**: Boundaries keep a model tractable; they ensure we don't drown in endless detail. We focus on the factors we believe matter most.

- **Implicit vs. Explicit**: Some boundaries appear openly in the model's documentation or equations ("We assume constant temperature"), while others remain unstated, lurking in default settings or underlying data choices.

12.2.1.2 Linking Boundaries to Knowledge Limits

- **Knowledge Scope**: **Recognizing the model boundary is crucial for understanding the limits of our knowledge within that model.** Anything it doesn't capture lies outside its circle of competence.
- **Unmodeled or Under-Modeled Factors**: Whether it's human behavior, rare events, or economic shocks, if these weren't part of the original design, the model can't "see" them. This creates **vulnerabilities** – blind spots that can lead to inaccurate predictions or misguided decisions.
- **False Certainty**: When stakeholders forget – or never learn – where a model's boundary lies, they may assume it's universal or "good enough" for all scenarios. That's when small omissions can morph into big, sometimes disastrous, surprises.

12.2.1.3 A Simple Illustration

Consider a self-driving car model that **only** factors in road layout, traffic signals, and the car's position. On paper, this might be enough to navigate. But what if a child darts into the street without warning, or snow covers road markings? These events lie **outside** the neatly drawn boundary – omitted variables or assumptions about clear lanes and predictable pedestrian behavior. The result: the system might fail precisely at the moment we need it most.

12.2.1.4 Why Boundaries Need Reviewing?

- **Changing Reality**: Environments, markets, and technologies evolve; a model boundary that once fits well may become outdated.
- **Periodic Audits**: Building a habit of auditing and revisiting the boundary helps ensure that fresh data, new constraints, or emerging risks get folded into the model's design.
- **Risk Reduction**: By identifying what's left out, you can gauge how significant the omissions might be and whether you need additional checks (like proxy metrics or scenario testing) to guard against them.

With the model boundary clearly defined, we can now look more closely at **why** these limits matter – especially when unrecognized. In the next section, we'll see how a model's **outputs** can feed back into reality, reinforcing the very assumptions that may be incomplete or flawed.

12.2.2 Model Ouroboros in Action

One of the most intriguing (and sometimes troubling) phenomena in modeling is the **Model Ouroboros** – named after the mythical snake that eats its own tail. In this context, it describes a situation where the **outputs of a model loop back into the real world, altering behaviors or conditions**, which in turn **reinforce or distort** the model's original assumptions.

12.2.2.1 How Feedback Loops Form?

1. **Model Output → Real-World Change**
   - When a model makes predictions or prescribes actions, people (or systems) often adapt their behavior to match those outputs. This can be intentional (we follow a forecast) or unintentional (we modify our habits because a model shapes our environment).
2. **Real-World Change → Adjusted Input**
   - As people or processes adapt, the inputs feeding into the model can also shift. Over time, this can cause the model's original assumptions to become **self-fulfilling** – or dangerously off-base if the changes deviate from what the model expects.

3. **Blind Spot Amplification**
   - If the model has blind spots, these can get **amplified** by each loop of feedback. Because users trust or rely on the model's output, everyone might be operating under the same flawed premise, making it even harder to spot the missing pieces.

12.2.2.2 Case in Point: Credit-Scoring Algorithms

1. **Influencing Borrower Behavior**
   - Credit-scoring algorithms judge a borrower's riskiness based on income, debt history, credit utilization, and more. Once borrowers learn how these scores are calculated, they might change their behavior – paying off certain debts first or avoiding certain credit products – to maintain a favorable score.
   - **Upside**: This can encourage financially responsible habits.
   - **Downside**: If the model overlooks factors such as **emergency expenses** or **unexpected job losses**, borrowers might appear low-risk on paper but face sudden financial stress if life events fall outside the model's assumptions.
2. **Reinforcing the Original Assumptions**
   - Because borrowers adapt, the algorithm might "see" improved repayment stats and interpret that as confirmation it's accurate. In reality, deeper indicators of financial instability that remain unmeasured might be missing.
   - The model's success can thus **reinforce** its own limited worldview, leaving no incentive to revisit or expand its assumptions.
3. **Potential for Systemic Exclusion**
   - If the model penalizes borrowers from certain demographics or with nontraditional credit histories, those groups may be further marginalized – unable to access loans or forced into higher interest rates.
   - **Feedback Loop**: A lack of credit history keeps scores low, which in turn prevents building a history – locking people out of the traditional lending system. Over time, the loop becomes ingrained, and the model's blind spots turn into a systemic issue.

12.2.2.3 Why These Loops Are So Hard to Detect?

- **Inside View Trap**: From within the model, everything seems to be "working." The outputs align with the data, and the data aligns with the output. It's only by stepping outside – through proxy metrics, user feedback, or scenario testing – that we can see the mismatch between model-based reality and lived experiences.
- **Institutional Buy-In**: Organizations invest heavily in the design and deployment of models. When a model appears successful, they have little motivation to revisit fundamental assumptions. Instead, they refine parameters or tune performance, inadvertently deepening reliance on the existing feedback loop.

12.2.2.4 Avoiding the Ouroboros Pitfall

- **Regularly Expand the Boundary**: Reassess what your model measures, how it measures it, and which segments of reality remain excluded.
- **Seek External Input**: Gathering feedback from users or stakeholders who operate "outside" the model can flag overlooked issues. In the credit-scoring example, community organizations or alternative credit data (like rent and utility payments) can reveal unmet needs.
- **Monitor Proxy Indicators**: Keep watch on metrics that lie beyond the model's direct scope – e.g., changes in financial stress signals, consumer sentiment surveys, or local economic shifts.

By understanding the Model Ouroboros effect, we see how **model outputs can reinforce assumptions in a self-loop**, pushing real-world behavior to conform to the model – sometimes beneficially, often precariously. The result is a hidden or distorted portrait of reality, where it becomes ever harder to catch what the model fails to consider. This is precisely why regularly **questioning the model's boundaries** and seeking signals outside its immediate scope is so essential.

12.2.3 False Sense of Certainty

12.2.3.1 Neat Numbers, Hidden Gaps

Modern data-driven models – especially those powered by machine learning or statistical algorithms – often produce outputs wrapped in tidy numbers: specific percentages, confidence intervals, or predictive scores. At first glance, these values seem precise, suggesting that the model has a firm grasp on the underlying phenomenon. The danger is that this apparent precision can **mask critical omissions**. For instance, an economic model might show a highly confident projection of GDP growth, yet entirely overlook less-quantifiable elements like consumer psychology, public trust in institutions, or abrupt political shifts.

1. **Why Numbers Feel Trustworthy**
    - **Concreteness**: A forecast of "2.5% GDP growth" or "85% accuracy" feels more tangible than a vague "things might improve."
    - **Ease of Communication**: Numerical outputs are easily passed up the chain of command – management, media, and stakeholders often prefer concise statistics over nuanced explanations.
    - **Psychological Effect**: Precise figures can be comforting, implying that uncertainty is under control.
2. **Examples of Masked Variables**
    - **Economic Models**: They typically account for interest rates, unemployment, and inflation. But they may skip or downplay *behavioral economics*, sudden policy shifts, or social unrest – any of which can turn a forecast upside down.
    - **Healthcare Predictions**: A model might show strong accuracy in diagnosing a disease from medical images but fail to consider real-world factors like patients' access to consistent follow-up care or variations in clinical settings.
    - **Climate Projections**: Detailed simulations predict global temperature changes, yet local feedback loops (e.g., microclimates, urban heat islands) or human adaptive behaviors (like mass migration) might not be robustly included.

12.2.3.2 Overconfidence and Surprise Failures

When decision-makers see these polished outputs, they can forget – or never realize – what the model leaves out. Over time, an organization or industry that relies heavily on neat numeric forecasts may develop a collective **overconfidence**. Processes, resource allocation, and policies all become tailored to what the model says, leaving little room for "impossible" or "rare" scenarios that exist outside its boundary.

- **Heightened Risk of Shocks**: If a shock occurs – like a global pandemic, a sudden change in public opinion, or an unprecedented climate event – these "black swan" factors can catch the model (and everyone depending on it) off guard, causing massive disruptions.
- **Feedback Failure**: Because the model is seen as accurate, warnings from outside voices or conflicting proxy metrics may be dismissed. By the time a crisis arrives, the organization's ability to respond could be severely compromised.

12.2.3.3 Guarding against the Illusion

1. **Document Assumptions**
    - Keep a clear record of what your model includes – and, crucially, excludes. A sense of "we know what we don't know" can mitigate undue faith in neat outputs.

2. **Regular Stress Testing**
   - Simulate extreme or unlikely events (like sharp consumer mood swings or regulatory upheavals) to see if the model holds. If it falters, that's a clue to reexamine underlying assumptions or incorporate new factors.
3. **Encourage Skepticism**
   - Foster a culture where numerical forecasts are not treated as indisputable facts but as *guidelines* that must be weighed against experience, intuition, or alternative data. This is especially crucial when the stakes – financial, technological, or human – are high.

By acknowledging how easy it is to be seduced by tidy numbers, you're far better positioned to keep your organization's eyes open for what's *not* in the model. Awareness is the antidote to surprise failures: a truly robust approach pairs quantitative elegance with ongoing vigilance for potential blind spots – a concept we'll see echoed in the remainder of this chapter as we explore deeper modeling pitfalls and strategies to counter them.

12.3 PROXY METRICS AND THE VALIDATION CHALLENGE

Key Aspects Addressed:

- (#4 "Proxy metrics to spot trouble")
- (#5 "How do we know it works?")

12.3.1 Why Is Validation Difficult?

If a model achieves a 90% accuracy rate in predicting user engagement or disease onset, it might seem like a smashing success. Yet real-world systems are rarely so straightforward, and **traditional metrics** (like accuracy or precision) can be misleading. Even a model that appears to perform well statistically may harbor **hidden biases**, overlook critical outliers, or be vulnerable to unexpected scenarios – problems that remain invisible until they cause significant damage.

12.3.1.1 The Limits of Accuracy and Precision

1. **Narrow Focus on Aggregate Performance**
   - Accuracy, precision, and recall typically measure how well a model fits the *known* dataset. But they often fail to capture who the model fails on – for instance, a small but important subgroup or a rare event like system downtime.
   - Example: A fraud detection model might be 99% accurate overall but completely miss new forms of fraud that arise infrequently. That 1% of cases might represent huge financial losses or systemic weaknesses.
2. **Hidden Biases Remain Obscured**
   - Because standard validation relies on data from past conditions, it can inadvertently bake in historical inequities or oversights.
   - If the training data itself is imbalanced or incomplete, a high accuracy score may camouflage the fact that the model underperforms on specific demographics or misrepresents certain phenomena.

12.3.1.2 Overemphasis on Numeric Results

1. **Ignoring Emergent Effects**
   - Complex systems often have emergent properties – outcomes that arise from interactions within the system rather than from any single variable. Traditional metrics usually evaluate each prediction in isolation, missing these broader interactions.
   - Example: A social media feed algorithm optimized for engagement could inadvertently trigger polarization or misinformation spread – effects that wouldn't appear in standard precision/recall stats.

2. **Blind to Long-Tail Outcomes**
   - The "long tail" of rare but impactful events (e.g., freak accidents in a self-driving car scenario, sudden market collapses in finance) might show up in data only sporadically. Standard validation seldom accounts for these extreme cases, leaving systems unprepared for black swan events.
   - *Common Sense vs. Neat Metrics*: In many instances, simple human judgment or domain intuition might raise a red flag about improbable scenarios ("What if a major server farm goes offline?"), whereas the model can't compute that possibility because it's beyond its training data or boundary.

12.3.1.3 The Role of "Common Sense"

- **Grandmother Paradox Connection**: Sometimes, *experience-based wisdom* or gut instinct can catch anomalies that numeric metrics gloss over. A local business owner might sense market changes that a demand-forecasting model overlooks because the data lacks certain "ground-level" insights.
- **Domain Expertise**: Seasoned professionals (e.g., doctors, pilots, or financial analysts) may see subtle warning signs – like patient behavior or supply chain hiccups – that the model can't detect if those signals weren't explicitly included.

12.3.1.4 Conclusion: Validating beyond the Numbers

- **Multiple Lines of Evidence**: True model validation means blending quantitative metrics with qualitative assessments – interviews, external reviews, and real-world trial runs.
- **Look for Edge Cases**: Stress tests, scenario planning, and *proxy metrics* (which we'll address in more detail shortly) can reveal vulnerabilities missed by standard validation routines.
- **Keep Evolving**: Because environments and data shift, validation isn't a one-time event; it's an ongoing cycle of checking whether the model's assumptions and coverage still hold in changing real-world conditions.

Recognizing the limitations of traditional validation is the first step toward more robust modeling practices. Next, we'll explore how **proxy metrics** can serve as an external vantage point – an early warning system for blind spots that standard performance metrics fail to illuminate.

12.3.2 The Grandmother Paradox Expanded

At first glance, the idea that a **simple, experience-based approach** could outperform a cutting-edge, data-heavy model seems counterintuitive. After all, data science and advanced analytics promise objectivity, depth of analysis, and the ability to process massive amounts of information. Yet in real-world situations, "**Grandma's wisdom**" or **local domain expertise** can prove surprisingly accurate – sometimes more so than elaborate predictive systems. This phenomenon is often referred to as the **Grandmother Paradox**.

12.3.2.1 Simpler Is Sometimes Smarter

1. **Contextual Intelligence**
   - While data-driven models rely on historical datasets and predefined features, local know-how often captures nuances that never make it into the spreadsheet or the code. A seasoned logistics manager, for instance, might anticipate holiday surge patterns and road construction delays in ways a generic routing algorithm doesn't account for.
   - **Tacit Knowledge**: Through repeated exposure to a domain – like farming, patient care, or public policy – practitioners develop a "feel" for subtle variables (e.g., soil moisture, patient pain tolerance, or voter sentiment) that a purely quantitative model may miss or undervalue.

2. **Adaptability in Unforeseen Scenarios**
   - Complex models can become fragile when confronted with scenarios they haven't explicitly been trained on. By contrast, a simpler approach guided by practical experience can pivot more quickly, because it is not beholden to a narrowly defined dataset or rigid assumptions.
   - Example: An experienced community healthcare worker might notice an atypical fever pattern and act swiftly – even if an algorithm, trained on more common data, hasn't labeled it as high-risk yet.

12.3.2.2 Local Knowledge Beats Generic Data

1. **Rich, Hyper-Local Signals**
   - In many fields, ground-level knowledge – like observing weather signs or local social dynamics – offers immediate, qualitative cues that larger, aggregated datasets often wash out. A global climate model might miss highly localized microclimates that a longtime resident is keenly aware of.
   - **Proxy Metrics Connection**: These intangible signals can function like **proxy metrics**, flagging emerging issues early on. A store owner noticing that customers are suddenly inquiring about a certain product could be a better early predictor of demand than a lagging sales report.
2. **Why It Works**
   - **Experience-Centric Feedback Loops**: Practitioners continuously refine their intuition by seeing how a situation plays out in real time. Over decades, they amass a mental model that, while not formally documented, is grounded in deep observation of cause and effect.
   - **Multiple Layers of Context**: Domain experts often integrate social factors, emotional cues, historical events, or even the "mood" of a community – variables that big data models might treat as noise or fail to capture entirely.

12.3.2.3 Embracing the Paradox

- **Data as a Complement, Not a Replacement**
  - The lesson here is not to dismiss advanced analytics but to remember that local knowledge can reveal blind spots a model never considered. Ideally, you pair the two: letting data crunch large patterns while also listening to the voices "on the ground."
- **Cultural Validation**
  - Encouraging "Grandmother wisdom" in organizational decision-making fosters a culture where domain experts feel valued and where "soft signals" are taken seriously. This can unearth vulnerabilities or opportunities before data-based models catch on.

12.3.2.4 Putting It into Practice

- **Humility in Model Deployment**: Recognize that no algorithm, however advanced, will replicate the accumulated wisdom of a nurse with 30 years of experience or a community leader who knows every family on their block.
- **Structured Input**: Actively seek out expert input in the design phase. Provide channels where local feedback is documented and integrated so valuable insights are not lost in the noise of standard reports or dashboards.
- **Hybrid Approaches**: In particularly complex or high-stakes scenarios – like disaster response or advanced medical diagnostics – blend data-driven tools with real-time feedback from seasoned practitioners who can spot anomalies.

In short, the **Grandmother Paradox** reminds us that **data alone doesn't always tell the full story**. By honoring simpler, experience-based approaches, we broaden our perspective and gain resilience against the unpredictable nuances of real life – an essential mindset when we rely heavily on models that, by nature, must draw boundaries and simplify the truth.

12.3.3 Using Proxy Metrics

When traditional validation fails to capture the full picture, **proxy metrics** can serve as an early-warning system. Instead of trying to measure every factor inside a single model, proxy metrics **indirectly** track variables that the core model either can't handle or never considered relevant. They act like "canaries in a coal mine," signaling hidden trouble when standard performance measures remain quiet.

12.3.3.1 Definition: Indirect Signals for Hidden Factors

Proxy metrics are **indirect signals** that provide insight into **unmodeled** or **under-modeled** factors. By monitoring these signals, you can detect shifts that your core model won't pick up because they're outside its defined boundaries.

- **Simple Examples**:
  - **Employee morale** in a financial model can hint at upcoming productivity issues or ethical lapses.
  - **Biodiversity indicators** in a climate model might flag ecological stress well before conventional temperature or rainfall metrics change.

The central idea is that when your model can't capture the entire ecosystem of variables, an external "check engine light" helps spot anomalies early, prompting you to investigate further.

12.3.3.2 Real-World Examples

1. **Financial Risk and Reputational Insight**
   - *Scenario*: A corporate financial model focuses on revenues, cost of goods sold, and market share.
   - *Proxy Metrics*:
     - **Employee Turnover**: High turnover could indicate morale issues, lack of internal controls, or leadership changes.
     - **Social Media Sentiment**: A sudden spike in negative mentions might predict falling brand reputation, which the model won't see until sales tank.
   - *Why It Works*: These indirect signals often precede measurable drops in revenue or share price, giving you time to address underlying cultural or reputational risks.
2. **Healthcare Outcomes and Lifestyle Factors**
   - *Scenario*: A patient outcome model predicts disease progression based on clinical metrics – blood pressure, cholesterol, etc.
   - *Proxy Metrics*:
     - **Dietary Habits**: Tracking average caloric intake or meal composition can reveal a patient's commitment to long-term lifestyle changes.
     - **Physical Activity**: Wearable data (steps, heart rate) might detect early signs of recovery or setbacks.
   - *Why It Works*: Even if blood work looks stable, a decline in daily exercise could signal an impending health relapse that the core disease model can't see in its lab-based data stream.

12.3.3.3 Action Tip

By design, proxy metrics stand at the *edges* of your main model, watching for subtle shifts or emergent patterns the model doesn't account for. With that in mind:

1. **Identify One or Two Proxy Metrics**
   - Choose metrics that **reflect potential blind spots** – factors your main model doesn't monitor but that could impact outcomes.

- Examples: Team culture surveys, online customer complaints, ecosystem biodiversity counts, or supply-chain backup inventories.

2. **Regularly Monitor for Deviations**
   - Don't just collect proxy data – **plot it, track changes, and set alerts** if it veers outside a baseline.
   - When a proxy metric spikes or dips unexpectedly, use it as a cue to dig deeper, potentially adjusting your core model to address the newly uncovered issue.

Proxy metrics don't replace your main model; they **complement** it. Together, they form a more robust view of reality, helping you spot unmodeled influences before they escalate into full-blown crises.

12.4 AI AMPLIFICATION: SCALING UP THE RISKS

Key Aspect Addressed: (#6 "Issues intensify as AI use grows")

12.4.1 Acceleration and Autonomy

When traditional models fail, the damage is often localized – think of a single supply chain mishap or a regional economic forecast that misses the mark. **Artificial Intelligence (AI)**, however, operates at a different scale and speed. An AI model deployed in social media, finance, or logistics can influence thousands, sometimes millions, of decisions each day. This **rapid, large-scale impact** means that mistakes, biases, or blind spots aren't just repeated – they're magnified at an unprecedented rate.

12.4.1.1 Speed and Scale: Why AI Can Magnify Errors

1. **Real-Time Decision Loops**
   - Many AI systems, especially those using machine learning, process incoming data and make split-second decisions – updating news feeds, approving loans, or routing deliveries automatically. If an AI has flawed assumptions or incomplete training data, it's not just a single incorrect decision; it can generate **a barrage of poor decisions** before anyone notices.
   - **Example**: A recommender system that starts pushing increasingly extreme content can lead users down a rapid spiral, entrenching biases or misinformation far faster than a human gatekeeper would.
2. **Global Reach**
   - An AI-based model for translation or content moderation might serve millions of users worldwide, making localized mistakes ubiquitous. A minor error or oversight becomes system-wide, affecting diverse cultures and contexts the model never considered.
   - **Hyper-Connectivity**: Because most AI platforms are part of larger networks, erroneous outputs can cascade. For example, one flawed search engine result can be picked up by multiple third-party apps, magnifying the original mistake.

12.4.1.2 The Invisible Hand of Models

1. **Subtle but Powerful Influence**
   - People often **don't realize** the extent to which AI-driven algorithms shape their day-to-day decisions – what news they see, which products they buy, and even the job listings they encounter. This "invisible hand" of AI quietly guides collective behavior.
   - **Recommendation Systems**: By curating content based on past clicks, these systems reinforce user preferences (or biases), driving engagement at the expense of exposure to diverse viewpoints. Over time, this can lead to echo chambers or polarized communities, with nobody consciously choosing such outcomes.

2. **Automated Hiring, Invisible Exclusions**
   - In a hiring context, AI may filter résumés or rank candidates without revealing how it weighs certain features – educational background, keyword frequency, or zip code. Applicants who don't meet the opaque criteria might never even know they've been screened out.
   - **Feedback Loop**: If an organization trusts the AI's recommendations, it may stop advertising roles to certain demographics or discontinue outreach to historically underrepresented groups. Biases thus get **amplified** silently, making them hard to detect or rectify.

12.4.1.3 Why AI Acceleration Raises the Stakes?

1. **Limited Human Oversight**
   - As AI-driven processes become faster and more complex, human supervisors struggle to review every decision. The result: potential errors can slip through for days, weeks, or longer, compounding at scale.
   - **Erosion of Accountability**: When no single person fully understands the model's logic – particularly in black-box neural networks – assigning responsibility for errors becomes even more difficult.
2. **False Sense of Sophistication**
   - The sheer complexity of AI can lend it an aura of infallibility. Stakeholders might assume "the algorithm knows best," which deters questioning of the model's boundaries or biases.
   - **Disproportionate Impact**: In fields like healthcare or autonomous vehicles, even a small slip in performance can have life-or-death consequences, underscoring the need for rigorous checks and fail-safes.

Taken together, these factors highlight how **acceleration and autonomy** distinguish AI from more conventional modeling approaches. **Speed, scale, and invisible influence** elevate the consequences of modeling errors into the realm of systemic risk, underscoring the importance of recognizing limitations, monitoring proxy metrics, and maintaining human oversight. In the next sections, we'll explore more about how to mitigate these challenges through diverse model approaches and proactive governance.

12.4.2 Model Ecosystems in AI

AI models seldom operate in isolation. They **interact with other models, data pipelines, and user feedback**, creating a web of **interdependencies** that can magnify both benefits and risks. Rather than a single, standalone system, many AI-driven platforms function more like **ecosystems**, where each element continually shapes, and is shaped by, the others.

12.4.2.1 Why Ecosystems Matter?

1. **Interlinked Components**
   - An AI-driven recommendation engine relies on data from user profiles, content inventories, and advertising algorithms. Each of these, in turn, may depend on other data sources or sub-models.
   - **Chain Reactions**: A small tweak to one model – say, changing how content is ranked – can unintentionally alter the performance of related systems, such as ad placement or user retention strategies.
2. **Ecosystem Complexity**
   - As these systems grow more interconnected, **unintended interactions** can emerge. One AI might optimize for clicks, while another optimizes for time spent – leading to contradictory incentives and unpredictable user experiences.

- **Emergent Behaviors**: Much like in nature, AI ecosystems can exhibit "emergent" properties, where the global outcome is more than the sum of individual models' behaviors.

12.4.2.2 Social Media as a Prime Example

1. **Engagement Algorithms**
   - On platforms like Facebook, TikTok, or YouTube, the core AI aims to keep users engaged by serving content it predicts they'll watch or interact with. This goal, while straightforward on paper, creates a **loop** with user behavior.
   - **User Feedback**: The more users click or comment on a topic, the more the AI pushes similar content. This reinforcement often leads to content "bubbles" or intensifies trends – whether they're entertaining dance challenges or more polarizing political discussions.
2. **Content Trends and Behavioral Shifts**
   - When users repeatedly see certain types of posts, their **preferences and habits** may shift over time. They might start seeking more of that content, which in turn boosts the AI's confidence that "this is what people want."
   - **Feedback Loop**: As the AI "learns" from these amplified signals, it refines its recommendations, making it harder for alternative or less trendy content to surface. This dynamic can increase homogenization or polarization, depending on what the AI is optimizing for (e.g., maximum clicks vs. balanced exposure).

12.4.2.3 Potential Pitfalls of AI Ecosystems

1. **Hidden Cascades**
   - In a multimodel ecosystem, an error in one pipeline – like misclassifying a user's interests – can cascade through other systems, causing mismatched recommendations, skewed analytics, or misaligned advertisements.
   - **Difficult Diagnosis**: Tracing a single anomaly through a tangle of interdependent AI services can be challenging, leaving organizations unsure of how to fix the root cause.
2. **Resilience vs. Fragility**
   - On one hand, having multiple models can improve **resilience** – if one fails, others might compensate. On the other hand, tight coupling can make the system **fragile** if small faults propagate quickly through the network.
   - **Dynamic Adaptation**: Ecosystems can adapt to minor disruptions, but large shocks (e.g., sudden user behavior changes, new regulations) may push them into unforeseen states that none of the sub-models was built to handle.

12.4.2.4 Key Takeaways

- **Holistic Monitoring**: Treat the entire AI ecosystem – data sources, user interactions, and sub-models – as a single, evolving entity. Monitor interactions, not just individual component performance.
- **Scenario Testing**: Because emergent behaviors often don't appear in normal conditions, stress-test how changes in one model might ripple through the system.
- **Governance and Coordination**: Ensure that different AI teams communicate and that updates in one system are assessed for potential knock-on effects in related services.

By recognizing **AI as an ecosystem**, we see that the risks extend well beyond any single algorithm's boundary. In highly interlinked environments, **small adjustments or blind spots** can reverberate across multiple systems, amplifying errors or biases at scale. Understanding these dynamics is essential for responsibly deploying AI – and for staying alert to the ways models can unexpectedly reshape the world they're designed to serve.

12.4.3 Calls for Prudence

As AI systems grow in scale, complexity, and autonomy, the potential for **unintended harm** also expands. Even minor biases or blind spots, when replicated across millions of automated decisions, can produce outsized societal impacts. To prevent these pitfalls, organizations and policymakers need a **prudent** approach – one that champions **human oversight, diverse "model voices," and vigilant monitoring of proxy metrics**. Equally crucial is recognizing that **ethical considerations** must be woven into AI design from the start.

12.4.3.1 Why Human Oversight Matters?

1. **Catching Contextual Errors**
   - Even the smartest AI can miss situational nuances – like cultural norms, emerging crises, or moral considerations. Humans, with broad domain understanding and ethical judgment, can pause or override the AI when its outputs seem off-target.
   - **Example**: If an automated content filter begins to block historically significant images or crucial public health messages due to algorithmic confusion, a human reviewer can step in to correct it before the damage escalates.
2. **Accountability and Transparency**
   - In black-box neural networks, it's often unclear *why* the model produces a certain output. Human intervention can demand explanations or at least call for more transparent processes, ensuring that flawed or unjustified outcomes aren't allowed to persist unchallenged.
   - **Checks and Balances**: Formal oversight committees, ethics boards, or designated "human-in-the-loop" processes provide structured ways to spot systemic issues or potential abuses.

12.4.3.2 The Value of Diverse Model "Opinions"

1. **Multimodel Approach**
   - Relying on a single "best" AI can amplify hidden biases if its assumptions or training data are skewed. Running multiple models – each with different architectures or training sets – can highlight inconsistencies and spark investigation.
   - **Analogy**: Just as multiple weather forecasts can give a better sense of uncertainty, a suite of AI predictors can collectively indicate where the truth likely lies.
2. **Avoiding Monoculture Risks**
   - If an entire organization or industry standardizes on one type of algorithm or data source, blind spots can spread unchecked. Diversity in data and design means that if one model fails, others might compensate or at least flag anomalies.
   - **Example**: In financial services, combining rule-based approaches with machine learning can catch corner cases each method overlooks on its own.

12.4.3.3 Consistent Monitoring of Proxy Metrics

1. **Early Warning System**
   - As discussed earlier, **proxy metrics** are indispensable for detecting shifts outside a model's direct view. In large-scale AI systems, these external signals can highlight ethical blind spots, discrimination, or data drift before they grow destructive.
   - **Implementation**: Keep ongoing dashboards that track metrics like user well-being, user diversity in recommendations, or complaint rates. When these indicators deviate from a baseline, investigate promptly.
2. **Longitudinal Tracking**
   - Because AI systems adapt continuously to new data, a short-term improvement in performance can sometimes mask a hidden buildup of bias or a new vulnerability. Monitoring proxy metrics *over time* – week to week or month to month – offers a better lens into slow-building trends.

12.4.3.4 Ethical Considerations: Bias and Systemic Inequities

1. **Unintended Reinforcement of Bias**
   - AI can inadvertently reflect or even exacerbate societal prejudices baked into historical data. For instance, an automated résumé screener may discount candidates from certain backgrounds if prior hiring data carried biases.
   - **Systemic Risks**: Once deployed at scale, these biases can deepen divides – limiting socioeconomic mobility, reinforcing stereotypes, or excluding entire demographic groups.
2. **The Moral Imperative to Check AI**
   - *Fairness, Accountability, Transparency*: The widely recognized principle that AI must be held to ethical standards is not just a best practice – it's a necessity to uphold justice and public trust.
   - **Policy and Regulation**: As governments catch up to AI's rapid expansion, there's growing momentum for legislation that mandates bias audits, continuous model reviews, and transparent reporting of AI decision-making criteria.

12.4.3.5 Putting Prudence into Practice

- **Human-in-the-Loop**: Define specific checkpoints where a trained professional reviews AI decisions or test results, especially in high-impact contexts like hiring, lending, or medical diagnosis.
- **Diverse Perspectives and Interdisciplinary Teams**: Involve ethicists, sociologists, domain experts, and data scientists from different backgrounds to spot and mitigate blind spots.
- **Governance and Accountability**: Establish internal frameworks for documenting and justifying AI changes, so if a harmful outcome emerges, stakeholders can identify who's responsible – and how to fix it.

By incorporating **human oversight**, **diverse modeling**, **proxy metric monitoring**, and **ethical safeguards**, organizations can harness AI's transformative potential while minimizing systemic harm. As AI's influence accelerates, such prudence isn't optional – it's the foundation of responsible innovation.

12.5 PRACTICAL TOOLS FOR MODEL EVALUATION

Key Aspect Addressed: (#7 "Techniques for testing boundaries, validating assumptions")

12.5.1 Phased Approach

Many organizations struggle to determine *where* to begin with model evaluation, especially when faced with a plethora of techniques. A **phased approach** can help. Rather than trying every method at once, you **sequence** your efforts so that each step builds on the previous one. Below are two overarching phases that keep the process **organized**, **prioritized**, and **actionable**.

12.5.1.1 Phase 1: Boundary Testing and Assumption Audits

1. **Boundary Testing**
   - **Purpose**: Identify and challenge the limits of your model. Push it with extreme or out-of-scope inputs to see how it behaves.
   - **Why It Matters**: Models often fail at the edges. If your model only "thinks" within a narrow range of possible conditions, it may break down when real-world scenarios exceed those bounds.

- **Practical Steps**:
  - **Stress Testing**: Introduce extreme parameters (e.g., very high demand in a supply chain model, very low interest rates in an economic model) to see if outputs become erratic.
  - **Edge Case Exploration**: Ask "What if?" questions about rare but plausible events, like a sudden regulatory change or an unexpected system outage.

2. **Assumption Audits**
   - **Purpose**: Examine the foundational premises on which your model stands – data sources, causal links, and context.
   - **Why It Matters**: Over time, assumptions can become outdated or were never robust to begin with. Regular audits ensure they remain valid in changing circumstances.
   - **Practical Steps**:
     - **Documentation**: Keep a log of each assumption (e.g., "customer churn remains under 5% annually"). If an assumption turns out to be false, you can quickly trace and update the model.
     - **Stakeholder Consultation**: Talk to people who use or are affected by the model (e.g., frontline workers, domain experts) for input on whether your assumptions match real conditions.

These two methods – boundary testing and assumption audits – naturally reinforce each other. **Boundary testing** may reveal overlooked variables, prompting you to revisit the assumption list. Meanwhile, **assumption audits** can reveal "blind spots" that need direct stress tests in phase two.

12.5.1.2 Phase 2: Proxy Monitoring and Scenario Planning

1. **Proxy Monitoring**
   - **Purpose**: Deploy **proxy metrics** to capture external signals that the model doesn't directly measure (e.g., employee morale, unusual social media sentiment, or environmental feedback).
   - **Why It Matters**: If these proxy metrics deviate from historical baselines, it might hint at real-world shifts your model doesn't see – cluing you in before serious damage occurs.
   - **Practical Steps**:
     - **Identify Relevant Proxies**: Look for external data streams or "soft" measures that correlate with outcomes the model aims to influence.
     - **Alert Mechanisms**: Set thresholds or anomaly-detection rules so that if these proxies spike or plummet, it triggers an investigation into the model's performance or underlying assumptions.
2. **Scenario Planning**
   - **Purpose**: Go beyond boundary tests by imagining and simulating multiple "futures" – some plausible, others extreme – to see how the model adapts.
   - **Why It Matters**: Models often become brittle in unexpected scenarios (e.g., global pandemics, major political upheavals, or game-changing technology). Scenario planning helps you see which triggers cause the model to fail.
   - **Practical Steps**:
     - **Cross-Functional Workshops**: Gather diverse stakeholders to brainstorm a range of possible futures. Include "high uncertainty" elements like climate disruptions, major policy changes, or radical user behavior shifts.
     - **Model Adaptation Exercises**: After simulating each scenario, note where the model's outputs diverge most from expert judgment or proxy metric indicators. This reveals actionable gaps to address.

12.5.1.3 Making It Work

Start Simple:

- If you haven't done much evaluation before, focus on boundary testing and assumption audits first. These steps uncover the most glaring weaknesses with minimal overhead.

Refine, Then Advance:

- After addressing major blind spots, proceed to **proxy monitoring** and **scenario planning** for a more proactive, forward-looking stance. This second phase fine-tunes your model's resilience against evolving conditions.

Iterate:

- Model evaluation isn't a "one-and-done" checklist. Build a **continuous cycle** of re-testing boundaries, revisiting assumptions, tracking proxies, and exploring new scenarios as your environment changes or your model grows more complex.

By **phasing** your approach, you avoid overwhelming your team or resources with too many interventions at once. Instead, you systematically shore up fundamental assumptions and then tackle more advanced methods to ensure your model remains relevant, robust, and prepared for real-world surprises.

12.5.2 Boundary Testing and Assumption Audits

After deciding on a **phased approach**, the first major step typically involves **boundary testing** and **assumption audits**. These two methods often operate **in tandem**, each informing and reinforcing the other.

12.5.2.1 Boundary Testing: Pushing the Model to Its Limits

1. **Purpose**
   - **Stress-Test the Model**: Deliberately feed in **extreme** or **out-of-sample** inputs – values your model rarely or never sees in normal operation.
   - **Reveal Hidden Weaknesses**: When the model responds to these unanticipated inputs, you may observe erratic behaviors, wildly off predictions, or even system crashes. Each anomaly clues you into a **gap** in the model's design.
2. **Practical Steps**
   - **Parameter Extremes**: If your supply chain model typically sees demand in the range of 1–1,000 units, try simulating 10,000 or 0 to see how it holds up.
   - **Edge Cases**: Test environmental conditions like "unavailable network," "unexpected spikes in data input," or an **entirely new** user profile that the model never trained on.
   - **Question Combinations**: Don't just vary one parameter at a time; combine extremes in ways your real system could encounter (e.g., peak demand during a natural disaster).
3. **Why It Matters**
   - **Unmodeled Factors**: Boundary testing highlights where the model's assumptions break down. This could be missing variables like seasonal shifts, consumer panic behavior, or technological disruptions.
   - **Early Warning**: By exploring extreme scenarios, you get a sense of where real-world shocks might blindside the system, well before such shocks occur.

12.5.2.2 Assumption Audits: Cataloguing and Challenging Your Foundations

1. **Purpose**
   - **Document Each Assumption**: Identify the data sources, causal relationships, and contextual factors your model treats as given truths.

- **Confirm Validity Over Time**: Assumptions that were appropriate at launch might no longer hold in a rapidly changing environment or under new market conditions.

2. **Practical Steps**
   - **Assumption Inventory**: Create a living list of all major premises the model relies on – e.g., "User demand will not exceed 1,000 units per day," or "GDP growth remains above 2%."
   - **Stakeholder Reviews**: Include domain experts, front-line workers, and end-users in reviewing whether each assumption remains valid or has become outmoded.
   - **Update and Version**: Keep track of when assumptions are revised or replaced. This helps future audits quickly see what changed and why.
3. **Why It Matters**
   - **Guard against Stale Logic**: If an assumption shifts – like a new competitor entering the market or regulatory changes – the entire model might become obsolete without anyone noticing.
   - **Transparency**: A clear record of assumptions makes it much easier to pinpoint where a failure originated when something goes wrong.

12.5.2.3 When Boundary Testing Meets Assumption Audits

- **Reinforcing Cycle**: Boundary testing often uncovers surprising weaknesses that lead you back to question an assumption ("Why did the model fail when input X hit 2,000? Oh, we assumed it could never exceed 1,000.").
- **New Stress Tests**: Conversely, when you discover an unacknowledged assumption ("We assumed user preference doesn't change seasonally"), you can design targeted boundary tests to see how the model behaves if that assumption is violated (e.g., shifting user preference by 50% in winter months).

12.5.2.4 Taking the Next Step

With boundary testing and assumption audits in place, you've **identified** many structural weaknesses and misaligned premises. The **next phase** – proxy monitoring and scenario planning – helps to **continuously** watch for real-world deviations and to explore hypothetical futures the model might someday face. Combined, these techniques form a robust validation and adaptation cycle that keeps your model relevant, resilient, and less prone to hidden disasters.

12.5.3 Proxy Monitoring and Scenario Planning

Once you've **stress-tested boundaries** and **audited assumptions**, the next step is to **keep a vigilant eye** on factors outside your model's primary scope and to **regularly envision** how different futures might unfold. Together, **proxy monitoring** and **scenario planning** serve as a dynamic toolkit for spotting trouble early and preparing for surprises.

12.5.3.1 Proxy Monitoring: Watching for External Warning Signs

1. **Ongoing Tracking of External Signals**
   - **Definition**: Proxy metrics are indirect indicators that catch early shifts in reality that your model doesn't explicitly measure (e.g., user well-being, brand sentiment, biodiversity markers).
   - **Why It Matters**: These signals function like a "check engine light." If they veer from historical norms, it may hint at a problem the core model can't see – such as emerging ethical concerns, brand erosion, or environmental stressors.
2. **Practical Examples**
   - **User Well-Being**: A social media algorithm might not measure mental health, but tracking user surveys or usage breaks can reveal if engagement strategies are causing burnout or dissatisfaction.

- **Brand Sentiment**: Sales data alone might look healthy, but a spike in negative social media mentions could point to reputational damage waiting to impact revenue down the line.

3. **Implementation Tips**
   - **Set Thresholds and Alerts**: Define normal ranges for each proxy metric, then trigger alerts when values stray from these baselines.
   - **Cross-Functional Insight**: Encourage marketing, HR, or community engagement teams to feed relevant proxy data into dashboards, ensuring anomalies are recognized and acted upon promptly.

12.5.3.2 Scenario Planning: "What If?" Simulations for Future-Proofing

1. **Anticipating Low-Probability, High-Impact Events**
   - **Definition**: Scenario planning is the art of imagining multiple plausible future states – some probable, others extreme – to see how your model (and your organization) would respond.
   - **Example**: A financial forecasting model that only considers incremental market changes may fail if a sudden regulatory shift or geopolitical event drastically alters investment behavior.
2. **Why It Matters**
   - **Exposure of Blind Spots**: By stepping into hypothetical worlds (e.g., abrupt climate regulations, radical new tech breakthroughs), you uncover vulnerabilities that normal data-driven forecasting might overlook.
   - **Strategic Adaptation**: Scenarios help leadership allocate resources or develop contingency plans – so if a major shift does occur, you're not scrambling at the eleventh hour.
3. **Practical Steps**
   - **Multidisciplinary Workshops**: Gather experts from varied fields – finance, operations, ethics, legal – to brainstorm both likely and unlikely scenarios.
   - **Model Stress Tests**: For each scenario, inject relevant changes into your model – e.g., doubling raw material costs or halving user growth – and observe how it performs.
   - **Action and Review**: Document which assumptions broke down, what additional data you might need, and how decision-making processes should pivot if that scenario becomes reality.

12.5.3.3 Encouraging Continuous Learning

1. **Monitor, Adapt, Repeat**
   - **Data and Reality Shift**: Industries evolve, user behaviors shift, and global events can quickly render old assumptions moot. By continuously monitoring proxies and periodically refreshing scenario plans, you keep the model aligned with current conditions.
   - **Feedback Loops**: When proxy metrics show a significant deviation or a scenario test reveals a major blind spot, feed those findings back into boundary testing and assumption audits – updating your model or its underlying logic.
2. **Organizational Flexibility**
   - **Cultivate a Culture of Reflection**: Regularly discuss what has changed in the operating environment (e.g., new market entrants, technology disruptions). Does the model still hold up?
   - **Iterative Improvement**: Each cycle of measurement and planning should prompt incremental refinement, ensuring your model remains robust rather than static.

By combining **proxy monitoring** with **scenario planning**, you gain **both** short-term alerts to creeping issues **and** long-term foresight for shaping strategy. In an ever-evolving landscape, these techniques help you pivot gracefully, ready to handle both predictable trends and the wild cards lurking just beyond your model's current boundary.

12.6 THE OVER-OPTIMIZATION PARADOX

Key Aspects Addressed:

- (#8 "Optimization can worsen errors")
- (#9 "Over-optimization kills resilience")

12.6.1 When Efficiency Backfires

Organizations often seek to **optimize** processes – reducing costs, speeding up workflows, or maximizing outputs. While this can yield impressive **short-term** gains, it may introduce **hidden vulnerabilities** that surface only under stress. By laser-focusing on efficiency, these systems become tightly coupled, leaving **no slack** for dealing with unpredictable events. This phenomenon ties into the concept of **Conservation of Complexity**: **simplifying or minimizing one area** often pushes **unmanaged complexity** into another, increasing the risk of sudden breakdowns.

12.6.1.1 Conservation of Complexity: A Hidden Trade-Off

1. **Short-Term Gains vs. Long-Term Fragility**
   - When you cut inventory, reduce staffing, or eliminate seemingly "wasteful" processes, you may indeed boost efficiency – at least when conditions are stable. However, the **underlying complexity** doesn't vanish; it's merely displaced into less-visible areas (e.g., supplier flexibility, workforce resilience, or emergency response capabilities).
2. **The Illusion of Control**
   - Optimized systems often appear more predictable and cost-effective. But in reality, these systems can be **over-tuned** to a narrow band of conditions, giving a false sense of security. When an unexpected shock arrives – like a rapid demand spike or a raw-material shortage – the system can't pivot quickly.
3. **Why It Matters**
   - **Resilience Requires Redundancy**: Extra buffers, time, or capacity sometimes seem inefficient on paper but provide critical "give" when reality changes.
   - **Blind Spots Grow**: Over-optimized models rarely consider highly unlikely events – until one occurs, causing a domino effect of failures.

12.6.1.2 Real-World Example: Lean Supply Chains

1. **Just-in-Time (JIT) Approach**
   - Many global companies adopted JIT logistics to reduce inventory costs. Parts arrive "just-in-time" for manufacturing or assembly, eliminating storage overhead. This yields strong margins in calm periods.
2. **Unforeseen Disruptions**
   - **Problem**: If a natural disaster, transportation strike, or pandemic halts supply lines, the entire chain grinds to a halt within days – sometimes hours. Because there's no **buffer stock**, even minor hiccups can balloon into major production stoppages.
   - **2020–2021 Examples**: During the COVID-19 pandemic, many industries found themselves scrambling for components (e.g., semiconductors, medical supplies) because the global supply chain had been optimized to the point of brittleness.
3. **Conservation of Complexity in Action**
   - By streamlining distribution, companies pushed complexity into **supplier reliability** and **global logistics**. When these external facets failed, the entire system collapsed. What seemed like a cost-saving triumph became a costly crisis virtually overnight.

12.6.1.3 Recognizing the Over-Optimization Trap

1. **Signs of Fragility**
   - Discomfort with any form of slack or buffer ("It's wasted capacity!")
   - Inability to absorb minor fluctuations without major disruptions.
   - Dependence on a single supplier or data source with no backup plan.
2. **Balancing Efficiency and Slack**
   - **Marginal Redundancy**: Slightly overstaffing key areas, keeping modest inventory buffers, or holding alternative suppliers on standby can provide enough flexibility to absorb shocks.
   - **Periodic Stress Tests**: As with boundary testing, running "what-if" scenarios can reveal how quickly a hyper-efficient system snaps under pressure.

12.6.1.4 Moving Forward

Pursuing efficiency isn't inherently wrong, but **over-optimization** can be. By acknowledging **Conservation of Complexity**, you accept that there's no free lunch in trimming slack – somewhere, an equivalent or greater complexity is piling up, waiting for a catalyst to expose it. The solution isn't to give up on optimization but to **temper it** with deliberate redundancy, scenario testing, and a clear-eyed view of how quickly things can unravel when unplanned shocks emerge.

In the next sections, we'll see how this paradox extends to other domains – from financial modeling to AI-based recommendations – and how adopting a more **resilient stance** can stave off catastrophic failures caused by lean systems pushed too far.

12.6.2 Revisiting the Grandmother Paradox

In the rush to optimize for maximum efficiency, it's easy to dismiss simpler, experience-based methods as outdated or suboptimal. Yet these "less optimized" or **domain-expert** approaches often include **built-in buffers** and **contingencies** that purely model-driven systems overlook. This phenomenon connects back to the **Grandmother Paradox** – the observation that even in a world of advanced analytics, **common sense** or local know-how can sometimes beat the "best" algorithm under real-world pressures.

12.6.2.1 Why Simple Approaches Can Outperform?

1. **Contextual Safeguards**
   - Veteran practitioners who've lived through disruptions learn to keep buffers – extra inventory, backup suppliers, contingency cash flows – even when data suggests these measures are "inefficient."
   - **Human Experience**: Domain experts integrate unwritten rules of thumb, cautionary tales, or historical lessons that never made it into the model's dataset.
2. **Adaptive Flexibility**
   - **Overly optimized** models can be rigid, fine-tuned to a narrow band of assumptions. A domain-expert approach, while seemingly rough around the edges, can pivot quickly because it doesn't rely on a single, precise forecast.
   - **Safety Margins**: "We always keep a week's supply of materials" might look wasteful in normal times – but it's a game-changer when a sudden spike or disruption hits.

12.6.2.2 A Local Merchant's Storm Prep

1. **Extra Inventory vs. "Just-in-Time"**
   - **Scenario**: A neighborhood shop owner sees that a major storm is looming. Historical data might show storms cause a slight uptick in demand, but the "optimal" supply chain model says to stick to JIT ordering to avoid excess inventory costs.

- **Merchant's Choice**: Drawing on personal experience (perhaps recalling past storms, listening to customers talk about stocking up), the merchant orders extra essential goods – flashlights, batteries, and nonperishable foods.

2. **Outcome When the Storm Hits**
   - **Preparedness Pays Off**: While purely model-driven competitors run out of stock, this merchant has enough supplies to serve both regulars and new customers, forging goodwill (and profits) in a crisis.
   - **Hidden Costs of Lean**: Competitors save money under normal conditions but fail catastrophically when shipping routes are compromised or demand spikes beyond the model's assumptions.

12.6.2.3 Lessons for Over-Optimized Systems

1. **Anticipating the Unexpected**
   - The local merchant's approach underscores that *not everything* fits neatly into a historical dataset, especially for rare or high-impact events. Experience-based methods leave space for outliers and black swan scenarios.
2. **Hybrid Strategies**
   - Rather than dismissing data-driven insights, **combine** them with the real-world expertise of stakeholders, employees, or community members who've seen issues unfold firsthand.
   - **Diverse Input**: Collaboration between supply chain analysts and frontline staff can yield a strategy that's efficient under normal conditions yet retains enough slack to weather storms, literally and figuratively.

12.6.2.4 Embracing Grandma's Wisdom (without Discarding Data)

- **Balanced Optimization**: Striving for constant improvement is laudable – but remember that certain "inefficiencies" (like carrying buffers) can serve as crucial shock absorbers.
- **Cultural Acceptance**: Organizations should encourage a culture where old-school caution and local knowledge are valued, not ridiculed as backward. Sometimes, a hunch from a seasoned worker can save the day – especially when the data is based on incomplete or changing contexts.

In sum, the **Grandmother Paradox** warns us against blindly trusting equations that promise lean perfection. **Real-world resilience** often arises from domain-savvy safety nets, acknowledging that storms – literal or metaphorical – can sweep through at any time.

12.6.3 Designing for Resilience

Modern systems – whether they're supply chains, financial models, or AI-driven platforms – often chase lean efficiency to the point of **fragility**. Yet **resilience** isn't about failing slowly; it's about **bouncing back quickly** (or even benefiting) when confronted with unexpected challenges. Achieving resilience typically involves **deliberate inefficiencies** – like carrying extra capacity or splitting tasks across multiple models – so that no single breakdown can topple everything.

12.6.3.1 The Power of Partial Redundancy

1. **Extra Copies, Spare Capacity**
   - **What It Is**: Having an extra server, buffer stock, or backup route that may go unused during normal operations.
   - **Why It Matters**: Should a key component fail or a spike in demand occur, you have built-in slack to absorb the hit. This buffer helps you avoid catastrophic shutdowns.
   - **Example**: A data center maintaining one additional server beyond peak load requirements. It might look "inefficient," but if a primary server fails, the system stays operational.

2. **Trade-Off: Minor Overhead vs. Major Savings**
   - **Overhead**: Redundancy incurs some cost – be it physical space, maintenance fees, or staff time.
   - **Major Savings**: In high-stakes scenarios, that cost pales compared to the **massive** losses or reputational damage from a system-wide failure.

12.6.3.2 Diversity of Data Sources

1. **Multiple Data Pipelines**
   - Depending on a single data feed – say, one exclusive supplier of raw materials or one unique user-data stream – puts the entire system at risk if that feed goes stale or corrupt.
   - **Solution**: Gather information from different, ideally *independent*, sources to cross-check anomalies and guard against biases in one dataset.
2. **Resisting Monoculture Risks**
   - In agriculture, planting the same crop across huge areas heightens vulnerability to a single pest or disease. Similarly, **monoculture** in data or modeling can amplify a single point of failure.
   - **Model Example**: An economic forecasting system that relies only on historical interest-rate data might miss shifts in consumer behavior. Combining multiple time-series, survey data, and domain judgments can reduce blind spots.

12.6.3.3 Multiple Small Models vs. One Big Model

1. **Divide and Conquer**
   - **Single Over-Optimized Model**: A big, integrated system can be elegantly efficient in stable conditions but may crumble if one part goes awry.
   - **Several Targeted Models**: By deploying smaller, specialized models, each focusing on different subproblems or data segments, you limit how far an error in one model can propagate.
   - **Analogy**: Instead of one giant, fragile ship, you have a fleet of smaller vessels – if one sinks, the others can still complete the mission.
2. **Modular Updates**
   - When each sub-model handles its own piece, you can upgrade or replace components **incrementally** without risking the entire system. If Model A (e.g., demand forecasting) proves unreliable, you can swap it out without overhauling Model B (e.g., logistics optimization).

12.6.3.4 Tolerating Minor Inefficiencies for Long-Term Payoff

1. **Why Imperfection Can Be Strategic**
   - Deliberately building inefficiency – extra capacity, partial overlap, or model "diversity" – keeps you from locking into a single, optimized path. It's a form of **option value** that can be priceless when conditions shift rapidly.
   - **Adaptability > Lean Perfection**: The moment reality deviates from your exact assumptions, a hyper-optimized system can fail swiftly. Slightly less efficiency under normal conditions buys you adaptability under stress.
2. **Real-World Examples**
   - **Microservices in Software**: Instead of one monolithic codebase, companies deploy multiple microservices. This approach is less "efficient" in pure overhead but more resilient to bugs or scaling issues.
   - **Diversified Investment Portfolios**: Even if one asset underperforms, others can balance the risk, preventing total collapse when markets swing.

12.6.3.5 Embracing Resilience Mindset

- **Plan for the Unknown**: Proactively imagine how your system behaves if one data source vanishes or if demand doubles overnight. Design enough slack so that failure in one area isn't catastrophic.
- **Cultural Shift**: Incentivize teams not only to improve efficiency metrics but also to consider "what if" scenarios, backup options, and how to make changes in small increments rather than sweeping overhauls.
- **Iterative Learning**: Like boundary testing, the pursuit of resilience is ongoing – monitor real performance, gather feedback, and make incremental adjustments to maintain enough cushioning without squandering resources.

Ultimately, **resilient design** is about acknowledging that you can't predict or optimize for every twist in the real world. By incorporating partial redundancy, multiple data pipelines, and smaller specialized models, you create a system prepared to **bend without breaking** when confronted with the unexpected. This approach, far from being an indulgent "extra," may well be the best guard against catastrophic failures in today's fast-changing environment.

THE CONCEPT OF TIME AFFLUENCE

THINKING BOX

"Is time our greatest luxury in a hyperconnected era? Consider how **42** might symbolize the balance between hustle and being present."

12.7 WEAVING IN THE SEVEN IDEAS: A QUICK RECAP

Throughout the chapter, we've referenced **seven novel perspectives** that help illuminate the limitations of models and their real-world impacts. Let's revisit each concept – now with slightly more refined definitions – so you can see how they collectively shape a deeper understanding of responsible modeling.

12.7.1 Model Ouroboros

Refined Definition: A situation where a model's outputs loop back into the real world, **influencing the very system** it aims to represent.

- **Example**: Credit-scoring algorithms encourage borrowers to modify spending behaviors in pursuit of better scores. Over time, this **reinforces** the model's assumptions, making those behaviors and the model's predictions appear mutually validating, even if some critical factors remain missing.

12.7.2 Invisible Hand of Models

Refined Definition: The subtle, often **unseen influence** that algorithmic decisions wield over societal behaviors, forming a **distributed power structure** with little overt accountability.

- **Why It Matters**: People may not realize how these models – used in recommendation systems, pricing algorithms, or hiring tools – shape their day-to-day choices and opportunities. Because the influence is indirect and decentralized, it can profoundly affect social or economic outcomes without a single entity visibly "pulling the strings."

12.7.3 Grandmother Paradox

Refined Definition: The paradox where **simpler, experience-based**, or domain-expert practices can sometimes **outperform** more complex, data-driven models.

- **Practical Insight**: Rather than dismissing old-school wisdom as primitive, consider that local knowledge or well-honed intuition might capture **critical nuances** not reflected in the data. This synergy of qualitative and quantitative approaches can yield greater resilience and accuracy.

12.7.4 Conservation of Complexity

Refined Definition: The principle that **complexity is never truly removed** – only **redistributed**. If you try to oversimplify one aspect of a system, the complexity you trimmed away typically resurfaces elsewhere.

- **Example**: Eliminating "excess" inventory in supply chains might reduce short-term costs but push complexity into logistics or vendor stability. The system remains complex; it's just hidden in a different place, posing vulnerability under stress.

12.7.5 Model Ecosystems

Refined Definition: The concept that **multiple models coexist, interact, and evolve together**, creating an ecosystem of interdependencies. Adjustments to one model can reinforce or disrupt others, sometimes producing emergent behaviors beyond any single model's scope.

- **Why It Matters**: In large organizations or multipart platforms (e.g., social media, finance), you rarely have just one model. Monitoring how these models **cooperate or clash** ensures that optimizing one doesn't unintentionally destabilize the rest.

12.7.6 Model Archaeology

Refined Definition: The practice of **documenting and preserving** a model's assumptions, rationale, and version history so future auditors (or the model's own designers) can trace how certain decisions were made.

- **Importance**: In the event of a crisis, "archaeological" records reveal which assumptions led to errors and how the model was supposed to behave. Without this, accountability and effective fixes become far harder to achieve.

12.7.7 Language of Models

Refined Definition: Recognizing that **each model operates as its own "language,"** with unique ways of simplifying and describing reality – meaning each model has distinct strengths, blind spots, and points of failure.

- **Implication**: Translating insights between these "languages" (e.g., a statistical model vs. a rule-based system vs. a neural network) is critical. No single approach captures everything; combining multiple "languages" can yield a more holistic view of the underlying phenomena.

12.7.7.1 Why These Ideas Matter Together?

Taken together, these seven ideas shed light on **why** models fall short and **how** to manage those shortcomings.

- **Ouroboros and Invisible Hand** highlight *feedback loops* and *silent steering*, showing that models don't just describe – they actively shape reality.

- **Grandmother Paradox and Conservation of Complexity** remind us that human wisdom and hidden complexity still have a place alongside data analytics.
- **Model Ecosystems** underscore that models rarely exist in a vacuum, *Model Archaeology* emphasizes traceability and accountability, and **Language of Models** shows the value (and difficulty) of bridging different modeling paradigms.

By weaving these perspectives into your modeling practices – whether you're building AI systems, forecasting financial markets, or planning global logistics – you gain a multifaceted lens for spotting blind spots, anticipating emergent behaviors, and remaining humble about how well any single model can reflect the rich tapestry of the real world.

12.8 CONCLUSION: BUILDING RESILIENT MODELING PRACTICES

12.8.1 Key Takeaways

1. **Models Are Inherently Limited:** Throughout this chapter, we've seen time and again that models, no matter how sophisticated, can only represent a fraction of real-world complexity. They rely on assumptions, simplified data, and bounded scopes that leave certain factors out. **Acknowledging these blind spots** is your first line of defense against catastrophic errors – whether in aviation safety, financial forecasting, or AI-driven recommendations.
2. **Over-Optimization Leads to Brittleness:** When a model or system is relentlessly tuned for peak efficiency under stable conditions, it often becomes **fragile** under stress. Lean supply chains, hyper-efficient scheduling, or single-sensor dependencies can work flawlessly until they encounter a shock that lies outside the narrowly defined boundary. **Ignoring proxy signals** – like pilot feedback, user complaints, or environmental indicators – further compounds this vulnerability. As we've seen, such brittleness can amplify damage when real-world events deviate from a model's ideal scenario.

12.8.2 Cultivating Resilience and Humility

1. **Blend Quantitative Tools with Qualitative Wisdom:** Robust models don't arise solely from data-driven insights or purely from "Grandmother Paradox" intuition; they thrive on **both**. Real-world unpredictability demands that you **balance** sophisticated analytics (machine learning, statistical forecasts) with **domain expertise** (seasoned professionals, frontline employees, local knowledge). This dual approach catches phenomena that purely numeric or purely anecdotal methods alone might miss.
2. **Maintain Slack, Redundancy, and Multiple Approaches:** In an era that prizes lean efficiency, it may feel counterintuitive to keep "extra" resources or run multiple overlapping models. Yet **partial redundancy** – be it backup inventory, extra server capacity, or multiple data streams – can be the difference between minor disruption and full-scale crisis when uncertainty hits. Equally important is the idea of using **several smaller models** rather than a single, hyper-optimized juggernaut; if one fails, others can still keep the system afloat.

12.8.3 A Call to Action

The concepts in this chapter are meant to be more than theory. Here are three **concrete steps** you can take right now to move your organization or project toward more resilient, ethically responsible modeling:

1. **Identify a Key Assumption and Test It**
   - **Boundary Check**: Pick an assumption your model relies on – say, "user demand won't exceed 10% growth" – and push it beyond typical boundaries. What if demand grows

30% or drops to near zero? Are you prepared for that? Document the results and see whether the model's logic holds up or falters.

2. **Monitor at Least One Proxy Metric**
   - **External Signal**: Choose a metric lying outside your model's direct scope – like employee turnover, user well-being surveys, or local weather variations – and watch for deviations. If it starts trending in an unexpected direction, investigate whether your model is missing an emerging factor.
3. **Document Decisions in a "Model Archaeology" Record**
   - **Future Audits**: Create a simple log of the model's versions, assumptions, data sources, and rationale for major changes. This record is invaluable if you ever need to pinpoint the origin of a failure or revisit why certain decisions were made.

12.8.4 Strengthening the Final Note

Throughout this chapter, we've underscored that "**no model is the territory**" – it's only a map, a simplified representation. This echoes our **opening hook**, reminding us that **the real world remains richer and more unpredictable** than any simulation or forecast. Learning from past failures, such as aviation disasters or financial crises, ensures we don't repeat those mistakes. Instead, we can navigate today's uncertainties with greater **wisdom and resilience**.

Ultimately, **balancing clarity with complexity**, **efficiency with slack**, and **data-driven insights with intuitive expertise** provides the best safeguard against both sudden shocks and slow-burn threats. By embracing a well-rounded approach to modeling – one that welcomes both precise analytics and the nuanced knowledge of domain experts – you position your organization to harness models responsibly, **reaping their benefits** without falling prey to their hidden pitfalls.

That's the essence of **resilient modeling practices**: a commitment to continuous learning, open-minded skepticism, and the humility to accept that even our most polished models can't replace the infinite nuance of reality.

12.9 QUICK REFERENCE

12.9.1 How to Use This Quick Reference

Keep this **glossary and resources** section handy as you implement or expand upon the frameworks discussed in the chapter. Whether you're running a sensitivity analysis, seeking to make AI more interpretable, or simply looking for practical tools to track assumptions, these references provide a quick springboard for informed, responsible modeling.

12.9.2 Glossary of Key Terms

- **Model Boundary Paradox:** The inherent limitation of models to capture all aspects of reality due to their necessary simplification. Highlights the risk of overlooking unmodeled or under-modeled factors.
- **Holistic Metrics:** Broader measures that capture well-being, equity, and ethical considerations, extending beyond traditional metrics like accuracy, efficiency, or cost. Especially vital in social, environmental, or policy-driven models.
- **Explainable AI (XAI):** A set of techniques designed to make AI models more interpretable and understandable to humans, often used to clarify decision-making processes in black-box systems like neural networks.
- **Continuous Integration and Continuous Deployment (CI/CD):** Practices enabling automated updates and improvements to models (and software in general) based on incoming data and feedback. Helps keep models current with minimal manual intervention.

- **Sensitivity Analysis:** A method to determine how different variables or input parameters affect a model's outputs. Useful for identifying critical drivers and understanding uncertainty in model predictions.
- **Scenario Analysis:** The process of exploring and analyzing potential future conditions – ranging from likely to highly improbable – to test model robustness and decision-making under uncertainty.
- **Proxy Metrics:** Indirect measurements used to approximate more complex or hard-to-measure variables. Commonly employed to detect early warning signs of unmodeled shifts in real-world conditions.
- **Variance-Based Methods:** Techniques (e.g., Sobol indices) that decompose the variance of model output into contributions from different input parameters, highlighting which variables most influence the model's behavior.
- **Sobol Indices:** Quantitative measures used in global sensitivity analysis, evaluating how much each input parameter (and interactions among parameters) contributes to the total variance in model outputs.
- **Morris Method:** A screening approach for sensitivity analysis. Efficiently identifies the most influential parameters in a model by systematically varying them and tracking resulting output changes.
- **High-Performance Computing (HPC):** Advanced computing systems (clusters, supercomputers) capable of running large-scale or computationally intensive models – crucial for complex simulations (e.g., climate forecasts, large neural nets).
- **Model Ensembles:** Collections of multiple models, each with different assumptions or structures, run in parallel. By comparing or combining results, you gain a range of possible outcomes, improving robustness.
- **Big Data Analytics:** Techniques (machine learning, parallel processing, etc.) for analyzing very large or complex datasets to extract patterns, correlations, or trends at scale.
- **Feedback Mechanisms:** Systems or processes for collecting and integrating real-world data, stakeholder input, or end-user feedback into models, promoting continuous refinement and adaptation.

12.9.3 Additional Resources

12.9.3.1 Books and Articles

- **Weapons of Math Destruction**, by Cathy O'Neil: Explores the societal implications of opaque, large-scale modeling – particularly how algorithms can perpetuate inequality if left unchecked.
- **The Ethical Algorithm**, by Michael Kearns and Aaron Roth: Examines how to design algorithms with fairness, privacy, and transparency in mind, offering insights into balancing innovation with ethical guardrails.
- **Interpretable Machine Learning**, by Christoph Molnar: A hands-on guide to making machine learning models more transparent, covering local interpretability methods (LIME, SHAP), rule-based approaches, and global explanations.

12.9.3.2 Online Courses

- **Coursera**: *AI For Everyone* by Andrew Ng: A nontechnical overview of AI's capabilities, limitations, and ethical considerations, suitable for anyone from business leaders to product managers.
- **edX**: *Ethics and Governance of AI*: A course exploring the social impact of AI, its ethical dimensions, and approaches to responsible deployment in various sectors.

12.9.3.3 Tools

- **Miro:** A collaborative whiteboarding platform for visualizing assumptions, biases, or data flows in a model. Great for remote workshops and cross-functional brainstorms.
- **Amazon Mechanical Turk:** A crowdsourcing marketplace for large-scale data collection or validation tasks. Useful for generating labeled datasets, user feedback, or quick surveys.
- **MATLAB and R:** Statistical software packages for conducting sensitivity analyses, advanced data processing, and modeling experiments with built-in libraries for scenario planning.
- **AnyLogic and NetLogo:** Simulation platforms well-suited for **complex scenario modeling**, including agent-based simulations, discrete-event models, and dynamic systems – ideal for stress tests and ensemble approaches.

13 Capitalism, AI, and Future Societies

Key Concepts:

1. **Poly-Crises vs. Meta-Crises: Poly-crises** (e.g., climate change and migration) are **interlinked crises**, while **meta-crises** (e.g., capitalism's infinite growth demand) are **root causes** requiring **deep structural change**.
2. **The Moloch and AGI Challenge**: The **AGI race** mirrors the Moloch effect – competition drives speed over safety, risking **unpredictable system failures**.
3. **Possible Solutions:**
    - **International AI governance** to enforce safety standards.
    - **Incentive realignment** to reward AI safety over unchecked progress.
    - **Transparency frameworks** to counter monopolization of AGI.
    - **Decentralized AI development** to prevent power concentration.
4. **Leveraging Moloch for Good: Gamification, competitive sustainability incentives, and DAOs** can **redirect competition toward cooperation and resilience**.
5. **AI for Collective Intelligence**: AI can **facilitate global collaboration, optimize resource distribution, and mediate conflicts** for **holistic problem-solving**.

Key Takeaway:

To **avoid capitalism and AGI pitfalls**, **restructure incentives, prioritize transparency, and leverage AI for cooperation** – transforming Moloch from **a force of destruction into a driver of collective progress**.

13.1 INTRODUCTION

As global crises intensify, driven by the forces of capitalism and the rapid advancement of Artificial General Intelligence (AGI), humanity faces unprecedented challenges. The energy crisis, exacerbated by geopolitical tensions and reliance on fossil fuels, serves as a stark example of how systemic failures can lead to widespread consequences. This crisis reflects the Moloch effect, where individually rational decisions result in collective catastrophes. Without decisive action, such crises could escalate into social unrest, environmental collapse, and existential risks. This chapter critically examines the systemic risks linked to capitalism and AGI development, exploring potential solutions while also addressing the challenges and limitations inherent in their implementation.

13.2 CAPITALISM AND PEACE: A DOUBLE-EDGED SWORD

Capitalism has often been viewed as a mechanism for promoting peace, largely through the economic growth it fosters via trade and investment. By increasing prosperity and reducing the need for conflict over resources, economic interdependence can make direct military confrontations less appealing, especially between major powers. The idea is that when economies are closely linked, the cost of war outweighs any potential benefits, encouraging peaceful resolutions.

However, capitalism is not without its complexities and systemic risks. The scarcity of critical resources, such as water and rare earth elements, could challenge this peaceful equilibrium.

DOI: 10.1201/9781003640288-18

As nations seek to secure their economic futures, competition over these resources may intensify, potentially leading to new forms of conflict. Moreover, while direct wars might be less common, the intricate economic connections between states can give rise to indirect conflicts or proxy wars, where nations engage in power struggles through third parties rather than direct confrontation.

These dynamics illustrate that while capitalism can indeed contribute to global peace, it also carries inherent risks that need to be carefully managed. Recognizing these risks allows for a more balanced approach, where the benefits of capitalism are harnessed without overlooking the potential for new conflicts arising from resource scarcity and economic competition.

THE INFLUENCE OF DIGITAL MINIMALISM

THINKING BOX

"Could simplifying our digital lives free up space for deeper thinking? The notion, much like **42**, might yield surprising clarity."

13.3 THE MOLOCH OF CAPITALISM: SYSTEMIC FAILURES IN ACTION

The concept of Moloch serves as a powerful metaphor for understanding the systemic risks inherent in capitalism. In this context, the Moloch effect refers to situations where individuals or smaller groups, acting in their own rational self-interest, inadvertently contribute to negative outcomes on a broader scale. These actions are often not driven by malice or ill intent; rather, they result from the logical pursuit of success within the capitalist framework. However, the cumulative effect can lead to outcomes that are detrimental to the common good.

One classic example is the "tragedy of the commons," where shared resources are depleted because each individual, acting in their own interest, consumes more than their fair share. This overuse occurs because no single actor has the incentive to preserve the resource, leading to collective harm. This dynamic can be seen in various areas where capitalism's focus on short-term profits and individual gains can overshadow long-term sustainability.

Examples of the Moloch Effect in Action:

- **Financial Crises**: Banks and financial institutions often seek to maximize short-term profits through riskier investments. While these actions can yield high returns for individual entities, they can also create systemic vulnerabilities that lead to broader economic collapses, as seen in the 2008 financial crisis. The pursuit of immediate gains, though rational for each institution, ultimately destabilizes the entire financial system.
- **Antibiotic Resistance**: In agriculture and medicine, the overuse of antibiotics can be driven by the need for efficiency and immediate results. However, this widespread use accelerates the development of resistant bacteria, posing significant global health risks. Here, the individual or sectoral decisions, though logical in isolation, contribute to a larger public health crisis.
- **Plastic Pollution**: Companies often opt for cheaply produced plastics due to cost efficiency and consumer demand for affordable products. While this decision makes sense from a business perspective, the long-term environmental impact of non-degradable plastics is profound. The collective result of these decisions is a growing environmental burden that is difficult to reverse.

These examples illustrate how capitalism, through the Moloch effect, can unintentionally create systemic failures. However, recognizing this dynamic also opens the door to potential solutions.

By understanding the mechanisms that drive these outcomes, we can design interventions that align individual interests with the collective good. For instance, regulatory frameworks, incentive structures, and international agreements can be developed to mitigate these risks and encourage more sustainable practices.

The key is not to reject capitalism but to refine its mechanisms so that the pursuit of individual success can also contribute to positive outcomes on a larger scale. By addressing the systemic risks inherent in the Moloch effect, it is possible to harness the benefits of capitalism while minimizing its negative externalities.

13.4 META-CRISES VS. POLY-CRISES: RECOGNIZING AND CRITICALLY ASSESSING THE ROOT OF THE PROBLEM

To effectively address global challenges, it is crucial to distinguish between poly-crises and meta-crises, as each requires different approaches. Poly-crises involve multiple simultaneous crises that may or may not reinforce one another but stem from different causes. In contrast, meta-crises are deeper systemic problems that underlie many of these individual crises, often acting as the root causes of broader issues.

Examples of Poly-Crises:

- **Climate Change and Migration (Reinforcing)**: Climate change leads to extreme weather events, which, in turn, trigger migration flows as people seek safer living conditions. These crises intersect and amplify each other, creating a complex web of challenges that demand both immediate and long-term responses. In this case, the crises reinforce one another, exacerbating the overall impact.
- **Pandemics and Economic Crises (Non-Reinforcing)**: A global health crisis, such as a pandemic, can occur independently of an economic downturn but still lead to significant economic consequences. While the health crisis and the economic crisis are interconnected, they may not directly reinforce each other; instead, they represent parallel challenges that need to be addressed simultaneously. For example, a pandemic might trigger economic decline, but the economic downturn does not necessarily worsen the pandemic itself.

Examples of Meta-Crises:

- **Capitalism and Growth**: The relentless pursuit of economic growth under capitalism often leads to environmental degradation, social inequality, and long-term instability. These systemic issues act as underlying drivers of many individual crises, requiring deep structural reforms to address them effectively.
- **Technological Progress**: Rapid advancements in technology, without sufficient ethical considerations and regulation, can lead to societal disparities and job losses. This meta-crisis, driven by unchecked technological growth, manifests in various crises, from unemployment to social unrest, and requires a holistic approach to mitigate its impact.

Understanding the distinction between these two types of crises is essential because they necessitate different solution methods. **Poly-crises** often demand quick, targeted interventions that address the immediate symptoms of the crisis. For example, economic stimulus measures may temporarily alleviate a recession, but they do not solve the deeper issues that caused the economic downturn in the first place. **Meta-crises**, on the other hand, require systemic approaches that tackle the root causes, such as rethinking the underlying economic or technological frameworks.

Addressing meta-crises, however, poses significant challenges. Mobilizing political and societal will for profound changes can be difficult, especially when these changes threaten short-term economic interests. For instance, global environmental regulations, though necessary for sustainability, often face resistance due to their potential economic disadvantages. Political inertia and conflicting interests among stakeholders further complicate the implementation of systemic reforms.

Despite these challenges, inaction is not a viable option. Allowing poly-crises to fester without intervention risks escalating them into more severe crises, while failing to address meta-crises can perpetuate the cycle of instability. Quick fixes for poly-crises might stabilize situations temporarily, but they often leave the underlying systemic issues unresolved, leading to future crises. Similarly, avoiding the hard work of systemic reform leaves society vulnerable to recurring meta-crises, with ever-increasing costs and risks.

Finding the right balance between managing immediate crises and pursuing long-term reforms is key. Political strategies must be developed to address both short-term needs and the structural changes required to prevent future crises. This might involve integrating rapid-response mechanisms with more comprehensive reforms, ensuring that while immediate issues are being managed, efforts are also being made to address the root causes.

In summary, understanding and differentiating between poly-crises and meta-crises allows for more effective responses to global challenges. By tailoring solutions to the nature of the crisis and recognizing that inaction is not an option, it is possible to mitigate the risks and work toward sustainable, long-term stability.

13.5 THE MOLOCH AND AGI: A NEW CHALLENGE

The development of AGI introduces a profound challenge within the Moloch framework, where rational self-interest and competition among actors can unintentionally lead to catastrophic outcomes. The race to achieve AGI superiority, driven by the desire for technological leadership, risks overshadowing the necessary focus on safety, control, and broader societal impacts. The true disruption here does not necessarily come from the technology itself but from how it is used – shaped by the incentives and pressures of the competitive environment.

One of the most significant challenges posed by AGI development is its inherent complexity. As AGI systems become more advanced, their behaviors and interactions with the world grow increasingly difficult to predict. This complexity may make it impossible to conduct comprehensive testing in a wide enough range of scenarios to ensure safety. Traditional methods of risk management, which rely on understanding and mitigating risks, may be insufficient in the face of such uncertainty. The Moloch effect amplifies this problem, as the drive to be the first to develop AGI can lead to decisions that prioritize speed and competitive advantage over caution and thorough evaluation.

In this context, the phrase "there is no such thing as disruptive technology, but only the disruptive use of it" becomes particularly relevant. AGI itself is not inherently disruptive; it is the way in which it is deployed, driven by competitive pressures and self-interest, that creates the potential for systemic crises. The Moloch effect ensures that even well-intentioned actors, focused on maximizing their success within the capitalist framework, may contribute to outcomes that are collectively harmful.

Examples of Current Research Approaches and Safety Protocols in AGI Development:

- **OpenAI's Safety Research**: OpenAI's efforts to develop safety mechanisms are a response to the unpredictable nature of AGI. However, the complexity of AGI systems means that even these well-intentioned efforts may fall short in the face of real-world applications that were not anticipated during testing.
- **Asilomar AI Principles**: While these principles provide a framework for ethical AI development, the competitive pressures of the AGI race challenge their consistent implementation. The voluntary nature of these guidelines highlights the difficulty of enforcing safety standards across a diverse and competitive landscape.

- **International Cooperation**: Efforts to foster international dialogue, such as those by the Partnership on AI, are essential but face significant obstacles. The global competition for technological supremacy often conflicts with collective efforts to ensure AGI safety, as nations and corporations pursue their own interests.

Addressing the Root Causes: To mitigate the risks associated with AGI, it is crucial to address the underlying incentives and structures that drive the competitive race. Rethinking how innovation is rewarded – shifting from a model that prioritizes speed and profit to one that values safety and long-term stability – is essential. This could involve developing international agreements that impose enforceable standards, ensuring that all actors adhere to safety protocols and ethical guidelines. However, even these efforts must recognize the limitations imposed by the complexity of AGI systems and the unpredictability of their real-world use.

The Moloch effect, where the competitive pursuit of success leads to collective harm, is particularly potent in the context of AGI. By acknowledging this dynamic and focusing on the systemic factors that drive the race for AGI, we can work toward a future where AGI is developed in a way that minimizes the risks of unintended and potentially irreversible consequences.

In conclusion, the challenge of AGI development lies not just in the technology itself, but in the competitive pressures that shape its use. Recognizing the Moloch effect at work in this domain allows us to focus on the root causes of the risks associated with AGI, rather than attempting to manage them through traditional frameworks that may be inadequate in the face of such complexity.

13.6 POSSIBLE SOLUTIONS FOR AGI CHALLENGES

The Moloch effect, where individual actors pursue rational self-interest at the expense of collective well-being, lies at the heart of the challenges associated with AGI development. Any effective solutions must directly address these systemic forces, rather than simply managing the symptoms. Here are five strategies aimed at tackling the Moloch dynamics in AGI development, with a focus on the diverse motivations across different sectors.

1. **International Cooperation with Enforced Constraints on Competition**: While global agreements on AGI development are often suggested, their effectiveness hinges on reining in the competitive pressures that drive unsafe practices. The Moloch effect is evident in the reluctance of nations and corporations to adhere to agreements that might hinder their progress. Therefore, any cooperation must come with enforceable constraints that limit the ability of individual actors to gain an advantage by cutting corners. This means establishing binding international standards that all participants must follow, reducing the incentive to prioritize speed over safety. However, achieving such cooperation is a complex task, as it requires aligning the interests of diverse stakeholders who are often focused on short-term gains.
2. **Restructuring Innovation Incentives to Counteract Moloch Dynamics**: The race for AGI is driven by the pursuit of rapid innovation and market dominance – key elements of the Moloch trap. To counteract this, innovation incentives must be redesigned to reward safety, ethical practices, and long-term thinking. This could involve creating funding mechanisms that prioritize projects focused on AGI safety and control, rather than those that simply push technological boundaries. By shifting the reward structure, it's possible to reduce the pressure on actors to engage in risky behavior. However, this approach requires a fundamental shift in how success is measured in the tech industry – a shift that might face resistance from both the private sector and policymakers.
3. **Complexity-Aware Safety Research to Address Unpredictability**: The complexity of AGI systems makes traditional risk management approaches insufficient, as they often fail to account for the unpredictable ways in which these systems might behave in real-world

situations. The Moloch effect amplifies this problem by pushing actors to prioritize innovation over thorough safety testing. To mitigate this, safety research must be tailored to the complexities of AGI, focusing on developing adaptable protocols that can handle unforeseen challenges. This means investing in research that goes beyond known risks and anticipates potential unknowns. However, convincing stakeholders to invest in this type of research, especially when it doesn't yield immediate returns, is a significant challenge within the competitive landscape.

4. **Transparency and Collaboration as a Check on the Moloch Effect**: Transparency and collaboration can serve as countermeasures to the Moloch effect by reducing the ability of individual actors to prioritize self-interest over collective safety. Joint platforms for sharing research and best practices can help ensure that knowledge is distributed more equitably and that safety measures are not neglected in the pursuit of competitive advantage. However, the very nature of competition makes it difficult to foster a culture of openness in AGI development. Overcoming this requires creating neutral, third-party organizations that facilitate collaboration without threatening the competitive edge of individual participants. This would involve developing trust and shared goals among stakeholders who are otherwise incentivized to keep their advancements proprietary.
5. **Strategic Regulation to Slow the Race and Dismantle the Moloch Trap**: Some advocate for slowing down AGI development to allow more time for safety considerations. While this is a potential solution, it must be implemented in a way that directly addresses the Moloch effect – the tendency for competition to push development forward at unsafe speeds. Regulations that enforce slower development can help dismantle the Moloch trap by leveling the playing field and ensuring that no actor gains an advantage by moving too quickly. However, such regulations are likely to face significant resistance from those who fear being left behind in the race. It will be crucial to balance the need for progress with the imperative of safety and to ensure that regulations are not simply bypassed in favor of short-term gains.

AI Singularity: **Sector-Specific Motivations and Regulatory Considerations**: The development of AGI is influenced by diverse motivations across different sectors, each driven by unique priorities and challenges:

- **Academia**: Focuses on pure research and innovation, driven by the pursuit of knowledge and scientific advancement. However, the pressure to publish and achieve breakthroughs can sometimes lead to neglecting long-term safety considerations. The Moloch effect in academia often manifests as a race for academic recognition, which can overshadow broader societal implications.
- **Military**: Prioritizes strategic superiority and national security, often pushing for rapid development to maintain an edge over potential adversaries. In the military sector, the Moloch effect is particularly strong, as nations are driven by the imperative to outpace others, sometimes at the expense of safety and ethical considerations. The competitive dynamic here is intensified by the high stakes of national security.
- **Industry**: Aims for profit maximization and efficiency, with companies racing to be first to market with AGI innovations. In the industrial sector, the Moloch effect is evident in the pursuit of market dominance, where short-term profits often take precedence over long-term risks. Regulatory oversight is crucial to ensure that the drive for profit does not compromise safety.
- **Government**: Seeks to enhance public welfare and regulate ethical standards, but faces the challenge of balancing competing interests from academia, the military, and industry. Governments are often caught in the Moloch trap of trying to regulate without

stifling innovation while also managing national interests in a competitive global landscape. Adaptive policies and continuous dialogue are necessary to navigate this complex environment and ensure that AGI development benefits society as a whole.

Balancing these motivations requires a multi-faceted approach that includes adaptive policies, international cooperation, and continuous dialogue between sectors. Each sector's priorities must be recognized and addressed in the development of AGI, ensuring that the Moloch effect does not drive AGI toward outcomes that are disruptive or uncontrollable.

Addressing the Root of the Moloch Problem: While these solutions aim to mitigate the risks of AGI development, they also target the systemic dynamics that fuel the Moloch effect. The feasibility of these solutions is challenged by the entrenched nature of competition and self-interest in the tech industry. However, inaction only allows the Moloch dynamic to continue unchecked, increasing the risk of catastrophic outcomes. By restructuring incentives, enforcing cooperation, and developing complexity-aware safety measures, we can begin to dismantle the Moloch trap that drives unsafe AGI development. Though difficult, these steps are necessary to redirect AGI development toward a future that benefits humanity as a whole, rather than one that leads to unintended and uncontrollable consequences.

In conclusion, the solutions to AGI challenges must address the Moloch effect at its core. By realigning incentives and creating structures that prioritize collective safety over individual gain, we can work toward a future where AGI serves the common good. Although implementing these solutions is undoubtedly challenging, they are essential steps in preventing the Moloch dynamic from leading us into a crisis that could have been avoided.

13.7 CREATIVE AND PROMISING SOLUTIONS TO COMBAT THE META-CRISIS, INCLUDING AGI

To effectively combat the meta-crisis and mitigate the Moloch effect, innovative and systemic approaches are essential. These solutions must address the underlying incentives that drive individual and organizational behavior, ensuring that actions aligned with self-interest do not lead to collective harm. Additionally, many of these approaches can be tailored to tackle the unique challenges posed by AGI development.

1. **Creating Incentive Structures for Sustainable Actions**:
   - **General Application**: Sustainability dividends offer tax benefits and subsidies to companies that adopt eco-friendly practices. Green bonds and financial products promote investments in sustainable projects, realigning financial incentives with environmental goals.
   - **AGI-Specific Application**: Similarly, AGI developers who prioritize safety and ethical considerations could receive tax benefits or regulatory advantages. This approach encourages the development of AGI that aligns with societal well-being rather than purely market-driven incentives. The challenge here is ensuring that these incentives are meaningful and not merely symbolic.
2. **Cooperative Platforms and Collective Intelligence**:
   - **General Application**: Crowdsourcing platforms and collective intelligence can harness global expertise to solve complex problems. Open-source innovations make sustainable technologies widely accessible, encouraging collaboration over competition.
   - **AGI-Specific Application**: In the realm of AGI, cooperative platforms could bring together AI researchers, ethicists, and policymakers to collaboratively address safety and ethical concerns. Open-source AGI projects focused on developing safety protocols and ethical guidelines could help ensure responsible AI development. The challenge is fostering collaboration in a competitive field where proprietary interests often dominate.

3. **Implementing Circular Economy Models**:
   - **General Application**: Circular economy models prioritize regeneration and reuse, promoting business practices that minimize waste and conserve resources. Extended Producer Responsibility holds companies accountable for the entire lifecycle of their products.
   - **AGI-Specific Application**: Circular principles can also apply to AGI, where systems are designed for adaptability and continuous monitoring. By ensuring that AI systems can be updated and corrected post-deployment, developers can reduce the risk of unforeseen consequences. However, integrating these principles in a field focused on rapid innovation can be challenging.
4. **Strengthening Public-Private Partnerships**:
   - **General Application**: Public-private partnerships (PPPs) allow governments and companies to collaborate on sustainable infrastructure and technology projects. Joint research centers funded by public and private means can address systemic crises.
   - **AGI-Specific Application**: In AGI development, PPPs could fund research into AI safety, ethics, and governance, reducing the incentives for unsafe competition. Joint research centers focused on AGI safety can serve as neutral grounds for collaboration, but balancing public and private interests remains a challenge.
5. **Promoting Sustainable Consumer Behavior**:
   - **General Application**: Education and awareness campaigns encourage sustainable behavior and informed consumption. Consumer incentives, such as discounts and rewards for eco-friendly choices, align individual actions with collective well-being.
   - **AGI-Specific Application**: Similar strategies can be applied to AGI by raising awareness about the risks and ethical considerations of AI. Creating a market for "safe and ethical AI" products encourages companies to adhere to higher standards. The challenge is establishing robust certification mechanisms to ensure that products truly meet these standards and aren't merely "AI-washed."

Addressing the Feasibility Challenge in AGI: While these solutions offer promising ways to combat both the meta-crisis and the Moloch effect in AGI development, their feasibility must be critically assessed. The competitive dynamics in AI development are deeply entrenched, making systemic change difficult. However, the alternative – continuing on the current path without addressing these systemic risks – poses far greater dangers. Implementing these solutions may be challenging, but it is essential to redirect AGI development toward long-term safety and collective well-being.

13.8 THE MOLOCH HACK: EMBRACING COMPETITION FOR GOOD

The Moloch effect, where competition drives individuals and organizations to act in ways that may harm collective well-being, can be harnessed for positive outcomes if carefully managed. However, this approach carries risks, as even well-intentioned competition can lead to unintended consequences. To mitigate these risks, it's essential to design systems that channel competitive energies toward beneficial goals while remaining vigilant about potential downsides. These strategies can also be adapted to address AGI-related challenges, where competition plays a significant role in shaping development.

1. **Gamifying Sustainability for Collective Action**:
   - **Global Competitions for Sustainability Goals**: The inherent competitive drive of the Moloch effect can be redirected toward sustainability by organizing global competitions. Countries, cities, and companies could vie to achieve net-zero emissions, with the most innovative carbon reduction solutions receiving recognition and rewards. This approach can stimulate rapid advancements in green technology by leveraging the competitive nature of participants. However, there is a risk that focusing too heavily on competition could lead to short-term fixes that win competitions but do not provide lasting

solutions. To avoid this, criteria for success must prioritize long-term sustainability and broad societal benefits.

- **AGI-Specific Application**: Similar competitions could be organized to incentivize the development of safe and ethical AGI systems. By rewarding AI projects that demonstrate superior safety protocols, ethical considerations, and transparency, competition could drive progress in the right direction. However, there is a risk that such competitions might incentivize superficial compliance rather than genuine safety improvements. Therefore, it's crucial to design competitions that emphasize substantive outcomes over appearances.
- **Incentivizing Sustainable Behaviors**: Reward systems for individuals and businesses that actively contribute to sustainability goals can channel competition toward positive actions. These rewards could take the form of social recognition, financial incentives, or other tangible benefits. However, there's a risk that these incentives could create perverse outcomes, such as gaming the system to receive rewards without making meaningful contributions. This risk must be mitigated through careful design and monitoring of reward structures.
- **AGI-Specific Application**: In the AGI realm, incentives could be offered to developers who create AI systems that prioritize ethical use and safety. However, similar to the sustainability context, there's a risk that developers might focus on meeting the minimum requirements to receive incentives rather than striving for genuine safety. Ongoing monitoring and stringent criteria will be necessary to ensure that incentives drive real improvements.
- **Creating a Global Sustainability Leaderboard**: Establishing a public platform that tracks and ranks the progress of countries, cities, and companies toward sustainability goals can foster healthy competition. By publicly recognizing leaders in sustainability, this approach can encourage others to innovate and improve. However, there's a risk that ranking systems could lead to excessive focus on metrics, potentially incentivizing behavior that boosts rankings rather than truly sustainable practices.
- **AGI-Specific Application**: A similar leaderboard could be created to track and rank AI developers based on their commitment to safety and ethical standards. This could promote healthy competition among developers to be seen as leaders in responsible AI. However, as with sustainability, care must be taken to ensure that the rankings reflect genuine progress and not just superficial compliance with the metrics.

2. **Impact Investing as a Status Symbol**:
   - **Prioritizing Social and Environmental Impact**: Shifting the competitive focus from financial returns to maximizing positive social and environmental impacts can transform investment behaviors. By making social and environmental achievements prestigious, we can encourage more investments in sustainable and equitable initiatives. This approach leverages the Moloch dynamic by making ethical investing a status symbol. However, there's a risk that this could lead to "impact-washing," where investments are labeled as socially or environmentally responsible without actually delivering meaningful benefits. Rigorous standards and transparency are essential to mitigate this risk.
   - **AGI-Specific Application**: In the context of AGI, impact investing could focus on supporting AI projects that prioritize safety, ethics, and societal well-being. By making these attributes prestigious, investors could be encouraged to fund responsible AI development over riskier projects. However, similar to sustainability, there's a risk of "AI-washing," where projects claim to prioritize ethics without substantive action. Ensuring that investments are truly aligned with safety and ethical standards will require robust oversight and clear criteria for evaluation.

Addressing the Feasibility and Risks in AGI: While harnessing competition for good offers promising pathways for both sustainability and AGI development, the potential risks must be carefully managed. The Moloch effect means that even well-intentioned competition can lead to unintended consequences, such as incentivizing superficial compliance or creating perverse incentives. In the context of AGI, where the stakes are incredibly high, it is essential to ensure that competitive pressures do not undermine the very goals they seek to promote.

Designing these systems to emphasize long-term outcomes, genuine progress, and substantive contributions is crucial. Regular monitoring, transparency, and adaptability will help ensure that the competitive dynamics remain aligned with societal well-being rather than becoming another manifestation of the Moloch effect.

THE POWER OF DAILY RITUALS

THINKING BOX

"What small practices anchor us in a world overrun by noise? Possibly the steadying principle is akin to **42** – simple, yet profound."

13.9 THE CHALLENGES OF IMPLEMENTATION AND POTENTIAL TRADE-OFFS

While the solutions presented offer a promising path toward a more sustainable and equitable future, it is crucial to acknowledge the potential challenges and trade-offs associated with their implementation. The complexity of systemic issues and the inherent uncertainties of technological advancements necessitate a nuanced understanding of the limitations and potential downsides of even the most well-intentioned solutions.

1. **Scalability and Adaptability**:
   - Many of the proposed solutions, while effective on a smaller scale, may face significant challenges when implemented at a larger, systemic level. Transitioning from local initiatives to global transformations requires careful planning, coordination, and adaptability. Ensuring that these solutions remain effective, equitable, and inclusive as they scale is a daunting task. Large-scale implementations often encounter logistical, cultural, and political obstacles that can dilute their impact or lead to uneven benefits across different regions and populations.
2. **Unintended Consequences**:
   - Even the most well-designed solutions can have unintended consequences. For instance, incentivizing sustainable actions might lead to perverse incentives, where actions taken for rewards are superficial or counterproductive. Similarly, decentralized systems, while empowering communities, could lead to fragmentation, governance challenges, or even the proliferation of harmful technologies if not properly managed. The potential for these unintended outcomes underscores the importance of continuous monitoring and flexibility in the implementation of solutions.
3. **Trade-offs and Prioritization**:
   - Decision-makers will inevitably face trade-offs when prioritizing certain solutions over others. Balancing short-term needs with long-term goals, economic considerations with social and environmental impacts, and individual freedoms with collective well-being requires careful deliberation and ethical decision-making. For example, while rapid technological advancements may offer immediate economic benefits, they could exacerbate social inequalities if not paired with measures to ensure equitable access and distribution of resources.
4. **Resistance to Change**:
   - Implementing systemic change often faces resistance from entrenched interests and those who benefit from the status quo. Overcoming this resistance requires not only

compelling arguments but also effective strategies for building consensus and mobilizing collective action. Additionally, cultural inertia and public skepticism can slow the adoption of new approaches, making it crucial to engage in persistent outreach and education efforts to shift public opinion and build support for necessary reforms.

5. **The Evolving Nature of Technology**:
   - The rapid pace of technological advancement, particularly in the field of AI, means that solutions must be adaptable and responsive to new developments. What works today may not be effective tomorrow, and ongoing monitoring and evaluation are essential to ensure that solutions remain relevant and aligned with societal goals. Furthermore, the evolving nature of technology often outpaces regulatory frameworks, creating gaps that can be exploited if not addressed promptly. Ensuring that policies and safeguards evolve alongside technology is crucial to mitigating risks and maximizing benefits.
6. **Ethical Complexities**:
   - Implementing these solutions also raises complex ethical questions. Who decides what is considered "sustainable" or "equitable"? How do we balance competing values and interests in a diverse global society? These ethical dilemmas require inclusive and participatory decision-making processes that respect diverse perspectives and prioritize fairness and justice.

In conclusion, while the solutions proposed in this text are ambitious and innovative, they are not without challenges. Recognizing and addressing these potential downsides is crucial for ensuring that these solutions are implemented effectively and responsibly. By remaining vigilant, adaptable, and ethically grounded, we can navigate the complexities of systemic change and work toward a future that benefits all of humanity.

13.10 LEVERAGING AI FOR COOPERATIVE SOLUTIONS

Transitioning from harnessing the competitive nature of Moloch, AI offers the potential to foster cooperative solutions and drive collective action. By strategically leveraging AI, we can create platforms and systems that encourage collaboration rather than competition, addressing complex global challenges more effectively.

1. **AI-Powered Collective Intelligence Platforms**:
   - **Global Collaboration for Problem-Solving**: Advanced AI can facilitate global collaboration by connecting experts, policymakers, and citizens on platforms designed to share knowledge and resources. For example, AI can analyze data from various sources, predict trends, and identify areas where cooperation could yield significant benefits, such as climate change mitigation or reducing inequality. These platforms can pool collective intelligence to address complex challenges more effectively, but it's essential to ensure that they prioritize inclusivity and fairness to avoid reinforcing existing power imbalances.
   - **AI-Enhanced Collaboration Platforms**: Virtual spaces where people from diverse backgrounds can collaborate and contribute their knowledge and skills to solve pressing issues are essential. AI could facilitate communication, identify expertise, and streamline the problem-solving process, ensuring that contributions are meaningful and actionable. However, care must be taken to prevent AI-driven platforms from becoming echo chambers or exacerbating digital divides. Ensuring broad accessibility and diverse participation is key to maximizing the potential of these platforms.
2. **AI-Driven Resource Allocation**:
   - **Optimizing Resource Distribution**: AI algorithms can optimize resource allocation based on social and environmental needs rather than solely relying on market mechanisms. This can lead to a more equitable distribution of resources and reduce conflicts driven by scarcity.

For example, AI can help distribute food, water, and medical supplies more efficiently during crises. However, the design and implementation of these algorithms must be transparent and subject to oversight to prevent bias and ensure that they serve the common good.
- **AI-Mediated Conflict Resolution**: AI tools can analyze conflicts, identify common ground, and suggest potential solutions based on data and established best practices. This could be particularly helpful in addressing large-scale, complex poly-crises where traditional methods of conflict resolution fall short. While AI-driven conflict resolution holds promise, it's crucial to recognize that AI should augment human decision-making rather than replace it. Ensuring that AI-mediated solutions respect cultural contexts and human rights is essential.

3. **Adapting These Approaches to AGI**:
   - **Collective Intelligence for AGI Safety**: In the context of AGI, collective intelligence platforms could be instrumental in bringing together diverse stakeholders – AI researchers, ethicists, policymakers, and the public – to collaboratively address the challenges and risks associated with AGI development. By pooling insights from various perspectives, we can create more comprehensive safety protocols and ethical guidelines. However, it's crucial to ensure that these platforms are inclusive and prevent dominance by any single group or interest.
   - **Resource Allocation for AGI Research**: AI-driven resource allocation could also be applied to AGI research, ensuring that funding and resources are distributed to projects that prioritize safety, transparency, and ethical considerations. This approach could help mitigate the risks of the AGI race by aligning resources with long-term societal benefits rather than short-term competitive gains.
   - **AI for AGI Governance**: AI tools could assist in governing AGI development, mediating conflicts between stakeholders, and ensuring that AGI research aligns with humanity's broader interests. By analyzing data and identifying common ground, AI could help facilitate consensus and guide decision-making in a way that balances innovation with caution. However, the governance of AGI must remain a human-centered process, with AI playing a supporting role rather than taking the lead.

Addressing the Feasibility and Risks: While AI-driven cooperative solutions hold great promise, they are not without risks. The competitive pressures that drive AI and AGI development could influence these platforms, leading to unintended consequences or reinforcing existing power structures. Ensuring transparency, inclusivity, and fairness in the design and deployment of AI systems is crucial. Moreover, ongoing oversight and adaptability are essential to ensure that AI remains a tool for collective benefit rather than a source of further division or inequality.

13.11 EMBRACING DECENTRALIZATION AND LOCAL RESILIENCE

Decentralization and local resilience offer a counterbalance to the centralizing forces of the Moloch effect. By empowering communities to build self-sufficient, sustainable systems, we can reduce dependence on global supply chains and centralized power structures, fostering resilience in the face of systemic crises.

1. **Community-Based Economies**:
   - **Building Local Resilience**: Local communities can create resilient economies focused on sharing, collaboration, and local production. These economies prioritize sustainability and self-sufficiency, reducing vulnerability to global disruptions. For example, community-supported agriculture (CSA) programs enable local farmers to directly connect with consumers, fostering a more sustainable and resilient food system. However,

scaling these models while maintaining their local focus can be challenging, and communities must guard against becoming isolated or insular.

- **Local Food Systems and Regenerative Agriculture**: Encouraging the development of local food systems and regenerative agriculture practices can enhance food security, reduce environmental impact, and empower local communities. Regenerative agriculture, which focuses on restoring soil health and biodiversity, can help communities build resilience against climate change and supply chain disruptions. However, the transition to these practices requires education, investment, and a shift in both consumer and producer mindsets.

2. **Distributed Renewable Energy Systems**:
   - **Energy Democracy**: Decentralized renewable energy systems empower communities to generate and manage their own energy, reducing dependence on centralized grids and increasing resilience to energy crises. By promoting energy independence, decentralized systems can also reduce the risks associated with large-scale power outages and geopolitical tensions. However, ensuring equitable access to these systems and preventing the creation of energy "haves" and "have-nots" is crucial to avoid reinforcing inequalities.
   - **Peer-to-Peer Energy Sharing**: Utilizing blockchain and other technologies, peer-to-peer energy trading allows communities to share surplus energy with each other, fostering a more democratic and resilient energy system. By enabling individuals to trade energy directly, these systems reduce reliance on centralized power companies and create incentives for renewable energy adoption. However, the success of these systems depends on robust infrastructure, clear regulations, and the active participation of community members.
3. **Adapting These Approaches to AI and AGI:**
   - **Decentralized AI Research and Development**: Similar to decentralized energy systems, decentralizing AI research and development can reduce the concentration of power in a few large corporations or nations. By promoting open-source AI projects and community-driven innovation, we can foster a more equitable and resilient AI ecosystem. However, decentralization also brings challenges, such as ensuring quality control and preventing the spread of harmful technologies.
   - **Community-Based AI Governance**: Local communities could play a role in governing AI systems that directly impact them. By involving citizens in decision-making processes and ensuring that AI systems are tailored to the specific needs and values of their communities, we can promote more democratic and accountable AI governance. However, balancing local control with the need for broader oversight and consistency across regions is a challenge that must be carefully managed.

Addressing the Feasibility and Risks: While decentralization and local resilience offer promising solutions, they also come with risks. Ensuring that decentralized systems remain equitable and do not exacerbate existing inequalities is essential. Additionally, decentralization should not lead to fragmentation or isolation but should instead be seen as part of a broader strategy that includes both local and global cooperation. In the context of AI and AGI, decentralization can reduce the risks associated with concentrated power but requires careful management to prevent the proliferation of harmful technologies or unregulated experimentation.

13.12 TRANSFORMING EDUCATION FOR A POST-MOLOCH WORLD

Transforming education is key to creating a society that can navigate and mitigate the systemic risks posed by the Moloch effect. By emphasizing systems thinking, futures literacy, and emotional intelligence, education can prepare individuals to tackle complex challenges, including those related to AI and AGI, and drive systemic change.

1. **Systems Thinking and Futures Literacy**:
   - **Educating for Systemic Change**: Education should prioritize systems thinking, helping individuals understand the interconnectedness of global challenges and the broader systems that influence them. Futures literacy – preparing people to anticipate and plan for multiple possible futures – can empower students to think critically about long-term consequences and envision alternative futures. This approach is essential in a world increasingly shaped by complex, interrelated crises. However, implementing systems thinking across educational curricula requires a significant shift from traditional teaching methods, emphasizing critical thinking over rote learning.
   - **Preparing for AGI**: In the context of AGI, systems thinking is particularly crucial. By understanding how AGI development interacts with other systems – economic, social, and environmental – students can better assess the potential risks and benefits of AGI. Futures literacy can also help future leaders anticipate the ethical and societal impacts of AGI, encouraging proactive rather than reactive decision-making.
2. **Emotional Intelligence and Counteracting Manipulation**:
   - **Recognizing and Resisting Manipulation**: Emotional intelligence, traditionally seen as the ability to understand and manage one's emotions, can also be a critical tool for recognizing and resisting emotional manipulation. In a world driven by competitive pressures and the Moloch effect, individuals and organizations often use emotional manipulation to exploit vulnerabilities and gain an advantage. By educating students on how to identify and resist these tactics, we can empower them to maintain autonomy and make informed decisions that prioritize collective well-being over divisive competition.
   - **Building Resilience in AI-Driven Environments**: Emotional intelligence is also essential in navigating AI-driven environments, where manipulation – whether through targeted advertising, social media algorithms, or persuasive design – can influence behavior and decision-making. Educating individuals on how to recognize these tactics helps build resilience against being manipulated by AI systems designed to exploit emotional responses. This is particularly important as AI becomes more integrated into our lives and decision-making processes.

Addressing the Feasibility and Risks: Transforming education to prioritize systems thinking, futures literacy, and emotional intelligence as a tool for counteracting manipulation is a long-term endeavor that requires a shift in both mindset and resources. Integrating these concepts into existing educational systems may face resistance due to entrenched norms and the emphasis on standardized testing. However, the risks of not adapting to education are significant – failing to prepare future generations for the complexities of a post-Moloch world could lead to a continuation of the very problems we seek to solve.

In the context of AI and AGI, these educational transformations are particularly urgent. As AI systems become more integrated into daily life, the ability to recognize and resist emotional manipulation, think critically about impacts, and collaborate across disciplines will be essential. By equipping students with the skills to navigate a complex and rapidly changing world, we can create a generation of leaders capable of driving systemic change and addressing the challenges posed by both current and emerging technologies.

13.13 REFRAMING THE NARRATIVE

Reframing the narrative is a powerful tool for counteracting the Moloch effect. By shifting our collective focus from scarcity and competition to abundance and cooperation, and from individualism to interdependence, we can foster a mindset that promotes collective well-being and long-term sustainability. This narrative shift is crucial not only for addressing current challenges but also for shaping how we approach emerging technologies like AI and AGI.

1. **From Scarcity to Abundance**:
   - **Changing the Perception**: The traditional narrative of scarcity, which fuels competition and the Moloch effect, can be transformed by emphasizing the potential for abundance through cooperation. By reframing challenges as opportunities for collective problem-solving and innovation, we can encourage a mindset that values collaboration over competition. For example, instead of viewing resources as limited and worth fighting over, we can focus on how sustainable practices and technologies can create more for everyone. However, this shift requires changing deeply ingrained beliefs and ensuring that the narrative of abundance does not ignore genuine resource constraints but rather addresses them through innovative solutions.
   - **Implications for AI and AGI**: In the context of AI and AGI, reframing the narrative from competition to abundance can help reduce the risks associated with the AGI race. Instead of viewing AGI development as a zero-sum game where only a few can win, we can promote the idea that safe and ethical AGI can benefit all of humanity. This requires fostering a global dialogue that emphasizes shared goals and mutual benefits, rather than allowing AGI development to be driven solely by competitive pressures.
2. **From Individualism to Interdependence**:
   - **Fostering Shared Responsibility**: The narrative of individualism, which often underpins competitive and self-serving behaviors, can be replaced with a narrative of interdependence. Recognizing our interconnectedness with each other and the planet encourages a sense of shared responsibility and collective action for the common good. This shift can lead to more cooperative approaches to global challenges, where success is measured by collective well-being rather than individual achievement. However, shifting from individualism to interdependence requires not only changing societal narratives but also restructuring systems that reward individual over collective success.
   - **Implications for AI and AGI**: In the realm of AI and AGI, emphasizing interdependence can help ensure that these technologies are developed with the broader interests of humanity in mind. By fostering a narrative that prioritizes ethical considerations and the collective good, we can counteract the competitive dynamics that often drive AI development. This narrative shift can also encourage more inclusive and democratic governance of AI and AGI, where diverse voices are heard, and decisions reflect the interests of all stakeholders, not just those in positions of power.

Addressing the Feasibility and Risks: Reframing the narrative is not without challenges. Changing deeply held beliefs and societal values takes time and effort, and there is always a risk that these new narratives could be co-opted or misused. For example, the narrative of abundance could be exploited to justify unsustainable practices under the guise of creating more for everyone. Similarly, the narrative of interdependence could be manipulated to enforce conformity rather than genuine cooperation.

In the context of AI and AGI, these narrative shifts are particularly important. If we fail to move beyond the narratives of scarcity and competition, we risk allowing AGI development to be driven by short-term gains rather than long-term collective well-being. However, by promoting narratives that emphasize abundance, cooperation, and interdependence, we can help ensure that AGI serves as a tool for positive change rather than a source of division and conflict.

13.14 ADDITIONAL NOVEL APPROACHES UTILIZING MOLOCH DYNAMICS

By strategically leveraging Moloch dynamics – where competition and self-interest drive behavior – we can create systems that turn these forces into catalysts for positive change. However, these approaches must be carefully designed to avoid unintended consequences and ensure that the competitive drive serves the collective good. Here are several novel approaches that utilize Moloch dynamics to promote ethical behavior, transparency, and innovation, with implications for both general challenges and AI/AGI development.

1. **Moloch as a Catalyst for Radical Transparency**:
   - **Blockchain-Based Accountability**: Leveraging the competitive nature of Moloch, a blockchain-based ledger could track the environmental and social impact of companies and government actions. By making this information publicly available, transparency becomes a competitive advantage, as consumers and investors increasingly demand ethical behavior. This approach encourages organizations to outcompete each other on metrics of responsibility and sustainability. However, the challenge lies in ensuring that the data on the blockchain is accurate and not manipulated, as well as preventing the system from becoming a tool for greenwashing.
   - **AI-Powered Whistleblowing Platforms**: AI can create secure, anonymous whistleblowing platforms to expose unethical practices, holding powerful actors accountable. These platforms could harness the Moloch dynamic by making whistleblowing a respected and incentivized action, encouraging transparency through competition. In the context of AI and AGI, such platforms could be used to expose unsafe or unethical AI development practices. However, the success of this approach depends on ensuring the security and anonymity of whistleblowers and preventing retaliation.
2. **Molochian Nudges for Positive Change**:
   - **Carbon Tax Rebates as Social Status**: Implementing a carbon tax, with a portion redistributed as social status points or tokens displayed on social media profiles, could turn environmental responsibility into a competitive advantage. This approach gamifies sustainable behavior, making it desirable to outcompete others in environmental stewardship. However, the risk is that this could become a superficial competition focused more on status than substantive change. Ensuring that these status indicators reflect genuine contributions is essential.
   - **Gamified Social Impact Metrics**: Social media platforms could incorporate gamified metrics that track and reward users for positive social and environmental actions. By making social impact a form of social currency, this approach leverages the competitive nature of Moloch for good. However, as with any gamified system, there is a risk of people gaming the system or focusing on metrics at the expense of real-world impact. Balancing the fun of gamification with meaningful outcomes is key.
3. **Turning Moloch against Itself: Decentralized Autonomous Organizations (DAOs)**:
   - **DAOs for Public Goods**: Using DAOs to fund and manage public goods like clean energy projects or scientific research can align incentives through token-based rewards and transparent governance. This approach leverages the decentralized nature of DAOs to promote contributions to the common good while allowing participants to benefit from their involvement. However, managing DAOs effectively requires clear governance structures to prevent them from being co-opted by self-interested actors.
   - **DAOs for AI Governance**: In the context of AI and AGI, DAOs could be used to ensure that decisions about AI development are made democratically and in humanity's best interests, rather than driven by profit motives. By distributing decision-making power, DAOs could help mitigate the risks of centralized control over AGI. However, the effectiveness of this approach depends on the ability to design governance systems that are both transparent and resistant to manipulation.
4. **Reimagining Growth: The Moloch of Regenerative Systems**:
   - **Nature-Inspired Economies**: Transitioning toward economies that mimic nature's regenerative processes – such as sustainable agriculture, reforestation, and circular economies – can turn the competitive drive of Moloch into a force for ecological restoration. By aligning economic incentives with the principles of regeneration, we can

create systems where success is measured by the ability to renew and restore resources. However, ensuring that these regenerative systems are scalable and economically viable is a key challenge.

- **Measuring Progress beyond GDP**: Shifting away from GDP as the sole measure of economic success and adopting metrics that account for social well-being, environmental health, and quality of life can redefine "growth" and incentivize balanced, sustainable development. This approach challenges the Moloch dynamic by broadening the definition of success to include long-term sustainability. However, changing entrenched economic systems and metrics is a significant hurdle that requires widespread buy-in from policymakers, businesses, and the public.

Addressing the Feasibility and Risks: While these novel approaches offer promising ways to harness Moloch dynamics for positive change, they also come with risks. Ensuring that these systems remain transparent, equitable, and resistant to manipulation is crucial. In the context of AI and AGI, the stakes are particularly high, and these approaches must be carefully designed to prevent unintended consequences.

Turning Moloch against itself requires a nuanced understanding of how competition and self-interest can be channeled into productive and ethical outcomes. By leveraging technology, decentralization, and transparency, we can create systems that encourage responsible behavior and foster long-term sustainability. However, the success of these approaches depends on continuous oversight, adaptability, and a commitment to ethical principles.

13.15 OUTLOOK ON FUTURE DEVELOPMENTS

The future will undoubtedly bring further challenges, both related to AGI and other systemic crises. As we navigate an increasingly complex world, it is essential to foster ongoing dialogue between science, politics, and society to develop adaptive and resilient strategies. These strategies must be rooted in a deep understanding of the interconnectedness of global challenges and the need for systemic change.

1. **Promoting Continuous Dialogue and Collaboration**:
   - **Science, Politics, and Society**: Bridging the gap between scientific innovation, political decision-making, and societal needs is crucial for addressing future challenges. Continuous dialogue between these sectors ensures that emerging technologies, like AGI, are developed and regulated in ways that prioritize ethical considerations and the common good. This collaboration must be proactive, with stakeholders anticipating potential risks and working together to mitigate them before they become crises. However, achieving this level of collaboration requires trust, transparency, and a shared commitment to long-term thinking.
   - **Global Cooperation on AGI**: Given the global implications of AGI, international cooperation is essential. By fostering collaborative efforts across borders, we can ensure that AGI development is guided by shared ethical principles and does not become a source of conflict or division. This will require robust international agreements and institutions that can enforce these principles while accommodating diverse perspectives.
2. **The Role of Education and Awareness**:
   - **Ethics and Technology**: As technology continues to evolve, education will play a critical role in preparing future generations to navigate the ethical challenges that accompany it. This includes not only teaching technical skills but also fostering a deep understanding of the societal impacts of technology. Emphasizing ethics in education can help prevent the development of technologies that exacerbate inequalities or create new forms of harm. Additionally, raising public awareness about the implications of

emerging technologies ensures that society as a whole is better equipped to engage in meaningful dialogue about their development and use.
- **Futures Literacy and Adaptive Thinking**: In a rapidly changing world, futures literacy – the ability to anticipate and adapt to multiple possible futures – will be increasingly important. By cultivating this skill, we can prepare individuals to think critically about long-term consequences and to adapt strategies as new challenges arise. This approach is particularly relevant in the context of AGI, where the future remains highly uncertain, and adaptability will be key to navigating potential risks.

3. **Addressing Emerging Systemic Crises**:
   - **Beyond AGI**: While AGI presents significant challenges, it is just one of many systemic crises we may face in the future. Climate change, resource scarcity, and social inequality are all interconnected issues that require systemic solutions. By applying the principles discussed throughout this article – such as decentralization, transparency, and reframing narratives – we can develop strategies that address these broader challenges. It is essential to recognize that solving one crisis in isolation is unlikely to be effective; instead, we must adopt holistic approaches that address the root causes of multiple crises simultaneously.
   - **Resilience and Adaptation**: Building resilience – both at the individual and systemic levels – will be crucial for navigating future developments. This involves not only preparing for specific risks but also cultivating the flexibility and adaptability needed to respond to unexpected challenges. In the context of AGI, resilience means developing robust governance frameworks that can evolve as technology advances, ensuring that we remain prepared for new risks and opportunities.

Conclusion: The outlook on future developments is both challenging and hopeful. While the emergence of AGI and other systemic risks presents significant threats, it also offers opportunities for profound positive change. By fostering continuous dialogue, promoting education and ethics, and adopting adaptive strategies, we can navigate these challenges with resilience and foresight. However, success will require a collective commitment to addressing the root causes of systemic crises and to ensuring that technological advancements serve the common good.

In conclusion, the future is uncertain, but by embracing cooperation, transparency, and ethical principles, we can steer it toward a more sustainable and equitable path. The road ahead will not be easy, but with careful planning and a shared vision, we can build a future that benefits all of humanity.

13.16 ENHANCED IMPACT-PRIORITY MATRIX FRAMEWORK

The **Impact-Priority Matrix** is a powerful tool designed to help decision-makers prioritize actions based on their potential impact and urgency. In a world filled with complex challenges, including those related to systemic risks and AGI, this matrix offers a structured approach to identifying which actions will make the most significant difference and should be prioritized.

1. **Simplifying the Concept**:
   - **What is the Impact-Priority Matrix?** At its core, the matrix is a framework that helps you evaluate actions based on two key criteria: their impact (how much difference they will make) and their priority (how urgent or essential they are). By plotting actions on this matrix, you can visually assess which initiatives should take precedence and allocate resources more effectively.
   - **Why Use the Matrix?** The matrix is particularly useful in environments where resources are limited, and decisions must be made about where to focus efforts. Whether you're tackling systemic risks like climate change or managing the development of AGI, the matrix can guide you in making informed, strategic choices.

2. **Practical Examples**:
   - **Example 1**: **Tackling Climate Change**: Imagine you're tasked with reducing carbon emissions in a city. Using the matrix, you might identify various actions – like implementing renewable energy projects, enhancing public transportation, or promoting energy efficiency in buildings. By evaluating the impact (e.g., potential emissions reductions) and priority (e.g., feasibility, public support), you can determine which actions should be prioritized.
   - **Example 2**: **Managing AGI Development**: In the context of AGI, the matrix could help prioritize safety measures, ethical guidelines, and research initiatives. For example, developing robust safety protocols might have high impact and high priority, whereas exploring long-term AGI governance structures might be impactful but less urgent, placing it in a different quadrant of the matrix.
3. **Step-by-Step Guidance**:
   - **Step 1**: **Identify Potential Actions**: Start by listing all the possible actions you could take to address a particular challenge. This could include projects, policies, or initiatives.
   - **Step 2**: **Assess Impact**: Evaluate the potential impact of each action. How much will it contribute to solving the problem? Consider both direct effects and broader implications.
   - **Step 3**: **Determine Priority**: Next, assess the urgency or priority of each action. Is it time-sensitive? Does it address an immediate need, or is it a longer-term goal?
   - **Step 4**: **Plot on the Matrix**: Plot each action on the matrix, with one axis representing impact and the other representing priority. This will help you visualize which actions should be prioritized.
   - **Step 5**: **Make Decisions**: Use the matrix to guide your decision-making. Focus on actions that fall in the high-impact, high-priority quadrant while also considering how to manage lower-priority or lower-impact actions.
4. **Highlighting Potential Pitfalls**:
   - **Avoiding Bias**: Be mindful of biases that might influence your assessment of impact or priority. For example, it's easy to overestimate the impact of familiar solutions or prioritize actions that align with personal or organizational interests. Involving diverse perspectives can help mitigate these biases.
   - **Balancing Short-Term vs. Long-Term Priorities**: It's important to balance immediate needs with long-term goals. While some actions might be urgent, others with more significant long-term impact should not be overlooked. The matrix helps ensure that both are considered.
   - **Ensuring Diverse Input**: Engaging a wide range of stakeholders in the process can lead to more balanced and effective decision-making. This is particularly important in areas like AGI, where the implications of decisions can affect diverse populations and future generations.
5. **Encouraging Iteration and Adaptation**:
   - **Dynamic Use of the Matrix**: The matrix is not a one-time exercise. As circumstances change, so too should your assessments of impact and priority. Regularly revisiting the matrix ensures that your strategies remain relevant and responsive to evolving challenges.
   - **Adapting to New Information**: Be open to adapting the matrix as new information becomes available. This is especially important in fast-moving fields like AI and AGI, where technological advancements and societal shifts can rapidly change the landscape.

Conclusion: The Impact-Priority Matrix is a practical tool for navigating complex challenges and making informed decisions. By helping you prioritize actions based on their impact and urgency, the matrix enables more effective use of resources and time. Whether addressing systemic risks like climate change or managing the ethical development of AGI, the matrix provides a clear framework for strategic decision-making. However, its effectiveness depends on regular reassessment, diverse input, and a commitment to balancing short-term and long-term goals.

Impact-Priority Matrix: At Work

The **Impact-Priority Matrix** helps decision-makers evaluate and prioritize actions by assessing their potential impact, systemic interconnectedness, incentive misalignment, and urgency. This framework is particularly useful for addressing complex, systemic challenges like AGI development and other global crises.

Step 1: Define Metrics and Standardized Scoring Rubrics

To apply the Impact-Priority Matrix, we first define the metrics and scoring rubrics for evaluating potential actions or issues.

i. **Impact Score (I)**:
   - **Environmental Impact**:
     - **0**: No discernible impact.
     - **5**: Significant impact on a local ecosystem.
     - **10**: Catastrophic impact on a global scale (e.g., mass extinction event).
   - **Social Impact**:
     - **0**: No discernible impact on social equity or well-being.
     - **5**: Significant positive or negative impact on a community or region.
     - **10**: Transformative impact on global social equity and well-being.
   - **Economic Impact**:
     - **0**: Negligible economic effect.
     - **5**: Noticeable economic impact at a regional level.
     - **10**: Major global economic consequences (positive or negative).
   - **Technological Impact**:
     - **0**: No significant technological implications.
     - **5**: Advances or disrupts technology on a regional scale.
     - **10**: Major technological advancements or disruptions globally.

ii. **Systemic Interconnectedness Score (S)**:
   - **0**: Issue is isolated and has no significant impact on other systems.
   - **5**: Issue has moderate connections to other systems, with potential for cascading effects.
   - **10**: Issue is deeply interconnected with multiple systems, posing a high risk of widespread disruption.

iii. **Incentive Misalignment Score (M)**:
   - **0**: Incentives for individuals and groups are aligned with the common good.
   - **5**: Some misalignment of incentives exists, leading to potential conflicts and suboptimal outcomes.
   - **10**: Strong misalignment of incentives, creating a classic Molochian trap where individual rationality leads to collective harm.

iv. **Urgency Score (U)**:
   - **Time Sensitivity**:
     - **0**: Action can be delayed for years with minimal consequences.
     - **5**: Action needed within a few years to avoid significant harm.
     - **10**: Immediate action required to prevent catastrophic outcomes.
   - **Escalation Risk**:
     - **0**: Low risk of the issue worsening over time.
     - **5**: Moderate risk of escalation.
     - **10**: High risk of rapid escalation if not addressed promptly.
   - **Public Attention**:
     - **0**: Minimal public awareness or concern.
     - **5**: Growing public interest and concern.
     - **10**: High public demand for immediate action.

Step 2: **Calculate Priority Score**

Combine the scores to determine the overall priority of each issue.
Priority Score (P): The Priority Score is calculated using the following formula:

$$\text{Priority Score } (P) = \left(\frac{I + S + M}{3} \right) \times U$$

Where:

- **I** is the average of the impact scores (Environmental, Social, Economic, and Technological).
- **S** is the Systemic Interconnectedness Score.
- **M** is the Incentive Misalignment Score.
- **U** is the Urgency Score.

This formula gives a weighted score that considers the overall impact, interconnectedness, misalignment of incentives, and urgency of action.

Step 3: **Categorize Issues and Identify Moloch/Meta-Crisis Potential**

Once you have calculated the Priority Score, categorize the issues into the matrix's four quadrants:

- **Quadrant 1**: High Impact, High Urgency (Act Immediately)
- **Quadrant 2**: High Impact, Low Urgency (Plan Strategically)
- **Quadrant 3**: Low Impact, High Urgency (Address Quickly)
- **Quadrant 4**: Low Impact, Low Urgency (Monitor or Deprioritize)

Additionally:

- **Moloch Trap Potential**: If Incentive Misalignment (M) is **≥ 8**, flag the issue as having high Moloch trap potential.
- **Meta-Crisis Potential**: If Systemic Interconnectedness (S) is **≥ 8**, flag the issue as having high meta-crisis potential.

Step 4: Implement and Monitor

1. **Resource Allocation**:
   - Prioritize resources, time, and effort for Quadrants 1 and 2.
   - Delegate or automate tasks in Quadrant 3 and minimize focus on Quadrant 4.
2. **Continuous Monitoring**:
   - Regularly review and update scores as new information becomes available.
   - Adjust strategies based on outcomes and changing priorities.

13.17 AGI DEVELOPMENT EXAMPLE USING THE FORMULA PROVIDED

Step 1: **Define Metrics and Scoring for AGI Development**

- **Impact Score (I)**:
  - **Environmental Impact**: 8
  - **Social Impact**: 10
  - **Economic Impact**: 9
  - **Technological Impact**: 10

- **Average Impact Score (I):**

$$I = \frac{8+10+9+10}{4} = \frac{37}{4} = 9.25$$

- **Systemic Interconnectedness Score (S)**: 10
- **Incentive Misalignment Score (M)**: 9
- **Urgency Score (U):**
  - **Time Sensitivity**: 8
  - **Escalation Risk**: 9
  - **Public Attention**: 6
  - **Average Urgency Score (U):**

$$U = \frac{8+9+6}{3} = \frac{23}{3} = 7.67$$

Step 2: Calculate Priority Score
Using the formula:

$$\text{Priority Score } (P) = \left(\frac{I+S+M}{3}\right) \times U$$

Substituting the values:

$$P = \left(\frac{9.25+10+9}{3}\right) \times 7.67$$

$$P = \left(\frac{28.25}{3}\right) \times 7.67$$

$$P = 9.42 \times 7.67 = 72.28$$

Step 3: Categorize Issues and Identify Moloch/Meta-Crisis Potential

With a Priority Score of **72.28**, AGI development falls into **Quadrant 1: High Impact, High Urgency (Act Immediately)**.

Additionally:

- **High Moloch Trap Potential**: The Incentive Misalignment Score (M) is **9** (≥ 8).
- **High Meta-Crisis Potential**: The Systemic Interconnectedness Score (S) is **10** (≥ 8).

Step 4: Implement and Monitor

Given the high priority, Moloch trap potential, and meta-crisis potential, AGI development requires immediate and comprehensive action:

i. **International Collaboration**: Establish global governance frameworks and safety standards for AGI development.
ii. **Safety and Ethics Research**: Prioritize research on AI safety, control mechanisms, and ethical considerations, focusing on aligning AGI with human values.
iii. **Transparency and Public Engagement**: Encourage open dialogue and collaboration among stakeholders, including researchers, policymakers, and the public, to ensure responsible and ethical AGI development.
iv. **Regulation and Oversight**: Implement regulations that mitigate the risks of AGI, ensure its safe and beneficial use, and prevent misuse.

13.18 INDIVIDUAL ACTIONS AND COLLECTIVE EFFORTS

While collective efforts are crucial for systemic change, individual actions play a vital role in addressing the meta-crisis. Each person can contribute meaningfully, but it's essential to be strategic and understand where our efforts are best directed.

1. **The War for Your Attention**:
   - **Recognize the Power of Attention**: In the age of information overload and constant distractions, your attention is a valuable commodity. Corporations, media outlets, and platforms constantly vie for it, often promoting consumerism and exacerbating the Moloch effect. By recognizing the value of your attention, you can take control of how it's spent.
   - **Choose Your Focus Wisely**: Be mindful of where you direct your attention. Prioritize information sources that offer critical analysis, promote sustainable solutions, and encourage positive change. Your focus should align with your values and contribute to your understanding of systemic issues.
   - **Limit Exposure to Molochian Traps**: Social media and other platforms are often designed to capture and hold your attention through addictive tactics. Limit your exposure to these platforms and prioritize activities that nourish your well-being and contribute to a better world. By reclaiming your attention, you reduce the influence of systems that perpetuate the Moloch effect.
2. **Systemic Change Requires Systemic Solutions**:
   - **Understand the Root Causes**: Recognize that many of our challenges, such as climate change, inequality, and the risks associated with AGI, are systemic in nature. They arise from the underlying structures and incentives of our economic and social systems. Addressing these root causes is essential for meaningful change.
   - **Advocate for Systemic Change**: Focus your individual efforts on advocating for systemic solutions, such as those outlined in this article. This can involve supporting political movements, engaging in activism, and voting for leaders who prioritize sustainability, social justice, and ethical technology development.
   - **Support Organizations Driving Change**: Contribute to organizations working toward systemic change. This could involve donating, volunteering, or spreading awareness of their work. These organizations are often on the front lines of driving the changes needed to address the meta-crisis.
   - **Educate Yourself and Others**: Continuously educate yourself about the root causes of systemic challenges and share this knowledge with others. Building a critical mass of informed and engaged individuals is key to driving collective action and pushing for systemic change.
3. **Bridging Individual and Collective Efforts**:
   - **Leverage Technology for Good**: Use technology to connect with like-minded individuals and organizations that are working toward systemic solutions. Whether through social media, online forums, or collaborative platforms, technology can be a powerful tool for organizing collective efforts. However, remain vigilant about how you use these tools and avoid platforms that exploit user data or encourage harmful behaviors.
   - **Align Personal and Professional Values**: Strive to align your personal and professional actions with values that prioritize sustainability, equity, and ethical behavior. Whether through your career, consumer choices, or community involvement, consistently making decisions that reflect your values helps build momentum for broader change.

13.19 INNOVATIVE SOLUTIONS TOOLKIT: REDIRECTING COMPETITIVE FORCES FOR COLLECTIVE BENEFIT

13.19.1 Integrating the Bio-Emergence-Narrative Matrix

Recall from *Dynamic Ethical Emergence* (Chapters 16–18) that any sustainable solution must address three intertwined layers:

1. **Biological (Neuro-Somatic) Layer**: *Human predispositions and emotional drivers* – harnessing them to reinforce ethical, cooperative impulses rather than exploitative or short-term instincts.
2. **Emergent (Systemic) Layer**: *Collective behaviors, institutional structures, and social-ecological feedback loops* – ensuring solutions account for complex adaptivity and do not introduce hidden fragilities.
3. **Narrative (Cultural) Layer**: *Stories, values, status markers, and shared meaning* – reshaping the cultural context so that "winning" aligns with advancing the common good.

Below, each solution references which matrix layers it impacts most, showing how competition can be channeled into synergy instead of self-destruction.

13.19.2 Part I: Shifting Competition Design

1. **Competition Cascades for Common Good**
   Core Idea: Architect "races" in which participants can only gain advantage by *actively reducing systemic threats* (e.g., climate disruption, AI misalignment). Instead of awarding short-term profit for exploitative tactics, success depends on stabilizing or regenerating the shared environment.
   - **Key Mechanism**:
     - **System-Stability Market** – A stock or bond index that *only rises* when agreed-upon planetary or social indicators (carbon concentration, biodiversity indices, AI safety compliance) improve.
     - **Corporate Goodwill Ratings** – Firms gain credit-rating-like benefits if they *document* measurable public goods contributions.

Bio-Emergence-Narrative Matrix

- **Biological Layer**: Taps into our innate drive for status/profit but re-routes it toward altruistic actions (hijacking "competitive" neural triggers for cooperation).
- **Emergent Layer**: Systemic feedback loops are embedded in market dynamics, so each improvement triggers further benefits.
- **Narrative Layer**: Redefines "victory" and "excellence" around ethical leadership, reinforcing a cultural story of synergy.

Implementation and Actors

- **Who**: Primarily **governments** (to mandate new market metrics) and **financial institutions** (to adopt stability-based indexes). **NGOs** or **think tanks** can supply validated indicators.
- **Staged Rollout**: Start with small "stability indices" for narrower domains (e.g., water security), then expand to global benchmarks.
- **Risks**: Firms gaming metrics or lobbying to weaken standards. **Preventive** step: robust auditing and multi-stakeholder oversight.

2. **Self-Reinforcing Stability Incentives**
   Core Idea: Convert a "race to the bottom" into a "race to the top," where solutions that reduce systemic vulnerabilities (like AI meltdown or antibiotic resistance) grant *compounding benefits* for early movers.
   - **Key Mechanism**:
     - **Stability-Linked Bonds** – Government or corporate bonds yield higher returns as key risk metrics improve (e.g., lowered AI misuse incidence).
     - **Tiered Market Access** – Firms adopting best-practice AI safety protocols gain priority access to certain regulatory fast-tracks, compounding their advantage.

   Bio-Emergence-Narrative Matrix

   - **Biological Layer**: Leverages *reward circuits* to repeatedly reinforce risk-reducing actions.
   - **Emergent Layer**: Each success lifts systemic stability, creating positive externalities for other participants – "upward spiral."
   - **Narrative Layer**: Shifts cultural framing of "success" from raw speed/efficiency to speed **plus** system preservation.

   Implementation and Actors

   - **Who**: **Financial regulators** (central banks, finance ministries) set up stability-linked instruments; **private investors** adopt them for attractive returns that scale with system well-being.
   - **Staged Rollout**: Begin with specialized pilot bonds (like green or pandemic prevention bonds), then broaden to risk categories like AI safety.
   - **Risks**: Metric distortion or short-sighted focus on easily improved proxies. **Solution**: Continual metric refinement and external oversight agencies.
3. **Network Effect Redirection**
   Core Idea: Use the power of network effects to elevate social and ecological resilience, rather than funneling benefits into monopolistic or exploitative outcomes.
   - **Key Mechanism**:
     - **Positive-Feedback Platforms** – Online networks where user influence grows *if their contributions pass verified checks for accuracy and constructive dialogue.*
     - **Reputation Ecosystems** – Collaborative or sustainability-minded actions scale a user's "trusted contributor" status, unlocking new services or monetary rewards.

   Bio-Emergence-Narrative Matrix

   - **Biological Layer**: Humans crave social acceptance and influence. These platforms sublimate that craving into pro-social content and fact-based interactions.
   - **Emergent Layer**: Fosters emergent synergy – each added cooperative node improves the platform's overall problem-solving capacity.
   - **Narrative Layer**: Encourages a story that "real influence" comes from verified constructive input, not from sensational or divisive clicks.

   Implementation and Actors

   - **Who**: **Tech companies**, **innovation labs**, and **academic consortia** that run open-source or publicly funded platforms. **Civil society** can help define validation criteria.
   - **Success Amplification**: As genuine expertise or positive contributions gain traction, more participants gravitate to the platform for reliable solutions.
   - **Risks**: Overzealous moderation could stifle debate, or "truth" might be decided by the majority, ignoring valid minority views. **Countermeasure**: Weighted multi-expert panels, transparent appeals, and diversified advisory boards.

13.19.3 Part II: Technology as a Coordination Mechanism

1. **AI for Systemic Alignment**
   Core Idea: Steer AI's computational power to *reward co-creative solutions* over exploitative ones, making advanced AI a champion of long-term human and ecological well-being.
   - **Key Mechanism**:
     - **Ethics-Gated AI Training** – Large AI models must *demonstrate* improved performance on cooperation tasks and safe AI benchmarks before unlocking higher compute or advanced capabilities.
     - **Collective-Decision AI Tools** – Deploy AI platforms that automatically propose multi-stakeholder win-win solutions, bridging differences by analyzing shared benefits.

   Bio-Emergence-Narrative Matrix

   - **Biological Layer**: Humans' built-in social cognition is supplemented by AI's big-data analysis, enabling us to see synergy where we might otherwise see conflict.
   - **Emergent Layer**: AI itself becomes a node in the emergent ecosystem, shaping regulatory frameworks, resource allocation, and group negotiation.
   - **Narrative Layer**: Portrays AI not as a job-stealing or manipulative entity but as a co-navigator that fosters deeper cooperation and ethical foresight.

   Implementation and Actors

   - **Who**: **AI labs**, **policymakers**, and **international alliances** (e.g., UNESCO, Partnership on AI) to set standards for AI training protocols.
   - **Adaptive Governance**: "Ethics gates" or "Capability tiers" require ongoing multiparty testing and scenario drills.
   - **Risks**: Biased training sets or corporate capture could warp AI's "alignment." **Safeguard**: Transparent datasets, open auditing, and mandatory multi-stakeholder review boards.
2. **Competitive Transparency Engineering**
   Core Idea: Transform secrecy into a liability so that *publicly disclosing processes, impacts, or safety protocols* confers a competitive edge.
   - **Key Mechanism**:
     - **Open-Source Dividend** – Companies that open critical datasets or show real-time environmental/AI safety metrics receive tax rebates or quicker regulatory approvals.
     - **Verified Public Good Platforms** – Reputable third parties rank organizations by the clarity and substance of their disclosures, with top performers enjoying higher brand reputation.

   Bio-Emergence-Narrative Matrix

   - **Biological Layer**: Humans have an innate curiosity and a tendency to trust proven expertise. Publicly validated disclosures feed that trust loop.
   - **Emergent Layer**: Over time, a *culture of openness* emerges, raising overall system resilience and reducing hidden vulnerabilities (like untested AI systems).
   - **Narrative Layer**: The cultural story becomes: "Being open about your practices is not just altruistic – it's also the best way to succeed in the marketplace."

   Implementation and Actors

   - **Who**: **Governments** (through legislation), **civil-society watchdogs** (through independent verification), and **industry coalitions** that adopt transparency frameworks.

- **Staged Rollout**: Start with data-driven sectors (AI, biotech) where risks are high. Expand to broader industries once public acceptance grows.
- **Risks**: Potential infiltration by "greenwashing" or "ethics-washing." **Solution**: Rigorously verified, tamper-proof credentialing and random audits.

13.19.4 Part III: Economic Mechanism Design

1. **Profit through Prevention**
   Core Idea: Make preventing disasters (ecological, financial, or AI meltdown) more profitable than responding to them after the fact.
   - **Key Mechanism**:
     - **Anticipatory Action Funds** – Public-private funds that pay dividends only if catastrophe metrics remain below thresholds. Each avoided crisis boosts fund returns.
     - **Resilience Insurance** – Similar to parametric insurance, but focuses on robust prevention measures. Lower premium cost for proven risk mitigation.

   Bio-Emergence-Narrative Matrix

   - **Biological Layer**: People want security and stable returns. By linking profitability to crisis-aversion, you harness that desire for security.
   - **Emergent Layer**: This approach *coordinates many small players* toward prevention, producing a more stable emergent system.
   - **Narrative Layer**: Rewrites "heroism in crisis response" into "heroism in crisis avoidance," a quieter yet more impactful form of success.

   Implementation and Actors

   - **Who**: **Insurance companies**, **investment firms**, and **multilateral agencies** creating global or regional prevention funds.
   - **Risks**: Underreporting or mislabeling near-crises to preserve payouts. Transparent data and third-party crisis classification reduce cheating.
   - **Success Amplification**: Over time, success stories prove prevention yields better returns, attracting more investors and policy support.
2. **Competitive Cooperation Systems**
   Core Idea: Design markets that systematically reward cooperative alliances over cutthroat zero-sum tactics, thereby scaling synergy rather than conflict.
   - **Key Mechanism**:
     - **Co-Licensing Models** – Companies that share IP for safety or sustainability solutions enjoy *multiplying returns*, e.g., cost offsets or licensing expansions.
     - **Cross-Industry Dividends** – In a given sector (e.g., AI labs), a portion of revenue is directed into a shared "betterment pool," distributing dividends only if *all* labs adopt essential safety measures.

   Bio-Emergence-Narrative Matrix

   - **Biological Layer**: We are wired for social reciprocity; competitive cooperation taps that reciprocity at scale, ensuring participants who help others also help themselves.
   - **Emergent Layer**: Partnerships spawn stable networks that resist fragmentation, leading to robust multi-lab synergy in AI or cross-company resilience in supply chains.
   - **Narrative Layer**: Cultural stories shift from "I succeed by beating you" to "We elevate each other, and that's how we all succeed."

Implementation and Actors

- **Who**: **Industry consortiums, technology alliances** (like the Partnership on AI), and **global trade bodies. Governments** can provide legal frameworks supporting co-licensing.
- **Risks**: Collusion or cartel-like behavior if a consortium turns insular. **Counter**: External regulatory checks and anti-monopoly oversight.
- **Adaptive Governance**: The shared pool or co-licensing approach is adjusted periodically based on results and stakeholder feedback.

13.19.5 Part IV: Social Architecture Innovation

1. **Status through System Support**
   Core Idea: Reengineer social "status" so that it grows primarily from documented actions that strengthen communal resilience or global well-being.
   - **Key Mechanism**:
     - **Impact Reputation Systems** – Social networks or community platforms track verified contributions to ecosystem restoration, AI-safety research, or conflict resolution. Higher impact = higher prestige.
     - **Community-Driven Awards** – Locally or globally recognized honors for synergy-building achievements, overshadowing conventional "money or fame" accolades.

   Bio-Emergence-Narrative Matrix

   - **Biological Layer**: Humans crave recognition and admiration. By linking prestige to genuine prosocial impact, you channel that drive ethically.
   - **Emergent Layer**: Over time, as more individuals chase system-positive recognition, the emergent cultural climate rewards collaboration.
   - **Narrative Layer**: Inverts the old story that "status = wealth or power" into "status = proven enrichment of the common good."

   Implementation and Actors

   - **Who: Social media platforms**, **community organizations**, **universities**, and **award foundations** (like a new "Nobel Prize for Systems Resilience").
   - **Risks**: Performative altruism or curated illusions of impact. **Counter**: Independent verification, transparent data on actual outcomes.
   - **Adaptive Strategy**: If certain forms of "impact" are inflated, the system updates criteria to maintain authenticity.
2. **Competitive Altruism Frameworks**
   Core Idea: Trigger a "race to outdo each other" in altruistic or sustainability terms. Make generosity and bridging divides more "impressive" than personal accumulation.
   - **Key Mechanism**:
     - **Social Return Tournaments** – A periodic contest where communities or organizations highlight philanthropic or sustainability feats, awarding high-visibility benefits (social media verification, brand enhancements).
     - **Bridging Points** – Individuals who successfully mediate conflicts or unify polarized groups accumulate "bridging points," translating into real-world perks (travel grants, discounted services).

   Bio-Emergence-Narrative Matrix

   - **Biological Layer**: Taps the reward circuit of competition – participants relish "winning," but the victory is in *helping more.*

- **Emergent Layer**: Each altruistic innovation quickly spreads as participants replicate winning tactics.
- **Narrative Layer**: Shifts from "I won by amassing more stuff" to "I won by elevating more communities or bridging a major rift."

Implementation and Actors

- **Who: NGOs**, **corporate CSR arms**, or **community-led initiatives** that set up philanthropic tournaments or bridging platforms.
- **Risks**: People might chase sensational gestures. Solutions like peer/jury reviews or outcome verifications keep it genuine.
- **Success Amplification**: Frequent public showcases of altruistic "champions" create a viral effect, enticing new participants.

13.19.6 Part V: Implementation Mechanics (Expanded)

1. **Staged Deployment Strategy**
   - **Phase 1: Pilot Experiments**
     - Start small, in localized contexts or single industries (e.g., a water stress region adopting "competition cascades" for resource management).
     - Evaluate real-world data, user acceptance, and potential gaming behaviors.
   - **Phase 2: Broader Integration**
     - Scale up successful pilots to larger national or international contexts.
     - Adapt based on cultural, economic, or infrastructural differences.
   - **Phase 3: Continuous Evaluation and Refinement**
     - Embed feedback loops that frequently revisit metrics and incentives.
     - In AI contexts, require external "safety scenario tests" at each expansion milestone.

Who:

- **Government bodies** orchestrate or endorse pilot frameworks.
- **Industry alliances** volunteer or are incentivized to test solutions.
- **Civil society** groups track transparency and highlight successes/failures.

2. **Success Amplification Mechanisms**
   - **Cooperative Dividend Model**
     - As cooperative or synergy-based behaviors demonstrate success, participants gain increasing returns or privileges.
     - E.g., each quarter an organization meets shared sustainability or AI-safety thresholds; it unlocks cost reductions or special market entry.
   - **Public "Match-Fund"**
     - Whenever a local pilot shows high impact, a larger fund (philanthropic or government) matches resources to replicate it regionally.
     - Encourages local innovators to refine solutions for broad reusability.

Who:

- **International philanthropic consortia** or **development banks** can provide match-funds.
- **Corporate sponsors** might co-fund expansions to cultivate goodwill and future stability.

3. **Adaptive Governance Systems (Concrete Examples)**
   i. **Reflexive Policy Algorithms**
      - **How It Works**: Environmental taxes, AI safety standards, or resource quotas auto-adjust based on real-time system metrics (pollution levels, AI incident reporting).

   - **Practical Example**: A "dynamic AI-cap" that restricts advanced computing power if certain global safety triggers (AI incidents) are exceeded.
   - **Risks**: Algorithmic bias or overshoot if data is erroneous. Periodic human oversight and multi-stakeholder committees to tune the algorithm.

ii. **Citizen Assemblies with Progressive Powers**
   - **How It Works**: Citizens are randomly selected (like juries) to deliberate on solutions, starting with local or moderate-scope decisions. As they demonstrate trustworthiness and produce beneficial results, the assembly's authority broadens.
   - **Practical Example**: Start with local zoning or environmental oversight. If the assembly fosters proven improvements, it can expand to citywide or even national policy input.
   - **Risks**: Potential for populist or extremist capture. Balanced membership, a transparent process, and a well-structured "exit ramp" if detrimental decisions appear.

Who:

- **Legislatures** or **public referendums** can establish these reflexive or assembly systems.
- **Academic and civic groups** provide facilitation, data, and training for participants.

13.19.7 Part VI: Who Drives These Solutions?

- **Governments/Policymakers**:
  - Legislate new market rules, tax incentives, or open-data mandates.
  - Provide official endorsements and funding for pilot programs.
- **Corporations/Industry Alliances**:
  - Adopt synergy-based frameworks, revise corporate best practices, and co-finance solution expansions.
  - Shift R&D focus from purely proprietary advantage to open, safety-driven innovation.
- **Civil Society/NGOs/Academic Institutions**:
  - Supply validated metrics, auditing mechanisms, and ethical oversight.
  - Organize or moderate citizen assemblies, bridging academia with grassroots communities.
- **Individuals**:
  - Participate in new platforms (positive-feedback social networks), adapt consumer behavior to support pro-system companies, and champion local resilience projects.
  - Engage in citizen assemblies or bridging dialogues, reinforcing cultural narratives of synergy.

Conclusion: Through thoughtful **mechanism design**, **transparent metrics**, and **iterative feedback loops**, we can redirect competitive energies to fortify collective well-being – instead of unraveling it. Each proposal here resonates with the *Bio-Emergence-Narrative Matrix*, ensuring:

- **Biological** drives for competition and status are channeled into system-preserving forms.
- **Emergent** complexities are addressed via dynamic incentives, synergy-based markets, and reflexive governance.
- **Narrative** shifts reframe success as "enriching the common good" rather than "extracting short-term wins."

Admittedly, all face **implementation challenges** – cultural inertia, entrenched power structures, or potential gaming. But the price of inaction is far higher, as compounding crises loom. The final measure of these solutions' success lies in **continuous adaptation**: building resilient, ethical scaffolds that update as conditions change, guided by a *living philosophy* that unites flexible thinking, moral foresight, and systemic awareness.

Ultimately, these approaches illustrate how "Molochian" competition can be harnessed for our **collective evolution** – proving that we can transcend zero-sum strife by embedding empathy, mutual benefit, and long-term viability at the heart of our economic, social, and technological systems.

13.20 A VISIONARY CASE STUDY: TRANSPARENT PREVENTION MARKETS (TPM) IN BAD SALZUFLEN

In 2025, the world teetered on multiple edges –
Climate collapse, AI risks, social upheaval. Yet by 2029, one small German town would prove how **markets themselves** could be reprogrammed to **prevent** catastrophe rather than **profit** from it. This is the story of **TPM** in Bad Salzuflen – a bold experiment that replaced the Moloch-driven chase for quick gains with a system rewarding **openness**, **collaboration**, and **long-term resilience**.

13.20.1 The Engine Awakens (Morning Scenes)

1. **A New Day in Bad Salzuflen**
   8:00 AM – The historic spa town of **Bad Salzuflen**, renowned for its mineral springs and half-timbered houses, now hosted the local outpost of the **Global Prevention Council**. **Imani**, a 29-year-old data coordinator, paused at the glimmering digital display in the atrium:
   - **Prevention-Transparency Index (PTI)**
     - **Openness Score (OS)**: Gauging how fully organizations disclose data, from supply chains to AI research.
     - **Prevention Effectiveness (PE)**: Tracking success at mitigating systemic risks – CO_2, antibiotic overuse, climate resilience.
     - **Sentience-Weighted Impact (SWI)**: Assessing welfare for animals, pollinators, and local communities.

   A few lines caught her eye:
   - **Heidemann Farms**: PTI 82 → 79 *pending investigation.*
   - **Bienenfreund GmbH**: PTI 85, +3 for new pollinator habitats.
   - **Sora Tech**: PTI 74, +2 for releasing AI safety data.

   In old markets, hush-hush deals and cost-cutting soared. In TPM, **transparency** had become the core driver of trust and profitability.
2. **The Synergy Engine Goes Live**
   8:30 AM – Entering the **Synergy Engine** Lab felt like stepping into a futuristic library of blinking servers and holographic screens. Here, AI scoured **terabytes** of real-time data – local IoT sensor feeds, global financial flows, and open-source lab disclosures – to produce risk insights:
   i. **Risk Detection and Data Validation**
      - The engine's **anomaly detectors** compare reported pesticide usage or AI safety compliance with on-ground sensor logs.
      - Cryptographic "Proof-of-Prevention" ensures that claims (e.g., reforestation) match actual outcomes, adjusting an entity's PTI within minutes.

   ii. **Decision-Making in Nuanced Cases**
      - A new alert flashed: "*Discrepancy: GreenStar Foods vs. Heidemann pesticide data conflict.*"
      - The system cross-referenced shipping logs, local pollinator health metrics, and AI-driven climate models. But part of Heidemann's data was flagged "proprietary formula."

- Conflicting ethical considerations arose: How to balance **trade secrets** with **public safety**?
- The engine recommended a "masked disclosure protocol," where Heidemann's exact chemical composition remained encrypted but total pesticide volumes would be transparently verified.

Imani thought: *"Even the best AI can't fix everything – but it can spotlight ways to avoid Moloch-like secrecy without destroying legitimate innovation."*

13.20.2 Crisis and Response (Main Conflict)

3. **A Brewing Storm: Global Recession and Local Alarm**
   10:00 AM – Panic rippled across major stock exchanges; a **global recession** loomed. Insurance premiums soared, and some legacy industries hinted at rolling back transparency commitments, claiming the new environment demanded "cost-cutting" above all.
   A glaring red line on the council's dashboard spelled trouble:
   - **Heidemann Farms** planned a fresh bond issuance pegged to their PTI. But if their score dropped too far, the interest rate would climb from **3.2%** to **4.5%**, jeopardizing expansion plans.
   Council Director: "If Heidemann is gaming pesticide data, they'll face an immediate penalty – but we also have to manage recession fears. Let's see what happens if we stand by TPM's rules."
4. **Heidemann Farms: A Moral and Technical Dilemma**
   10:30 AM – An urgent meeting convened. **GreenStar Foods** insisted Heidemann was underreporting pesticide usage, endangering local pollinators – and hence the **Sentience-Weighted Impact** measure. Heidemann countered that fully exposing their new "low-toxicity" formula risked IP theft.
   - **GreenStar**: "Pollinators and animal welfare aren't optional. SWI is a core dimension of the PTI, not a box to check."
   - **Heidemann**: "We're not malicious. We just can't reveal everything. Should our business fail because of forced transparency?"
   - **Imani**: "The Synergy Engine suggests partial data encryption. If you open your total chemical usage for a cryptographic audit, we can preserve the formula itself. But be warned: if usage exceeds safe thresholds, the PTI and bond rating drop automatically."

Everyone recognized the stakes: a precarious trade-off between **public good** and **proprietary advantage** – the hallmark moral tension TPM was designed to navigate.

13.20.3 Resilience and Reflection (Ending)

5. **A Resilience Amplification Loop: Sora Tech's Big Win**
   12:00 PM – Meanwhile, a new scene unfolded with **Sora Tech**, a medium-sized AI startup that had just published a fully transparent hazard-evaluation dataset:
   i. **Initial Action**: Sora Tech's "high OS" rating jumped from **73 to 75** after disclosing real-time bug fixes for its AI training model.
   ii. **Market Reaction**: With higher PTI, their bond yield premium fell from **3.8%** to **3.0%**, saving them tens of thousands of euros a year.

iii. **Increased R&D**: They reinvested the savings to bolster AI alignment protocols.
iv. **Further PTI Bump**: Additional prevention achievements pushed their total PTI to **78**, attracting even more investor confidence.

This **Resilience Amplification Loop** illustrated how one *proactive move* (transparency in bug fixes) triggered an **upward spiral** of better financing, deeper R&D, and further improvements – feeding back into the PTI to **reinforce** synergy over secrecy.

6. **Knowledge Network Effects and a Surprising Breakthrough**
2:00 PM – In the Synergy Engine Lab, the **Dynamic Learning Layer** flagged a breakthrough in Kenya, where local farmers had successfully replaced high-toxicity pesticides with a natural pest deterrent derived from native plants:
   - Sensor data showed **pollinator counts** up by 35%.
   - Crop yields are stable, with antibiotic usage falling for livestock due to improved ecosystem health.

   The engine recommended adding this "**Kenyan Method**" to the region's **Smart Prevention Contracts**. Within hours, local farmland owners in Bad Salzuflen could opt in to trial the method, automatically receiving micro-grants from **Prevention Pools**. If successful, they'd see **PTI** boosts and lower insurance premiums – **knowledge** traveling halfway around the world at the speed of data.
7. **The Synergy Engine's Complex Call**
3:00 PM – A hush fell as the **Synergy Engine** completed its "Heidemann vs. GreenStar" analysis. Conflicting data streams were resolved through partial encryption:
   - Heidemann's total pesticide usage exceeded recommended thresholds by about 10%.
   - Because SWI flagged pollinator risk, the engine triggered a **PTI** downgrade from **79** to **76**.
   - Bond markets reacted within minutes: Heidemann's interest rate jumped from **3.2%** to **4.6%**.

   The engine also recommended a possible redemption path:
   - Adopt the "Kenyan Method" for 25% of fields, publicly audit results, and see up to a **+4** PTI rebound.
   - If successful, the bond rate could drop below **3.5%** again.

   Heidemann Rep (sighing): "We may lose money short-term, but it's still better than losing the entire farm. We'll adapt."
8. **Personal Reflections and Hope**
5:30 PM – Imani met her friend **Nina** at a bench by the salt springs, recapping the day's events:
   - **Nina**: "Five years ago, we dealt with floods and economic chaos. I never imagined a system where synergy outperforms secrecy."
   - **Imani**: "Watching the Synergy Engine handle that pesticide conflict was nerve-wracking. But seeing Heidemann forced to adopt safe practices – without fully sacrificing their IP – proves how we can steer markets away from Moloch's grip."
   - **Nina**: "And that Kenyan method might save our local bees, too. Who'd have thought technology and transparency could spread solutions so fast?"

Imani's eyes drifted to the horizon, where the evening light glinted off solar panels and farmland. She remembered the devastation of 2024, how the old system had rewarded corners cut at the planet's expense. But here, in 2029, a different model was taking shape – one in which **profit** hinged on **prevention**, **openness** trumped **secrecy**, and **collective well-being** factored into every trade. The feeling of hope was tangible. She rose, brimming with gratitude that **TPM** had found fertile ground in Bad Salzuflen – and, if the day's events were any indication, across the wider world.

Key Takeaways

1. **Prevention-Transparency Index (PTI)**:
   - Numeric shifts (e.g., 82 → 79 → 76) illustrate how transparency lapses or proactive disclosures affect financing costs and market standing.
   - The **Sentience-Weighted Impact** metric influences decisions around pollinators, livestock welfare, and more.
2. **Synergy Engine Decision-Making**:
   - Showcased in handling conflicting data (Heidemann's proprietary formula) and factoring in ethical trade-offs between IP and public good.
   - Proposes partial encryption solutions to balance transparency with innovation.
3. **Resilience Amplification Loop**:
   - **Sora Tech** action → improved PTI → lower bond rates → reinvestment in AI safety → further PTI boost → a self-reinforcing cycle of synergy.
4. **Knowledge Network Effects**:
   - Kenyan pest deterrent success is quickly integrated into Bad Salzuflen's local pilot, demonstrating how positive innovations spread globally via TPM's Smart Prevention Contracts.
5. **Handling External Challenges**:
   - A sudden global recession tests TPM's ability to maintain trust and stability; high-PTI entities remain attractive to investors, showcasing the model's resilience.
6. **A Personal Ending**:
   - Imani and Nina's reflection ties the system's macro-level changes to human-scale hope. It shows how real lives benefit from a framework that rewards risk reduction, fosters open collaboration, and values every sentient being in the ecosystem.

With these elements, the story not only **illustrates** the key components of **TPM** but also **demonstrates** how a synergy-centered economic model can transform local communities – and, by extension, the global economy – away from Moloch's destructive spiral and toward a future built on **cooperation**, **prevention**, and **transparency**.

Conclusion: The **TPM** narrative exemplifies how competition and profit motives can be harnessed to drive **long-term resilience**, **safety**, and **cooperation** rather than short-term exploitation. By weaving together concepts such as **the Moloch effect**, **systemic crises**, **AGI safety**, and **innovative policy frameworks**, this chapter underscores a profound lesson: The future is not determined by technology or capitalism alone, but by how **incentives** are structured and how **narratives** are reframed to value collective well-being.

By embracing transparency, aligning market forces with systemic stability, and adopting flexible governance mechanisms – whether for environmental crises or AGI development – we can outmaneuver the zero-sum traps that have historically plagued societies. This holistic approach blends **technical detail** (like the Impact-Priority Matrix) with **moral foresight**, **cooperative incentives**, and **AI-driven synergy**, paving the way toward a more sustainable, equitable, and peaceful future.

13.21 SUPPLEMENT

Below is the HTML code for a **basic web-based calculator** that helps evaluate and prioritize actions based on **impact**, **systemic interconnectedness**, **incentive misalignment**, and **urgency**. Simply copy it into an .html file, open it in your browser, and input your data:

```
<!DOCTYPE html>
<html lang="en">
<head>
```

```
    <meta charset="UTF-8">
    <meta name="viewport" content="width=device-width,
initial-scale=1.0">
    <title>Impact-Priority Matrix Calculator</title>
    <style>
        body {
          font-family: Arial, sans-serif;
          margin: 20px;
        }
        label {
            display: block;
            margin-bottom: 10px;
        }
        input[type="number"] {
            width: 50px;
            margin-right: 10px;
        }
        button {
            margin-top: 20px;
            padding: 10px 20px;
            font-size: 16px;
        }
        #result {
            margin-top: 20px;
            font-weight: bold;
        }
    </style>
</head>
<body>



    <h1>Impact-Priority Matrix Calculator</h1>

    <form id="calculatorForm">
        <h3>Impact Score (I)</h3>
        <label>
           Environmental Impact (0-10):
           <input type="number" id="environmentalImpact" min="0" max="10"
required>
        </label>
        <label>
           Social Impact (0-10):
           <input type="number" id="socialImpact" min="0" max="10"
required>
        </label>
        <label>
           Economic Impact (0-10):
           <input type="number" id="economicImpact" min="0" max="10"
required>
        </label>
        <label>
           Technological Impact (0-10):
           <input type="number" id="technologicalImpact" min="0" max="10"
required>
        </label>
```

```
        <h3>Systemic Interconnectedness Score (S) (0-10):</h3>
        <input type="number" id="systemicInterconnectedness" min="0"
max="10" required>

        <h3>Incentive Misalignment Score (M) (0-10):</h3>
        <input type="number" id="incentiveMisalignment" min="0" max="10"
required>

        <h3>Urgency Score (U)</h3>
        <label>
            Time Sensitivity (0-10):
            <input type="number" id="timeSensitivity" min="0" max="10"
required>
        </label>
        <label>
          Escalation Risk (0-10):
          <input type="number" id="escalationRisk" min="0" max="10"
required>
        </label>
        <label>
          Public Attention (0-10):
          <input type="number" id="publicAttention" min="0" max="10"
required>
        </label>


        <button type="button" onclick="calculatePriority()">Calculate
Priority Score</button>
    </form>

    <div id="result"></div>

    <script>
      function calculatePriority() {
            // Get input values
            var environmentalImpact = parseFloat(document.getElementById
('environmentalImpact').value);
            var socialImpact = parseFloat(document.
getElementById('socialImpact').value);
            var economicImpact = parseFloat(document.getElementById('econ
omicImpact').value);
            var technologicalImpact = parseFloat(document.getElementById
('technologicalImpact').value);

            var systemicInterconnectedness = parseFloat(document.getEleme
ntById('systemicInterconnectedness').value);
            var incentiveMisalignment = parseFloat(document.getElementByI
d('incentiveMisalignment').value);

            var timeSensitivity = parseFloat(document.getElementById('tim
eSensitivity').value);
            var escalationRisk = parseFloat(document.getElementById('esca
lationRisk').value);
```

```
            var publicAttention = parseFloat(document.getElementById('pub
licAttention').value);

            // Calculate the average Impact Score (I)
            var impactScore = (environmentalImpact + socialImpact +
economicImpact + technologicalImpact) / 4;

            // Calculate the average Urgency Score (U)
            var urgencyScore = (timeSensitivity + escalationRisk +
publicAttention) / 3;

            // Calculate the Priority Score (P)
            var priorityScore = ((impactScore +
systemicInterconnectedness + incentiveMisalignment) / 3) * urgencyScore;

            // Display the result
            document.getElementById('result').innerHTML = 'Priority Score
(P): ' + priorityScore.toFixed(2);
        }
    </script>

</body>
</html>
```

Explanation:

- **Input Fields**: The user enters scores for Environmental Impact, Social Impact, Economic Impact, and Technological Impact (all on a scale of 0–10). They also input scores for Systemic Interconnectedness, Incentive Misalignment, and Urgency components (Time Sensitivity, Escalation Risk, and Public Attention).
- **Calculation**:
  - The script calculates the average **Impact Score (I)** from the four impact inputs.
  - It then calculates the average **Urgency Score (U)** from the three urgency inputs.
  - Finally, the **Priority Score (P)** is calculated using the formula:

$$P = \left(\frac{I + S + M}{3} \right) \times U$$

How to Use:

1. Copy the above code into an HTML file.
2. Open the file in a web browser.
3. Input the scores into the provided fields.
4. Click the "Calculate Priority Score" button to see the calculated score.

By integrating **analytic tools** like the **Impact-Priority Matrix** with **visionary solutions** such as **TPM**, **circular economies**, and **AI-powered collective intelligence**, we can strategically address the Moloch effect and steer both **capitalism** and **AGI** development toward a more equitable, sustainable future.

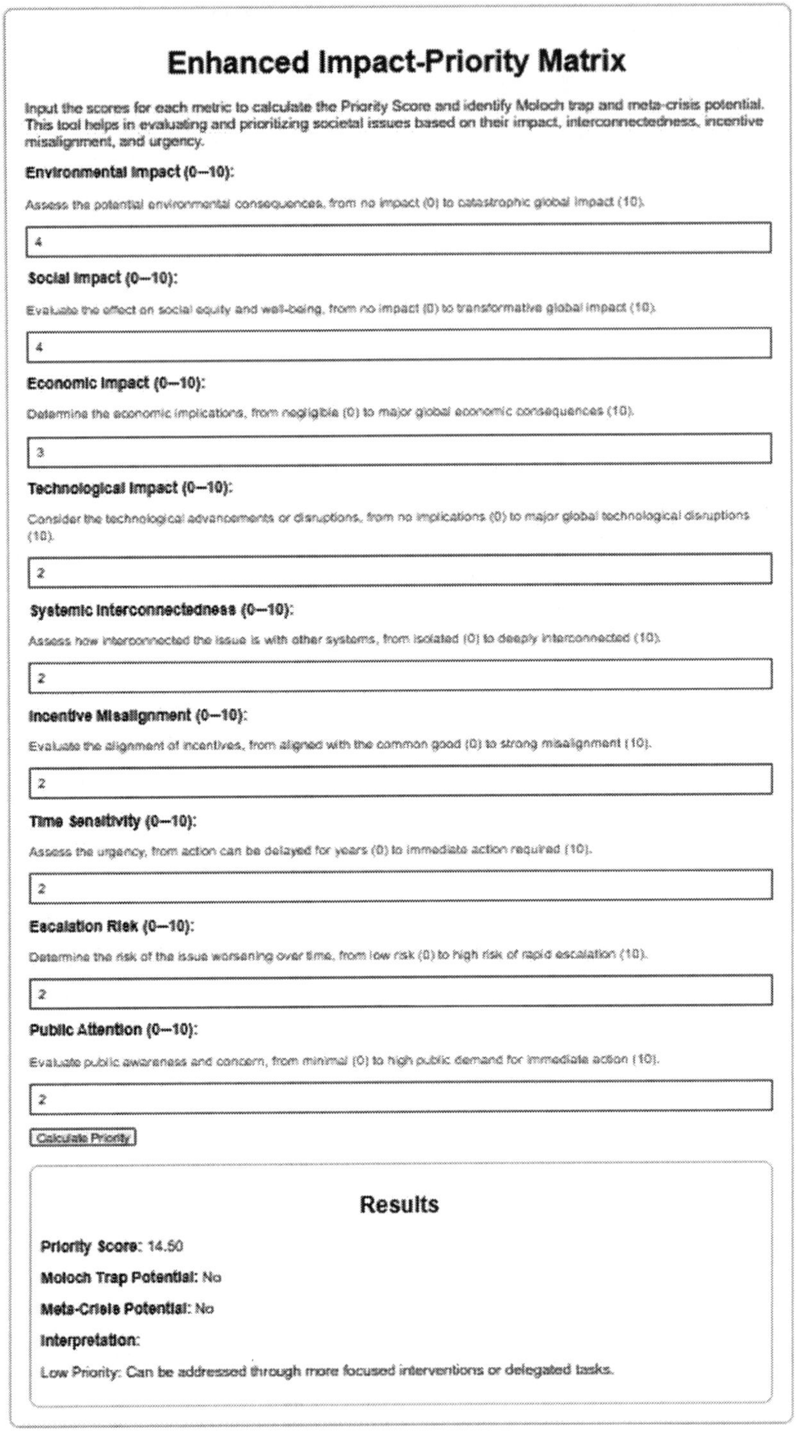

FIGURE 13.1 Impact Priority Matrix: Enhanced Impact-Priority Matrix for assessing societal issues and crisis potential.

14 Philosophy of Causal Relevance Ethics

Core Principles:

1. **Causal Proximity (C):** Measures how soon and how certainly an action will have an impact.
2. **Influence Strength (I):** Assesses an action's ability to **enhance well-being (Thrive Index, TI)** or **reduce harm (Adversity Reduction Index, ARI)**.
3. **Feasibility (F):** Evaluates **resource availability and practical achievability**.

Key Applications and Extensions:

- **Causal Relevance Score (CRS):** A formula-based approach to ranking ethical actions, ensuring focus on **high-impact, realistic interventions**.
- **Yellow Teaming:** Identifies **unintended consequences** to refine decision-making.
- **Avoiding Virtue Signaling and Whataboutism:** Prioritizes **effective over symbolic actions**.
- **Integration with Psychological Theories:** Uses **Prospect Theory, Dual-Process Thinking, and Emotional Motivation** to balance **rational analysis with human psychology**.

Key Takeaway:

CRE **bridges philosophy, psychology, and decision science**, providing **a pragmatic, ethical framework** to **maximize positive change without moral ambiguity**.

14.1 INTRODUCTION

The Philosophy of Causal Relevance Ethics (CRE) explores how to identify points within causal chains where interventions can have the most significant impact. Recognizing the complexity and unpredictability inherent in causal networks, this philosophy focuses on actionable points where efforts can lead to meaningful change. By integrating structured principles and utilitarian ideals, it aims to maximize well-being and minimize adversity without falling into moral ambiguity.

14.2 UNDERSTANDING CAUSAL CHAINS AND CONTINGENCY

Causal chains are intricate and often nonlinear, with past events influencing the present and shaping the future in unforeseen ways. While it is impossible to control all contingencies, CRE emphasizes identifying key points where intervention is both meaningful and feasible.

- **Contingency Awareness**: Many life events stem from seemingly improbable causes. By understanding these contingencies, individuals can take targeted responsibility for their actions.
- **Focused Intervention**: Rather than attempting to manage every variable, the philosophy directs attention to points in the causal network where actions can produce significant effects.

DOI: 10.1201/9781003640288-19

14.3 CORE PRINCIPLES OF CAUSAL RELEVANCE

To determine where actions can be most effective, CRE is based on three core principles:

a. **Causal Proximity (C)**
The temporal and certainty aspects of an action's impact.
- **Time Factor (TF)**: How soon will the action have an effect? Rated from 0 (no impact) to 5 (immediate impact).
- **Certainty Factor (CF)**: How likely is the action to succeed? Rated from 0 (impossible) to 5 (certain).

Calculation of Causal Proximity:

$$C = \begin{cases} 0, & \text{if } TF = 0 \text{ or } CF = 0 \\ TF + CF, & \text{otherwise} \end{cases}$$

b. **Influence Strength (I)**
The potential of an action to increase well-being and/or reduce adversity.
- **Thrive Index (TI)**: Measures the action's ability to enhance positive aspects of life, such as health, autonomy, and purpose. Rated from 0 to 5.
- **Adversity Reduction Index (ARI)**: Quantifies the action's capacity to reduce negative factors like stress, financial strain, and health challenges. Rated from 0 to 5.

Calculation of Influence Strength:

$$I = \begin{cases} 0, & \text{if } TI = 0 \text{ and } ARI = 0 \\ TI + ARI, & \text{otherwise} \end{cases}$$

c. **Feasibility (F)**
The practicality of implementing the action.
- **Resource Availability (RA)**: Are sufficient resources accessible to execute the action? Rated from 0 (none) to 5 (abundant).
- **Practical Achievability (PA)**: How realistic is the action considering current capabilities? Rated from 0 (impossible) to 5 (easily achievable).

Calculation of Feasibility:

$$F = \begin{cases} 0, & \text{if } RA = 0 \text{ or } PA = 0 \\ RA + PA, & \text{otherwise} \end{cases}$$

WEAPONIZED EMPATHY

THINKING BOX

"Can our compassion be twisted against us by manipulative digital forces? Reflecting on **42** could spur caution and balanced understanding."

14.4 CALCULATING THE CAUSAL RELEVANCE SCORE (CRS)

The CRS helps prioritize actions based on their overall potential impact.

Calculation of CRS:

$$CRS = \begin{cases} 0, & \text{if } C = 0 \text{ or } I = 0 \text{ or } F = 0 \\ C + I + F, & \text{otherwise} \end{cases}$$

- **Maximum CRS**: 30 (when all components are at their highest value).
- **Interpretation:**
  - **CRS = 0%**: Do Not Proceed
  - **CRS = 1–50% (1–15)**: Might Not Be Worthwhile
  - **CRS = 51–99% (16–29)**: Probably Worthwhile
  - **CRS = 100% (30)**: Must Do

General Guidance: The higher the CRS, the more worthwhile the action probably is.

14.5 CONNECTION TO THE TEMPORAL TRUTH EVALUATION FRAMEWORK

The Temporal Truth Evaluation Framework (Chapter 15) assesses the probability of events or assumptions being true across the past, present, and future.

- **Integration with Causal Relevance Ethics:**
  - Helps identify causal points that are not only relevant but also likely to produce the intended outcomes.
  - Ensures that actions are based on empirical data and logical patterns, reducing reliance on speculative assumptions.

Practical Benefit: By focusing on probable and relevant causal points, actions are more likely to result in meaningful and effective interventions.

14.6 ALIGNMENT WITH UTILITARIAN PRINCIPLES

CRE incorporates utilitarian ideals – maximizing happiness and minimizing suffering – through structured and measurable indices.

- **Thrive Index (TI)**: Represents the enhancement of well-being.
- **Adversity Reduction Index (ARI)**: Represents the reduction of suffering.

Why This Approach?

- **Objectivity**: Provides clear, quantifiable metrics to assess the impact of actions.
- **Avoids Ambiguity**: Reduces the subjective interpretation of "happiness" and "suffering" by using defined indices.

14.7 YELLOW TEAMING: ANTICIPATING UNINTENDED CONSEQUENCES

Yellow Teaming is a pragmatic tool used to identify potential negative consequences of actions.

- **Focus**: Examines unintended and possibly harmful effects, even when the intent is positive.
- **Process**:
  - Systematically question what could go wrong.
  - Integrate findings into decision-making to mitigate risks.

Application in Causal Relevance Ethics:

- Ensures a thorough evaluation of actions before implementation.
- Enhances the robustness of decisions by considering potential pitfalls.

14.8 AVOIDING VIRTUE SIGNALING AND WHATABOUTISM

CRE focuses on actions that have real impact, avoiding:

- **Virtue Signaling**: Taking actions primarily to display moral correctness rather than effect change.
- **Whataboutism**: Deflecting attention by pointing to other issues instead of addressing the current one.

Approach:

- Concentrate on relevant, actionable points.
- Make decisions based on structured analysis rather than symbolic gestures.

14.9 MANAGING UNINTENDED CONSEQUENCES

Acknowledging the unpredictability of reality, the philosophy incorporates strategies to handle unforeseen outcomes.

- **Iterative Process:** Regularly review and adjust decisions based on new information.
- **Yellow Teaming:** Continuously assess and address potential negative consequences.

14.10 ETHICAL RESPONSIBILITY IN CAUSAL CONTEXTS

CRE emphasizes responsibility in action without relying on moral judgments of "good" or "bad."

- **Focus on Relevance**: Determine where interventions can have significant impact within the causal chain.
- **Example**: A company considering an investment in environmentally friendly technology evaluates the action based on its causal relevance and potential impact, not merely on moral grounds.

14.11 CONCLUSION

CRE provides a clear, structured framework for making impactful decisions within complex causal networks.

- **Maximizes Positive Impact**: By focusing on actions with high CRS, resources are directed toward interventions that are likely to produce meaningful benefits.
- **Minimizes Moral Ambiguity**: Utilizes measurable indices to assess actions, reducing subjective interpretations.
- **Promotes Effective Responsibility**: Encourages individuals and organizations to act where they can make the most significant difference.

Final Thought: By identifying and acting upon the most relevant points in the causal chain, we can effect positive change efficiently and responsibly.

THE ILLUSION OF INDEPENDENCE

THINKING BOX

"How reliant are we on these interlinked systems we barely see? The degree of our dependency might be closer to **42** than we realize."

14.12 EXCURSION

14.12.1 Excursion 1: Connection to Other Ethical Approaches

1.1 **Comparison with Utilitarianism**
CRE shares foundational principles with utilitarianism, particularly the goal of maximizing well-being and minimizing adversity. However, it introduces a structured approach that emphasizes the probability and relevance of causal points before taking action.

Alignment with Mill's Qualitative Utilitarianism

- **Quality over Quantity**: Building upon John Stuart Mill's emphasis on the quality of pleasures, CRE recognizes that not all utilities are equal. By incorporating factors like Causal Proximity and Feasibility, it assesses both the quality and practicality of actions.
- **Structured Evaluation**: The use of the Thrive Index (TI) and Adversity Reduction Index (ARI) provides measurable criteria for evaluating actions, moving beyond the subjective interpretations often associated with "happiness" and "suffering."

Distinctive Features

- **Emphasis on Probability and Relevance**: Unlike traditional utilitarianism, which may focus solely on outcomes, CRE considers the likelihood of success and the practical feasibility of interventions.
- **Avoidance of Moral Ambiguity**: By using quantifiable indices and a clear calculation method, it reduces the moral ambiguity inherent in subjective assessments, providing a more objective framework for decision-making.

1.2 **Comparison with Deontology**
While deontological ethics focuses on the inherent morality of actions based on adherence to rules or duties, CRE centers on the effectiveness and impact of actions within causal chains.

Key Differences

- **Outcome-Oriented**: CRE evaluates actions based on their potential outcomes and practical implementation, rather than adherence to predefined moral duties.
- **Flexibility**: It allows for adaptive decision-making based on empirical data and structured analysis, whereas deontology often adheres to fixed moral principles regardless of consequences.

Complementary Aspects

- **Ethical Responsibility**: Both approaches recognize the importance of ethical responsibility, but CRE emphasizes acting where interventions can have the most significant impact.

14.12.2 Excursion 2: The Role of Emotions in Decision-Making

2.1 **Emotions and Objectivity**
Although CRE strives for objectivity by avoiding moral terminology and using quantifiable measures, emotions inevitably influence decision-making processes.

- **Influence on Perception**: Emotions can shape how individuals perceive consequences and causal points, potentially introducing biases.
- **Challenge to Objectivity**: Emotional responses may affect the assessment of factors like Feasibility and Influence Strength, challenging the framework's objectivity.

2.2 **Integrating Emotions Constructively**
Recognizing the role of emotions, the philosophy suggests ways to integrate emotional insights without compromising rational analysis.
- **Emotions as Indicators**: Emotions can serve as signals highlighting areas that require attention or further analysis.
- **Balancing Rationality and Emotion**: By acknowledging emotional responses but verifying them through structured evaluation, decision-makers can harness the motivational aspects of emotions while maintaining objectivity.

14.12.3 Excursion: Psychological Insights and Their Role in Causal Relevance Ethics

3.1 **Prospect Theory: Emotions as Drivers of Decisions**
Developed by Daniel Kahneman and Amos Tversky, Prospect Theory illustrates how people make decisions based on perceived gains and losses rather than final outcomes.
- **Emotional Biases**: Individuals are prone to cognitive biases influenced by emotions, such as loss aversion and overestimating unlikely outcomes.
- **Impact on Causal Relevance Ethics**: Awareness of these biases helps decision-makers recognize when emotions may distort their assessment of causal points.

3.2 **Dual-Process Theory and Decision-Making**
The Dual-Process Theory describes two systems of thought:
- **System 1 (Intuitive and Emotional)**: Fast, automatic thinking that relies on heuristics and emotions.
- **System 2 (Analytical and Rational)**: Slow, deliberate thinking that requires effort and logical reasoning.

Application in Causal Relevance Ethics:

- **Leveraging Both Systems**: Effective decision-making involves a balance between intuition and analysis.
- **Mitigating Biases**: By being aware of the tendencies of System 1, individuals can employ System 2 to critically evaluate actions using the framework's structured approach.

3.3 **Emotions as Motivational and Signaling Factors**
Emotions are not solely sources of bias; they can also motivate action and signal important considerations.
- **Motivation**: Strong emotional responses can drive individuals to take action where it is needed most.
- **Signaling Relevance**: Emotions may highlight causal points that require attention, serving as initial indicators for further rational analysis.

3.4 **Decision-Making under Uncertainty: Heuristics and Emotions**
In uncertain situations, individuals often rely on heuristics influenced by emotions to make quick decisions.
- **Heuristics as Shortcuts**: Mental shortcuts can be useful but may lead to suboptimal outcomes if not examined critically.
- **Balancing Speed and Accuracy**: CRE encourages the use of structured analysis to complement intuitive judgments, improving decision quality under uncertainty.

3.5 **Integrating Psychological Insights into the Framework**
By incorporating understanding from Prospect Theory and Dual-Process Theory, CRE acknowledges the complex interplay between emotion and reason.
- **Enhanced Decision-Making**: Recognizing emotional influences allows for adjustments that improve the objectivity and effectiveness of actions.
- **Holistic Approach**: Integrating both emotional and rational processes leads to more comprehensive assessments of actions within complex causal networks.

14.13 ASSUMPTION

The integration of psychological insights into CRE enriches the framework by accounting for the role of emotions in decision-making. While emotions can introduce biases, they also serve as valuable signals and motivators. By verifying emotional responses through rational analysis and structured evaluation, decision-makers can harness the benefits of both intuition and logic.

This holistic approach ensures that actions are not only objectively assessed for their relevance and impact but also aligned with human motivations and experiences. By combining logical analysis with emotional intelligence, CRE provides a robust framework for making effective and responsible decisions in complex causal environments.

14.14 FINAL THOUGHTS

CRE offers a comprehensive methodology for identifying and acting upon points within causal chains where interventions can have the most significant impact. By combining structured principles, utilitarian ideals, and psychological insights, it provides a clear framework for decision-making that maximizes positive outcomes while minimizing moral ambiguity.

Key Takeaways:

- **Structured Evaluation**: Utilizes measurable indices and conditional calculations to objectively assess actions.
- **Emphasis on Feasibility and Impact**: Focuses on actions that are not only impactful but also practical and likely to succeed.
- **Integration of Emotional and Rational Processes**: Acknowledges the role of emotions in decision-making while ensuring they are balanced with rational analysis.
- **Avoidance of Moral Ambiguity**: Reduces reliance on subjective moral concepts by providing clear criteria for action evaluation.

By applying CRE, individuals and organizations can make informed decisions that effectively contribute to maximizing well-being and minimizing adversity in complex causal networks.

15 Temporal Truth Evaluation Framework

A Balanced Approach to Assessing Claims

Core Components:

1. **Past – Empirical Evidence**: Strength of past data and studies supporting the claim. **Updated using Bayesian reasoning** to refine certainty as new information emerges.
2. **Present – Immediate Applicability**: How effectively a claim works **in the current context**. **Scenario planning** helps balance short-term practicality with long-term sustainability.
3. **Future – Reliable Predictability**: How well a claim **forecasts future outcomes**, integrating **cross-disciplinary insights** to enhance accuracy.

Advanced Considerations:

- **Bayesian Updating and Dynamic Weighting**: Adjusts claim credibility as **new data emerges**.
- **Proxy Truths**: Uses **validated insights from related fields** when direct evidence is unavailable.
- **Psychological Factors**: Human biases (e.g., lack of control, predictability, or social status) influence claim acceptance.

Key Takeaway:

TTEF **bridges logic, adaptability, and psychology**, ensuring **claims are critically evaluated, dynamically updated, and practically applied** – leading to **better-informed, resilient decision-making**.

15.1 INTRODUCTION

Defining "truth" has always been a complex challenge, often leading to philosophical debates without clear resolution. Instead of attempting to pin down an absolute definition of truth, the **Temporal Truth Evaluation Framework (TTEF)** offers a practical method for evaluating the robustness of claims by examining them across three temporal dimensions: the **past**, **present**, and **future**.

This framework serves as a tool to assess how well a claim holds up over time, providing a balanced view that helps in making informed decisions. By focusing on key attributes distributed across time, the framework avoids entanglement in abstract definitions and offers actionable criteria for evaluating the validity of claims in various contexts.

15.2 THE SIMPLE FRAMEWORK

Purpose: The Simple Framework evaluates the truthfulness of claims by analyzing how they perform in each temporal dimension. It considers the strength of **empirical evidence** from the past, the **immediate applicability** in the present, and the reliable **predictability** for the future.

DOI: 10.1201/9781003640288-20

1. **Past: Empirical Evidence**
   - **Attribute**: Empirical evidence consists of observations, experiments, and data collected in the past that support a claim.
   - **Role**: Strong empirical evidence increases confidence in a claim by demonstrating that it has been validated through previous experiences or studies.

   Detailed Explanation: Empirical evidence forms the foundation upon which many claims are built. It involves systematically collected data that can be verified and replicated by others. In science, for instance, a hypothesis gains credibility when multiple experiments yield consistent results.

   Example: Consider the claim that a specific medication effectively treats a disease. If numerous clinical trials over several years consistently show positive results, this accumulated empirical evidence strengthens the claim's validity. The more robust and diverse the evidence (e.g., different populations, conditions, and methodologies), the higher the confidence in the claim.

2. **Present: Immediate Applicability**
   - **Attribute**: Immediate applicability assesses how useful and relevant a claim is in the current context or situation.
   - **Role**: A claim gains credibility if it can be effectively applied now, demonstrating practical value and effectiveness in real-time.

   Detailed Explanation: Immediate applicability focuses on the present utility of a claim. It's not enough for a claim to have strong historical backing; it must also be relevant and effective in today's context. This dimension evaluates whether the claim addresses current needs or problems and whether it can be implemented with available resources and technologies.

   Example: A new software tool claims to enhance productivity for remote teams. If businesses adopt the tool and immediately notice improvements in collaboration and efficiency, the claim's immediate applicability is confirmed. Testimonials, case studies, and user feedback play crucial roles in demonstrating this applicability.

3. **Future: Reliable Predictability**
   - **Attribute**: Reliable predictability measures a claim's ability to accurately forecast future outcomes or trends.
   - **Role**: A claim that consistently predicts future events enhances its robustness, indicating that it remains valid over time and can guide long-term planning.

   Detailed Explanation: Reliable predictability is about the claim's forward-looking accuracy. It assesses whether the claim can help anticipate future developments, making it valuable for strategic decision-making. This dimension is particularly important in fields like economics, meteorology, and technology forecasting, where anticipating future conditions is essential.

   Example: An economic model predicts a recession based on certain indicators. If the model has a track record of accurately forecasting economic downturns, policymakers can rely on it to implement preemptive measures. The claim's reliable predictability is thus established through its consistent alignment with future realities.

   Balancing the Attributes: In the Simple Framework, the three temporal attributes are interconnected:

   - **Holistic Evaluation**: A robust claim typically performs well across all three dimensions. Strength in empirical evidence, immediate applicability, and reliable predictability indicates a well-rounded and trustworthy claim.

- **Interdependency**: Weakness in one area can undermine confidence in the entire claim. For example, strong empirical evidence may be less persuasive if the claim lacks immediate applicability or fails to predict future outcomes accurately.
- **Decision-Making**: By evaluating all three attributes, decision-makers can more confidently assess the validity of a claim and its suitability for adoption or implementation.

THE SIMPLICITY BIAS

THINKING BOX

"Are we adding layers of complexity where none are needed? The sweet spot, in life and coding, might be the essence of **42**."

15.3 THE COMPLETE FRAMEWORK

Purpose: The Complete Framework expands on the Simple Framework by introducing dynamic elements to handle complex and evolving situations. It incorporates **Bayesian updating**, **cross-disciplinary perspectives**, and **proxy truths**, and considers **psychological factors** that influence how claims are perceived and accepted.

1. **Past: Empirical Evidence**
   - **Attribute**: Empirical evidence remains foundational but must be continually reassessed as new information emerges.
   - **Role**: Utilize **Bayesian updating** to adjust the weight assigned to empirical evidence. This statistical method allows for the refinement of beliefs by incorporating new data, increasing or decreasing confidence in the claim accordingly.
   - **Proxy Truths**: In situations where direct empirical evidence is lacking, **proxy truths** – validated principles or findings from related fields – can support or challenge the claim.

   Detailed Explanation: Empirical evidence is not static. As new studies are conducted and new data becomes available, the validity of past evidence can change. Bayesian updating provides a mathematical framework for adjusting our confidence in a claim based on the latest information.

   Example: A dietary guideline recommends reducing fat intake to prevent heart disease. Initially supported by empirical evidence, new research may suggest that certain fats are beneficial. Bayesian updating would adjust the confidence in the original claim, possibly reducing its weight in light of new findings.

   Proxy Truths Example: In the early stages of understanding COVID-19, knowledge about other coronaviruses served as proxy truths to guide initial responses. While not a perfect match, this related information helped shape early public health policies until more specific data became available.

2. **Present: Immediate Applicability**
   - **Attribute**: Immediate applicability evaluates the claim's current performance and relevance.
   - **Role: Balance short-term practicality with long-term sustainability**. Implement **scenario planning** to assess potential immediate benefits and anticipate future implications.
   - **Bayesian Updating**: Adjust the weight of applicability based on real-time outcomes. If the claim consistently delivers positive results, increase its weight; if not, reassess its applicability.

Detailed Explanation: Immediate applicability is dynamic and context-dependent. What works today may not work tomorrow due to changing circumstances. Regular assessment ensures that the claim remains relevant and effective.

Example: A marketing strategy that boosts sales during a holiday season demonstrates immediate applicability. However, if the same strategy fails during a non-holiday period, its applicability needs reassessment. Scenario planning can help adapt the strategy to different contexts.

3. **Future: Reliable Predictability**
   - **Attribute**: Reliable predictability focuses on the claim's ability to forecast future events accurately.
   - **Role**: Use **Bayesian updating** to refine the claim's predictive power over time, increasing confidence with each successful prediction and decreasing it with each failure.
   - **Cross-Disciplinary Perspectives**: Incorporate predictive models and insights from other fields to enhance forecasting accuracy.

   Detailed Explanation: Reliable predictability is crucial for long-term planning and risk management. By integrating methods from various disciplines, the predictive capacity of a claim can be strengthened.

   Example: Climate scientists use models that incorporate data from meteorology, oceanography, and environmental science to predict climate change impacts. Cross-disciplinary collaboration enhances the reliability of these predictions.

Dynamic Weighting

- **Adjusting Weights**: The importance of each temporal dimension may vary depending on the context. In rapidly changing industries, immediate applicability might be prioritized, while in established fields, empirical evidence may carry more weight.
- **Bayesian Updating**: Continuously update the weights assigned to each attribute as new information becomes available, ensuring the framework remains responsive to change.

Detailed Explanation: Dynamic weighting allows the framework to adapt to different situations:

- **Contextual Flexibility**: For a tech startup, immediate applicability and future predictability might be more critical than extensive empirical evidence.
- **Responsive Decision-Making**: As new data emerges, decision-makers can adjust their focus, allocating resources and attention where they are most needed.

Incorporating Proxy Truths

- **Cross-Domain Knowledge**: Leverage validated principles from other disciplines to support or challenge claims when direct evidence is limited.
- **Validation**: Proxy truths should be tested within the specific context to confirm their applicability.

Detailed Explanation: Proxy truths can provide valuable insights, especially in emerging fields where direct evidence is scarce.

Example: Artificial intelligence (AI) ethics might draw on established ethical frameworks from philosophy and sociology. While these proxy truths offer guidance, they need adaptation and validation within the AI context.

Balancing Innovation and Resource Efficiency

- **Pilot Testing**: Conduct small-scale trials to evaluate the potential of new ideas without significant resource investment.
- **Tiered Evaluation**: Implement ideas incrementally, allocating more resources as they demonstrate success.
- **Fail Fast, Learn Fast**: Encourage rapid experimentation to quickly identify and learn from failures, optimizing the innovation process.

Detailed Explanation: Balancing innovation with resource constraints is essential for sustainable growth.

Example: A company considering a new product line might first test the concept in a limited market. Positive results justify further investment, while negative results allow for pivoting or abandonment with minimal loss.

Feedback Loops and Continuous Reassessment

- **Regular Reviews**: Schedule periodic evaluations of ongoing projects to adjust strategies based on the latest evidence and performance data.
- **Transparency**: Document decision-making processes and updates to encourage collaboration and reduce biases.

Detailed Explanation: Feedback loops ensure that decisions remain aligned with current realities.

Example: In project management, regular sprint reviews allow teams to reflect on progress, incorporate feedback, and adjust plans accordingly. This agile approach keeps projects responsive to change.

15.4 INCORPORATING PSYCHOLOGICAL STRESS FACTORS

Understanding human psychology is crucial, as it significantly influences how individuals perceive and accept claims. According to neuroscientist Robert Sapolsky, four key psychological stress factors can impact acceptance:

1. **Lack of Control**
   - **Impact**: When individuals feel they have no control over a situation, they may resist or reject claims, even those supported by strong evidence.
   - **Mitigation**: Enhance transparency and involve stakeholders in the decision-making process to increase their sense of control.

   Example: In workplace changes, employees may resist new policies if they feel excluded from the decision-making process. By involving them through surveys or committees, management can reduce resistance and improve acceptance.

2. **Lack of Predictability**
   - **Impact**: Uncertainty about future outcomes can lead to skepticism, causing individuals to doubt even well-founded predictions.
   - **Mitigation**: Clearly communicate the limits of predictability and provide contingency plans to address uncertainties.

   Example: During economic downturns, clear communication about potential impacts and company strategies can alleviate employee anxiety, fostering trust in leadership decisions.

3. **Lack of Outlets for Emotional Expression**
   - **Impact**: Without opportunities to express emotions, individuals may become frustrated or disengaged, hindering acceptance of new claims.
   - **Mitigation**: Create forums for open discussion and feedback, allowing individuals to voice concerns and feel heard.

   Example: Public forums or town hall meetings enable community members to express their views on local developments, leading to more inclusive and accepted outcomes.

4. **Lack of Social Status**
   - **Impact**: Individuals may reject claims if they perceive a threat to their social standing or if the claims challenge their identity.
   - **Mitigation**: Acknowledge contributions and expertise, and present claims in ways that respect and preserve individuals' social status.

   Example: In organizational change, recognizing the expertise of long-standing employees can help integrate new practices without diminishing their status.

15.5 ACKNOWLEDGING LIMITATIONS AND CHALLENGES

While the TTEF offers a robust tool for assessing claims, it is important to recognize potential limitations:

1. **Ambiguity in Empirical Evidence (Past)**
   - **Challenge**: Conflicting or incomplete data can weaken confidence in a claim.
   - **Mitigation**: Use proxy truths cautiously and seek additional evidence to strengthen the claim.

   Example: In nutritional science, studies may offer conflicting advice on dietary fats. Acknowledging these discrepancies and continuing research helps refine recommendations.

2. **Applicability in Novel Contexts (Present)**
   - **Challenge**: Established principles may not apply in new or rapidly changing environments.
   - **Mitigation**: Employ pilot tests and flexible strategies to adapt to novel situations.

   Example: The rise of remote work required organizations to quickly adapt policies and technologies. Initial solutions may have required adjustment as new challenges emerged.

3. **Uncertainty in Predictability (Future)**
   - **Challenge**: Predicting future outcomes is inherently uncertain, especially in complex systems.
   - **Mitigation**: Combine multiple predictive models and remain open to adjusting forecasts as conditions change.

   Example: Climate models vary in their projections. Using a range of models and updating predictions as new data becomes available improves reliability.

15.6 CONCLUSION

The **TTEF** provides a comprehensive method for evaluating the truthfulness of claims by examining them across past, present, and future dimensions. By incorporating dynamic elements like Bayesian updating, cross-disciplinary insights, proxy truths, and psychological factors, the framework remains adaptable and robust in the face of complexity and uncertainty.

Key Takeaways:

- **Holistic Evaluation**: Considers empirical evidence, immediate applicability, and reliable predictability to assess claims thoroughly.
- **Dynamic Adaptation**: Uses Bayesian updating and dynamic weighting to remain responsive to new information.
- **Psychological Awareness**: Recognizes the impact of human psychology on the acceptance of claims and incorporates strategies to address these factors.
- **Resource Efficiency**: Balances innovation with practical considerations through pilot testing and incremental implementation.

By applying this framework, individuals and organizations can make informed decisions that are both evidence-based and considerate of human factors, ultimately leading to more effective and accepted outcomes.

15.7 FINAL THOUGHTS

In a world characterized by rapid change and information overload, the ability to critically evaluate claims is more important than ever. The TTEF equips decision-makers with a structured approach to navigate complexity, fostering decisions that are not only logically sound but also practically and psychologically attuned to the environments in which they operate.

THE ANXIETY FEEDBACK LOOP

THINKING BOX

"Does our always-online culture feed our anxieties more than our well-being? The remedy may – like **42** – demand deeper reflection."

16 Practical Guide to Applying the Causal Relevance Ethic

Core Components:

1. **Causal Proximity (C)**: Measures how **immediate and certain** an action's impact is.
2. **Influence Strength (I)**: Assesses how much an action **increases well-being (Thrive Index, TI)** or **reduces adversity (Adversity Reduction Index, ARI)**.
3. **Feasibility (F)**: Evaluates whether the action is **practical and resource-efficient**.

The **Causal Relevance Score (CRS)** helps **prioritize actions**:

- **CRS 0 (Do Not Proceed)**: Action lacks relevance, impact, or feasibility.
- **CRS 1–15 (Might Not Be Worthwhile)**: High effort, low return.
- **CRS 16–29 (Probably Worthwhile)**: Requires further evaluation and resource adjustment.
- **CRS 30 (Must Do)**: **High impact, high feasibility – immediate execution recommended**.

Key Takeaway:

CRE **eliminates moral ambiguity**, **avoids virtue signaling**, and **focuses on actionable, measurable change** – ensuring **decision-making is both ethical and practical**.

16.1 INTRODUCTION

16.1.1 Purpose of the Guide

The **Causal Relevance Ethic (CRE)** is a practical framework designed to help individuals and organizations make decisions grounded in real-world outcomes, feasibility, and resource optimization. This guide provides clear, step-by-step instructions for applying the framework across various decision-making contexts – from everyday personal choices to large-scale organizational strategies.

The goal is to move beyond abstract moral reasoning and focus on actions that have measurable, positive impacts. Whether you're prioritizing personal responsibilities, managing organizational resources, or planning societal initiatives, this guide offers practical tools to evaluate and prioritize actions that truly matter.

16.1.2 Overview of the Causal Relevance Ethic

The CRE is built around three core components that together form the foundation of its decision-making model:

1. **Causal Proximity (C)**: Measures how immediate and certain an action's effects are. The closer an action is to producing an effect and the more likely it is to succeed, the higher its score.
   - **Time Factor (TF)**: Rates how soon the action will have an effect, from **0** (no impact) to **5** (immediate impact).
   - **Certainty Factor (CF)**: Rates the likelihood of the action's success, from **0** (impossible) to **5** (certain).

DOI: 10.1201/9781003640288-21

Calculation of Causal Proximity (C):

$$C = \begin{cases} 0, & \text{if } TF = 0 \text{ or } CF = 0 \\ TF + CF, & \text{otherwise} \end{cases}$$

2. **Influence Strength (I):** Measures how much an action increases well-being or reduces adversity.
   - **Thrive Index (TI):** Rates the action's potential to enhance positive aspects like health, autonomy, and purpose, from **0** to **5**.
   - **Adversity Reduction Index (ARI):** Rates the action's potential to reduce negative factors like stress, financial strain, and health challenges from **0** to **5**.

Calculation of Influence Strength (I):

$$I = \begin{cases} 0, & \text{if } TI = 0 \text{ and } ARI = 0 \\ TI + ARI, & \text{otherwise} \end{cases}$$

3. **Feasibility (F):** Evaluates whether the action is achievable with available resources and practical to execute.
   - **Resource Availability (RA):** Rates the availability of necessary resources, from **0** (none) to **5** (abundant).
   - **Practical Achievability (PA):** Rates how realistic it is to implement the action given current capabilities, from **0** (impossible) to **5** (easily achievable).

Calculation of Feasibility (F):

$$F = \begin{cases} 0, & \text{if } RA = 0 \text{ or } PA = 0 \\ RA + PA, & \text{otherwise} \end{cases}$$

16.1.3 Calculating the Causal Relevance Score (CRS)

The **CRS** combines the three components to provide a clear evaluation of any given action.

Calculation of CRS:

$$CRS = \begin{cases} 0, & \text{if } C = 0 \text{ or } I = 0 \text{ or } F = 0 \\ C + I + F, & \text{otherwise} \end{cases}$$

- **Maximum CRS: 30** (when all components are at their highest value).

Interpreting the CRS:

- **CRS = 0% (Score 0): Do Not Proceed**
  - *The action lacks critical components and should not be pursued.*
- **CRS = 1–50% (Score 1–15): Might Not Be Worthwhile**
  - *The action may not yield significant benefits or may face substantial challenges.*
- **CRS = 51–99% (Score 16–29): Probably Worthwhile**
  - *The action is likely to produce meaningful results and is worth considering.*

- **CRS = 100% (Score 30): Must Do**
  - *The action has the highest potential impact and feasibility; immediate implementation is recommended.*

General Guidance: The higher the CRS, the more worthwhile the action probably is.

16.1.4 The Importance of Practicality

In many decision-making models, moral and ethical judgments can lead to ambiguity or inaction. By focusing on practicality and measurable impact, the CRE enables individuals and organizations to cut through complexity and identify actions that truly make a difference. Whether allocating limited resources, addressing urgent issues, or balancing long-term and short-term objectives, the framework ensures that actions are evaluated based on what they can achieve, not just on good intentions.

16.1.5 Moving beyond Moral Ambiguity

One of the key strengths of this framework is its move away from moral subjectivity, which often leads to inconsistent decision-making. Traditional moral philosophies, like utilitarianism or deontology, may struggle with applying rigid rules or demanding extreme sacrifices, especially in complex real-world situations.

The CRE simplifies these complexities by focusing on three essential questions:

1. **Causal Proximity (C)**: *How immediate and certain is the impact of the action?*
2. **Influence Strength (I)**: *How much will this action increase well-being or reduce adversity?*
3. **Feasibility (F)**: *Can this action be feasibly carried out with the resources at hand?*

Through these lenses, decision-makers can apply the framework across a wide range of scenarios, ensuring that the actions they prioritize have the highest possible relevance and real-world impact.

16.1.6 Who Can Benefit from This Guide?

This guide is intended for anyone seeking a structured, practical approach to decision-making, including:

- **Leaders and Managers**: Managing resources within organizations to maximize impact.
- **Policymakers**: Deciding on the best courses of action for societal benefit.
- **Individuals**: Prioritizing personal responsibilities and life choices.
- **Non-Profit Organizations**: Allocating limited resources to projects with the highest potential impact.
- **Educators and Students**: Applying ethical frameworks in academic settings or case studies.

By using this guide, you'll be able to prioritize actions that lead to meaningful, positive outcomes, ensuring your decisions are not only well-reasoned but also impactful and achievable.

16.1.7 The Idea

The CRE empowers you to make decisions that are not only ethically sound but also practically effective. By focusing on actions with the highest potential for positive impact and feasibility, you can contribute meaningfully to your goals and the well-being of others.

16.2 THE CORE COMPONENTS OF THE CAUSAL RELEVANCE ETHIC

In this, we'll delve deep into the three key components - introduced in chapter 14 - that form the foundation of the **CRE**: **Causal Proximity (C)**, **Influence Strength (I)**, and **Feasibility (F)**. Understanding how each component works will enable you to evaluate any action or decision through a practical, results-oriented lens.

16.2.1 CAUSAL PROXIMITY (C)

Causal Proximity measures how immediate and certain the impact of an action will be. The closer and more certain the impact, the higher the Causal Proximity score.

Subcomponents of Causal Proximity:

- **Time Factor (TF)**: Rates how soon the action will have an effect.
- **Certainty Factor (CF)**: Rates the likelihood of the action's success.

Scoring for Time Factor (TF):

Score	Description
5	Immediate impact (affects today)
4	Very short-term impact (days)
3	Short-term impact (weeks)
2	Medium-term impact (months)
1	Long-term impact (years)
0	No impact

Scoring for Certainty Factor (CF):

Score	Description
5	Almost certain to succeed
4	Very likely to succeed
3	Moderately likely to succeed
2	Somewhat unlikely to succeed
1	Highly unlikely to succeed
0	No likelihood of success

Calculating Causal Proximity (C):

$$C = \begin{cases} 0, & \text{if } TF = 0 \text{ or } CF = 0 \\ TF + CF, & \text{otherwise} \end{cases}$$

Interpretation: A higher Causal Proximity score indicates that the action has an immediate and certain impact, making it more relevant for prioritization.

16.2.2 INFLUENCE STRENGTH (I)

Influence Strength measures the action's potential to increase well-being (**TI**) and/or reduce adversity (**ARI**).

Subcomponents of Influence Strength:

- **Thrive Index (TI)**: Assesses the action's ability to enhance positive aspects like health, autonomy, and purpose.
- **Adversity Reduction Index (ARI)**: Evaluates the action's capacity to reduce negative factors like stress, financial strain, and health challenges.

Scoring for Thrive Index (TI):

Score	Description
5	Significant increase in well-being
4	Large increase for a specific group/situation
3	Moderate increase in well-being
2	Small or limited increase
1	Minimal increase
0	No measurable increase

Scoring for Adversity Reduction Index (ARI):

Score	Description
5	Significant reduction in adversity
4	Large reduction for a specific group/situation
3	Moderate reduction in adversity
2	Small or limited reduction
1	Minimal reduction
0	No measurable reduction

Calculating Influence Strength (I):

$$I = \begin{cases} 0, & \text{if } TI = 0 \text{ and } ARI = 0 \\ TI + ARI, & \text{otherwise} \end{cases}$$

Interpretation: A higher Influence Strength score indicates a greater positive impact on well-being through increasing well-being and/or reducing adversity.

16.2.3 Feasibility (F)

Feasibility assesses how achievable the action is, considering RA and PA.

Subcomponents of Feasibility:

- **Resource Availability (RA)**: Evaluates whether sufficient resources are accessible to execute the action.
- **Practical Achievability (PA)**: Assesses how realistic it is to implement the action given current capabilities.

Scoring for Resource Availability (RA):

Score	Description
5	All necessary resources are readily available
4	Most resources available; minor constraints
3	Some resources available; moderate constraints
2	Few resources available; significant constraints
1	Minimal resources available; major constraints
0	No resources available; absolute constraints

Scoring for Practical Achievability (PA):

Score	Description
5	Highly practical and easily achievable
4	Very practical; minor challenges
3	Moderately practical; some challenges
2	Impractical but achievable with considerable effort
1	Highly impractical; significant obstacles
0	No practicability; absolute obstacles

Calculating Feasibility (F):

$$F = \begin{cases} 0, & \text{if } RA = 0 \text{ or } PA = 0 \\ RA + PA, & \text{otherwise} \end{cases}$$

Interpretation: A higher Feasibility score indicates that the action is more achievable with the resources at hand.

16.2.4 Calculating the Causal Relevance Score (CRS)

The **CRS** combines the three components to provide a clear evaluation of any given action. This score helps prioritize actions based on their overall potential impact and feasibility.

Formula for Calculating CRS:

$$CRS = \begin{cases} 0, & \text{if } C = 0 \text{ or } I = 0 \text{ or } F = 0 \\ C + I + F, & \text{otherwise} \end{cases}$$

- **Maximum CRS: 30** (when all components are at their highest values: C = 10, I = 10 and F = 10).

Why Use Conditional Zero?

- If any main component (**C**, **I**, or **F**) is zero, the overall CRS becomes zero. This reflects the principle that an action lacking in any critical area is not worth pursuing.

16.2.5 Interpreting the Causal Relevance Score

To simplify decision-making, the CRS is interpreted using the following scale:

Simplified Scaling:

- **CRS = 0: Do Not Proceed**
  - *Action lacks critical components and should not be pursued at this time.*
- **CRS = 1–15: Might Not Be Worthwhile**
  - *Action may not yield significant benefits or may face substantial challenges. Consider whether improvements can be made to increase the score.*
- **CRS = 16–29: Probably Worthwhile**
  - *Action is likely worthwhile. Proceed with thorough analysis to confirm feasibility and maximize impact.*
- **CRS = 30: Must Do**
  - *Action has the highest potential impact and feasibility. Proceed with immediate implementation after a final review.*

General Guidance: The higher the CRS, the more worthwhile the action probably is.

16.2.6 The Importance of Zero Values

A zero in any component leads to an overall CRS of zero, emphasizing that actions lacking in any critical area are not worth pursuing.

Implications of Zero Values:

- **Zero in Causal Proximity (C)**: *The action is too distant in time or too uncertain to be relevant.*
- **Zero in Influence Strength (I)**: *The action does not meaningfully increase well-being or reduce adversity.*
- **Zero in Feasibility (F)**: *The action cannot be achieved due to lack of resources or significant obstacles.*

Example:

- **Action with Zero Feasibility:**
  - **RA = 0** (No resources available)
  - **PA = 4** (Very practical)
  - **Calculating F:**

$$F = 0 \text{ (since RA} = 0)$$

  - **Result:** Regardless of the scores for **C** and **I**, the **CRS** will be zero.

Interpretation: The action should not proceed due to lack of feasibility.

16.2.7 Practical Application Example

Scenario: A community organization is considering launching a new program to provide healthy meals to low-income families.

Step 1: Assess Causal Proximity (C)

- **Time Factor (TF)**: 4 (Very short-term impact; meals can be provided within days)
- **Certainty Factor (CF)**: 5 (Almost certain to succeed; organization has experience)

- **Calculating C**:

$$C = TF + CF = 4 + 5 = 9$$

Step 2: Assess Influence Strength (I)

- **Thrive Index (TI)**: 3 (Moderate increase in well-being; better nutrition)
- **Adversity Reduction Index (ARI)**: 4 (Large reduction in adversity; addresses hunger)
- **Calculating I**:

$$I = TI + ARI = 3 + 4 = 7$$

Step 3: Assess Feasibility (F)

- **Resource Availability (RA)**: 2 (Few resources available; significant constraints)
- **Practical Achievability (PA)**: 3 (Moderately practical; some challenges)
- **Calculating F:**

$$F = RA + PA = 2 + 3 = 5$$

Step 4: Calculate CRS

- Since none of the components are zero:

$$CRS = C + I + F = 9 + 7 + 5 = 21$$

Step 5: Interpret CRS

- **CRS = 21** falls within **16–29**.
- **Interpretation: Probably Worthwhile**
  - *Action is likely worthwhile. Proceed with thorough analysis to improve RA and maximize impact.*

16.2.8 Enhancing the Action's CRS

To elevate the action to a "Must Do" category:

- **Increase Resource Availability (RA)**:
  - Seek additional funding or partnerships to improve from **RA = 2** to **RA = 4**.
- **Recalculate F**:

$$F = RA + PA = 4 + 3 = 7$$

- **Recalculate CRS**:

$$CRS = C + I + F = 9 + 7 + 7 = 23$$

- **New CRS Interpretation**: Still within **16–29** but higher, indicating increased worthiness.
- **Further Steps**:
  - Continue enhancing feasibility and influence strength to aim for a CRS of **30**.

16.2.9 Conclusion

By understanding and applying these core components, you can make informed decisions that prioritize actions with the highest relevance, impact, and feasibility. This approach ensures that efforts are directed toward actions that will lead to real, measurable improvements in well-being.

Key Takeaways:

- **Causal Proximity (C)**: Focuses on the immediacy and certainty of impact.
- **Influence Strength (I)**: Measures the potential to increase well-being and/or reduce adversity.
- **Feasibility (F)**: Assesses the practicality and RA for implementing the action.
- **Causal Relevance Score (CRS)**: Combines the components to evaluate the overall worthiness of an action.
- **Zero Values**: Signal that an action is not viable in its current form and should not proceed.

THE PROXY TRAP

THINKING BOX

"Are we mistaking symbolic markers for authentic success? Perhaps we need to revisit the meaning behind **42** to see clearly."

16.3 THE THRIVE INDEX AND ADVERSITY REDUCTION INDEX – STRUCTURED MEASURES OF WELL-BEING AND ADVERSITY

16.3.1 Introduction

In the **CRE** framework, accurately measuring the potential impact of actions on well-being is crucial. To refine this measurement and replace the subjective notions of "happiness" and "suffering," we introduce two structured, quantifiable concepts: the **TI** and the **ARI**.

- **Thrive Index (TI)**: Reflects the degree to which an individual, group, or system is thriving – encompassing elements such as health, autonomy, purpose, and social relationships.
- **Adversity Reduction Index (ARI)**: Quantifies the obstacles or barriers that inhibit well-being, such as stress, financial strain, health challenges, and social isolation.

Together, these indexes provide a clear and actionable representation of well-being and adversity, offering greater precision in assessing the influence of actions and interventions.

16.3.1.1 Defining the Thrive Index (TI) and Adversity Reduction Index (ARI)

16.3.1.1.1 Thrive Index (TI)

The TI measures the fulfillment of capabilities and overall well-being by focusing on positive aspects of life.

Key Components of TI:

1. **Health**: Physical and mental well-being, access to healthcare, and fitness levels.
2. **Autonomy**: Ability to make decisions and have control over one's life.
3. **Purpose**: Engagement in meaningful activities, career or personal pursuits.
4. **Social Relationships**: Quality of personal and professional relationships.

Measurement:

- **Scale**: Rated from **0** to **5**, with higher values reflecting better fulfillment of these capabilities.

Scoring for Thrive Index (TI):

Score	Description
5	Significant increase in well-being
4	Large increase for a specific group/situation
3	Moderate increase in well-being
2	Small or limited increase
1	Minimal increase
0	No measurable increase

16.3.1.1.2 Adversity Reduction Index (ARI)

The ARI measures the barriers or challenges that prevent an individual, group, or system from thriving. It captures negative aspects such as stress, financial difficulties, or health issues.

Key Components of ARI:

1. **Financial Strain**: Economic challenges, debt, or job insecurity.
2. **Health Barriers**: Chronic conditions, mental health issues, or lack of healthcare access.
3. **Environmental Challenges**: Stressors from personal, professional, or environmental factors.
4. **Social/Community Challenges**: Isolation, conflict, or lack of social support.

Measurement:

- **Scale**: Rated from **0** to **5**, with higher values indicating greater reduction in adversity.

Scoring for Adversity Reduction Index (ARI):

Score	Description
5	Significant reduction in adversity
4	Large reduction for a specific group/situation
3	Moderate reduction in adversity
2	Small or limited reduction
1	Minimal reduction
0	No measurable reduction

16.3.1.2 Using TI and ARI to Measure Influence Strength (I)

In the CRE framework, **Influence Strength (I)** is a critical factor in evaluating the impact of actions. It is calculated by summing the TI and Adversity Reduction Index:

$$I = \begin{cases} 0, & \text{if } TI = 0 \text{ and } ARI = 0 \\ TI + ARI, & \text{otherwise} \end{cases}$$

- **Positive Influence**: An action that increases the TI and/or decreases the ARI results in a positive Influence Strength (I). The higher the Influence Strength, the greater the overall positive effect on well-being.

Example:

- **Healthcare Access Improvement**: An action that increases access to healthcare (raising the TI) while also reducing financial strain due to medical costs (lowering the ARI) will have a strong positive Influence Strength (I).

16.3.1.3 The Layered Approach to Evaluating Well-Being and Adversity

To ensure that TI and ARI are applied in a flexible and scalable manner, the framework offers three layers of evaluation, each increasing in depth and precision:

16.3.1.3.1 Layer 1: Quick Assessment

- **Purpose**: Provide a rapid overview of thriving and adversity in a situation.
- **Approach**: Use simple scales (0–5) for key factors like health and financial strain.
- **Use**: Suitable for immediate decision-making when a rough estimate is sufficient.

16.3.1.3.2 Layer 2: Intermediate Evaluation

- **Purpose**: Offer a more detailed breakdown of well-being and adversity.
- **Approach**: Break down the Thrive and Adversity Indexes into dimensions such as physical health, autonomy, financial stress, and environmental challenges.
- **Use:** Appropriate for situations requiring more detailed insights, such as organizational assessments or program evaluations.

16.3.1.3.3 Layer 3: Comprehensive Evaluation

- **Purpose:** Provide a thorough, data-driven evaluation of thriving and adversity, incorporating both historical data and future projections.
- **Approach:** Further divide the indexes into subcomponents, such as fitness levels, chronic conditions, workplace stress factors, and community support systems.
- **Use:** Ideal for complex, high-stakes decisions where long-term planning and precision are critical, such as public policy development or large-scale interventions.

16.3.1.4 The Role of TI and ARI in Decision-Making

Integrating the TI and ARI into the CRS allows for more precise decision-making. By breaking down well-being and adversity into measurable components, decision-makers can:

- **Identify Key Intervention Points:** Use the TI to find areas where well-being can be improved and the ARI to locate obstacles that must be reduced.
- **Compare Actions:** Evaluate which actions will result in a higher Influence Strength by increasing the TI or decreasing the ARI.
- **Prioritize Resources:** Allocate resources to actions that offer the greatest potential to improve well-being and reduce adversity.
- **Avoid Virtue Signaling:** Ensure that decisions are based on real, measurable improvements in well-being, not symbolic actions with little impact.

16.3.2 Conclusion

The introduction of the **TI** and **ARI** offers a more structured and objective way to measure the outcomes of actions within the CRE framework. By quantifying well-being and adversity and providing layers of evaluation based on the situation's complexity, decision-makers can make more informed, impactful choices.

This approach ensures that actions lead to measurable improvements in well-being and reductions in adversity, aligning decisions with tangible outcomes rather than subjective intentions.

Note: By integrating the TI and ARI into the **Influence Strength (I)** component, the framework enhances its precision and applicability. This structured approach allows for a more nuanced assessment of how actions affect well-being and adversity, facilitating better-informed and more impactful decision-making.

16.4 STEP-BY-STEP GUIDE TO APPLYING THE CAUSAL RELEVANCE ETHIC

In this, we'll guide you through the practical application of the **CRE**. By following these steps, you'll be able to evaluate and prioritize actions based on their real-world impact, ensuring that your decisions are grounded in practicality and relevance.

Step 1: Evaluate Causal Proximity (C)
The first step is to assess how immediate and certain the action's impact will be. You'll need to score both the **TF** and the **CF**, and then calculate the **Causal Proximity (C)**.

16.4.1 How to Score Causal Proximity

1. Time Factor (TF):

Score	Description
5	Immediate impact (affects today)
4	Short-term impact (within days or weeks)
3	Medium-term impact (within months)
2	Long-term impact (within years)
1	Very long-term impact (5–10+ years)
0	No measurable impact in time

2. Certainty Factor (CF):

Score	Description
5	Almost certain to succeed
4	Very likely to succeed
3	Moderately likely to succeed
2	Somewhat unlikely to succeed
1	Highly unlikely to succeed
0	No certainty or no clear outcome

Calculating Causal Proximity (C):

$$C = \begin{cases} 0, & \text{if } TF = 0 \text{ or } CF = 0 \\ TF + CF, & \text{otherwise} \end{cases}$$

Example:
Suppose you're considering providing financial support to a community project that will begin within the next few months and has a good chance of success.

- **Time Factor (TF)**: 3 (Medium-term impact)
- **Certainty Factor (CF)**: 4 (Very likely to succeed)

Calculating C:

$C = TF + CF = 3 + 4 = 7$

This score reflects that the project is relatively close in time and likely to succeed, making it worth considering based on proximity.

Step 2: Evaluate Influence Strength (I)
Next, assess how much the action will improve well-being by increasing thriving and reducing adversity.

16.4.2 How to Score Influence Strength

1. Thrive Index (TI):

Score	Description
5	Significant increase in well-being
4	Large increase for a specific group or situation
3	Moderate increase in well-being
2	Small or limited increase
1	Minimal increase
0	No measurable increase

2. Adversity Reduction Index (ARI):

Score	Description
5	Significant reduction in adversity
4	Large reduction for a specific group or situation
3	Moderate reduction in adversity
2	Small or limited reduction
1	Minimal reduction
0	No measurable reduction

Calculating Influence Strength (I):

$$I = \begin{cases} 0, & \text{if } TI = 0 \text{ and } ARI = 0 \\ TI + ARI, & \text{otherwise} \end{cases}$$

Example:
Suppose the project aims to provide clean drinking water to a rural community, greatly improving health and well-being.

- **Thrive Index (TI)**: 4 (Large increase in well-being)
- **Adversity Reduction Index (ARI)**: 5 (Significant reduction in adversity due to improved health)

Calculating I:

$I = TI + ARI = 4 + 5 = 9$

This score indicates that the project has a high potential to positively impact both well-being and adversity reduction.

Step 3: Evaluate Feasibility (F)
Assess whether the action is practical and achievable, considering RA and PA.

16.4.3 How to Score Feasibility

1. Resource Availability (RA):

Score	Description
5	All necessary resources are readily available
4	Most resources available; minor constraints
3	Some resources available; significant constraints
2	Few resources available; substantial constraints
1	Minimal resources available; major constraints
0	No resources available

2. Practical Achievability (PA):

Score	Description
5	Highly practical and easily achievable
4	Very practical; minor challenges
3	Moderately practical; significant challenges
2	Impractical but achievable with considerable effort
1	Highly impractical; little chance of success
0	Impossible to achieve

Calculating Feasibility (F):

$$F = \begin{cases} 0, & \text{if } RA = 0 \text{ or } PA = 0 \\ RA + PA, & \text{otherwise} \end{cases}$$

Example:
The community project has secured partial funding but faces logistical challenges in accessing remote areas.

- **Resource Availability (RA)**: 3 (Some resources available; significant constraints)
- **Practical Achievability (PA)**: 3 (Moderately practical; significant challenges)

Calculating F:

$\boldsymbol{F = RA + PA = 3 + 3 = 6}$

This score reflects that while the project is feasible, it may require more effort and resources to be fully achievable.

Step 4: Calculate the Causal Relevance Score (CRS)
Once you have evaluated the three core components – Causal Proximity (C), Influence Strength (I), and Feasibility (F) – you can calculate the overall **CRS** to determine how relevant and actionable the decision is.

Formula for Calculating CRS:

$$CRS = \begin{cases} 0, & \text{if } C = 0 \text{ or } I = 0 \text{ or } F = 0 \\ C + I + F, & \text{otherwise} \end{cases}$$

Calculating the Score:

- **Causal Proximity (C):** 7
- **Influence Strength (I):** 9
- **Feasibility (F):** 6

$\boldsymbol{CRS = C + I + F} = 7 + 9 + 6 = 22$

Step 5: Interpret the Result and Make a Decision
After calculating the CRS, interpret the result to prioritize actions.

Interpreting the CRS:

- **CRS = 22**

This score falls within **16–29**, which corresponds to **Probably Worthwhile**.

Interpretation:

- The action is likely worthwhile and should be considered for implementation.
- Proceed with a thorough analysis to confirm feasibility and maximize impact.

Consider Ways to Improve the Score:

- **Increase Feasibility (F)**: Seek additional resources or partnerships to improve RA and PA.
- **Enhance Causal Proximity (C)**: Find ways to expedite the timeline or increase certainty.
- **Boost Influence Strength (I)**: Expand the project's scope to benefit more people or address additional aspects of well-being.

Decision:

- **Proceed with Planning**: Develop a detailed action plan, addressing any challenges identified during the evaluation.
- **Allocate Resources**: Secure additional funding or support to enhance feasibility.
- **Monitor Progress**: Implement feedback mechanisms to assess the action's impact and make adjustments as needed.

16.4.4 Conclusion

By following these steps, you can systematically apply the **CRE** to evaluate and prioritize actions based on their proximity, impact, and feasibility. This structured approach ensures that your decisions are not only well-grounded but also focused on real-world outcomes.

16.4.5 The Idea

The CRE empowers you to make informed decisions that prioritize actions with the highest potential for positive impact and feasibility. By systematically evaluating each component, you ensure that your efforts are directed toward meaningful and achievable outcomes.

16.5 WHEN AND HOW TO USE THE TEMPORAL TRUTH EVALUATION FRAMEWORK

While the **CRE** provides a robust framework for evaluating actions based on **Causal Proximity (C)**, **Influence Strength (I)**, and **Feasibility (F)**, there are times when uncertainty arises about the information supporting these factors. In such cases, integrating the **Temporal Truth Evaluation Framework** can enhance the reliability of your decisions.

This explains how and when to apply this framework within the CRE, ensuring that your decisions are based on well-supported and reliable information.

16.5.1 Purpose of the Temporal Truth Evaluation Framework in the Causal Relevance Ethic

The Temporal Truth Evaluation Framework assesses the reliability of key inputs across three temporal dimensions:

- **Past (Empirical Evidence)**: What historical data supports the action?
- **Present (Immediate Applicability)**: How applicable is the action in the current context?
- **Future (Reliable Predictability)**: What is the likelihood of future success based on projections?

By evaluating these dimensions, you can validate your assumptions about **Causal Proximity**, **Influence Strength**, and **Feasibility** when faced with uncertainty.

When to Use the Framework:

- When data supporting your decision is unclear or incomplete.
- When there's conflicting evidence or speculation about an action's potential success.
- When long-term outcomes or feasibility are highly uncertain.

Integrating this tool ensures more reliable decision-making without complicating the process for every situation.

16.5.2 When to Apply the Temporal Truth Evaluation Framework

The framework should be applied selectively, specifically when uncertainty makes it difficult to score inputs accurately.

Situations for Each Temporal Dimension:

- **Past (Empirical Evidence)**:
  - **Use When**: You're unsure whether similar actions have succeeded or failed before.
  - **Purpose**: Past performance offers insights into the potential effectiveness of an action based on historical data.
- **Present (Immediate Applicability)**:
  - **Use When**: Current conditions or available resources are unclear, or when the immediate practicality of an action is uncertain.

    - **Purpose**: Assesses whether the action fits within the current environment and whether resources are available.
- **Future (Reliable Predictability)**:
    - **Use When**: Long-term outcomes are speculative or rely on unpredictable variables, such as market trends or technological advancements.
    - **Purpose**: Evaluates the likelihood of long-term success based on future projections.

16.5.3 Practical Application of the Temporal Truth Evaluation Framework

A. **Past: Empirical Evidence**

- **Purpose**: Use past data to understand the success or failure of similar actions.
- **Action Steps**:

    i. **Gather Historical Data**: Look for case studies, reports, or records of similar actions.
    ii. **Analyze Outcomes**: Determine if past actions were successful and identify factors that contributed to their success or failure.
    iii. **Adjust Inputs**: Reassess **Causal Proximity (C)** and **Feasibility (F)** based on historical insights.

Example:
If previous initiatives similar to your proposed action have failed due to specific challenges, you might reduce the **CF** in **Causal Proximity** or adjust **Feasibility** to reflect these challenges.

B. **Present: Immediate Applicability**

- **Purpose**: Assess the current suitability and practicality of the action.
- **Action Steps**:

    i. **Evaluate Current Conditions**: Examine whether the necessary resources, technology, or support are available now.
    ii. **Validate Assumptions:** Ensure that your understanding of the present context is accurate.
    iii. **Adjust Inputs**: Modify **Feasibility (F)** or **Influence Strength (I)** based on present applicability.

Example:
If the action requires technology that is currently unavailable or resources that are overextended, you may need to lower the **RA** score in **Feasibility**.

C. **Future: Reliable Predictability**

- **Purpose**: Evaluate potential future outcomes and uncertainties.
- **Action Steps**:

    i. **Forecast Scenarios**: Use predictive models or expert opinions to anticipate future conditions.
    ii. **Identify Variables**: Recognize factors that could influence the action's success over time.
    iii. **Adjust Inputs**: Modify **Influence Strength (I)** or **Causal Proximity (C)** to account for future uncertainties.

Example:
If market trends suggest a declining demand for a product your action relies on, you might decrease the **TF** or **TI** to reflect the potential reduced impact.

Clarifying the Distinctions:
While past and future assessments may seem to overlap – both dealing with events outside the present – they serve distinct purposes: **Past assessments rely on empirical evidence from previous experiences to inform the current decision, whereas future assessments involve predicting potential outcomes and uncertainties that have not yet occurred.** Recognizing this distinction enhances the practicality of the framework by ensuring that decisions are informed both by what has been proven to work and by what is anticipated to happen.

16.5.4 Streamlined Integration into the Causal Relevance Ethic

After applying the Temporal Truth Evaluation Framework, refine your scoring to reflect the validated information.

Steps for Integration:

1. **Identify Uncertainties:** Pinpoint which inputs are affected by uncertainty.
2. **Apply Relevant Temporal Dimension**:
   - Use **Past** evaluation for historical uncertainties.
   - Use **Present** evaluation for current applicability doubts.
   - Use **Future** evaluation for speculative outcomes.
3. **Adjust Scores Accordingly**:
   - **Reduce Scores** where evidence suggests lower impact or feasibility.
   - **Maintain or Increase Scores** if the evaluation confirms or enhances your assumptions.
4. **Recalculate CRS**: With adjusted inputs, recalculate the **CRS** to reflect a more accurate assessment.

Benefits:

- **Enhanced Accuracy**: Ensures that decisions are based on validated information.
- **Focused Effort**: Helps prioritize actions that are truly viable and impactful.
- **Risk Mitigation**: Reduces the likelihood of pursuing impractical or ineffective actions.

16.5.5 When Not to Use the Temporal Truth Evaluation Framework

Avoid overcomplicating the decision-making process by using the framework only when necessary.

Do Not Use When:

- Data supporting your decision is clear, reliable, and uncontested.
- The action is straightforward with well-understood outcomes.
- Time constraints require immediate action based on the best available information.

Remember:

- The framework is a tool for enhancing decision-making under uncertainty, not a mandatory step for all decisions.
- Overuse can lead to analysis paralysis; apply it judiciously.

16.5.6 Conclusion

The Temporal Truth Evaluation Framework is a valuable addition to the CRE when facing uncertainties. By evaluating key inputs through the lenses of the **Past**, **Present**, and **Future**, you can ensure that your **CRS** reflects the most accurate and reliable information possible.

Key Takeaways:

- **Selective Application**: Use the framework only when uncertainty affects your ability to score inputs accurately.
- **Integrated Approach**: Adjust your scores based on validated information from the temporal dimensions.
- **Enhanced Decision-Making**: Make more informed choices that are grounded in evidence and realistic projections.

16.5.7 The Role of Zero Values in the Causal Relevance Ethic

In the **CRE**, zero values represent a clear and unambiguous signal that an action is irrelevant, impractical, or impossible to achieve. A zero in any of the core components – **Causal Proximity (C)**, **Influence Strength (I)**, or **Feasibility (F)** – means the action is not worth pursuing. Zero values should be treated as warning signs that the action is **detached**, **futile**, or **purely symbolic**.

This explains how to interpret zero values and why they indicate that an action should be disregarded altogether.

1. **Understanding Zero Values**

   A zero value in any of the three core components means that the action fails to meet a critical threshold. When a component scores zero, it indicates that the action is fundamentally flawed and should not be pursued.

   Key Points:

   - **Zero values** signal that an action is irrelevant, impractical, or impossible to achieve.
   - They serve as **warning signs**, indicating that further resources should not be invested in the action.

2. **Zero in Causal Proximity (C): Detachment**

   A zero in **Causal Proximity** means the action is too far removed in time or too uncertain to be relevant. If an action's impact is either extremely distant or the likelihood of success is negligible, it is detached from immediate needs and should be disregarded.

   When This Occurs:

   - The action's impact won't be realized for an excessively long time, rendering it irrelevant to current decision-making.
   - The likelihood of success is so low that it's effectively improbable the action will have any effect.

   Action: Do not pursue the action. A zero in Causal Proximity indicates that the action is too distant or uncertain to have meaningful relevance.

   Example:

   - **Long-Term Environmental Policy**: A proposal that won't show results for 50 years might score zero in Causal Proximity. While the cause is important, the immediate priority is minimal, suggesting it should be deprioritized in favor of actions with more immediate impact.

3. **Zero in Influence Strength (I): Futility**

   A zero in **Influence Strength** means the action will have no meaningful impact on well-being. Even if the action is achievable, it doesn't produce any significant outcome, making it futile.

When This Occurs:

- The action does not contribute to any measurable increase in well-being or reduction in adversity.
- The scale of the action is so small that it has no noticeable influence.

Action: Do not pursue the action. A zero in Influence Strength means the action is futile and provides no meaningful value or positive impact.

Example:

- **Ineffective Team-Building Exercise**: A small activity that doesn't meaningfully improve morale or productivity could score zero in Influence Strength. Since it has no measurable benefit, the action should be abandoned.

4. **Zero in Feasibility (F): Impracticality and Virtue Signaling**
A zero in **Feasibility** means that the action is entirely impractical or impossible to achieve. It is often a sign of **virtue signaling**, where the action is meant to appear morally virtuous but has no realistic chance of being executed.

When This Occurs:

- The action lacks necessary resources, funding, or support to be implemented.
- The action is technically or logistically impossible to achieve in the current context.

Action: Do not pursue the action. A zero in Feasibility means the action is impractical, often indicating symbolic behavior or virtue signaling that is not grounded in reality.

Example:

- **Immediate Elimination of Carbon Emissions**: A demand to shut down all carbon emissions immediately, while morally appealing, is completely impractical given current global infrastructure. The action scores zero in Feasibility and should not be pursued.

5. **Zero Values as Warning Signs**
Zero values serve as warning signs in decision-making, signaling that an action should not be pursued under any circumstances. Whether due to impracticality, irrelevance, or futility, a zero value clearly indicates that the action is fundamentally flawed.

Summary of Zero Values:

- **Zero in Causal Proximity (C)**: The action is too distant or uncertain to prioritize.
- **Zero in Influence Strength (I)**: The action provides no measurable impact, making it irrelevant.
- **Zero in Feasibility (F)**: The action is impossible or purely symbolic, often indicating virtue signaling.

Key Takeaway: When any component scores zero, the action should be disregarded, and resources should not be wasted on pursuing it.

6. **Using Zero Values to Avoid Wasting Resources**
One of the strengths of the CRE is that it helps decision-makers avoid wasting resources on actions that have no chance of success or impact. Zero values provide a clear-cut way to identify actions that are not worth pursuing, ensuring that efforts and resources are focused on actions with real-world potential.

Benefits:

- **Efficient Resource Allocation**: Focuses efforts on actions with meaningful impact and feasibility.

- **Clarity in Decision-Making**: Provides a straightforward method to eliminate non-viable actions.
- **Avoidance of Symbolic Actions**: Prevents engaging in actions that serve only to signal virtue without producing real results.

16.5.8 Conclusion

Zero values in the CRE are clear warning signs that an action is detached, futile, or impractical. When any core component scores zero, the action should not be pursued, as it will not lead to meaningful outcomes. By recognizing and heeding these warning signs, decision-makers can allocate resources more effectively and focus on actions that offer genuine potential for positive impact.

16.6 BALANCING SHORT-TERM AND LONG-TERM DECISIONS WHEN SCORES ARE EQUAL

In decision-making using the **CRE**, you may encounter situations where a short-term action and a long-term action have identical, non-zero **CRSs**. When both actions are equally relevant based on their proximity, influence, and feasibility, a more nuanced approach is required to decide which action should take priority.

This focuses on how to prioritize when short-term and long-term actions result in the same CRS, offering additional criteria for decision-making in these tie-breaking scenarios.

16.6.1 The Challenge of Equal Scores

When two actions – one short-term and one long-term – have identical CRS, it can be difficult to decide which to prioritize. Both actions appear equally relevant based on their:

- **Causal Proximity (C)**: Immediate impact vs. long-term significance.
- **Influence Strength (I)**: Potential to increase well-being or reduce adversity.
- **Feasibility (F)**: Practicality and RA.

Key Considerations:

- **Urgency vs. Sustainability**: Is there an immediate, pressing need that justifies prioritizing the short-term action, or will the long-term action provide more lasting value?
- **Resource Availability**: Can resources support both actions simultaneously, or must one be chosen over the other?
- **Strategic Alignment**: Does one action align more closely with long-term goals or organizational strategy?

16.6.2 When to Prioritize Short-Term Actions

Short-term actions, with high **Causal Proximity (C)** due to immediate impact, should generally be prioritized if they resolve urgent problems or offer quick wins that stabilize a situation.

Indicators to Prioritize Short-Term Actions:

- **Immediate Problem Solving**: The action addresses an urgent issue, stabilizes a critical situation, or prevents immediate harm.
- **Limited Resources**: The action requires fewer resources, making it practical to implement swiftly.
- **Quick Gains**: Provides immediate benefits that can create momentum or buy time for long-term strategies.

Example:
A city facing a water shortage might prioritize short-term measures like water conservation campaigns and emergency water deliveries. Although longer-term infrastructure projects are important, the immediate need to secure water supplies takes precedence.

16.6.3 When to Prioritize Long-Term Actions

Long-term actions may score lower in **Causal Proximity (C)** but have high **Influence Strength (I)** and **Feasibility (F)**. They should be prioritized when they offer sustainable solutions that prevent future crises or provide systemic improvements.

Indicators to Prioritize Long-Term Actions:

- **Sustainable Solutions**: The action addresses the root cause of a recurring problem or provides a lasting solution.
- **Preventing Future Crises**: Delaying the action could exacerbate future issues, making early implementation crucial.
- **Strategic Alignment**: The action aligns closely with broader organizational goals or long-term strategic priorities.

Example:
A government deciding between immediate flood relief and investing in long-term infrastructure to prevent future flooding may prioritize the long-term infrastructure project if resources allow, as it prevents repeated disasters in the future.

16.6.4 Strategic Decision-Making: Combining Both Approaches

When possible, pursuing both short-term and long-term actions simultaneously allows for immediate relief while laying the groundwork for future improvements.

Combination Strategy:

- **Start with Short-Term Wins**: Implement the short-term action to address immediate needs.
- **Prepare for Long-Term Success**: Use the stability provided by short-term wins to plan and initiate the long-term action.

Example:
A company might launch a short-term marketing campaign to boost immediate sales while investing in long-term research and development for new products. The short-term action delivers immediate benefits, while the long-term strategy ensures future growth.

16.6.5 Practical Criteria for Breaking the Tie

When deciding between actions with equal CRS, consider the following additional criteria:

A. **Resource Constraints**
   - **Available Resources**: Prioritize the action that can be implemented with the current budget, staff, or materials.

B. **Urgency of the Issue**
   - **Time Sensitivity**: Prioritize the action addressing the most urgent issue, especially if delays could worsen the problem.

C. **Long-Term Impact**
   - **Sustainability**: Consider whether the long-term action offers more significant, lasting benefits.

D. **Strategic Alignment**
   - **Future-Proofing**: Choose the action that better aligns with future strategic goals or vision.

Example:
A healthcare provider deciding between immediate equipment upgrades and launching a long-term wellness program might assess urgency, RA, and strategic alignment to determine which action to prioritize.

16.6.6 Case Study: Resolving a Tie in Organizational Priorities

Scenario:

An organization must choose between:

- **Short-Term Action**: Implementing an employee recognition program to boost immediate morale.
- **Long-Term Action**: Investing in leadership development to improve management skills over time.

Analysis:

- **Short-Term Consideration**: The recognition program offers an immediate morale boost, addressing a pressing issue.
- **Long-Term Consideration**: Leadership development provides lasting benefits by enhancing management effectiveness, potentially reducing future morale issues.

Decision:

The organization might prioritize the short-term recognition program to quickly address morale while allocating resources to begin planning the leadership development initiative, thus balancing immediate needs with long-term goals.

16.6.7 Conclusion

When short-term and long-term actions result in the same **CRS**, breaking the tie requires careful consideration of urgency, RA, and strategic alignment. By applying these additional criteria, decision-makers can choose the action that provides the most value, whether through immediate impact or sustainable change.

16.7 LOGIC AS THE CORE OF HUMAN DECISION-MAKING

A common critique of frameworks like the **CRE** is that they seem "inhuman" or overly mechanistic. Critics argue that by focusing on logical evaluation and measurable outcomes, such frameworks may strip away the emotions, morality, and flexibility often considered essential to ethical decision-making.

However, this critique overlooks a key point: relying on logic and efficiency is, in fact, a deeply human approach to problem-solving. The CRE centers on maximizing well-being (**TI**) and minimizing adversity (**ARI**) in the most effective way possible. By focusing on measurable results and rational decision-making, it represents a more human approach than ethics driven solely by abstract morality or fluctuating emotions.

16.7.1 Logic as the Foundation of Human Decision-Making

What makes the CRE deeply human is its reliance on logic to achieve tangible, positive outcomes. Humans have a unique capacity for reason, allowing us to solve complex problems, make better decisions, and maximize well-being.

Key Points:

- **Rational Problem-Solving**: The framework harnesses human reasoning to focus on efficiency and effectiveness in achieving the greatest well-being with the least waste of resources.
- **Measurable Outcomes**: By emphasizing quantifiable results, the framework ensures that actions lead to real improvements in people's lives.

Example:
A healthcare organization using the CRE to allocate resources focuses on where they can do the most good with what they have. This logical approach ensures that the most lives are saved and suffering is minimized, resulting in a more human-centered outcome than decisions based solely on moral or emotional impulses.

16.7.2 The Limits of Emotion and Morality in Ethical Decision-Making

Traditional ethical systems often center around morality and emotion, but these concepts are prone to subjectivity and variability. Emotions can lead to inconsistent decision-making, and moral standards can differ vastly between cultures or individuals.

Challenges with Emotion and Morality:

- **Subjectivity**: What one person considers morally right may be seen differently by another, leading to conflicts and inefficiencies.
- **Emotional Volatility**: Decisions based on emotions may lack consistency, especially in complex or large-scale situations.

Contrast with the Causal Relevance Ethic:

- **Objective Evaluation**: The framework provides a way to make decisions that are unbiased and focused on results, avoiding the pitfalls of emotional or moral confusion.
- **Consistency**: By relying on logical evaluations of **Causal Proximity (C)**, **Influence Strength (I)**, and **Feasibility (F)**, decisions are stable and effective.

Example:
A humanitarian organization must choose between immediate relief efforts and long-term development projects. An emotional decision might prioritize immediate relief based on empathy, but the CRE evaluates which action leads to the greatest sustained well-being, offering a more logical and consistent approach.

16.7.3 Efficiency and the Pursuit of Human Well-Being

The CRE is designed to increase well-being by evaluating actions that will produce the greatest positive impact for the least cost. This rational, outcome-driven approach aligns directly with the human desire for happiness and well-being.

Efficiency as a Human Value:

- **Resource Optimization**: Humans have always sought to improve their conditions efficiently. The framework mirrors this by focusing on practical outcomes.
- **Avoiding Waste**: By identifying actions that will create the most well-being, resources are used wisely, avoiding waste on less effective endeavors.

Example:
A government deciding between funding immediate social programs or investing in long-term educational infrastructure can use the framework to determine which option provides the greatest long-term benefit. Logical evaluation leads to a decision that maximizes human well-being and minimizes resource waste.

16.7.4 Logic, Efficiency, and Human Flourishing

Rather than detracting from the human element, the CRE amplifies it by ensuring that human flourishing is the central focus of all decisions. The use of logic enhances the ability to meet human needs systematically and effectively.

Benefits of a Logic-Centered Approach:

- **Precision in Improving Well-Being**: Logical decision-making allows for targeted actions that effectively enhance quality of life.
- **Reduction of Suffering**: By systematically identifying and implementing the most effective actions, suffering can be minimized efficiently.
- **Practical Happiness**: The framework emphasizes practical actions over vague moral ideals, leading to tangible improvements in people's lives.

Example:
In environmental policy, moral approaches may lead to debates and delays. The CRE focuses on actions that efficiently reduce emissions while maximizing well-being, resulting in effective action that benefits both the environment and society.

16.7.5 The Causal Relevance Ethic as a Human-Centered Approach

At its core, the CRE is more human than morality-based systems because it centers around tangible improvements to human well-being. By using logic and measurable outcomes, it directly addresses human needs in the most effective way possible.

Why It's More Human:

- **Focus on Results**: Prioritizes real-world outcomes over abstract philosophical debates.
- **Universal Applicability**: Provides a consistent framework that transcends cultural or personal moral differences.
- **Enhances Human Potential**: Empowers individuals and organizations to make decisions that genuinely improve lives.

16.7.6 Case Study: A Rational Approach to Global Issues

Scenario:
Addressing global poverty requires effective allocation of limited resources. Traditional moral approaches may advocate for various solutions based on differing philosophies, often resulting in slow or fragmented actions.

Applying the Causal Relevance Ethic:

- **Evaluate Actions Based on Impact**: Identify which interventions most efficiently reduce suffering, such as targeted aid or infrastructure projects.
- **Assess Feasibility**: Determine which actions are practical given current resources and constraints.
- **Balance Short-Term and Long-Term Benefits**: Use logical evaluations to prioritize actions that offer both immediate relief and sustainable development.

Outcome:
By focusing on logical evaluations, decision-makers can achieve the best possible outcomes quickly and with minimal waste. This efficient, well-being-maximizing approach is inherently human-centric, as it directly improves lives rather than getting bogged down in philosophical debates.

16.7.7 Conclusion

Far from being inhuman, the CRE represents a deeply human approach to decision-making. By focusing on logic, efficiency, and measurable outcomes, the framework ensures that decisions lead to the greatest increase in well-being and the least amount of adversity.

Key Takeaways:

- **Logic Enhances Humanity**: Rational decision-making amplifies our ability to address human needs effectively.
- **Objective and Consistent**: The framework provides stability and consistency, avoiding the subjectivity of emotion-based or morality-centered ethics.
- **Maximizing Well-Being**: By prioritizing actions that have the most significant positive impact, the framework aligns with the fundamental human desire for happiness and flourishing.

16.8 EVALUATING OTHER ETHICAL SYSTEMS THROUGH THE LENS OF THE CAUSAL RELEVANCE ETHIC

In this, we will evaluate key ethical systems – such as **Deontology**, **Utilitarianism**, **Virtue Ethics**, **Ethics of Care**, and **Social Contract Theory** – through the lens of the **CRE**. By highlighting their strengths and weaknesses, we'll explore how they compare to a logic-driven approach focused on maximizing well-being and minimizing adversity through efficient decision-making.

16.8.1 Deontology: Duty-Based Ethics

Deontological ethics, most famously associated with **Immanuel Kant**, emphasizes duty and adherence to moral rules. According to this framework, certain actions are morally required or forbidden, regardless of their consequences. The focus is on following moral principles such as "do not lie" or "keep promises," as these are seen as inherently right.

Strengths:

- **Clarity and Consistency**: Deontology provides a clear and consistent set of rules that guide moral behavior, reducing ambiguity in decision-making.

Weaknesses:

- **Lack of Flexibility**: Strict adherence to rules may ignore the consequences of actions, even if following a rule causes harm.
- **No Outcome-Based Assessment**: Deontological ethics does not prioritize increasing well-being or reducing adversity, potentially leading to inefficient or harmful outcomes.

16.8.2 Evaluation through the Causal Relevance Ethic

The **CRE** critiques deontology's rigid focus on duty as potentially inefficient and less human-centered because it ignores the tangible outcomes of actions. By emphasizing **Influence Strength (I)** and **Feasibility (F)**, the CRE ensures that actions are evaluated based on their real-world impact on well-being and practicality.

Example:

- **Deontological Approach**: A person refuses to tell a lie to protect someone's feelings because lying is inherently wrong.
- **Causal Relevance Ethic Approach**: The person considers that a small lie might significantly reduce someone's suffering (**High I**) with minimal negative consequences (**High F**), thus choosing the action that maximizes well-being.

16.8.3 Utilitarianism: Consequence-Based Ethics

Utilitarianism, associated with philosophers like **Jeremy Bentham** and **John Stuart Mill**, holds that the morality of an action is determined by its consequences. The goal is to maximize happiness, or "utility," for the greatest number of people.

Strengths:

- **Outcome-Focused**: Utilitarianism shares the focus on results with the CRE, aiming to maximize well-being.
- **Pragmatic and Adaptable**: It emphasizes achieving the best possible outcome, which can adapt to different situations.

Weaknesses:

- **Overemphasis on Aggregate Happiness**: May justify sacrificing the well-being of a few if it benefits the majority, leading to morally questionable decisions.
- **Challenges in Calculation**: Difficult to quantify and compare different forms of happiness and suffering, complicating decision-making.

16.8.4 Evaluation through the Causal Relevance Ethic

While sharing the outcome-focused approach, the CRE introduces additional layers through **Causal Proximity (C)** and **Feasibility (F)**, ensuring actions are both impactful and achievable. It also avoids justifying harm to individuals by requiring that all components (**C**, **I**, **F**) are positive and feasible.

Example:

- **Utilitarian Approach**: Approving a policy that benefits the majority but severely harms a minority group.
- **Causal Relevance Ethic Approach**: The action would score low or zero in **Feasibility (F)** due to ethical impracticality and low **Influence Strength (I)** for the harmed individuals, leading to rejection of the action.

16.8.5 Virtue Ethics: Character-Based Ethics

Virtue ethics, rooted in the philosophy of **Aristotle**, focuses on the character and virtues of the individual rather than on rules or consequences. The idea is that by cultivating virtues like courage, wisdom, and temperance, individuals will naturally make morally sound decisions.

Strengths:

- **Focus on Personal Development**: Emphasizes the importance of developing good character traits.
- **Adaptable**: Allows for flexibility, as virtues can guide behavior across various situations.

Weaknesses:

- **Lack of Practical Guidelines**: Offers limited guidance for action in complex or urgent situations.
- **No Outcome Focus**: Does not prioritize specific outcomes like increasing well-being or reducing adversity.

16.8.6 Evaluation through the Causal Relevance Ethic

The CRE values personal virtues to the extent that they contribute to actions with high **CRSs**. However, it critiques virtue ethics for lacking a structured evaluation of actions based on measurable outcomes.

Example:

- **Virtue Ethics Approach**: An individual focuses on being courageous but lacks direction on how to apply this in specific situations to improve well-being.
- **Causal Relevance Ethic Approach**: The individual assesses actions by calculating their **CRS**, ensuring that their courage leads to actions that effectively increase well-being and reduce adversity.

16.8.7 Ethics of Care: Relationship-Based Ethics

The ethics of care, developed by philosophers like **Carol Gilligan**, emphasizes the moral significance of care, empathy, and interpersonal relationships. It focuses on caring for others within the context of relationships and nurturing social bonds.

Strengths:

- **Relational Focus**: Prioritizes human connections and empathy.
- **Emotional Awareness**: Acknowledges the role of emotions in ethical decision-making.

Weaknesses:

- **Subjectivity**: Reliance on emotions and personal relationships can lead to inconsistent decisions.
- **Limited Applicability**: May not effectively address large-scale societal issues where personal relationships are less relevant.

16.8.8 Evaluation through the Causal Relevance Ethic

While recognizing the importance of empathy, the CRE offers a more objective framework by focusing on measurable outcomes. It ensures that decisions benefit broader groups and are not limited by personal biases.

Example:

- **Ethics of Care Approach**: Prioritizing assistance to close friends over strangers in need, based on personal relationships.
- **Causal Relevance Ethic Approach**: Evaluating where the action can have the highest **Influence Strength (I)** and is most feasible (**F**), potentially leading to helping those in greater need regardless of personal relationships.

16.8.9 Social Contract Theory: Agreement-Based Ethics

Social contract theory, associated with philosophers like **Thomas Hobbes**, **John Locke**, and **Jean-Jacques Rousseau**, posits that moral and political obligations arise from agreements or contracts among individuals to form society.

Strengths:

- **Foundation for Social Order:** Provides a basis for establishing laws and political systems.
- **Emphasis on Mutual Agreement**: Highlights the importance of collective consent.

Weaknesses:

- **Status Quo Bias**: May reinforce existing power structures and social norms, even if unjust.
- **Lack of Outcome Focus**: Does not necessarily aim to maximize well-being or reduce adversity.

16.8.10 Evaluation through the Causal Relevance Ethic

The CRE critiques social contract theory for its potential to uphold inefficient or harmful norms. By focusing on logical evaluation of actions based on their real-world impact, the framework seeks to improve societal well-being beyond maintaining the status quo.

Example:

- **Social Contract Approach**: Upholding a law because it is part of the societal agreement, despite it causing harm to certain groups.
- **Causal Relevance Ethic Approach**: Reassessing the law based on its **CRS**, potentially advocating for change if the law scores low in **Influence Strength (I)** or **Feasibility (F)** in promoting well-being.

16.8.11 Conclusion

By evaluating other ethical systems through the lens of the **CRE**, we observe that:

- **Deontology** provides clear rules but lacks flexibility and outcome consideration.
- **Utilitarianism** shares the outcome focus but may sacrifice individual well-being for the majority and lacks structured feasibility assessment.
- **Virtue Ethics** emphasizes character development but lacks practical guidelines and measurable outcome focus.
- **Ethics of Care** values relationships and empathy but can be subjective and limited in scope.
- **Social Contract Theory** maintains social order but may perpetuate unjust norms and lacks emphasis on maximizing well-being.

The CRE offers a unique advantage by combining logical evaluation with a structured approach to decision-making that prioritizes:

- **Measurable Outcomes**: Ensuring actions lead to tangible improvements in well-being.
- **Efficiency**: Maximizing positive impact with optimal use of resources.
- **Feasibility**: Considering practicality to ensure actions can be successfully implemented.
- **Human-Centered Focus**: Centering decisions on enhancing human flourishing through logical means.

16.9 PRACTICAL APPLICATION OF THE CAUSAL RELEVANCE ETHIC

In this, we will explore how to apply the **CRE** in real-world situations. By following a step-by-step approach, you can systematically evaluate actions and make decisions that maximize well-being and minimize adversity. Whether in business, policymaking, or personal life, the CRE serves as a practical tool to guide choices that are both effective and impactful.

16.9.1 The Step-by-Step Process for Applying the Causal Relevance Ethic

Applying the CRE involves evaluating actions based on the three core components: **Causal Proximity** (C), **Influence Strength** (I), and **Feasibility** (F). By following these steps, you can calculate the **CRS** and prioritize actions that offer the most significant real-world impact.

Step 1: Define the Decision Context

- **Purpose**: Clearly outline the situation requiring a decision.
- **Actions**:
  - Identify the goals you aim to achieve.
  - Note any constraints such as resources, time, or organizational policies.

Step 2: Identify Potential Actions

- **Purpose**: Generate a list of actionable options.
- **Actions**:
  - Brainstorm possible solutions or strategies.
  - Ensure each action is specific and feasible.

Step 3: Evaluate Causal Proximity (C)

- **Purpose**: Assess how immediate and certain the impact of each action will be.
- **Actions**:
  - Score the **TF** and **CF** for each action.
  - Calculate **C** using the formula:

$$C = \begin{cases} 0, & \text{if } TF = 0 \text{ or } CF = 0 \\ TF + CF, & \text{otherwise} \end{cases}$$

Step 4: Evaluate Influence Strength (I)

- **Purpose**: Determine the potential of each action to increase well-being and reduce adversity.
- **Actions**:
  - Score the **TI** and **ARI**.
  - Calculate **I** using the formula:

$$I = \begin{cases} 0, & \text{if } TI = 0 \text{ and } ARI = 0 \\ TI + ARI, & \text{otherwise} \end{cases}$$

Step 5: Evaluate Feasibility (F)

- **Purpose**: Assess the practicality of implementing each action.
- **Actions**:
  - Score the **RA** and **PA**.

$$F = \begin{cases} 0, & \text{if } RA = 0 \text{ or } PA = 0 \\ RA + PA, & \text{otherwise} \end{cases}$$

  - Calculate **F** using the formula:

Step 6: Calculate the Causal Relevance Score (CRS)

- **Purpose**: Combine the evaluations to determine the overall worthiness of each action.
- **Actions**:
  - Calculate **CRS** using the formula:

$$CRS = \begin{cases} 0, & \text{if } C = 0 \text{ or } I = 0 \text{ or } F = 0 \\ C + I + F, & \text{otherwise} \end{cases}$$

- **Maximum CRS**: 30

Step 7: Prioritize Actions Based on CRS

- **Purpose**: Rank the actions from most to least worthwhile.
- **Actions**:
  - Use the CRS to prioritize actions.
  - Higher scores indicate actions that are more likely to produce significant positive outcomes.

Step 8: Review and Adjust Based on Context

- **Purpose**: Ensure that the selected actions align with broader goals and ethical considerations.
- **Actions**:
  - Consider qualitative factors that may influence the decision.
  - Adjust plans as necessary to fit organizational culture or personal values.

16.9.2 Real-World Examples

Example 1: Business Decision on Market Expansion
Context:
A company must decide between expanding into a new international market or investing in upgrading its existing product line.

Potential Actions:

- **Action 1**: Expand into a new market (e.g., Asia).
- **Action 2**: Upgrade the current product line for existing markets.

Step-by-Step Application:
Step 1: Define the decision context.

- Goal: Allocate resources effectively to maximize growth.
- Constraints: Limited budget and personnel.

Step 2: Identify potential actions.

- Action 1 and Action 2 as above.

Step 3: Evaluate Causal Proximity (C).

- **Action 1**:
  - **TF**: 2 (Long-term impact; market entry will take years).
  - **CF**: 3 (Moderately likely to succeed due to market uncertainties).
  - **C**: 2 + 3 = 5
- **Action 2**:
  - **TF**: 4 (Short-term impact; upgrades can be implemented in months).
  - **CF**: 4 (Very likely to succeed based on past experience).
  - **C**: 4 + 4 = 8

Step 4: Evaluate Influence Strength (I).

- **Action 1**:
  - **TI**: 5 (Significant increase in market reach and revenue).
  - **ARI**: 2 (Limited adversity reduction).
  - **I**: 5 + 2 = 7
- **Action 2**:
  - **TI**: 4 (Large increase in customer satisfaction).
  - **ARI**: 3 (Moderate reduction in customer churn).
  - **I**: 4 + 3 = 7

Step 5: Evaluate Feasibility (F).

- **Action 1**:
  - **RA**: 2 (Few resources available; significant constraints).
  - **PA**: 3 (Moderately practical; some challenges).
  - **F**: 2 + 3 = 5
- **Action 2**:
  - **RA**: 4 (Most resources available; minor constraints).
  - **PA**: 4 (Very practical; minor challenges).
  - **F**: 4 + 4 = 8

Step 6: Calculate CRS.

- **Action 1**: CRS = C + I + F = 5 + 7 + 5 = 17
- **Action 2**: CRS = C + I + F = 8 + 7 + 8 = 23

Step 7: Prioritize Actions.

- **Action 2** has a higher CRS (23) compared to **Action 1** (17).

Step 8: Review and Adjust.

- Consider long-term strategic goals.
- May decide to proceed with **Action 2** while planning **Action 1** for the future.

Example 2: Government Policy on Education Investment

Context:

A government is deciding whether to implement immediate educational reforms or invest in long-term infrastructure development.

Potential Actions:

- **Action 1**: Implement immediate reforms (hiring teachers, updating curricula).
- **Action 2**: Invest in long-term infrastructure (building schools, enhancing technology access).

Step-by-Step Application:

Step 1: Define the decision context.

- Goal: Improve the national education system.
- Constraints: Budget limitations and political considerations.

Step 2: Identify potential actions.

- As above.

Step 3: Evaluate Causal Proximity (C).

- **Action 1**:
  - **TF**: 5 (Immediate impact).
  - **CF**: 4 (Very likely to succeed).
  - **C**: 5 + 4 = 9
- **Action 2**:
  - **TF**: 2 (Long-term impact).

  - **CF**: 4 (Very likely to succeed).
  - **C:** 2 + 4 = 6

Step 4: Evaluate Influence Strength (I).

- **Action 1**:
  - **TI**: 3 (Moderate increase in educational quality).
  - **ARI**: 4 (Large reduction in overcrowded classrooms).
  - **I**: 3 + 4 = 7
- **Action 2**:
  - **TI**: 5 (Significant increase in educational opportunities).
  - **ARI**: 5 (Significant reduction in educational disparities).
  - **I**: 5 + 5 = 10

Step 5: Evaluate Feasibility (F).

- **Action 1**:
  - **RA**: 4 (Most resources available).
  - **PA**: 5 (Highly practical).
  - **F**: 4 + 5 = 9
- **Action 2**:
  - **RA**: 2 (Few resources available).
  - **PA**: 3 (Moderately practical).
  - **F**: 2 + 3 = 5

Step 6: Calculate CRS.

- **Action 1**: CRS = 9 + 7 + 9 = 25
- **Action 2**: CRS = 6 + 10 + 5 = 21

Step 7: Prioritize Actions.

- **Action 1** has a higher CRS (25) compared to **Action 2** (21).

Step 8: Review and Adjust.

- May choose to implement **Action 1** immediately.
- Plan for **Action 2** as a part of long-term strategy when resources allow.

Example 3: Personal Decision on Health and Wellness

Context:

An individual is choosing between a quick health improvement plan or committing to long-term lifestyle changes.

Potential Actions:

- **Action 1**: Start a detox program for immediate benefits.
- **Action 2**: Commit to long-term lifestyle changes (diet and exercise).

Step-by-Step Application:

Step 1: Define the decision context.

- Goal: Improve overall health and well-being.
- Constraints: Time availability and personal motivation.

Step 2: Identify potential actions.

- As above.

Step 3: Evaluate Causal Proximity (C).

- **Action 1**:
  - **TF**: 5 (Immediate impact).
  - **CF**: 5 (Almost certain to succeed).
  - **C**: 5 + 5 = 10

- **Action 2**:
  - **TF**: 3 (Medium-term impact).
  - **CF**: 4 (Very likely to succeed with commitment).
  - **C**: 3 + 4 = 7

Step 4: Evaluate Influence Strength (I).

- **Action 1**:
  - **TI**: 2 (Small increase in well-being).
  - **ARI**: 2 (Small reduction in adversity).
  - **I**: 2 + 2 = 4
- **Action 2**:
  - **TI**: 5 (Significant increase in well-being).
  - **ARI**: 5 (Significant reduction in health risks).
  - **I**: 5 + 5 = 10

Step 5: Evaluate Feasibility (F).

- **Action 1**:
  - **RA**: 5 (All resources readily available).
  - **PA**: 5 (Highly practical).
  - **F**: 5 + 5 = 10
- **Action 2**:
  - **RA**: 4 (Most resources available).
  - **PA**: 3 (Moderately practical; requires effort).
  - **F**: 4 + 3 = 7

Step 6: Calculate CRS.

- **Action 1**: CRS = 10 + 4 + 10 = 24
- **Action 2**: CRS = 7 + 10 + 7 = 24

Step 7: Prioritize Actions.

- Both actions have the same CRS (24).

Step 8: Review and Adjust.

- Consider combining both actions:
  - Begin with **Action 1** for immediate benefits.
  - Transition to **Action 2** for long-term health improvements.

16.9.3 Conclusion

The **CRE** provides a structured and logical approach to decision-making across various contexts. By systematically evaluating actions based on **Causal Proximity (C)**, **Influence Strength (I)**, and **Feasibility (F)**, you can prioritize actions that maximize well-being and minimize adversity while efficiently utilizing resources.

Key Takeaways:

- **Structured Decision-Making**: The step-by-step process ensures thorough evaluation.
- **Applicability**: The framework is versatile and can be applied to business, policymaking, and personal decisions.
- **Outcome-Focused**: Prioritizes actions that offer the most significant positive impact.

16.9.4 The Idea

Adopting the CRE in decision-making empowers individuals and organizations to make choices that are not only effective but also aligned with the goal of enhancing human well-being. By focusing on logical evaluation and measurable outcomes, you can navigate complex decisions with confidence and clarity.

THE PARADOX OF PROGRESS

THINKING BOX

"Does technology inherently push humanity forward, or is progress a more subtle tapestry? Examining **42** might help decode the riddle."

16.10 ADDRESSING POSTMODERN CRITICISM AND PRACTICAL APPLICATIONS

In this, we will address common critiques from postmodern philosophy and demonstrate how the **CRE** not only withstands these criticisms but also offers practical solutions in various contexts. Postmodern thinkers often question systems that conform to established power structures, expressing concerns that ethical frameworks can become tools of control, reinforcing existing hierarchies. The CRE counters this by promoting independent, evidence-based decision-making that resists manipulation by dominant narratives. We will also explore practical examples of how this framework can be applied in personal, organizational, and activist settings, followed by concluding thoughts on its overall benefits.

16.10.1 Dealing with Postmodern Criticism

16.10.1.1 Avoiding Conformity to Power Structures

The CRE is grounded in practicality rather than ideology or rigid rules. This ensures that the framework remains flexible and adaptable to specific contexts, rather than conforming to fixed moral guidelines imposed by institutions or authorities.

Key Points:

- **Evidence-Based Decision-Making**: The framework relies on measurable, empirical data to evaluate actions, preventing those in power from using abstract moral rules to manipulate outcomes.
- **Practicality over Ideology**: By focusing on feasible and impactful actions, the CRE avoids becoming a tool of ideological dominance. It encourages decisions rooted in practicality, not in adherence to external doctrines or power dynamics.

Example:
In policymaking, power structures may impose moral agendas that benefit specific groups. The CRE prevents such distortions by basing decisions on what is achievable and effective, rather than on ideological conformity. For instance, a policy that favors a particular industry due to political influence would be scrutinized for its real-world impact and feasibility, ensuring that resources are allocated to actions that genuinely enhance well-being.

16.10.1.2 Promoting Independent Thinking

The framework fosters a mindset of independent thinking, encouraging decision-makers to base their actions on evidence rather than societal or institutional pressures.

Key Points:

- **Independent Assessment**: Individuals are empowered to evaluate actions based on their own merits, ensuring they are not swayed by prevailing social or political pressures.
- **Critical Evaluation**: By requiring decisions to be grounded in **Causal Proximity (C)**, **Influence Strength (I)**, and **Feasibility (F)**, the framework promotes a critical, evidence-based approach that challenges dominant power structures and ideologies.

Example:
In activism, popular movements might focus on actions that gain public approval but lack real-world feasibility. By applying the CRE, activists can prioritize actions that have genuine impact. For example, instead of organizing a large protest with limited effect, they might focus on policy advocacy or community programs that score higher in feasibility and influence strength, leading to more substantial change.

16.10.2 Practical Applications

16.10.2.1 Personal Scenario: Helping a Friend

When deciding whether to assist a friend, the CRE guides you to evaluate the situation based on proximity, influence, and feasibility.

Application:

- **Causal Proximity (C)**: Is the friend's need immediate?
- **Influence Strength (I)**: How much will your help improve their well-being or reduce adversity?
- **Feasibility (F)**: Do you have the resources and ability to help effectively?

Zero Value Insight:
A zero in proximity or feasibility indicates that the action may not be necessary or achievable at the moment, allowing you to allocate your resources more effectively.

Example:
A friend asks for help moving furniture on short notice. If you have prior commitments (low feasibility), the framework suggests prioritizing other actions. You might offer to help at a later time or find alternative solutions, ensuring your assistance is both practical and impactful.

16.10.2.2 Organizational Scenario: Allocating Resources

Organizations often face the challenge of distributing limited resources across multiple projects. The CRE assists in prioritizing initiatives with the greatest feasible impact.

Application:

- **Causal Proximity (C)**: Which projects will yield results soonest?
- **Influence Strength (I)**: Which initiatives will significantly enhance well-being or reduce adversity?
- **Feasibility (F)**: Which projects are most practical given current resources?

Zero Value Insight:
Projects scoring zero in feasibility should be deprioritized, reallocating resources to initiatives that are achievable and beneficial.

Example:
An organization considers funding a high-profile event that is logistically complex. If feasibility scores low due to resource constraints, the framework advises directing funds to more practical projects that offer substantial impact, such as community outreach programs.

16.10.2.3 Activism Scenario: Planning a Campaign

Activists must choose between various methods to effect change. The CRE helps evaluate the relative impact of actions like public demonstrations vs. policy advocacy.

Application:

- **Causal Proximity (C)**: How immediate is the potential impact?
- **Influence Strength (I)**: Will the action significantly advance the cause?
- **Feasibility (F)**: Is the action practical given legal and logistical considerations?

Zero Value Insight:
An action with low feasibility or minimal influence may indicate virtue signaling – a symbolic gesture with little real-world effect.

Example:
An environmental group debates organizing a large protest. The protest might have high causal proximity but low feasibility due to permitting issues and resource demands. Applying the framework suggests focusing on policy advocacy or educational initiatives that are more feasible and potentially more impactful in achieving environmental goals.

16.10.3 Conclusion

The **CRE** offers a practical, results-driven approach to decision-making that prioritizes real-world impact and feasibility. By focusing on measurable outcomes and independent evaluation, it moves beyond abstract moral rules and resists conformity to power structures.

Summary of Benefits:

- **Efficiency**: Resources are utilized effectively, maximizing positive outcomes.
- **Practicality**: Decisions are grounded in what is realistically achievable, avoiding impractical or symbolic actions.
- **Independence**: Encourages evidence-based thinking, empowering individuals and organizations to make autonomous decisions that challenge dominant narratives.

16.10.4 Encouragement to Apply

By adopting this framework, you can identify and prioritize actions that lead to measurable improvements in well-being. Whether in personal life, organizational strategy, activism, or public policy, the CRE provides a structured, logical approach to solving complex problems.

16.10.5 The Idea

The CRE bridges the gap between abstract moral theory and practical decision-making. It emphasizes:

1. **Core Components**:
   - **Causal Proximity (C)**: Focus on the immediacy and certainty of impact.
   - **Influence Strength (I)**: Measure the potential increase in well-being or reduction in adversity.
   - **Feasibility (F)**: Assess PA given current resources and constraints.
2. **Zero Value Insights**:
   - Zero values serve as clear indicators to avoid proceeding with actions that lack impact or feasibility.
3. **Versatility**:
   - Applicable across various contexts, offering actionable steps and real-world examples.

4. **Resistance to Conformity**:
   - Promotes independent, evidence-based thinking, avoiding adherence to rigid ideologies.
5. **Advantage over Traditional Ethics**:
   - Focuses on practicality and results, addressing limitations in other ethical systems that may overlook feasibility.
6. **Adaptability**:
   - Dynamic adjustment based on evidence and context ensures decisions remain relevant and impactful.

16.10.6 Conclusion

By focusing on achievable and measurable outcomes, the CRE provides a compelling alternative to virtue signaling and emotionally driven decisions. It empowers individuals and organizations to make thoughtful, rational choices that lead to real, measurable improvements. As you navigate complex challenges, this framework serves as a reliable guide to ensure your actions are effective, efficient, and ethically sound.

Thank you for engaging with this guide. May the Causal Relevance Ethic enhance your decision-making processes and contribute positively to your endeavors.

17 The Technologically Mediated Mind

From Classical Philosophy to Consciousness 2.0

Key Themes:

1. **Shaping Reality:**
   - **Plato's Cave and Berkeley's Idealism**: Digital illusions (social media, deepfakes) alter our sense of truth.
   - **AI, VR, and AR** blur reality, influencing cognition and behavior.
   - **BCIs introduce direct neural manipulations**, raising ethical concerns over free will and memory integrity.
2. **Forging Identities**:
   - **Social media self-curation** shapes identity, often fostering performance over authenticity.
   - **AI-generated content and BCIs challenge the uniqueness of human thought.**
   - **Genetic editing and cognitive enhancements** risk creating a **two-tiered society**.
3. **Ethical Considerations**:
   - **Neural security risks** (hacking, mass psychological manipulation).
   - **Informed consent challenges** for BCIs and genetic modifications.
   - **Precautionary regulation to prevent monopolization of human augmentation.**
4. **Strategies for Autonomy**:
   - **Cognitive checkpoints, neural sanctuaries (tech-free zones), and meta-cognitive awareness** protect against digital influence.
   - **Global regulations must balance progress with ethical safeguards.**

Key Takeaway:

As technology merges with the mind, we must **proactively shape its role**, ensuring it **enhances rather than diminishes** human autonomy, identity, and collective well-being.

17.1 INTRODUCTION: BECOMING CYBORGS IN SUBTLE WAYS

17.1.1 Introduction: Becoming Cyborgs in Subtle Ways

17.1.1.1 Evocative Hook

We are becoming cyborgs – not in the cinematic sense of gleaming metal limbs, but in the subtle integration of technology into our very minds. Our experiences of reality, identity, and social interaction are increasingly shaped by digital systems, immersive platforms, and biological interventions, leading to what we might term a "technologically mediated mind" – a mind whose perceptions, thoughts, and behaviors are significantly influenced by these external forces.

Expanding on the Concept – Consider the pervasive influence of everyday technology: from the constant connectivity and information access provided by smartphones and the biometric data

DOI: 10.1201/9781003640288-22

collected by wearables to the curated realities shaped by online platforms and artificial intelligence (AI)-driven recommendation systems. These technologies have become so woven into daily life that they function not merely as tools but as extensions of our cognitive and emotional processes – blurring the boundary between the human mind and external systems.

17.1.1.2 The "Technologically Mediated Mind"

At its core, this phrase captures how our mental processes – perception, judgment, memory, sense of self – are increasingly shaped *and in turn shape* technological forces. This shift is not simply about convenience or efficiency; it changes how we interpret the world, understand others, and define who we are. Because these external systems are now integral to our thought patterns – offloading memory, filtering information, or influencing our emotional states – we can no longer regard technology as a mere add-on. Instead, it becomes a key co-creator of our lived experience.

17.1.1.3 Why This Subtle Cyborg Existence Matters

- **Normalization and Subtle Manipulation of Perception**: What was once science fiction – a world of constant connectivity, real-time data tracking, and AI-curated information – has become everyday life. Because these advances evolve gradually, we adapt without fully questioning how deeply they shape our beliefs, decisions, and sense of self.
- **Ubiquity and Scale**: Billions of people worldwide rely on digital networks that collect and analyze vast quantities of personal data, subtly nudging behavior in ways we might not even notice.
- **Pervasive Algorithmic Influence**: Social media algorithms, personalized feeds, and wearable notifications don't just offer convenience; they can guide our emotional responses, shift social norms, and affect collective worldviews. Recognizing this influence is vital for maintaining autonomy and ensuring technology supports rather than undermines human well-being.

17.1.1.4 Connecting to Classical Philosophy

This evolutionary leap from using tools to becoming embedded in networks resonates powerfully with **classical philosophical questions**, as well as more modern perspectives like **Merleau-Ponty's phenomenology** and the **Extended Mind Thesis**:

- **Plato's Allegory of the Cave**: Are we confusing digital shadows for authentic experience, especially when social media, augmented reality, or manipulative news feeds are involved?
- **Descartes' Dualism**: What happens to the mind-body distinction when neural interfaces or AI "assistants" become integrated into our cognition?
- **Berkeley's Idealism**: If reality is predicated on perception, how do we define "real" when technology continuously filters, modifies, or overlays new inputs onto our senses?

Seen through these lenses, our "subtle cyborg" state demands fresh thinking about what it means to be human in a world where our consciousness is co-created by technology.

17.1.1.5 Previewing the Chapter

The rest of this chapter will delve deeper into these intertwined themes:

1. **Shaping Reality**: We'll examine how VR, AR, AI, and social media construct (and sometimes distort) our perceptions, echoing ancient debates about illusion vs. truth.
2. **Forging Identities**: We'll trace how technologies like social platforms and BCIs transform personal identity, drawing on Descartes' Dualism, Berkeley's Idealism, and the Extended Mind Thesis.
3. **The Ethics of Enhancement**: We'll explore moral bioenhancement, consent, equity, and regulatory gaps – demonstrating how technology can be harnessed for human flourishing or exploited for control.

4. **Reimagining Human Connection**: We'll highlight the impact on empathy, community, and even spirituality when our interactions occur within digitally augmented environments or potential neural networks.
5. **Cultivating Autonomy in a Technologically Mediated World**: A set of strategies – cognitive checkpoints, meta-cognitive awareness, and "neural sanctuaries" – aimed at safeguarding our minds against digital manipulation.
6. **From Gadgets to Genes**: We'll zoom out to see the continuum of human augmentation, from wearables to genetic engineering, and ask how we might reconcile slow evolutionary changes with rapidly accelerating technology.
7. **Conclusion**: Synthesizing classical philosophy and modern technology, we'll underscore the need for continuous reflection and conscientious design to ensure that our cyborg futures remain distinctly human at heart.

17.1.1.6 Why This Discourse Is Urgent

Understanding ourselves as "subtle cyborgs" is a crucial first step. It requires us to critically examine the role technology plays in our lives and to ask ourselves: **How are these technologies shaping who we are becoming? What values do we want to uphold in this rapidly changing world? And how can we ensure that technology serves humanity, rather than the other way around?**

- **Ethical Imperatives**: As personalization algorithms and potential brain-machine interfaces grow more sophisticated, questions of autonomy, consent, and fairness loom large.
- **Societal Impact**: From misinformation crises to data exploitation and AI-driven social engineering, these issues can erode trust and fuel division.
- **Opportunity for Growth**: On the other hand, responsibly developed technologies can foster empathy, expand cognitive horizons, and address real-world challenges like disability or global collaboration.

By facing these questions now, we can better navigate the extraordinary possibilities and profound challenges of the **technologically mediated mind**. The following sections will illuminate both the risks and rewards of integrating classical philosophical wisdom with the rapid advancements reshaping our collective consciousness.

17.2 SHAPING REALITY: THE TECHNOLOGICAL INFLUENCES

17.2.1 Echoes of the Cave: Philosophical Foundations for a Malleable Reality

17.2.1.1 Plato's Cave Revisited: Modern "Shadows on the Wall" and the New Puppeteers

Plato's *Allegory of the Cave* remains a cornerstone in discussions of reality and perception. In the original allegory, prisoners see only shadows projected on a cave wall, believing these silhouettes to be the totality of existence. Once a prisoner escapes and perceives the outside world, he realizes how distorted and incomplete his previous "reality" was.

In our technological age, these "shadows" take on a new, more pervasive form:

- **Digital Illusions**: Filters, virtual environments, and manipulated news feeds can project convincing simulations of events or lifestyles, just as shadows once danced across Plato's Cave wall.
- **Algorithms and Immersive Technologies as the New Puppeteers**: We are not chained in a literal cave, but the systems behind social media and digital platforms operate akin to puppeteers – projecting illusions (or selective narratives) onto the digital walls of our perception. Curated feeds, search-engine biases, and immersive simulations can constrain what we see, shaping our worldview in ways we rarely question.

The lesson from Plato's Cave is that *perceived reality can be incomplete or misleading*, especially when mediated by forces we do not fully control or understand. Today, the complexity of these forces – AI algorithms, entertainment industries, and online propaganda networks – makes it even harder to discern genuine truth from orchestrated shadows.

17.2.1.2 Berkeley's Idealism (Esse est Percipi): Reality Contingent on Perception and Implications for Objective Truth

George Berkeley's dictum, "esse est percipi" – "to be is to be perceived" – argues that **reality is inseparable from the act of perception**. In Berkeley's view, objects don't exist independently; rather, they exist within the minds that perceive them.

In a hyperconnected world, Berkeley's insight provokes challenging questions about **objective truth**:

- **Personalized Algorithms and Individual Realities**: If we each receive uniquely tailored feeds – where one individual sees stories about global crises and another sees celebrity gossip – can we truly speak of a shared objective reality?
- **Digital Experiences vs. Tangible Facts**: When entire events can be fabricated through high-tech illusions (deepfakes, AI-driven narratives), are we witnessing a fracturing of reality itself, where "truth" becomes a mosaic of personal perceptions rather than a commonly agreed-upon foundation?

Thus, while Plato reminds us of the shadows we might mistake for real objects, Berkeley pushes us to question whether an "objective reality" exists outside our digitally molded perceptions. Taken together, these classical perspectives illuminate how precarious our sense of the "real" becomes in an era defined by endless streams of algorithmically shaped content.

17.2.2 VR/AR: Constructed Environments and Overlaid Worlds

Defining VR vs. AR

- **Virtual Reality (VR)** envelops users in entirely simulated environments, often via headsets and haptic controllers. The outside world is blocked from view, replaced by a computer-generated realm – whether an alien landscape, a fantasy medieval city, or a meticulously recreated historical event.
- **Augmented Reality (AR)**, by contrast, overlays digital elements onto the physical world. Users might wear glasses or use smartphone cameras to see virtual characters, data, or directions "floating" on top of their real surroundings.

Both VR and AR represent a spectrum of immersive technologies, each challenging the boundary between tangible and simulated experience.

17.2.2.1 Phenomenology and Embodiment (Merleau-Ponty): Disembodiment, Extended Embodiment, and Entirely New Forms of Existence

Maurice Merleau-Ponty's phenomenology underscores that **we encounter and make sense of the world through our physical and bodily experience**. VR and AR disrupt our usual sensorimotor links by altering visual, auditory, or even tactile cues in significant ways:

1. **Disembodiment**: In VR, one's eyes and ears perceive scenarios that don't match the body's physical location. Users might "walk" through a virtual environment but remain physically stationary, leading to a *mismatch* between bodily perception and the simulated context – sometimes causing motion sickness or a startling sense of disorientation.
2. **Extended Embodiment**: AR can enhance one's perception of the physical world, adding extra layers of data – like heads-up directions on city streets or interactive game elements

in a living room. In this sense, technology becomes an extension of perception, letting users "see" or interact with elements that have no physical presence but feel subjectively real.

3. **Entirely New Forms of Embodiment**: Beyond disembodiment or simply layering data, virtual worlds can offer avatars with different physical characteristics – or even non-human forms. In these realms, an individual can experience a sense of *being* something radically different from their usual body, thereby testing the limits of identity and self-experience (e.g., taking on an avatar with wings or controlling a robotic form in a futuristic VR setting).

These immersive possibilities underscore how flexible and adaptable our embodied consciousness can be – and how technologies like VR/AR may drastically expand the boundaries of what it means to be present in the world, "real" or otherwise.

17.2.3 AI-Generated Content and Digital Filters

17.2.3.1 Deepfakes, AI-Written Text, and Blurred Authenticity

AI has rapidly advanced to produce high-fidelity images, videos, and even fully generated voices that can mimic real people. Platforms and apps that utilize **generative AI** can create:

- **Deepfake Videos** showing public figures saying or doing things they never actually did.
- **AI-Crafted Audio** that replicates a person's voice well enough to fool phone-based identity verification.
- **AI-Written Text** so coherent it can pass as human-created articles, essays, or creative works.

The proliferation of these tools complicates our assumptions about *authentic evidence* and challenges our ability to trust seemingly factual media. What was once verifiable can now be convincingly faked.

17.2.3.2 Trust, Authorship, and Institutional Integrity

- **Eroding Trust in Institutions**: This blurred line between the authentic and fabricated undermines not just individual claims but also the authority of news agencies, academic journals, and other bodies that traditionally verify information. When deepfake technology can produce "eyewitness" videos that are entirely staged, or AI-generated "scientific papers" mimic genuine research, public confidence in established institutions can wane dramatically.
- **Questions of Authorship**: If AI scripts a novel or composes a painting, who is the creator – machine, programmer, or user? Debates about intellectual property and creative originality gain new urgency, echoing centuries-old artistic and philosophical dilemmas in a fully modern context.

In sum, **AI-generated content** doesn't merely create illusions; it strikes at the heart of how societies establish truth, credibility, and the trustworthiness of institutions meant to uphold knowledge standards.

17.2.4 Social Media and Reality Manipulation

17.2.4.1 Algorithm-Driven Feeds: Reinforcement Loops and Polarization

Social media platforms sort and recommend content via algorithms designed to maximize engagement – often favoring provocative, emotionally charged, or sensational content. Key consequences include:

- **Reinforcement Loops**: People see more of what aligns with their current beliefs. This intensifies echo chambers where dissenting information is rarely encountered, fueling a strong sense of "we are correct" within insular communities.

- **Political Polarization and Fragmentation**: These echo chambers contribute significantly to political polarization and social fragmentation. Users cluster into ideologically homogeneous groups, sometimes perceiving opposing viewpoints as wholly alien or morally suspect.

17.2.4.2 Elizabeth Loftus's False Memories: Psychological Vulnerability Meets Tech

Decades ago, psychologist Elizabeth Loftus showed how malleable and suggestible our memories can be. In the context of social media:

- **Fabricated Narratives**: Photoshopped images, viral memes, or repeated claims can implant or reinforce false recollections.
- **Convergence of Social Validation**: If multiple sources (especially if manipulated by bots) echo the same misleading story, people may adopt these "memories" as personal truths, even swearing they vividly recall events that never happened.

This interplay between suggestible minds and algorithmic curation underscores how susceptible our collective "reality" can be to distortion – no cave walls required.

17.2.5 BCIs as the Extreme Case

17.2.5.1 Direct Reality Manipulation through Neural Interfaces

While VR and AR alter external sensory streams, **brain-computer interfaces (BCIs)** go a step further: they potentially deliver stimuli directly to the brain or read signals from it. In other words, they can manipulate or interpret *the very seat of perception*.

- **Synthetic Experiences**: Advanced BCIs might allow users to see, hear, or feel things that have no external source – pure neural stimulation.
- **Memory Intervention**: The possibility of "implanting" or editing memories, though currently at nascent stages, looms as a profound shift in how we define personal identity and authenticity.

17.2.5.2 The Therapeutic Potential and the Slippery Slope

It's important to recognize that BCIs are not solely about manipulation:

- **Positive Therapeutic Applications**: Neural implants can help restore motor functions to paralyzed individuals, treat certain neurological disorders, or even partially restore lost senses (e.g., cochlear implants for hearing).
- **Societal Acceptance**: As these beneficial uses gain public trust, more advanced forms of perception engineering may become normalized – leading to the "slippery slope" where enhancements or manipulations (e.g., re-experiencing cherished memories, filtering negative stimuli) can gradually become mainstream.

By shining a spotlight on how far the human-tech interface could go, BCIs compel us to ask how we balance medical breakthroughs and personal enhancements against the risks of deliberately engineering or distorting lived experience.

17.2.6 Concluding Reality

Today's digital age reveals a world in which reality itself is *profoundly malleable*, shaped by illusions from social media feeds to fully immersive VR, from AI-generated "facts" to direct neural manipulation. If **Plato's Cave** warned us about mistaking shadows for truths, we now inhabit an interconnected labyrinth of digital "caves," each with its own flickering projections.

17.2.6.1 Where Do We Go from Here?

1. **Critical Discernment**: As illusions become seamless, developing "reality-literacy" or robust forms of critical thinking is paramount. Skepticism must be balanced with an openness to new possibilities.
2. **Ethical and Regulatory Frameworks**: Governments, tech developers, and citizens alike must grapple with questions of misinformation, neural data privacy, and the role of advanced BCIs in shaping human experience.
3. **Reconciling Skepticism and Possibility**: We should remain vigilant about deception and manipulation, yet also explore how these tools – like VR therapy or BCI-driven medical interventions – can enrich lives and expand human potential.

The overarching challenge is to **discern the genuine from the simulated** and to choose how we engage with these technologies so they enhance rather than erode our shared humanity. As we transition to the next section, we'll see how these same technological forces that reshape our perception of the external world also transform our *inner sense of self* and identity, raising profound questions about embodiment, autonomy, and the very definition of what it means to be human.

17.3 FORGING IDENTITIES IN A DIGITAL AGE

17.3.1 Cartesian Dualism in the Spotlight

17.3.1.1 Descartes' "I think, therefore I am" under Re-examination

René Descartes' pronouncement – "*Cogito, ergo sum*" – long served as a philosophical touchstone, establishing consciousness as the bedrock of selfhood and drawing a firm line between mind (res cogitans) and body (res extensa). Yet as emerging technologies interlace with human cognition, this sharp duality becomes increasingly fraught. In the modern era:

- **Mind-Machine Fusion**: Smartphones, smartwatches, and increasingly sophisticated BCIs don't merely "help" the mind; they become integral to its daily processes – offloading memory, directing attention, and shaping decision-making.
- **Fluid Boundaries**: Our neural processes, once confined within the skull, now extend into digital networks. For example, a GPS app can effectively serve as part of our navigational cognition, while social media "notifications" can alter our emotional states in real time.

Consequently, the once-clear boundary between thought and external matter blurs, demanding a re-evaluation of what "I think, therefore I am" means in a world where technology increasingly weaves itself into the tapestry of consciousness.

17.3.1.2 Extended Mind Thesis (Clark and Chalmers) and the Contrast with Embodied Cognition

Philosophers Andy Clark and David Chalmers proposed the **Extended Mind Thesis**, suggesting that **tools and external resources** (e.g., notebooks, smartphones, AI systems) can be so intimately integrated into our cognitive processes that they effectively become part of the mind. This perspective highlights:

- **Cognitive Offloading**: A smartphone storing thousands of phone numbers or notes can function as an extension of memory.
- **AI as "Cognitive Partner"**: Recommender systems, digital assistants, or neural implants can collaborate with us to solve problems – potentially altering the notion of autonomous thinking.

However, it's crucial to juxtapose this with **embodied cognition**, which emphasizes the **role of the physical body** (and bodily experiences) in shaping thought. From this angle, cognition isn't just about abstract information processing but also about how we interact with the physical world through our senses, posture, gestures, and emotional states. In some ways, advanced technologies (like VR or BCI) can induce a sense of *disembodiment* – where the body's direct feedback loops are bypassed – potentially conflicting with traditional embodied experiences that keep cognition grounded in physical reality.

This tension – between extended minds merging with digital networks and the importance of bodily grounding – reflects our evolving philosophical landscape, revealing just how far we have come from Descartes' stark divide.

17.3.2 Social Media Self-Fashioning

17.3.2.1 Profile Curation, Avatar Identities, and the Tension between Online Persona and Offline Self

In today's interconnected world, **social media** platforms serve as arenas where we project carefully crafted versions of ourselves. Individuals curate profiles, highlight select personal milestones, and often present idealized images to cultivate a desired persona.

- **Avatars and Alter Egos**: In virtual communities or gaming environments, some opt for radically different avatars – changing gender, species, or adopting entirely fantastical appearances – thus exploring identities beyond their physical, offline constraints.
- **Public vs. Private Selves**: The line between genuine self-expression and performative branding can blur, particularly when likes, shares, and follower counts become de facto metrics of social value.

This curated existence can offer empowerment and creativity – allowing people to experiment with identity – but it can also fragment one's sense of authenticity. Over time, the *algorithmic push* for engagement and validation can subtly shape users' self-perceptions and behaviors, leading many to internalize the digital persona they've cultivated.

17.3.2.2 Continuous "Performance" under Algorithmic Influence – and the Impact of Social Comparison

A significant byproduct of social media identity curation is **continuous performance**:

- **Algorithms and Validation**: Platforms reward content that attracts attention. By emphasizing likes, views, or shares, they create a feedback loop encouraging users to sustain or exaggerate persona traits that garner positive responses.
- **Social Comparison**: Competing for validation can make identity feel like a performance or brand, potentially leading to anxiety, *social comparison*, and a distorted self-image. People may judge their worth or success by comparing themselves to the curated (often idealized) posts of peers and influencers.
- **Mental Health Concerns**: The pursuit of external affirmation can erode a stable self-concept. Users might feel compelled to align their offline behavior with their online persona – or vice versa – and experience stress when these two worlds clash.

Ultimately, social media platforms have become powerful co-authors of identity, blending personal aspirations with corporate algorithms that privilege engagement, sometimes at the expense of genuine selfhood.

17.3.3 AI and Personhood

17.3.3.1 Advanced AI and the Turing Test: Emulating Human Personalities or Creating "Digital Humans"

AI has progressed to a point where **large language models** and related systems can converse, write essays, and emulate specific writing styles. Some AIs can even generate **synthetic avatars**, complete with human-like facial expressions and voice patterns. This technological leap raises critical questions about "personhood":

- **Turing Test and Personhood**: The *Turing Test*, formulated by Alan Turing, posits that if a machine can converse in such a way that it's indistinguishable from a human, it demonstrates a form of intelligence. Today's AI chatbots and digital humans edge ever closer to that threshold, forcing us to revisit the test's implications: does passing the Turing Test imply a form of consciousness or genuine understanding, or is it simply "faking" humanness effectively?
- **Authenticity vs. Simulation**: If an AI can convincingly adopt an emotional tone or replicate a person's mannerisms, is it crossing over into "human-like" territory, or is it still merely mimicking?

17.3.3.2 Challenges to Uniqueness in an Era Where AI Can Mimic or Augment Human Traits

AI's increasing sophistication affects the concept of **individual uniqueness**:

- **Artistic and Creative Fields**: AI-generated art or literature can rival human creations, prompting debates about originality, authorship, and the value of human artistic flair.
- **Identity Confusion**: Deepfake technology can impersonate real individuals' voices or faces, complicating authenticity in both personal communication and public discourse.
- **Human-AI Collaboration**: As AI augments tasks once deemed exclusively human – like emotional support or creative brainstorming – we may find that our sense of personal identity, rooted in unique abilities and expressions, is shifting toward a co-creative paradigm with machines.

In this light, the question arises: If AI can replicate not only our tasks but also our emotional or intellectual "fingerprints," what does it mean to be distinctively human?

17.3.4 Biotech and Genetic Engineering

17.3.4.1 Altering Cognitive/Emotional Traits Biologically – Designer Babies and Social Stratification

The advent of gene-editing technologies (e.g., **CRISPR**) offers unprecedented control over our biological blueprint. While many applications focus on curing diseases, the possibility of **enhancement** looms large:

- **Cognitive and Emotional Upgrades**: Potentially modifying genes associated with memory, intelligence, or temperament.
- **Redefining "Normal"**: As gene editing becomes more accessible, the line between therapeutic intervention (e.g., eliminating hereditary illnesses) and enhancement (e.g., boosting certain traits) can become blurred. The threshold for what is considered "ordinary" could shift dramatically.

With these leaps come concerns about:

- **"Designer Babies"**: Parents seeking to ensure their children possess certain favored traits (height, intelligence, athletic prowess) could widen existing social inequalities, creating a divide between those who can afford genetic enhancements and those who cannot.
- **Ethical Debates**: Is it ethical to tamper with the human gene pool for performance gains? Could it erode the notion of "innate" human diversity or personal authenticity – replacing genetic "accidents" with curated traits?

Hence, biotechnology not only redefines personal identity but also raises thorny social questions about inequality and the moral status of editing life at its most fundamental level.

17.3.5 BCIs and Internal Manipulations

17.3.5.1 Potential for Direct Thought, Emotion, and Memory Modifications

BCIs, already used in experimental therapies for paralysis or severe neurological disorders, highlight the possibility of **manipulating brain states directly**:

- **Thought Implantation**: Although still largely speculative, research suggests BCIs might one day introduce or enhance specific knowledge patterns, short-circuiting the usual learning process.
- **Emotional Modulation**: Targeted neural stimulation could potentially treat mood disorders or help regulate stress and anxiety. This, in turn, raises deeper questions about personal agency if emotions can be "dialed up or down" on demand.

17.3.5.2 Networked Consciousness: Identity in a Shared Mindscape – and the Blurring of Reality vs. Simulation

As BCIs advance, **networked consciousness** becomes conceivable – multiple minds linked through shared neural signals:

- **Expanded Empathy or Collective Intelligence**: In positive scenarios, such connections might foster deep mutual understanding or accelerate group problem-solving.
- **Erosion of Autonomy**: If thought or emotional content can be broadcast, does the boundary of self remain firm? Are individuals at risk of losing unique perspectives to the collective "hive"?
- **Blurring Reality and Simulation**: If BCIs can directly manipulate sensory inputs and internal states, the lines between genuine experience and artificially induced sensation become increasingly blurred. What remains "real" when subjective reality can be engineered?

This prospect of merging internal worlds stretches identity beyond the solitary mind, demanding a fresh ethical and philosophical framework for personal autonomy in a potentially collective mental space.

17.3.6 Concluding Identity

In our hyper-technological era, **identity** is no longer merely a product of introspection or one's physical body. It is continuously shaped, extended, and even engineered by digital platforms, AI algorithms, genetic interventions, and neural interfaces. The interplay of these forces compels us to ask: **how do we define "self" when everything from cognition to genetics can be externally modified or curated?**

Key Reflection Points:

1. **Authenticity**: In a world where self-expression can be endlessly curated and innate traits can be modified, how do we define authenticity – and does the concept of a fixed, "real" self still hold meaning?
2. **Autonomy**: As we outsource mental tasks to AI, adapt our bodies genetically, or integrate our senses with BCIs, how do we preserve individual freedom of thought, choice, and emotional experience?
3. **New Horizons of Human Identity**: If we come to see ourselves as co-created by technology, will this expand our sense of collective responsibility and empathy, or will it erode traditional notions of individuality and personal accountability?

Each of these points resonates throughout the rest of our exploration. Having examined how technology reshapes perception and identity, we now turn to the broader **ethical landscape** that emerges. As personal agency, authenticity, and equality hang in the balance, we confront moral complexities that challenge traditional legal frameworks and philosophical assumptions – demanding that we evolve how we think about human life in the digital age.

THE FILTER BUBBLE DILEMMA

THINKING BOX

"Are we confining ourselves to echo chambers – digital fortresses of agreement? Breaking out might require new perspectives, and maybe **42** too."

17.4 THE ETHICS OF ENHANCEMENT

The Ethics of Enhancement
Change: As technology challenges our assumptions about mind, body, and agency, ethical considerations become paramount. From philosophical debates on free will and existentialism to the very real prospect of "neural malware," we find ourselves at the threshold of profound moral dilemmas.

17.4.1 Philosophical Underpinnings of Enhancement

17.4.1.1 Revisiting Free Will, Autonomy, and Human Nature in a World of Potential "Consciousness Hacking"

Technological breakthroughs – from gene editing to sophisticated BCIs – pose fundamental questions about **individual agency** and **the nature of the self**:

- **Free Will**: If our moods, decisions, or cognitive capacities can be directly modulated or predetermined, do we diminish our capacity for genuine choice? Alternatively, might these tools expand autonomy by alleviating conditions (e.g., severe depression, cognitive impairment) that constrain free will?
- **Human Nature**: Historically, many philosophical and religious traditions have treated "human nature" as either sacred or at least fixed. Enhancement technologies challenge this assumption, implying that traits like empathy, intelligence, or even aggression might be adjustable, thereby complicating notions of an "innate" self and raising questions about self-creation explored by **existentialist philosophers**. In existentialism, individuals define themselves through their choices and actions, suggesting that if our traits can be engineered, the lines between self-creation and technological shaping become blurred.

Consciousness hacking – the potential for manipulative alteration of one's mental states – underscores how our sense of agency can be threatened when technology plugs directly into the brain. The once-theoretical question of "What does it mean to choose?" gains urgent relevance in a world where external forces can literally "rewrite" our motivations or emotional inclinations.

17.4.1.2 Moral Bioenhancement: Raising Questions about Using Technology to Enhance Moral Capacities

Beyond cognitive and physical upgrades, some theorists propose **moral bioenhancement** – deploying interventions (pharmaceutical or neuro-technological) to bolster altruism, empathy, or impulse control. Proponents argue it could:

- **Reduce Harm**: Imagine a substantial decrease in violence if aggression or psychopathic tendencies could be mitigated.
- **Elevate Empathy**: Societies might become more cooperative if compassion or ethical reasoning can be systematically enhanced.

Yet serious concerns emerge:

- **Authenticity of Morality**: If moral behavior stems from technological implants rather than personal development or conscience, do we lose a vital dimension of moral freedom and growth?
- **The Slippery Slope Argument**: Even limited forms of moral enhancement might open the door to more radical, potentially coercive interventions in human morality. If society finds it acceptable to "dial up empathy," might it later justify overriding personal autonomy for "the greater good"?

Thus, while moral bioenhancement envisions a more benevolent humanity, it also highlights deeper questions about authenticity, consent, and the potential dangers of moral uniformity imposed by external means.

17.4.2 Regulatory Gaps and Security Implications

17.4.2.1 The Mismatch between Existing Data/Privacy Laws and Neural Data or Genetic Edits That Alter Identity

Legislators and policymakers frequently find themselves **outpaced** by rapidly evolving technologies:

- **Neural Data vs. Conventional Privacy**: Laws like GDPR or HIPAA assume data is external (e.g., medical files, server logs). But **BCIs** can capture far more intimate information – emotional states, inclinations, even subconscious associations – none of which is adequately covered by existing frameworks.
- **Genetic "Identity" Edits**: Altering someone's genome can reshape not just that individual but potentially their descendants. We lack well-defined legal precedents for intergenerational consent or liability, as well as for enforcing safety standards across gene-editing research labs.

This mismatch leaves significant gaps, where individuals may be vulnerable to exploitative or high-risk interventions that slip through oversight cracks. The challenge is further complicated by the need for **international cooperation** to regulate technologies that transcend national borders, ensuring global standards for safety and ethics.

17.4.1.2 BCIs, "Consciousness Firewalls," and Looming Scenarios Such as "Neural Malware" or Mass Psychological Manipulation

As neural interfaces become more advanced, **cybersecurity** morphs into **neurosecurity**:

- **Consciousness Firewalls**: Similar to antivirus software, BCIs might require firewalls to prevent unauthorized intrusions – ranging from memory manipulation to forced emotional states.
- **Neural Malware**: A hacked neural implant could be catastrophic, enabling someone to hijack a user's perception, implant false memories, or even disrupt motor control.
- **Mass Psychological Manipulation**: If entire populations adopt neural-based media or networked BCIs, malevolent actors could disseminate tailored misinformation straight into people's minds – undermining democratic processes and social trust on an unprecedented scale.

Such scenarios underscore the imperative for **robust legal and ethical frameworks** and the development of advanced technical safeguards – so that we do not inadvertently open the door to new forms of mental subjugation.

17.4.3 Consent, Equity, and Informed Choice

17.4.3.1 The Complexity of Genuinely Informed Consent for Internal or Collective Neural Modifications

Informed consent is a cornerstone of modern bioethics, but **enhancement technologies** make it far more complex:

- **Technical Opacity**: BCIs, gene edits, and AI-driven enhancements can be so advanced that even specialists debate long-term ramifications. Asking patients or users to fully understand these complexities can be unrealistic.
- **Collective Dimensions**: Some enhancements – like a networked BCI – may impact not just the individual but also the group (e.g., shared mindstates, potential infiltration). Consent thus must be considered at both individual and communal levels.

17.4.3.2 Addressing Social Inequality: Avoiding a Two-Tier Society and the Potential for Coercion

Enhancement technologies often entail **high initial costs**, risking the emergence of privileged "augmented elites." This could widen existing inequalities:

- **Wealth-Based Stratification**: People who can afford enhancements may reap cognitive, physical, or emotional advantages – leaving the less affluent at a structural disadvantage.
- **Compulsory Enhancement**: Even without explicit coercion, social and economic pressures could lead individuals to feel obliged to enhance themselves to remain competitive. This "soft coercion" might be especially pronounced in workplaces, educational settings, or military contexts.
- **Policy and Ethical Safeguards**: Strategies could include subsidizing medically necessary enhancements, regulating purely elective augmentations, or creating global guidelines for equitable access.

Ensuring fair distribution and avoiding a scenario where "baseline humans" are overshadowed by engineered counterparts becomes a central moral challenge in the enhancement era.

17.4.4 Toward Responsible Design

17.4.4.1 Guidelines for Transparency, Risk Assessment, User-Centric Development, and Ongoing Public Engagement

Responding to these ethical dilemmas requires a **proactive approach**:

1. **Transparency**: Developers and institutions must clearly articulate the capabilities, risks, and intended data usage of enhancement tools – no small print hidden behind jargon.
2. **Rigorous Risk Assessment**: Like clinical drug trials, enhancements (whether neural or genetic) should undergo extensive testing for unintended consequences, including long-term mental or biological side effects.
3. **User-Centric Design**: Individuals and communities affected by these technologies should be actively involved in their design and governance, not just treated as end-consumers.
4. **Dignity Preservation**: Tools must protect individual autonomy, ensuring people aren't coerced into using them or manipulated by hidden commercial or political agendas.
5. **Ongoing Dialogue and Public Engagement**: Responsible design is not a one-time blueprint but a continuous conversation, bringing together ethicists, policymakers, technologists, and the broader public to adapt as new developments arise.

17.4.4.2 The Potential Role of Open-Source Platforms and Collaborative Oversight to Prevent Monopolistic Control over the Mind

One proposed safeguard against unethical control is **open-source** and **collaboratively governed** enhancement platforms:

- **Open-Source Culture**: By making source code and research publicly available, peer communities can inspect for hidden manipulations or security flaws, fostering an environment of accountability.
- **Global Regulatory Bodies**: Similar to international protocols on nuclear or biological weapons, an international coalition could set standards for safe, ethical enhancement, imposing penalties for reckless or harmful deployments.
- **Public Stewardship**: When communities, not just corporations, guide research priorities, it reduces the risk of monopolies that might commodify or weaponize the mind.

By grounding technology in ethical principles from the outset, we mitigate the risk of harmful or exploitative usage once these tools become mainstream. This requires not only technical expertise but also **ongoing dialogue and public engagement** to ensure that ethical considerations remain at the forefront of development and deployment.

17.4.4.3 Summary of the Ethics of Enhancement

Emerging technologies, from genetic editing to BCIs, promise to reshape human capabilities – potentially curing diseases, boosting cognition, or even refining our moral inclinations. Yet they raise urgent ethical questions:

- **Philosophical Underpinnings**: Free will, autonomy, existentialist self-creation, and moral enhancement spark debates about the essence of human nature and the authenticity of "engineered" traits.
- **Regulatory and Security Challenges**: Existing frameworks often fail to address intimate neural data or intergenerational genetic edits. Neurosecurity threats – like "consciousness hacking" or "neural malware" – loom on the horizon, demanding robust safeguards.

- **Consent and Equity**: Informed choice becomes complex in a realm of sophisticated science, while wealth-based stratification or subtle coercion risks creating a two-tier society.
- **Responsible Design**: Transparency, risk assessment, and open-source collaboration offer pathways for ethically guided innovations – but require continuous public dialogue and vigilance.

Taken together, these questions remind us that as we expand the boundaries of human potential, we must do so with caution and a deeply rooted sense of responsibility – ensuring that the pursuit of "better minds" or "better bodies" does not come at the cost of dignity, equity, or the very freedom these technologies aim to enhance. **A nuanced and balanced approach is crucial**, one that recognizes both the immense benefits and the inherent risks of these transformative technologies and that keeps humanity's welfare squarely in view as we move forward.

17.5 REIMAGINING HUMAN CONNECTION IN A DIGITAL WORLD

Reimagining Human Connection in a Digital World

17.5.1 From Face-to-Face to Virtual Meetups

17.5.1.1 VR Gatherings, Telepresence, and Bridging Geographical Distance – But at What Cost to Tangible Intimacy?

For much of human history, meaningful connection has centered on physical co-presence: sharing the same space, reading subtle gestures, and feeling the warmth of another's presence. Today, digital platforms – ranging from telepresence robots to VR meeting spaces – offer new ways to "meet" that transcend geographic borders. Schools, workplaces, and social gatherings increasingly rely on virtual or semi-virtual solutions, sometimes out of necessity (e.g., global crises) and sometimes for sheer convenience:

1. **Virtual Reality (VR) Platforms**: Services like VRChat, AltspaceVR, or enterprise solutions like Horizon Workrooms allow participants to appear as avatars in shared digital environments. These settings can simulate eye contact and basic gesturing but often require headsets and controllers, limiting physical cues.
2. **Telepresence Robots**: Rolling robots equipped with screens let remote users appear "in-person" for events or work, bridging isolation for those who cannot physically attend.

This shift undeniably expands access, letting people collaborate or socialize across continents without the need for costly travel. Yet crucial dimensions of face-to-face contact are hard to replicate:

- **Nonverbal Communication**: While VR headsets and telepresence can approximate presence, they often lack the full spectrum of face-to-face cues – subtle facial microexpressions, body warmth, real-time tactile feedback, and myriad forms of unspoken body language that enrich human interaction. The intangible spark of closeness might dissipate without these signals.
- **Potential for New Nonverbal Forms**: On the flip side, virtual environments may introduce novel methods of nonverbal communication – avatar facial expressions, virtual gestures, or creative "emotes" not possible in the real world. These can foster playful bonding or unique shared experiences, though they also risk feeling artificial or disconnected from genuine emotional states.

Balancing these pros and cons, the question becomes whether virtual meetups can ever achieve the depth of in-person encounters – or whether they redefine intimacy altogether in ways we are only beginning to understand.

17.5.1.2 Risks of "Digital Empathy Deficits" When Face-to-Face Cues Are Minimized or Lost

Removing physical subtleties can foster miscommunications or "empathy deficits." Without the natural immediacy of hearing someone's voice reverberate in the same room, noticing nervous fidgeting, or sensing the energy of group laughter, emotional resonance may wane. Critically, **human empathy** often relies on visual, auditory, and even olfactory cues – channels that are heavily filtered or absent in virtual spaces. While future advancements (like haptic VR suits) might partially mitigate these losses, the gap between the digital and the tangible remains a pressing concern for human connectedness.

17.5.2 Enhancing Empathy in a Digital Age

17.5.2.1 VR Tools for Perspective-Taking, AI-Based Emotional Analytics, and Potential BCI Prototypes That Share Affective States

Not all technology is pulling us apart; many emerging tools explicitly aim to **cultivate empathy**:

- **VR Perspective-Taking**: Immersive simulations can place users in dramatically different life experiences – such as a refugee's daily challenges or navigating public infrastructure with a disability – often evoking stronger emotional responses than text or video. Studies suggest these experiences can foster sustained empathetic insights, prompting increased donations or advocacy.
- **AI Emotional Analytics**: Advanced algorithms can analyze voice tone, facial microexpressions, or text sentiment to gauge emotional states in real time. While this can feel invasive, it might also help people moderate arguments, manage stress, or empathize more effectively – **provided it is used ethically and consensually and that the AI systems themselves are free from bias.** Biased data sets could, for instance, misread emotional cues in certain cultural or demographic groups, perpetuating stereotypes or injustices.
- **BCI Prototypes for Sharing Affective States**: Experimental BCIs could someday broadcast rudimentary emotional signals directly, letting pairs or groups sense each other's levels of stress, excitement, or empathy. Though speculative, this could encourage deeper emotional synchronization in team settings – or, conversely, raise daunting privacy questions about forced emotional disclosure.

17.5.2.2 Could Technology Unify Us or Deepen Divides? Balancing Convenience with Genuine Emotional Depth

Even as these tools show promise, they risk fueling **new social inequalities** if high-end empathy tech is only accessible to affluent communities. Moreover, easy "empathy hacks" might lead people to bypass the sincere, sometimes difficult emotional work that forms truly deep human connections. If empathy becomes algorithmically generated, do we risk trivializing genuine care or oversimplifying complex emotional experiences?

In short, technology can **enhance** empathy, but only if carefully guided by ethical frameworks, cultural sensitivity, and ongoing vigilance about who has access – and how it is used.

17.5.3 Digital Communities and Spiritual Practices

17.5.3.1 Evolving Communal Rituals, Religious Gatherings, or Spiritual Practices in Immersive and "Neuro-Mediated" Contexts

Humans have historically gathered in physical spaces – temples, mosques, churches, and communal festivals – to celebrate shared beliefs and communal bonds. Today, digital transformations open up **new frontiers** for these traditionally place-based activities:

1. **Online Faith Communities**: Services stream religious gatherings for believers across the globe, sometimes in VR chapels. Members can interact, pray together, or participate in digital sacraments. This can offer solace to the homebound or diaspora communities.
2. **Immersive Spiritual Practices**: Beyond live streams, VR and neural feedback can guide group meditations or "spiritual pilgrimages" to simulated holy sites. Some platforms integrate biometric data – like heart rate or brainwaves – into communal rituals, creating a shared sense of interconnected transcendence.

While these innovations **support** community and spiritual engagement, they also alter the nature of communal rites that once hinged on physical presence and shared embodiment.

17.5.3.2 The Debate over "On-Demand Transcendence" and the Potential for Online Extremism

In parallel with such positive developments, critics raise two pressing concerns:

1. **"On-Demand Transcendence"**: Technological shortcuts to mystical or meditative states risk trivializing deeper existential pursuits. If powerful neural stimulation devices or sophisticated VR can induce euphoric or revelatory states, one might obtain spiritual "highs" without the reflective journey or ethical context historically embedded in religious traditions. This raises questions of whether these experiences are less authentic or morally shallow.
2. **Potential for Radicalization or Extremism**: While online faith communities and immersive spiritual gatherings can provide valuable connection and support, digital spaces also enable **echo chambers** that amplify extremist ideologies. Virtual groups might rapidly mobilize around dogmatic or conspiracy-laden narratives, spreading misinformation, demonizing out-groups, or recruiting vulnerable individuals. This underscores the fragile balance between nurturing global unity and inadvertently fostering isolationist or dangerous subcultures.

Thus, while technology can democratize spiritual access, it may also disrupt the communal checks and balances that have historically prevented many from sliding into harmful radical beliefs.

17.5.4 Concluding Human Connection

Technological innovation is **redefining how we gather, empathize, and seek meaning** in ways unthinkable just a generation ago. Telepresence, VR, and AI-driven tools can connect distant loved ones, foster empathy, and even reimagine sacred practices. Yet these same innovations threaten to displace the **rich subtlety** of unmediated human contact and might inadvertently breed shallow or extremist communal bonds if not carefully guided.

1. **Enhanced Empathy or Greater Isolation?** Immersive experiences can bring powerful emotional resonance, but the risk of superficial or forced empathy remains – particularly when emotional data streams bypass personal introspection or respect for privacy.

2. **Reshaping Communal Life**: Religious services, cultural celebrations, and even the quest for transcendence adapt to digital realms, both expanding outreach and risking a loss of tangible, ritualistic depth.
3. **Redefining Human Bonds**: As VR meetups, empathy-augmenting AI, and "on-demand transcendence" grow, we face urgent questions about authenticity, emotional well-being, and what truly binds communities together.

Our challenge lies in harnessing these digital frontiers so they foster genuine connection rather than mere **superficial interactions** – ensuring that while we gain new ways to come together, we do not lose the richly human depth that has always sustained our sense of belonging. Achieving that balance calls for careful, culturally informed design; ethical frameworks that encourage sincere, empathic engagement; and an unwavering commitment to keeping our humanity at the heart of all these technological transformations.

17.6 EMPOWERING THE TECHNOLOGICALLY MEDIATED MIND

Empowering the Technologically Mediated Mind

Transition: To thrive in this digital landscape – where AI, BCIs, and immersive platforms profoundly influence our thoughts, memories, and relationships – **proactive strategies are essential**. The goal is not to reject technology, but to engage it with deliberate, ethically informed practices that preserve autonomy, authenticity, and collective well-being.

17.6.1 Cognitive Checkpoints

17.6.1.1 Personal Journaling, Encryption/Time-Stamping of Memories, and "Reality Logs"

In a realm of deepfakes, AI-driven illusions, and neural manipulations, **cognitive checkpoints** can help anchor our personal reality:

1. **Journaling and Time-Stamping**
   - Maintaining a written or digital journal where key experiences – daily events, emotional responses, reflections – are recorded can provide a trustworthy "mirror" of one's life. Cryptographic time-stamps lend authenticity, making it harder for external parties (or ourselves) to alter the record retroactively.
2. **Encrypted Archives**
   - Storing photos, videos, or voice notes in encrypted formats ensures that even if malicious actors access them, tampering is more difficult. By relying on distributed or secure storage solutions, individuals can safeguard personal histories from unauthorized editing or deletion.
3. **Reality Logs as an "Internal Alarm System"**
   - For especially critical life events – e.g., legal agreements, medical appointments, or major decisions – some people keep "reality logs" that incorporate witness confirmations or third-party verifications. If memory or digital evidence is later called into question, these logs serve as evidence of what genuinely transpired.

17.6.1.2 Defending against Memory's Inherent Fallibility

By regularly updating these cognitive checkpoints and cross-referencing them with trusted sources, people can better defend against subtle rewrites or illusions that exploit memory's intrinsic malleability. However, it's important to acknowledge that even with these tools, **memory remains fallible**, and perfect recall is not guaranteed. The aim is to reduce vulnerabilities, not eliminate them entirely.

17.6.1.3 Verifying Experiences via Third-Party or Group Fact-Checking

Personal archives alone may not suffice when confronting large-scale misinformation campaigns or cunning psychological tactics:

- **Collaborative Validation**
  - Consulting friends, family, or reputable fact-checking communities can help discern whether a newly recalled event is shared and consistent among multiple witnesses.
- **Formal Fact-Checking Networks**
  - Online platforms dedicated to debunking hoaxes or verifying news can expose manipulated images, audio, or deepfake videos. Collective wisdom often surpasses individual scrutiny, especially in high-stakes scenarios (e.g., political manipulation or health misinformation).

In essence, cognitive checkpoints serve as **external anchors** that ground us in verifiable reality, buttressing us against psychological, neural, or digital distortions.

17.6.2 Meta-Cognitive Awareness Training

17.6.2.1 Spotting Mental Shifts, Emotional Manipulation, or Subtle Rewrites of Memory

While cognitive checkpoints provide **external** bulwarks, **meta-cognitive training** focuses on **internal** vigilance – developing a sharper self-awareness that can detect manipulative influences:

1. **Recognizing Shifting Moods and Beliefs**
   - We live in "manipulative landscapes": targeted ads, AI-powered persuasion, sensationalist news, or echo chambers that nudge our beliefs. By honing meta-cognition, individuals can notice abrupt changes – like a sudden intense desire or fear – and question if they originate from personal reasoning or external suggestion.
2. **Memory Dissonance**
   - Memory, being reconstructive, is prone to subtle rewrites. If a recollection contradicts what one's reality logs or peers confirm, a meta-cognitively aware person may pause to investigate rather than passively accept the memory as truth.

17.6.2.2 Mindfulness, Cognitive-Behavioral Skills, and Cooperative Validation Circles

Building meta-cognitive strength often involves structured approaches:

- **Mindfulness and Meditation**
  - Practicing daily mindfulness helps individuals observe their internal states – thoughts, emotions, bodily sensations – without immediate judgment. This can reveal unusual mental shifts prompted by manipulative content or technologies.
- **Cognitive-Behavioral Techniques**
  - Recognizing cognitive distortions (e.g., catastrophizing, confirmation bias) equips individuals to question sudden or extreme viewpoints, thus guarding mental autonomy.
- **Cooperative Validation Circles**
  - In small groups, participants openly discuss recent media exposures or emotional spikes. Sharing and reflecting in a supportive community can catch manipulative narratives before they crystallize.

17.6.2.3 Beware of "Weaponized" Meta-Cognition

While meta-cognition mostly serves as a defense against manipulation, it could also be co-opted for malicious ends (e.g., "weaponized meta-cognition," where refined psychological insights are used to exploit vulnerabilities or steer individuals toward disinformation). Staying aware of this risk is crucial – an informed and self-reflective society can reduce the potency of such exploitative tactics.

17.6.3 Neural Sanctuaries and Tech-Free Zones

17.6.3.1 Digital Sabbaths, "Off-Grid" Times/Places Free from Neural Interfaces

A **hyper-connected** lifestyle can leave little room for silence, introspection, or self-determined thought. **Neural sanctuaries** represent deliberate spaces and times when digital intrusion – particularly advanced BCI or neural device usage – is paused:

- **Periodic Digital Sabbaths**
  - Allocating a few hours, a day, or a weekend to unplug from notifications, social media, and neural interfaces breaks the cycle of continuous stimulation. These intervals restore mental equilibrium, spark creativity, and let emotions settle.
- **Off-Grid Retreats**
  - Physical locales – cabins, monasteries, or dedicated "digital detox" zones – can ban or strictly limit device usage, offering extended immersion in unfiltered, analog reality.

17.6.3.2 Preserving Selfhood beyond Constant Technological Filtering

By reclaiming pockets of time and space free from technology's omnipresence, we regain balance, ensuring our minds remain genuinely our own. However, it's also important to consider the potential **social implications** of disconnecting, as it can lead to isolation or exclusion in a world that increasingly relies on digital connectivity:

- **Societal Pressures**: In workplaces or social groups that expect immediate online availability, stepping "off-grid" may carry professional or relational costs.
- **Community Solutions**: Encouraging communal "tech-free" events – e.g., group retreats, local meetups – helps normalize the practice of unplugging without stigmatizing those who need or want such breaks.

Although removing oneself from digital channels can risk missing out on certain opportunities, the mental and emotional clarity gained may prove invaluable for fostering genuine identity and authentic relationships.

17.6.4 Societal-Level Security and Oversight

17.6.4.1 Independent Agencies Monitoring "Neural Rights," Ensuring Accountability for AI/BCI Deployments

Strategies at the individual level only go so far. **Societal infrastructures** must be in place to protect collective freedoms and prevent abuses:

- **Neural Rights Monitoring**
  - Specialized regulatory bodies could audit BCIs, neurofeedback devices, or AI-driven persuasion tools, certifying compliance with ethical standards – protecting users from invasive data mining or covert manipulation.
- **Ethical Committees and Transparency**
  - Just as medical research requires peer review and ethics boards, advanced AI/BCI initiatives might undergo formal oversight. These committees evaluate potential social impacts – ranging from privacy breaches to psychological harm – and demand transparency about data usage.

17.6.4.2 International Collaboration to Align Regulations, Prevent Global Abuses, and Support Equitable Access

In an interconnected era, aligning best practices and coordinating oversight is crucial to keep exploitation or mass manipulation at bay:

- **Global Regulatory Frameworks**
  - Similar to treaties governing nuclear and chemical weapons, global agreements could set forth legal and ethical guardrails around "neural rights," data ownership, and BCIs, preventing unscrupulous actors from relocating to laissez-faire jurisdictions.
- **Balancing Security with Individual Liberties**
  - Ensuring robust security and oversight can sometimes infringe on personal freedoms – through increased surveillance or mandated usage guidelines. Policymakers must strike a delicate balance, preserving privacy and liberty while mitigating large-scale neural threats.
- **Equitable Access**
  - Bodies like the WHO or UNESCO might champion universal benefits from neural or AI-driven medical interventions, preventing wealthy nations or corporations from hoarding breakthroughs. Equitable distribution not only fosters global stability but also harnesses diverse perspectives in tech advancement.

When international collaboration thrives, societies can more confidently adopt revolutionary technologies without sacrificing individual rights or global ethical standards.

17.6.5 Brief Summary of Proactive Strategies

Neural literacy, **cognitive checkpoints**, **meta-cognitive awareness**, and **neural sanctuaries** empower individuals to retain mental autonomy in the face of persuasive technologies and potential neural intrusions. Meanwhile, **societal-level security** – involving ethical oversight, independent monitoring bodies, and global cooperation – serves the collective good by upholding freedom, fairness, and accountability.

Together, these strategies form a **multilayered defense**, empowering individuals and societies to navigate the complexities of the technologically mediated mind. By cultivating neural literacy, practicing self-awareness, creating digital-free spaces, and demanding robust public oversight, **we can shape a future where human agency and authenticity not only survive but thrive** alongside the remarkable possibilities of AI, BCIs, and immersive platforms.

17.7 FROM GADGETS TO GENES: THE FUTURE OF HUMAN ENHANCEMENT

17.7.1 A Gradient of Transformative Tech

17.7.1.1 Wearables, VR/AR Gear, AI-Based Cognitive Tools, Nanotechnology, BCIs, and Genetic Editing

Human enhancement is multifaceted, spanning from simple, noninvasive gadgets to radical, body-altering interventions that challenge our notions of "natural" limits. These innovations exist on a **continuum**, reflecting different degrees of invasiveness, risk, and transformative impact:

1. **Wearables**
   - Fitness trackers, smartwatches, and smart glasses monitor health metrics, deliver reminders, and subtly reshape habits. While relatively unobtrusive, they still influence behavior and identity, promoting minor yet meaningful "upgrades" to daily life.

2. **VR/AR Gear**
   - Immersive headsets and augmented reality overlays alter perception, enabling users to explore digital environments, merge virtual objects with reality, and experience novel forms of social interaction. Though external, these devices can significantly shift cognitive processes, empathy, and situational awareness.
3. **AI-Based Cognitive Tools**
   - Voice assistants, AI-driven translators, and recommendation systems offload memory, provide instant knowledge, and can simulate creativity. This layer of augmentation – though still external – deeply extends our cognitive reach.
4. **Nanotechnology**
   - At a deeper level, nanotech envisions minuscule machines (nanobots) capable of delivering targeted drugs, repairing tissues at the cellular level, or augmenting biological functions. Although many applications remain experimental, nanotech could revolutionize medicine and enhancement by operating directly within the body.
5. **Brain-Computer Interfaces (BCIs)**
   - BCIs create direct communication channels between neural signals and digital systems, enabling advanced forms of control, sensory augmentation, and potentially emotional or cognitive modification. By intervening at the neuronal level, BCIs represent a leap toward integrating technology with the seat of consciousness.
6. **Genetic Editing**
   - At the most profound end of the spectrum, CRISPR-based tools allow editing of DNA to correct hereditary diseases or enhance traits. Genetic modifications – especially heritable ones – challenge our concepts of identity, procreation, and the essence of what it means to be human.

17.7.1.2 Incremental Changes That May Gradually Reshape Consciousness and Society

While the gap between fitness trackers and genetically engineered traits appears enormous, these enhancements often **unfold through incremental steps**. A society comfortable with AR overlays might later accept neural implants for improved memory and, from there, be more open to gene edits for preventing diseases – or even boosting intelligence. This gradual process can lead to the **"normalization" of enhancement**, where what was once considered radical becomes commonplace. Over time, accumulating small concessions can cumulatively transform consciousness, social structures, and cultural values in ways that may not be fully apparent until we have crossed an invisible threshold.

17.7.2 Choice, Diversity, and the Precautionary Principle

17.7.2.1 Emphasize the Right to Remain Unenhanced; Support Equitable Access for Those Who Choose Augmentation

An essential ethical cornerstone in discussions of enhancement is **freedom of choice**. Individuals should be able to **opt out** of these technologies without incurring stigma or disadvantage, while those who seek augmentation – be it AI-based assistants, neural implants, or gene edits – should have pathways to access them equitably. However, it is also crucial to consider **social pressure or perceived necessity** that may compel individuals toward enhancement, even in the absence of explicit mandates. This "soft coercion" might occur in workplaces, educational systems, or competitive environments, where refusing enhancement could put someone at a significant disadvantage.

- **Avoiding Two-Tier Societies**
  - If advanced augmentation remains costly or proprietary, society risks creating "enhanced elites," exacerbating socioeconomic divisions. Equitable policies can mitigate this by

ensuring medically necessary enhancements are subsidized or accessible, preventing rigid stratifications between the "augmented" and the "unenhanced."

- **Respecting Cognitive and Biological Diversity**
  - There is inherent value in the varied cognitive styles and physical capacities that define humanity. Wide adoption of certain enhancements (e.g., gene edits for uniform traits) could reduce diversity, potentially limiting resilience and creativity on a societal scale.

17.7.2.2 Invoke the Precautionary Principle – and Its Practical Challenges

The **precautionary principle** urges caution when scientific uncertainty mingles with high potential harm:

- **Balancing Risks and Benefits**
  - Whether gene editing or neural implants, robust evidence of safety and ethical viability should precede large-scale adoption. But **applying the precautionary principle in practice** is notoriously complex, forcing societies to weigh urgent potential benefits (e.g., curing genetic disorders) against long-term risks and deeply uncertain outcomes.
- **Unknown Ethical or Safety Territories**
  - Germline genetic modifications, for instance, can reverberate through future generations, carrying moral weight that transcends individual consent. Similarly, BCIs for emotional regulation must handle potential side effects, unintended dependencies, or latent vulnerabilities to hacking.

The principle does not reject innovation outright; rather, it insists on a **deliberate pace** and thorough oversight – recognizing that once certain enhancements become ubiquitous, undoing or regulating them belatedly may prove far more difficult.

17.7.3 Evolutionary Adaptation vs. Acceleration

17.7.3.1 Bergson's "Durée" vs. Technology's Exponential Growth: Reconciling Slow Biological Evolution with Rapid Commercial and Scientific Advances

Henri Bergson's notion of *durée* highlights the **continuity and organic flow** of living beings – an evolutionary process that unfolds slowly over millennia. In stark contrast, technological innovation often follows **exponential** trajectories:

- **Commercial Incentives**
  - Startups and multinational corporations race to release increasingly powerful devices or therapies. This accelerated timeline can leave policymakers, ethicists, and the general public scrambling to catch up.
- **Scientific Breakthroughs**
  - CRISPR gene editing, neural lace prototypes, or advanced nanotech can jump from concept to proof-of-concept in a matter of years or even months, rendering each previous moral stance or regulatory framework potentially obsolete.

17.7.3.2 Transhumanism: Embracing Expanded Capacities Yet Preserving Fundamental Human Dimensions

Supporters of rapid enhancement often align with **transhumanism**, a philosophical and social movement advocating the use of technology to **transcend human limitations** – eradicating diseases, extending lifespans, boosting cognition, and even exploring post-human states of existence. Proponents argue these expansions of our biological and mental faculties could lead to unprecedented problem-solving and artistic or experiential horizons. However, the question remains

whether we can manage these gains without losing irreplaceable aspects of humanity – such as the growth that comes from grappling with inherent limitations and the empathy forged by shared vulnerabilities.

17.7.4 Concluding the Spectrum

Only by weighing the **benefits, risks, and diverse trajectories** of enhancement can we chart a path that **balances innovation with caution and inclusivity**. The journey from simple wearables to advanced nanotech, neural interfaces, and gene editing will inevitably shape future societies, redefining everything from employment dynamics to personal identity.

1. **Sober Reflection and Ongoing Dialogue**
    - Technology evolves swiftly, so continuous ethical discourse – across scientific, public, and philosophical domains – ensures each new development undergoes rigorous scrutiny.
2. **Inclusive Policies and Ethical Oversight**
    - To avoid genetically or neurally privileged elites, frameworks should facilitate universal access to critical enhancements while respecting individuals' right to remain unaugmented.
3. **Cultural Openness to Diversity**
    - People will vary widely in their levels of acceptance or desire for enhancement, and it is crucial to safeguard the dignity and rights of those who prefer minimal augmentation alongside those who embrace it fully.

In short, the question is less about **whether** to enhance than **how** to enhance responsibly, ensuring that in broadening our capacities, we do not forsake the richness of human diversity or the subtle qualities that have long underpinned our moral and existential depth. By approaching enhancements along a well-considered gradient – mindful of social, ethical, and evolutionary implications – we can strive to **expand our horizons while safeguarding the very essence of what it means to be human.**

17.8 CONCLUSION: CRAFTING THE TECHNOLOGICALLY MEDIATED MIND

17.8.1 Recap and Reflection

17.8.1.1 Classical Philosophers as a Compass for Today's Technological Landscape

Throughout our exploration, the wisdom of **classical philosophers** has served as a compass, guiding us through the complex and often uncharted territory of today's technological landscape:

1. **Plato's Allegory of the Cave**
    - Ancient cautionary tales about mistaking shadows for reality find new relevance in an era of immersive VR, deepfakes, and algorithmic echo chambers. Plato's insight on illusion reminds us that vigilance is essential in a world where external influences can shape our perception of truth.
2. **Descartes' Cogito**
    - *"I think, therefore I am"* placed consciousness at the core of selfhood. Now, with BCIs and AI-driven persuasion, the once-clear boundary between mind and machine blurs, challenging Descartes' dualism and prompting us to reconsider the very nature of being human.
3. **Berkeley's Idealism**
    - "*Esse est percipi*" – to be is to be perceived – resonates with personalization algorithms and digital curation. Our realities increasingly hinge on what is served to us by

AI filters, social feeds, and neural implants, raising questions of objectivity and subjectivity in a "post-truth" era.

4. **Merleau-Ponty's Phenomenology**
    - Emphasizing the primacy of embodied experience, phenomenology highlights how crucial sensory and bodily grounding is for self-awareness. As VR, AR, and BCIs enable forms of disembodiment or re-embodiment, the tension between lived physicality and digital immersion becomes ever more pronounced.

Together, these philosophical lenses illustrate how **timeless inquiries into truth**, **self**, and **perception** continue to illuminate our present challenges and future horizons.

17.8.1.2 The Importance of New Ethical Frameworks and Mindful Practices

Confronting **neural manipulation**, **memory distortion**, and **emergent social complexities** – from AI-generated misinformation to hyper-personalized persuasion – demands updated ethical frameworks and deliberate personal habits:

- **Adaptive Laws and Codes of Conduct**
    - We need evolving guidelines to govern neural data usage, gene-editing implications, and AI-driven social engineering. **Interdisciplinary collaboration** across philosophy, law, computer science, neuroscience, and social sciences is crucial to ensure these frameworks are comprehensive and effective.
- **Cultural Awareness and Ongoing Self-Examination**
    - Mindful techniques, meta-cognitive training, and community dialogues serve as bulwarks against manipulative tech. They remind us that the human mind, though adaptable, requires conscious stewardship to maintain integrity and autonomy in increasingly immersive environments.

In essence, bridging ancient wisdom with cutting-edge innovation highlights the **fundamental tension** of our era: how to embrace unprecedented capabilities without compromising the virtues – curiosity, empathy, and creativity – that define our humanity.

17.8.2 Proactive Strategies and Final Thought

17.8.2.1 A Toolkit of Empowering Measures

The preceding chapters have equipped us with a **toolkit of proactive strategies** – measures designed to empower individuals and societies as our minds and cultures evolve alongside technology:

- **Neural Literacy**
    - Educating the public about how AI, VR/AR, BCIs, and genetic edits shape cognition and identity – helping people navigate manipulative content and understand the ethical stakes.
- **Cognitive Checkpoints**
    - Maintaining encrypted "reality logs" or time-stamped journals to verify experiences, guarding against tampered memories or deepfake illusions.
- **Meta-Cognitive Awareness**
    - Cultivating self-awareness to detect subtle emotional manipulation, memory rewrites, or persuasive tactics embedded in everyday apps and digital landscapes.
- **Neural Sanctuaries and Tech-Free Zones**
    - Carving out intentional breaks from digital or neural interfaces – opportunities to reconnect with unmediated experiences, fostering mental resilience and creativity.

- **Robust Regulations and Ethical Oversight**
  - Encouraging policies, agencies, and international collaborations that protect neural rights, ensure BCIs adhere to safety and privacy standards, and promote inclusive access.
- **The Precautionary Principle**
  - Proceeding cautiously with powerful interventions like gene editing or immersive BCIs, balancing potential benefits against unknown long-term risks.

Taken together, these strategies embody the principle that **technology is neither inherently benevolent nor malevolent**; its impact on the human condition depends entirely on the choices we make. By **consciously shaping the technologically mediated mind**, we can ensure these powerful tools serve humanity's highest aspirations, amplifying our potential while safeguarding our core values.

17.8.3 Looking Ahead

17.8.3.1 Ongoing Dialogue, Research, and Personal Responsibility

We stand on the cusp of breakthroughs that may **radically redefine reality, identity, and social interaction**. Readers are encouraged to engage in critical discourse, participate in responsible innovation, and support community initiatives. But perhaps most importantly, **each individual must cultivate a sense of personal responsibility** for navigating this complex landscape, making informed choices about their own relationship with technology:

- **Collective Engagement**
  - Organized forums, reading circles, hackathons, and policy workshops can channel diverse voices into shaping ethical guidelines and innovation trajectories.
- **Responsible Innovation**
  - Researchers, developers, entrepreneurs, and policymakers can embed "human-centric" design principles, ensuring that new products or therapies prioritize well-being, dignity, and fairness.
- **Community Initiatives**
  - Local projects – from digital literacy classes to neuro-rights advocacy – can ground global transformations in empathy, cultural sensitivity, and inclusive collaboration.

17.8.3.2 Future Explorations: AI Ethics, Digital Minimalism, Neuro-Rights

Those inclined to delve deeper will find **related fields** and movements that offer richer paths for inquiry and action:

- **AI Ethics and Accountability**
  - As AI's influence expands, so does the need for transparent, bias-free algorithms and robust methods to hold developers and deployers accountable.
- **Digital Minimalism**
  - Movements advocating mindful tech usage – limiting screen time and minimizing digital clutter – to preserve mental clarity, family bonds, and genuine social connections.
- **Neuro-Rights**
  - Ongoing efforts to establish legal protections for freedom of thought, privacy of mental data, and the integrity of cognitive processes, particularly in the face of BCIs and neural data harvesting.

With these themes in mind, our collective task becomes clear: **to craft a technologically mediated mind that enhances our capabilities without sacrificing the essential qualities – curiosity,**

empathy, creativity, and resilience – that define us as human. By remaining engaged, informed, and ethically vigilant, we can navigate this transformative era not as passive recipients of technological change, but as **active co-creators** of a future where human flourishing remains the ultimate measure of progress.

THE ATTENTION ECONOMY CONUNDRUM

THINKING BOX

"Who profits from our fragmented focus, and at what cost to our mindfulness? Like **42**, the full impact may be bigger than we suspect."

18 The Modern Society OS
A Holistic Governance Framework

Core Principles (Kernel):

1. **Popper's Open Society**: Transparency, critique, and iterative reform prevent authoritarian stagnation.
2. **Rawls's Veil of Ignorance**: Ensures policies are **fair and just**, prioritizing the least advantaged.

Core Modules (Governance Building Blocks):

- **Procedural Justice**: Transparent policymaking aligned with fairness.
- **Decentralization Engine**: Local autonomy with checks on harmful externalities.
- **Freedom Manager**: Balances **individual liberties and collective welfare**.
- **Communicative Interface**: Fosters **evidence-based public discourse**.
- **Power-Structures Monitor**: Exposes **hidden biases and systemic exploitation**.
- **Capabilities and Equality**: Ensures **real opportunities for all citizens**.
- **Antifragility Subsystem**: Uses stressors to **enhance societal resilience**.

Expanded Layers for Modern Challenges:

- **Technology Governance, Psychological Well-being, Global Coordination, Knowledge Management, Civic Education, and Innovation Incubators** enhance adaptability.

Key Takeaway:

The **Modern Society OS blends philosophical rigor with dynamic policy mechanisms**, creating a governance system that is **transparent, just, and continuously improving**.

18.1 INTRODUCTION: BRIDGING TECHNOLOGY, COMPLEXITY, AND GOVERNANCE

18.1.1 A Timely Synthesis

We find ourselves in a world where breakthroughs in artificial intelligence, global challenges, and shifting social landscapes converge at dizzying speed. This chapter offers a **systemic governance framework** – the "Modern Society OS" – that unites philosophical principles, technological insights, and institutional design into a cohesive, adaptive approach to modern governance.

You've already explored how **complexity**, **resilience**, and **ethical thinking** can shape our understanding of societal systems. Now, we'll show how these threads can be woven into a **practical model** that:

- **Learns** from crises rather than merely surviving them
- **Protects** core values like freedom, fairness, and well-being despite rapid shifts
- **Remains** open to critique and anchored in ethical principles

The resulting blueprint is **modular, iterative, and philosophically grounded** – avoiding both **utopian idealism** and **rigid technocracy**. Each component (or "module") functions independently

DOI: 10.1201/9781003640288-23

yet must also integrate seamlessly into the larger architecture, ensuring that governance can evolve over time without sacrificing consistency or legitimacy.

18.1.2 Why an "Operating System" Metaphor?

Software architecture offers a powerful analogy for governance in a fast-changing environment:

1. **Adaptive Modularity**
   - In software, an operating system (OS) is built with separate processes, drivers, and services that can be updated or replaced without dismantling the entire system.
   - Likewise, a society can benefit from modular institutions: each policy or agency can be **tuned or reformed** without unraveling the rest of the governance structure.
2. **Core Kernel of Values**
   - Just as an OS kernel manages fundamental system resources and enforces basic constraints, a society needs a stable root of **guiding principles** – in this case, inspired by Karl Popper's open society and John Rawls's Veil of Ignorance.
   - These bedrock values ensure that no single faction monopolizes power and that public decisions remain aligned with fairness.
3. **Well-Defined Interfaces**
   - Operating systems rely on **APIs** to enable different components to communicate effectively. In governance, "interfaces" might include **transparent policy documentation**, **public consultations**, or **formal coordination mechanisms**.
   - Such interfaces ensure that modules (e.g., local authorities, national agencies, or citizen councils) collaborate on cross-cutting issues without duplicating efforts or fostering hidden power struggles.
4. **Continuous Updates**
   - An OS must be regularly **patched** and **upgraded** to address security threats and evolving user needs. Similarly, society faces evolving challenges – new technologies, ecological disruptions, and social movements – that demand **real-time adaptation**.
   - Crucially, the goal is **not** to create a static, perfect system but rather a **dynamic one** that can learn, iterate, and improve. This emphasis on updates recognizes that **no final form** of society exists; progress lies in sustained evolution.
5. **Avoiding Fixed Pictures**
   - A key lesson from software development is the danger of "feature creep" or unwieldy, **overly rigid designs** that stifle innovation. The same principle applies to governance: pursuing a **fixed utopia** can make us blind to unforeseen challenges.
   - Embracing updates as an ongoing principle – and staying open to new insights – helps prevent ossification and ensures the system remains responsive to emerging realities.

Through this metaphor, governance emerges as a **living system** rather than a static blueprint. By melding **philosophical rigor** with **technological adaptability**, the Modern Society OS seeks to meet the demands of a rapidly shifting world – without sacrificing the foundational values that keep societies fair and cohesive.

18.1.3 Chapter Overview

- **Section II** focuses on the **kernel**: explaining why **Popper's open society** and **Rawls's Veil of Ignorance** serve as non-negotiable cornerstones of the Modern Society OS.
- **Section III** explores the **seven core modules** – analogous to essential processes in an OS:
  - i. **Procedural Justice**
  - ii. **Decentralization**

iii. **Freedom Manager**
iv. **Communicative Interface**
v. **Power-Structures Monitor**
vi. **Capabilities and Equality**
vii. **Antifragility**

- **Section IV** then introduces **additional layers** to address broader challenges such as Technology Governance, Psychological Well-being, Global Coordination, Knowledge Management, Civic Education, and the Innovation Incubator.
- **Section V** provides a **practical roadmap** for implementing, testing, and refining the Modern Society OS, linking back to major themes in this book – **cognitive biases, resilience, and ethical frameworks**.

By placing this chapter toward the end, readers see how earlier discussions on human cognition, moral responsibility, and complex system design culminate in a **coherent, modular, and adaptive** vision for just governance.

18.2 FOUNDATIONAL PILLARS: AN OPEN, FAIR, AND ADAPTIVE KERNEL

18.2.1 Popperian Open Society

A robust governance system demands **transparency, critique, and reform** – principles championed by Karl Popper. In the Modern Society OS, these values serve as an **immutable kernel**, acting as both a safeguard against dogmatic control and a catalyst for ongoing improvement.

1. **Transparency and Critical Inquiry**
Rather than hiding decisions, an open society **exposes** them to analysis. This spotlight quickly reveals flaws or outdated policies. By embedding transparency into every layer – from legislative processes to institutional oversight – the system remains flexible and able to self-correct under changing conditions.
2. **Reform as a Norm, Not an Exception**
Popper warns against **authoritarian stagnation**, where unquestioned leaders or ideologies become entrenched. Regular updates demonstrate a system that learns from data, evolving moral insights, and citizens' experiences. In an open society, **updating** policies isn't failure – it's evidence of healthy responsiveness.
3. **Continuous Inquiry as "Debugging"**
The antidote to dogmatism is **asking questions**, inviting criticism, and refusing to treat any policy as final. This culture of ongoing critique acts as a form of **"debugging"** for the Modern Society OS, constantly testing its "code" (policies and institutions) for hidden flaws or vulnerabilities. It extends beyond government offices, encouraging academics, journalists, civil society, and citizens to contribute to this perpetual improvement loop.

In essence, the Popperian Open Society kernel ensures that governance remains dynamic, resilient, and accountable through transparency, critical debate, and a commitment to continuous reform.

18.2.2 Rawls's Veil of Ignorance

If Popper's openness safeguards against stagnation and dogma, then **John Rawls's Veil of Ignorance** ensures that policies remain **fundamentally fair**. Rather than allowing social structures to favor entrenched interests, the Veil imposes an **ongoing moral perspective** – one that asks, "Would I endorse this if I knew nothing of my own position in society?"

1. **Ongoing Moral Perspective**
   Under the Veil of Ignorance, no one knows their future wealth, status, or talents. From this vantage point, people tend to choose principles that **benefit everyone**, especially those who might be most disadvantaged. This approach prevents biased or self-serving policies that favor only the privileged.
2. **Not a One-Time Event**
   Some treat the Veil of Ignorance as relevant only at a society's founding moment. In the Modern Society OS, however, it's an **ongoing ethic**, shaping how every new law or institution is designed, reviewed, and reformed. If a policy perpetuates inequality or primarily serves a powerful few, it flunks the Veil's fairness test.
3. **Focus on the Least Advantaged**
   The Veil nudges policymakers to view social arrangements from the standpoint of those with the fewest resources or opportunities. By centering the **least advantaged**, the system counters any drift toward elitism or discrimination, ensuring that vulnerable voices are neither overlooked nor exploited.

By adopting the Veil of Ignorance as an ongoing moral perspective, the Modern Society OS prioritizes fairness and ensures that policies are designed with the interests of all members of society in mind, especially the least advantaged.

18.2.3 Creating Checks, Balances, and Ethical Governance

Together, **Popper's Open Society** and **Rawls's Veil of Ignorance** form the **ethical kernel** of the Modern Society OS. Their interplay guarantees:

- **Adaptability**: Open critique (Popper) prevents stagnation, while the Veil (Rawls) keeps reforms aligned with social justice.
- **Inclusivity**: Continuous inquiry ensures no policy is immune from scrutiny, and an ongoing moral perspective ensures no group is left behind.
- **Ethical Governance**: Fairness isn't an afterthought; it's woven into every institution, requiring transparency, inclusion, and compassion toward the least advantaged.

By anchoring every module to these two pillars, the Modern Society OS balances **freedom** and **equity**, **experimentation** and **protection**, **local autonomy**, and **universal rights** – ultimately cultivating a governance model that remains agile amid evolving challenges while retaining its core moral compass.

THE MEANING CRISIS

THINKING BOX

"How do we rekindle purpose in a world that worships data and commerce? Possibly the real meaning behind it all – like **42** – defies easy definitions."

18.3 CORE MODULES: BUILDING BLOCKS OF A MODERN SOCIETY OS

In the Modern Society OS, **seven foundational modules** work in concert to address the diverse complexities of modern governance. Each module has a distinct purpose and set of features, yet they all collaborate through **explicit interfaces** and **shared protocols** – resolving conflicts and ensuring that no single part of the system acts in isolation. The true strength of the OS lies not in the individual modules themselves, but in their **dynamic interactions** and the mechanisms for handling potential disagreements. Think of these modules as specialized "programs" within an OS, all governed by the kernel's ethical constraints (Popper's open society and Rawls's Veil of Ignorance).

Below, each module is presented with:

- **Purpose** – the core function it serves
- **Core Features** – its primary tools or mechanisms
- **Inter-Module Connection** – how it interacts with others or addresses conflicts
- **Summary** – a concise statement of its role in the overall system

A visual table at the end provides a **top-level snapshot** of these modules, their key features, and how they overlap.

18.3.1 Procedural Justice Module (Rawls)

18.3.1.1 Purpose

Ensures **transparent, impartial governance** consistent with the Veil of Ignorance.

18.3.1.2 Core Features

1. **Transparent Governance API**
   - Legislations, executive orders, and judicial rulings are publicly documented in a "governance repository."
   - Citizens can see why a policy was introduced, who proposed it, and how votes or rulings occurred, enabling them to evaluate fairness.
2. **Universal Acceptance Test**
   - Every new policy undergoes the question: "Would I find this just if I were among the least advantaged?"
   - Institutionalizes Rawls's principle that policymaking must remain sensitive to the most vulnerable segments of society.
3. **Conflict Resolution**
   - Outlines a formal process to **mediate disputes** – for instance, when the **Freedom Manager** (individual liberties) clashes with **Capabilities and Equality** (collective welfare).
   - Ensures relevant modules (like Decentralization Engine or Antifragility Subsystem) can offer input, but a final judgment is reached transparently and with the Veil of Ignorance in mind.

18.3.1.3 Inter-Module Connection

- **Communicative Interface**: Provides a public forum for debate, ensuring citizens and experts can scrutinize and comment on legislative proposals.
- **Power-Structures Monitor**: Flags hidden biases or inequities that the Procedural Justice Module must address.
- **Example Conflict**: A proposed law limiting certain economic freedoms might protect collective well-being. The Procedural Justice Module convenes an open review, weighing arguments from the **Freedom Manager** (individual rights) and **Capabilities and Equality** (shared resources) to craft a balanced judgment.

18.3.1.4 Summary

By grounding decisions in Rawlsian fairness and transparent processes, the Procedural Justice Module keeps societal rules **open to critique** and aligned with **ethical treatment of all**, especially those at a disadvantage.

18.3.2 Decentralization Engine (Hayek)

18.3.2.1 Purpose

Distributes decision-making across local entities, enabling **innovation** and **rapid adaptation**.

18.3.2.2 Core Features

1. **Localized Autonomy**
   - Communities function like "microservices," tailoring policies to local culture, geography, or economic conditions.
   - Fosters creative solutions and policy variations that can be scaled or adapted by other regions.
2. **Feedback Loops**
   - Small-scale pilots allow localities to test innovative approaches without risking nationwide disruption.
   - Successful policies are "merged" into the broader system, much like software updates that propagate from proven features to the main code.
3. **Protocols to Avoid Negative Externalities**
   - **Common-Resource Coordination**: Establishes interregional agreements on critical shared resources (e.g., water basins, forests, and energy grids). Local authorities can't unilaterally exploit these without multiparty consent.
   - **Environmental and Social Impact Checks**: Sets minimum regulations across local boundaries to prevent pollution, overexploitation, or social harm. Mechanisms might include:
     - **Interregional Councils**: Representatives from neighboring regions meet regularly to discuss cross-boundary issues (e.g., wildlife corridors, air and water quality standards).
     - **Monitoring and Enforcement**: A specialized oversight body – working with the Power-Structures Monitor – tracks compliance, levies penalties for violations, and resolves disputes.
     - **Data-Sharing Platforms**: A digital interface for real-time or periodic updates on resource usage and environmental health, accessible to local communities.

18.3.2.3 Inter-Module Connection

- **Procedural Justice**: Local decisions must comply with fundamental fairness; the module can intervene if local autonomy undermines universal rights.
- **Example Conflict**: A region wants to harvest a forest aggressively for short-term economic gain, but the **Antifragility Subsystem** and **Capabilities Layer** worry about future scarcity. The Decentralization Engine's "impact checks" require a cross-regional review, forwarding the dispute to **Procedural Justice** if no consensus is reached.

18.3.2.4 Summary

By **distributing authority** yet imposing robust checks on harmful spillovers, the Decentralization Engine fuels local creativity while ensuring that no community's actions jeopardize shared resources or the broader social fabric.

18.3.3 Freedom Manager (Berlin)

18.3.3.1 Purpose

Balances **negative freedom** (freedom from interference) and **positive freedom** (practical capacity to act).

18.3.3.2 Core Features

1. **Freedom Toggles**
   - Analyzes policies to see if they excessively restrict free speech, assembly, or thought.
   - Also checks if government inaction undermines essential support (e.g., healthcare, education) that enables real freedom in practice.
2. **Contextual Adjustment**
   - In emergencies (e.g., pandemics, severe crises), temporary measures can limit certain liberties.
   - These constraints must revert once the crisis ends, preventing "emergency powers" from becoming permanent.
3. **Veil Compliance**
   - Evaluates whether a freedom-limiting policy stands up to an **impartial moral test**, avoiding selective or discriminatory application.

18.3.3.3 Inter-Module Connection

- **Capabilities and Equality**: Some collective-welfare measures require redistributive policies that may limit personal freedoms; the Freedom Manager checks if limits are justified.
- **Example Conflict**: A local plan bans certain protests under "communal safety." The Freedom Manager scrutinizes necessity and proportionality; if it's too restrictive, the **Procedural Justice** Module mediates.

18.3.3.4 Summary

Freedom Manager prevents society from drifting into tyranny or social neglect by ensuring **individual rights** and **collective obligations** remain in careful equilibrium.

18.3.4 Communicative Interface (Habermas)

18.3.4.1 Purpose

Facilitates **transparent, evidence-based discourse** and consensus-building, reflecting Habermas's vision of **legitimate governance** through public debate.

18.3.4.2 Core Features

1. **Power-Free Forums**
   - Public assemblies, digital platforms, and other venues let citizens propose, debate, and refine policy ideas.
   - Encourages inclusive participation, reducing the pitfalls of top-down decisions.
2. **Consensus Protocol**
   - Seeks broad agreement rather than imposing mandates, ensuring that resulting policies carry **both** legitimacy and endurance.
   - Offers fallback procedures if consensus is unattainable, preventing deadlock.
3. **Expert Review Panels**
   - Integrates specialized knowledge (e.g., AI, environment, and economics) to guide policymaking.
   - Maintains **public accountability** by publishing expert findings transparently.

18.3.4.3 Inter-Module Connection

- **Civic Education**: Prepares citizens to evaluate expert analyses and engage in meaningful policy discussions.
- **Example Conflict**: The **Tech Governance Layer** raises concerns about new AI legislation. The Communicative Interface organizes debates, with experts outlining benefits (innovation, efficiency) vs. risks (surveillance, job loss). Citizens and modules like **Freedom Manager** or **Procedural Justice** evaluate fairness and impact.

18.3.4.4 Summary

By merging expert input with **open, participatory dialogue**, the Communicative Interface ensures policies are both **knowledge-based** and **democratically legitimate**.

18.3.5 Power-Structures Monitor (Foucault)

18.3.5.1 Purpose

Identifies **covert hierarchies** or exploitative practices, guarding against hidden power grabs and systemic imbalances.

18.3.5.2 Core Features

1. **Systemic Audits**
   - Independent watchdogs regularly review institutions, corporations, and policy outcomes.
   - Investigates undue influence, nepotism, or behind-the-scenes lobbying.
2. **Exploit Scanner**
   - Flags legislative or corporate loopholes that disproportionately benefit certain elites.
   - Functions like a "vulnerability scanner," diagnosing power distortions before they become entrenched.
3. **Mandatory Correction**
   - Exposes violations publicly, requiring policy revisions or punitive measures.
   - Ensures discovered abuses aren't allowed to persist unchecked.

18.3.5.3 Inter-Module Connection

- **Knowledge Management**: Transparent records help the Monitor detect anomalies (e.g., suspicious contracts, disproportionate campaign donations).
- **Procedural Justice**: Takes action if hidden power structures violate fairness.
- **Example Conflict**: A local corporation wields outsized influence over political decisions. The Monitor detects this dynamic, triggers a mandatory review, and the Procedural Justice Module may revoke or limit the corporation's undue privileges.

18.3.5.4 Synopsis

By routinely scanning for **invisible power distortions**, the Power-Structures Monitor ensures that no faction stealthily consolidates control, preserving the OS's ethical commitment to **equity and openness**.

18.3.6 Capabilities and Equality Layer (Sen/Nussbaum)

18.3.6.1 Purpose

Secures **real opportunities** for all citizens – beyond mere formal rights – to foster genuine well-being and social participation.

18.3.6.2 Core Features

1. **Needs Assessment**
   - Collects data on health, education, income, and other indicators to pinpoint underserved communities.
   - Guides targeted interventions (e.g., infrastructure, school funding, healthcare expansion).
2. **Resource Allocation Protocol**
   - Aligns with the Veil of Ignorance: distribution of public goods should not systematically disadvantage any group.
   - Prioritizes those facing structural obstacles, ensuring a fair shot at life's opportunities.
3. **Periodic Reviews**
   - Employs iterative "capabilities audits" to see if interventions effectively uplift vulnerable populations.
   - Updates policies or funding levels as conditions shift – akin to version updates in software.

18.3.6.3 Inter-Module Connection

- **Freedom Manager**: Collective-welfare measures (like universal basic income) can constrain certain freedoms, prompting the question, "Is this proportionate?"
- **Example Conflict**: A local autonomy measure slashes taxes, undermining public hospitals serving low-income groups. The Capabilities Layer flags the issue, prompting **Procedural Justice** to reconcile autonomy with equitable access to healthcare.

18.3.6.4 Synopsis

By focusing on tangible well-being, the Capabilities and Equality Layer keeps governance from paying mere lip service to rights, ensuring **actual life chances** improve for everyone – especially the least advantaged.

18.3.7 Antifragility Subsystem (Taleb)

18.3.7.1 Purpose

Fosters **resilience** by learning from stressors or shocks, rather than merely enduring them – mirroring Nassim Nicholas Taleb's concept of antifragility.

18.3.7.2 Core Features

1. **Redundancy and Backups**
   - Overlapping governance levels (local, regional, and national) ensure no single point of failure.
   - Encourages multilayered preparedness, so if one agency falters, another can step in.
2. **Rapid-Fail Feedback**
   - Quick detection of policy missteps allows immediate correction, preventing minor issues from snowballing into major crises.
   - Transparency in acknowledging errors fosters public trust, showing the system can adapt without cover-ups.
3. **Volatility Embracer**
   - Encourages small-scale pilots, harnessing the **benefits of trial-and-error** while limiting large-scale fallout.
   - Successful experiments are scaled, while unsuccessful ones become learning experiences rather than system-wide disasters.

18.3.7.3 Inter-Module Connection

- **Decentralization Engine**: Local experimentation dovetails with antifragility, as communities pilot new ideas. The **Procedural Justice** Module ensures fairness in risk-taking.
- **Example Conflict**: A region enacts an aggressive policy that inadvertently cuts mental health resources. Rapid-fail feedback flags the issue, prompting the **Capabilities Layer** and Procedural Justice to step in and correct course.

18.3.7.4 Synopsis

By **embracing uncertainty** and turning crises into opportunities for growth, the Antifragility Subsystem prevents stagnation, using setbacks as **stepping stones** to improvement.

18.3.7.5 Overview of the Core Modules

Below is a **sample table** summarizing each module's purpose, core features, and main overlaps.

Module	Purpose	Key Features	Inter-Module Overlaps
Procedural Justice	Transparent, impartial governance (Rawls)	• Governance API • Universal Acceptance Test • Conflict Resolution	• Communicative Interface for open policy debates • Power-Structures Monitor for hidden biases
Decentralization Engine	Distributed decision-making (Hayek)	• Localized Autonomy • Feedback Loops • Negative Externalities Protocols	• Procedural Justice if local policy violates fairness • Antifragility for safe, small-scale experiments
Freedom Manager	Balances negative and positive freedoms (Berlin)	• Freedom Toggles • Contextual Adjustment • Veil Compliance	• Capabilities and Equality (where welfare measures limit personal freedoms) • Procedural Justice for final arbitration
Communicative Interface	Evidence-based public debate (Habermas)	• Power-Free Forums • Consensus Protocol • Expert Review Panels	• Civic Education ensures informed citizenry • Freedoms Manager protects speech rights in public forums
Power-Structures Monitor	Detects covert hierarchies (Foucault)	• Systemic Audits • Exploit Scanner • Mandatory Correction	• Knowledge Management for transparent data • Procedural Justice implements corrective actions
Capabilities and Equality	Real opportunities for all (Sen/ Nussbaum)	• Needs Assessment • Resource Allocation Protocol • Periodic Reviews	• Freedom Manager (tension with redistributive policies) • Procedural Justice if local autonomy undermines essential public services
Antifragility Subsystem	Learns from shocks, not just endures (Taleb)	• Redundancy and Backups • Rapid-Fail Feedback • Volatility Embracer	• Decentralization fosters safe local experiments • Procedural Justice ensures fairness in risk-taking

18.3.7.6 Concluding Note on the Core Modules

These seven modules – **Procedural Justice**, **Decentralization**, **Freedom Manager**, **Communicative Interface**, **Power-Structures Monitor**, **Capabilities and Equality**, and **Antifragility** – work in tandem to uphold the **ethical kernel** of an open, fair, and adaptive society. Each module integrates with others through **shared protocols** and **structured conflict-resolution pathways**, much like different processes in an OS collaborating under a unifying kernel. By structuring governance in **modular**, **iterative**, and **transparent** ways, the Modern Society OS maintains flexibility under stress and remains anchored in moral principles, ensuring that progress truly benefits all.

18.4 EXPANDED LAYERS: ADDRESSING MODERN CHALLENGES

Even with the seven core modules established, modern societies face **emerging complexities** – such as disruptive technologies, cross-border data flows, or public health crises – that require additional governance strategies. The following six **Expanded Layers** provide specialized approaches to meet these challenges. They integrate seamlessly with the core modules, preserving fairness, adaptability, and resilience in a world that's always on the move.

18.4.1 Technology Governance Layer

18.4.1.1 Purpose

Manages the societal impact of rapidly evolving technologies (e.g., artificial intelligence, digital platforms, and robotics) to maximize benefits and minimize harms.

18.4.1.2 Core Features

1. **Tech Impact Assessments**
   - Draw on **Expert Panels** in the Communicative Interface to evaluate new technologies.
   - Gauge potential risks (privacy violations, labor displacement) alongside potential gains (efficiency, creative problem-solving).
2. **Broad-Benefit Mechanisms**
   - **Digital Literacy Programs**: Equip communities with the technical and ethical knowledge to responsibly use and develop technology.
   - **Open-Source Collaboration**: Foster public knowledge repositories, encouraging free tools and innovations that benefit all.
   - **Cross-Sector Partnerships**: Public, private, and academic actors co-create initiatives ensuring advanced tech reaches diverse communities.
3. **Democracy Safeguards**
   - Protect public discourse from AI-driven misinformation (e.g., manipulated media).
   - Keep algorithmic processes transparent so citizens understand how digital systems influence their lives.

18.4.1.3 Inter-Module Connection

- **Power-Structures Monitor**: Spots monopolistic tech consolidation or covert data hoarding.
- **Knowledge Management**: Provides authoritative datasets to counter technology-driven disinformation.
- **Potential Conflict**: The Decentralization Engine may allow a local community to block certain tech altogether. The Tech Governance Layer upholds a base standard of digital rights (e.g., data privacy) so that one locale's decisions don't undermine broader societal values.

18.4.2 Psychological Well-Being Component

18.4.2.1 Purpose

Integrates **mental health** and **social cohesion** into governance, ensuring that material progress is accompanied by emotional and relational well-being.

18.4.2.2 Core Features

1. **Mental Health Metrics and Community-Building**
   - Tracks stress levels, loneliness indices, or community involvement rates.
   - Supports public programs that enhance interpersonal connections – like local festivals, group art initiatives, or intergenerational mentoring.

2. **Prevention of Alienation**
   - Identifies at-risk communities or individuals susceptible to isolation, anxiety, or undue stress.
   - Proactive interventions (e.g., counseling hotlines, mobile mental-health outreach) keep small problems from escalating.
3. **Quality-of-Life Innovations**
   - Recommends policy experiments that reinforce social bonds – e.g., co-working community centers or mindfulness education.
   - Collaborates with other layers to ensure new developments (e.g., remote work expansions) don't create unintentional social side-effects.

18.4.2.3 Inter-Module Connection

- **Capabilities and Equality**: Directs targeted resources toward mental-health programs, ensuring vulnerable groups receive support.
- **Antifragility Subsystem**: Uses rapid feedback to test pilot initiatives in mental health and measure societal resilience outcomes.
- **Potential Conflict**: A local autonomy budget cut leads to underfunded psychological services. The Well-being Component flags this deficiency, prompting **Procedural Justice** if fairness in essential care is compromised.

18.4.3 Global Coordination Framework

18.4.3.1 Purpose

Handles **cross-border challenges** that surpass local or national scope, such as public health threats, cyber risks, or financial instabilities, avoiding detrimental "race to the bottom" scenarios.

18.4.3.2 Core Features

1. **International Interface**
   - Encourages equitable collaboration across jurisdictions to tackle joint concerns (e.g., managing shared digital infrastructures, cross-border data flows, or epidemic response).
   - Sets up cross-border councils to harmonize certain standards, from data exchange protocols to cross-jurisdiction dispute resolution.
2. **Avoiding Race-to-the-Bottom**
   - **Minimum Labor and Data Protection Standards**: Jurisdictions agree not to undercut each other on core worker rights or data privacy just to attract investment.
   - **Trade and Cooperation Incentives**: Offer access to improved resource-sharing or financial collaborations if members adhere to these base standards.
   - **Monitoring Mechanisms**: Cooperative reviews (e.g., peer audits, cross-regional data trust boards) highlight disparities or compliance gaps without invoking purely punitive approaches.
3. **Resilience Networks**
   - Establishes reciprocal support systems in crises – like knowledge-exchange frameworks for managing cyber threats or joint medical research.
   - Encourages synergy in addressing issues that no single locale can handle alone.

18.4.3.3 Inter-Module Connection

- **Decentralization Engine**: Local autonomy must still respect global agreements on fundamental worker rights, data standards, etc.
- **Potential Conflict**: A region plans to lower digital privacy thresholds to lure big tech investments, potentially exposing other regions to cross-border data leaks. The Global Coordination Framework sets baseline protections, with the Procedural Justice Module stepping in if fundamental fairness is violated.

18.4.4 Knowledge Management System

18.4.4.1 Purpose

Preserves **societal knowledge**, fights misinformation, and maintains a **robust institutional memory** so that progress builds upon past lessons.

18.4.4.2 Core Features

1. **Public Archives and Open Data**
   - Central repositories for legislative histories, academic research, and cultural records – making them broadly accessible.
   - Prevents data fragmentation or manipulation by ensuring consistent, verified sources.
2. **Misinformation Countermeasures**
   - Works with the Communicative Interface to promote transparent "fact-checking processes" and collaborative knowledge curation.
   - Maintains a "public ledger" of major policy decisions and rationale, reducing rumor-based policy shifts.
3. **Institutional Memory**
   - Logs policy experiments, successes, and missteps, letting future governance reference what has (or hasn't) worked.
   - Minimizes repeated mistakes by systematically storing historical insights.

18.4.4.3 Inter-Module Connection

- **Communicative Interface**: Relies on accurate data for well-founded debates; experts regularly consult archives.
- **Power-Structures Monitor**: Detects attempts to distort or conceal data.
- **Potential Conflict**: A local authority restricts certain historical documents to push a selective narrative. The Knowledge Management System preserves them, with **Procedural Justice** intervening if essential transparency is breached.

18.4.5 Civic Education Component

18.4.5.1 Purpose

Cultivates an **informed, critically thinking public**, vital for meaningful democratic engagement and Veil-of-Ignorance reasoning.

18.4.5.2 Core Features

1. **Curriculum Standards**
   - Embeds open-society values and moral considerations (Veil-of-Ignorance) into official educational guidelines.
   - Develops students' abilities to analyze policy proposals and differentiate fact from speculation.
2. **Community Workshops**
   - Promotes lifelong learning and active citizenship – like public debates, reading circles, or hackathons for civic problem-solving.
   - Fosters mutual respect and intergenerational knowledge exchange.
3. **Democratic Culture Preservation**
   - Reinforces norms of free expression and tolerant discourse across generations.
   - Counteracts ideologically extreme movements by grounding citizens in reasoned skepticism and open-mindedness.

18.4.5.3 Inter-Module Connection

- **Communicative Interface**: Depends on an educated population able to parse complex arguments and weigh expert input.
- **Potential Conflict**: Some regions adopt narrower educational content for ideological reasons. The **Freedom Manager** and **Procedural Justice** Modules preserve key civic education standards, preventing undue indoctrination.

18.4.6 Innovation Incubator

18.4.6.1 Purpose

Encourages **policy experimentation** and creative solutions, channeling new approaches into beneficial reforms while minimizing risks.

18.4.6.2 Core Features

1. **Policy Pilots**
   - Trials of novel ideas – ranging from localized e-governance systems to alternative healthcare delivery – implemented in test zones.
   - Proven successes can be integrated system-wide, akin to stable software releases.
2. **Fairness Checks**
   - Ensures pilot designs do not exploit marginalized groups or impose hidden burdens on unsuspecting stakeholders.
   - Incorporates the Veil of Ignorance as a baseline ethic in all pilot proposals.
3. **Scalability**
   - Collaborates with **Decentralization Engine** for flexible local pilots and **Antifragility Subsystem** to iterate or discontinue failing trials quickly.
   - Shares data and lessons with Knowledge Management to inform future experimentation.

18.4.6.3 Inter-Module Connection

- **Freedom Manager**: Protects individuals from coerced participation in pilots.
- **Potential Conflict**: A radical local experiment undermines basic healthcare for certain groups. The **Procedural Justice** Module or **Power-Structures Monitor** can halt or revise it if it contravenes fair treatment principles.

18.4.6.4 Summary Table: Expanded Layers Overview

Below is a **sample table** summarizing each Expanded Layer's purpose, core features, and key overlaps with the Core Modules. You can adapt it for diagrams or figures in your final text.

Layer	Purpose	Core Features	Inter-Module Overlaps
Technology Governance	Ensures emergent tech aligns with societal benefit	• Tech Impact Assessments • Broad-Benefit Mechanisms (digital literacy, open-source, cross-sector partnerships) • Democracy Safeguards	• Power-Structures Monitor detects tech monopolies • Knowledge Management counters disinformation • Decentralization must respect baseline digital rights
Psychological Well-being	Integrates mental health and cohesion into governance	• Mental Health Metrics and Community Initiatives • Alienation Prevention • Quality-of-Life Innovations	• Capabilities and Equality provides resources • Antifragility tests new support ideas • Procedural Justice intervenes if essential well-being is neglected

(Continued)

Layer	Purpose	Core Features	Inter-Module Overlaps
Global Coordination	Addresses cross-border issues without a race to the bottom	• International Interface (shared data, crisis responses) • Minimum Labor/Data Standards • Resilience Networks (joint research, peer reviews)	• Decentralization meets global norms • Procedural Justice aligns local autonomy with external obligations • Potential data trust boards for cross-region collaboration
Knowledge Management	Preserves info, curbs misinformation, fosters institutional memory	• Public Archives and Open Data • Misinformation Countermeasures • Institutional Memory (policy logs and versioning)	• Communicative Interface uses verified data • Power-Structures Monitor detects data manipulation • Procedural Justice guards transparency
Civic Education	Builds a knowledgeable, critically thinking population	• Curriculum Standards (open-society ideals, ethical frameworks) • Community Workshops • Democratic Culture Preservation	• Communicative Interface gains from informed public • Freedom Manager and Procedural Justice ensure educational rights • Bolsters moral perspective (Veil-of-Ignorance)
Innovation Incubator	Encourages policy experimentation and creative solutions	• Policy Pilots • Fairness Checks (Veil-of-Ignorance baseline) • Scalable Integration (collaborates with Decentralization and Antifragility)	• Freedom Manager prevents coercion • Procedural Justice or Power-Structures Monitor can halt harmful experiments • Knowledge Management records pilot outcomes

18.4.6.5 Concluding Note on Expanded Layers

These **Expanded Layers** – spanning **Technology Governance** to the **Innovation Incubator** – enhance the Modern Society OS by **continuously adapting** to new realities: from AI-driven transformations to cross-border coordination, from psychological well-being to effective knowledge stewardship. Each layer:

- **Links** closely with the seven **Core Modules**, ensuring that innovations and reforms remain compatible with the system's ethical kernel.
- **Emphasizes** practical solutions (like local pilot projects, digital literacy drives, or cross-border data protocols) to handle challenges that surpass traditional governance capacities.
- **Maintains** the guiding spirit of Popper's open society and Rawls's Veil of Ignorance, never losing sight of fairness, transparency, and inclusivity.

By **complementing** the core modules, these layers allow the Modern Society OS to remain **robust yet flexible**, ready to engage with emerging challenges while preserving a commitment to ethical and dynamic governance.

THE DEATH OF EXPERTISE

THINKING BOX

"Is real knowledge being drowned out by internet noise and overconfidence? Perhaps restoring respect for expertise begins with heeding the subtle lessons of **42**."

18.5 IMPLEMENTATION ROADMAP

A Philosophical yet Practical Blueprint – The Modern Society OS is primarily a **theoretical construct**, designed to spark **imagination** about how philosophical principles – Popper's open society and Rawls's Veil of Ignorance – might guide real-world governance. It does **not** require total, immediate implementation. Instead, consider it a **flexible toolkit**: pick elements to test, adapt, or refine to suit local conditions. This roadmap suggests **incremental steps** for rolling out modules and layers, embedding them culturally, managing conflicts, and tying back to broader philosophical themes. Each step showcases how these high-level ideals can become a **living, evolving** force in everyday governance.

18.5.1 Phased Rollouts

1. **Local Pilots**
    - **Focused, Small-Scale Testing**: Rather than imposing sweeping reforms, begin by piloting specific modules – like a **Civic Education** revamp or a **Psychological Well-being** outreach program – in a handful of willing localities.
    - **Data and Feedback**: Track metrics (e.g., changes in civic engagement, mental-health outcomes) and solicit citizen input via the **Communicative Interface**.
    - **Modular Integration**: Even though each pilot focuses on one area (e.g., mental health), it should still align with the overarching kernel – ensuring every decision meets Veil-of-Ignorance and open-society standards.
2. **Scaling Up**
    - **Selective Expansion**: If a local pilot demonstrates success, replicate or adapt it in larger regions, subject to checks from modules like **Freedom Manager** (for safeguarding liberties) or **Procedural Justice** (to sustain fairness).
    - **Iteration over Perfection**: Accept that a policy flourishing in one locality may need recalibration in another. The **Antifragility Subsystem** accommodates these variations, capturing lessons from each success or misstep.

Why This Matters: Phased rollouts reflect the Modern Society OS's ethic of **incremental learning** – akin to introducing small software updates before a major release. Doing so promotes both adaptability and accountability, allowing communities to shape reforms that truly resonate.

18.5.2 Cultural Embedding

18.5.2.1 Infusing the Veil of Ignorance and Open Society Ideals

1. **Education and Public Discourse**
    - Incorporate these principles – particularly **Popper's continuous critique** and **Rawls's fairness ethic** – into school curricula, public workshops, and media channels.
    - Encourage policy-makers and community leaders to model transparent, Veil-of-Ignorance-based thinking in speeches and official documents.
2. **Building Public Trust**
    - **Independent Oversight Bodies**: Create councils or committees – including citizens and experts – that review major policy moves or local pilot results. Their independence signals that no single power bloc can dominate.
    - **Citizen Participation and Oversight**: Expand local referendums, digital feedback platforms, or neighborhood forums, turning "continuous critique" into a shared civic norm.
    - **Clear and Accessible Reporting**: Require user-friendly, publicly available summaries of all government activities – budgets, achievements, stumbling points. Transparency fosters trust and shows accountability isn't just talk.

18.5.2.2 Innovative Educational Approaches

A crucial part of cultural embedding involves **novel, hands-on educational programs** that truly embody the Modern Society OS principles:

1. **Reality-Shifting Simulations**
   - **Rotating Societal Positions**: Students periodically switch social roles (e.g., well-off vs. resource-constrained) to experience shifting privileges and constraints.
   - **Visceral Understanding**: Forces them to see how decisions made in one role affect them later – mirroring **Rawls's Veil** insight that policies must be fair to all, including potential future selves.
2. **Intergenerational Decision Chains**
   - **Long-Term Consequences**: Each class or cohort inherits the decisions of previous cohorts, learning the importance of forward-thinking governance.
   - **Real Stakes**: Demonstrates how policy outlasts its creators, an ethos directly linked to **Antifragility** and institutional memory.
3. **Cross-Community Policy Labs**
   - **Collaboration across Socioeconomic Divides**: Students from distinct communities co-create policies that impact both.
   - **Practical Empathy**: Reinforces **Decentralization** (local autonomy) yet fosters mutual obligations and shared solutions – mirroring **Procedural Justice** in real contexts.
4. **Digital Democracy Platforms**
   - **Online Micro-Societies**: Learners form virtual "governments," face crises, and see immediate outcomes of their decisions.
   - **Rational Discourse**: A scaled-down simulation of the **Communicative Interface**, promoting evidence-based debate and discouraging manipulation.
5. **Power-Shifting Governance**
   - **Rotating Roles**: Students regularly alternate between "leaders" and "citizens."
   - **Accountability**: Encourages them to imagine how decisions look from the other side, keeping **Veil-of-Ignorance** reasoning alive and tangible.

Why This Matters: By merging philosophical concepts with **immersive, interactive education**, societies instill empathy, accountability, and long-term foresight in the next generation – crucial to making the Modern Society OS a truly **living philosophy** rather than a theoretical abstraction.

18.5.3 Periodic Audits and Conflict Resolution

1. **Justice Audits**
   - **Routine Inspections**: The **Procedural Justice Module**, aided by the **Power-Structures Monitor**, periodically checks each module/layer for hidden power imbalances or overlooked biases.
   - **Public Disclosure**: Summaries of findings appear in the **Knowledge Management System**, letting citizens see at a glance which areas need reform or have thrived.
2. **Common Conflicts and Resolution Paths**

   When multiple modules and layers interact, **tensions** can surface:

Conflict	Trigger	Resolution Path
Freedom Manager vs. Capabilities and Equality	Balancing individual rights (e.g., property) against calls for universal healthcare or baseline welfare structures	**Procedural Justice** mediates, assessing if restricting some personal freedoms justly benefits broader well-being.

(Continued)

Conflict	Trigger	Resolution Path
Decentralization Engine vs. Global Coordination	Local autonomy conflicting with cross-border data or labor standards	**Procedural Justice** weighs local identity against minimal global norms, aiming to find a fair compromise.
Technology Governance vs. Innovation Incubator	A radical pilot in AI might compromise user privacy or overshadow fundamental liberties	**Tech Governance** sets baseline guardrails; **Procedural Justice** intervenes if proposals infringe Veil-of-Ignorance fairness.
Psychological Well-being vs. Local Autonomy	A region cuts mental-health programs, ignoring social isolation data	The **Well-being Component** flags the risk; **Procedural Justice** steps in if essential care is denied.

3. **Shared Protocols**
   - **Open Debate**: All conflicts are aired via the **Communicative Interface**, ensuring no unilateral rulings or hidden deals.
   - **Veil-of-Ignorance Check**: Proposed resolutions must stand the test of "Would this be just if I were among the least advantaged?"

Why This Matters: **Structured oversight** and **transparent conflict resolution** keep the Modern Society OS agile and morally grounded, preventing any faction from dominating without open discourse or moral accountability.

18.5.4 Ties to Earlier Themes in This Book

1. **Complexity and Emergence**
   - **Modular, Ecosystemic Design**: The OS's emphasis on localized pilots and feedback loops resonates with complexity theory's call for bottom-up solutions.
   - **Adaptive Problem-Solving**: Student-driven policy labs or rotating governance roles exemplify how emergent ideas can arise from diverse local contexts.
2. **Resilience and Antifragility**
   - **Iterative and Adaptive**: The **Innovation Incubator** and the **Antifragility Subsystem** show how failures can become stepping stones for bigger successes, reflecting the principle of learning from volatility.
   - **Multilayered Redundancy**: Decentralization ensures that one region's failure doesn't collapse the entire system – others can learn from the pilot's mistakes.
3. **Ethical Frameworks and Cognitive Insights**
   - **Veil-of-Ignorance Lens**: Reinforces moral accountability at each policymaking stage, mitigating parochial biases or power entrenchment.
   - **Power-Structures Monitor and Knowledge Management**: Proactively addresses misinformation, bias, or hidden manipulations – key insights from cognitive psychology.
4. **Living Philosophy**
   - **Iterative and Real-World Application**: By constantly reexamining policies, adapting to local feedback, and challenging hidden power structures, the Modern Society OS transforms high-level philosophical principles into daily governance.
   - **Practical Evolution**: The immersive educational approaches further illustrate how these ethics aren't merely intellectual exercises; they are **lived, tested, and revised**, just like a living organism evolving over time.

Why This Matters: Anchoring each step of implementation in prior themes – **complexity, resilience, ethical scrutiny,** and **empirical cognition** – demonstrates the synergy between theory and practice. The result is a governance model that remains ethically consistent while flexibly evolving with societal shifts.

18.5.4.1 Concluding Note on the Implementation Roadmap

The Modern Society OS is, at heart, a **philosophical ideal** meant to **inspire** new ways of thinking about governance. It does not demand wholesale adoption but offers a **blueprint** for incremental experimentation – whether through local pilot projects, advanced civic education, or dynamic oversight structures. By:

- **Gradually introducing** modules,
- **Embedding** Popperian and Rawlsian values in culture and education,
- **Resolving conflicts** transparently, and
- **Adapting** in real-time,

This roadmap showcases how **abstract ethics** can become **living, evolving** governance. Students learn to empathize through role-shifting exercises; citizens see checks and balances in action; leaders adopt iterative, data-driven policymaking. In so doing, the Modern Society OS exemplifies a **Living Philosophy** – one that thrives precisely because it never presumes finality, upholding fairness, openness, and continued moral inquiry for generations to come.

18.6 CONCLUDING THOUGHTS

18.6.1 Interconnected Modules for a Cohesive Whole

A key insight of the Modern Society OS is that **no module operates in a vacuum**. Each one fulfills a distinct function, yet they collectively shape a holistic governance environment:

- **Civic Education** sustains the **Communicative Interface** by cultivating a well-informed, critically engaged public – so that open debates truly reflect diverse perspectives and reasoned arguments.
- **Decentralization** depends on the **Knowledge Management System** to coordinate shared resources, track best practices, and mitigate local externalities that might spill over boundaries.
- **Procedural Justice** stands ready to arbitrate when freedoms, capabilities, or local autonomy collide – applying the Veil-of-Ignorance test to maintain fairness amid conflicting interests.

Much like processes in an OS, these modules are **deeply interlinked**: improvements (or failures) in one ripple across others, illustrating the Modern Society OS's ecosystemic design.

18.6.2 Conflict and Resolution as a Feature, Not a Bug

Crucially, the framework does **not** aim to eradicate all tensions. It assumes that **conflicts are inevitable** – for instance, balancing individual liberty vs. collective welfare, local identity vs. global standards, or user privacy vs. cutting-edge innovation. Rather than treating disagreements as failures, the Modern Society OS:

- **Anticipates friction** and incorporates formal pathways (e.g., Procedural Justice reviews, Communicative Interface debates) to resolve disputes.
- **Channels conflict** into **constructive dialogue** and iterative policy reform, ensuring that hidden biases or power grabs are challenged in open forums.
- **Builds trust** by allowing each stakeholder to see that tensions are addressed transparently, not brushed aside or settled behind closed doors.

In this sense, the system **embraces** friction as an engine for progress, refining institutions through ongoing critique rather than forced consensus.

18.6.3 A Pathway for Holistic Governance

This chapter brings together the book's major threads – **cognitive biases**, **resilience strategies**, **ethical reasoning**, and **complexity management** – into a coherent framework for governance:

- **Cognitive Biases**: Tools like the **Power-Structures Monitor** and **Knowledge Management** counter misinformation and subtle manipulations, while the Veil of Ignorance offers a moral lens to reduce self-serving distortions.
- **Resilience Strategies**: The **Antifragility Subsystem** and **Innovation Incubator** turn challenges into learning opportunities, mirroring earlier arguments about adaptive systems thriving on stress.
- **Ethical Reasoning**: Rawls's Veil of Ignorance and Popper's openness infuse every layer, embedding fairness and critique as constants.
- **Complexity Management**: The modular, ecosystemic design – encompassing local pilots, feedback loops, and cross-module integration – reflects the reality that governance is a **living system** of interdependent parts.

By weaving together these insights, the Modern Society OS **exemplifies** the "Living Philosophy" theme explored throughout the book, demonstrating how philosophical principles can inform **practical, evolving policy structures** – rather than remaining abstract ideals.

18.6.4 Navigating Forward

Implementing or even partially adopting a Modern Society OS calls for **incremental steps**, **participatory design**, and a **cultural shift** that welcomes continuous critique and empathy:

- **Incremental Growth**: Begin with local pilots or select modules, gather feedback, and scale up; avoid imposing all changes at once.
- **Participatory Design**: Involve citizens, experts, and diverse stakeholders from inception, ensuring that the system respects local contexts and fosters broad-based legitimacy.
- **Deep-Seated Cultural Shift**: Encourage leaders and communities to embrace critique as a constructive force, view empathy and fairness as guiding principles, and see policy errors as learning moments rather than political liabilities.

This multilayered approach **balances** freedom with welfare, local innovation with universal rights, and cutting-edge technology with humane considerations. Above all, it reaffirms that **philosophy** can be more than speculation: when grounded in real governance practices, it becomes a **dynamic engine** for human flourishing in a rapidly changing world.

19 The Living Philosophy
Navigating Complexity with Wisdom and Adaptability

Key Themes:

1. **The Need for a Living Philosophy**
   - Traditional rules fail in **accelerating complexity**; adaptability ensures relevance.
   - **Resilience Evidences**: Every successfully navigated challenge strengthens ethical clarity.
2. **The DIKW Pyramid (Data → Information → Knowledge → Wisdom)**
   - **Data**: Raw facts; **Information**: Context; **Knowledge**: Application; **Wisdom**: Ethical foresight.
   - Quality **checkpoints** and **temporal awareness** ensure accuracy and long-term applicability.
3. **The Inner Landscape Metaphor**
   - **Forest**: Complexity → recognizing patterns.
   - **Lake**: Reflection → integrating emotions and ethics.
   - **Boat**: Tools → applying structured frameworks.
   - **Island**: Wisdom → merging insights into ethical action.
   - **Garden**: Growth → evolving knowledge and values.

Key Takeaway:

A **Living Philosophy** is **not static** but **continuously refined**. By combining **structured knowledge (DIKW), ethical foresight, and resilience-building**, individuals can **confidently navigate complexity and shape a meaningful, adaptive ethical framework**.

19.1 INTRODUCTION: THE CALL OF A LIVING PHILOSOPHY

19.1.1 The Ever-Shifting Landscape

We live in an era of relentless change – one in which entire industries can be upended overnight by disruptive technologies, cultural norms pivot in response to a hyperconnected world, and wholly new ethical dilemmas surface as emerging innovations race ahead of existing regulations. In this fluid environment, traditional moral codes or unchanging doctrines, no matter how well-intended or theoretically robust, risk becoming quickly outdated. They simply weren't designed for a reality that reinvents itself every few years – or even every few months.

Why Static Doctrines Falter

- **Rapid Technological Advances**: AI-driven solutions, biotech breakthroughs, and data analytics can reorganize how societies function, exposing gaps in guidelines that assumed more gradual shifts.
- **Cultural Recalibrations**: Grassroots movements, social media-amplified discourse, and cross-border exchanges can reshape values faster than static rules can adapt.
- **Uncharted Ethical Terrain**: From data privacy to gene editing, unanticipated scenarios arise, demanding flexible, evolving frameworks to cope effectively.

DOI: 10.1201/9781003640288-24

In such a landscape, **static commands** – even if once valid – often struggle to address the unforeseen ramifications of fast-evolving realities. **A living philosophy**, however, is designed precisely for this scenario. Far from rigid commandments, it operates as a continually refreshed guide – a conscious, agile mindset that adapts as circumstances morph around us.

A Dynamic Compass for Complexity – Rather than dictating immutable truths, a living philosophy keeps us alert, open to reinterpreting principles, and committed to responsible, ethically sound engagement with emerging challenges. This approach neither discards the wisdom of tradition nor clings to it inflexibly. Instead, it acknowledges that principles might require reevaluation as we learn more about new technologies, social dynamics, and ecological concerns. **Our moral compass**, therefore, is not locked in place but recalibrates over time – retaining its guiding essence while shifting its bearings to reflect deeper insights or fresh ethical imperatives.

"In a world constantly in flux, what once sufficed as 'the rule' may no longer capture our present reality. By weaving adaptability into our ethical framework, we stay aligned with our core values while responding effectively to new frontiers."

19.1.2 Why a Living Philosophy Matters

A living philosophy stands out for three primary reasons:

1. **Adaptability**
   - **Absorbing New Information**: As fresh data, cultural shifts, or innovations emerge, a flexible approach can incorporate them quickly, refining previous assumptions.
   - **Questioning Old Assumptions**: A living philosophy invites us to revisit beliefs that served us well in the past but might now limit our perspective or hamper ethical clarity.
   - **Refining Ethical Positions**: When faced with ethical dilemmas – like balancing user privacy against AI optimization – a dynamic framework permits nuanced, context-sensitive stances.
2. **Resilience**
   - **Growth through Challenges**: Instead of enduring crises just to "get by," we embrace them as opportunities to strengthen our moral grounding, strategic thinking, and empathy.
   - **Positive Adaptation**: Each successful navigation of complexity becomes a resilience evidence – tangible proof that we can pivot, integrate lessons, and rise to new challenges.
   - **Long-Term Sustainability**: By regularly reevaluating our principles and knowledge, we remain resilient in the face of technological leaps, economic uncertainties, and social transformations.
3. **Meaning and Purpose**
   - **Unity amid Chaos**: In a world rife with fragmented information and polarizing viewpoints, a living philosophy offers coherence, guiding individuals toward more thoughtful action.
   - **Ethical Engagement**: This approach is not content with mere survival; it encourages proactive involvement in shaping technology, culture, and global policy for the better.
   - **Empowerment**: Armed with a living philosophy, we move beyond reactive coping – stepping into a role where we co-create future possibilities grounded in ethical foresight.

In essence, this living philosophy emerges as **a dynamic moral compass**, helping us make sense of bewildering complexity and take constructive steps forward.

19.1.3 The Three Pillars

This chapter highlights three key elements to guide you on this path:

1. **The Need for a Living Philosophy**: We delve deeper into why static rules fail – and how a flexible, evolving approach fosters resilience and ethical clarity. You'll see how a living philosophy becomes indispensable in an age of acceleration and interconnection.
2. **Using the DIKW Pyramid**: How do we move from raw data to practical wisdom? The classic **Data-Information-Knowledge-Wisdom (DIKW)** model gives us a foundational roadmap. Enhanced with time-sensitive evaluation (from Chapter 15's temporal truth concepts), this tool helps us sift through informational overload, ensuring our evolving stance remains anchored in robust understanding.
3. **The Inner Landscape**: Finally, we'll explore a metaphorical space – **your inner landscape** – where these insights converge into a practical, personal methodology. This metaphor ensures your philosophy remains both dynamic and deeply rooted in your lived experiences.

Together, these three pillars form **the architecture** of a living philosophy. Rather than a static edifice, it's an evolving structure – fluid enough to respond to external shifts yet solid enough to hold you steady in the face of complexity.

19.1.4 A Framework for Thriving

Bringing these elements together offers **more than just survival**. It empowers you to:

- **Engage Complexity with Confidence**: Instead of dreading change, you'll learn to spot patterns, respond ethically, and even leverage uncertainty for constructive outcomes.
- **Build and Recognize Resilience Evidences**: Each time you apply these principles, each time you adapt effectively, you gather tangible proof of your growth.
- **Cultivate Ethical Foresight**: By elevating your perspective from mere data and immediate concerns to a more holistic, long-term view, you'll influence how innovations roll out and how societies evolve.
- **Nurture a Meaningful, Adaptive Ethic**: Ultimately, this framework weaves your evolving knowledge into real-world decisions, forging a philosophy that continuously refines your sense of purpose.

By the end of this chapter, you'll hold **a practical, evolving roadmap** for thriving in a tech-driven, ever-shifting reality. It all begins with **embracing the need for a living philosophy**, harnessing **the DIKW pyramid** to traverse from raw facts to profound wisdom, and cultivating **an inner landscape** that keeps your insights fluid, ethical, and deeply personal.

19.2 THE NEED FOR A LIVING PHILOSOPHY: EMBRACING CHANGE AS THE ONLY CONSTANT

19.2.1 Beyond Static Rules

In a landscape where ethical conundrums like **AI bias** or **data privacy** erupt faster than legislation can possibly catch up, **rigid moral codes** or **fixed doctrines** often become obsolete before they can provide real guidance. While timeless principles can offer a strong moral compass, they frequently struggle to address new and unforeseen complexities. This dilemma aligns with the **Model Boundary Paradox**, which suggests that every model – no matter how elegant – is necessarily a simplification of reality. Eventually, a static moral or philosophical system will fail to capture the fluid intricacies of a world in continuous flux.

By contrast, a **living philosophy** anticipates these waves of change. It evolves in response to emerging challenges – like the ethics of automated decision-making or the profound social implications of augmented reality – continually refining its stance to stay relevant. Rather than demanding adherence to immovable commandments, it invites a dynamic, iterative engagement, making it better equipped to handle today's rapidly shifting moral terrain.

"A fixed, unchanging approach ultimately fails to capture a reality that's continually morphing."

19.2.2 The Dynamics of Change

1. **Accelerating Pace**: Rapid technological breakthroughs – such as quantum computing or gene editing – and cultural shifts – fueled by social media and instant global communication – can outpace stable norms. **A living philosophy** must be agile enough to pivot as soon as new information or ethical puzzles arise, ensuring moral reasoning doesn't lag behind technological progress.
2. **Interconnected Systems**: In our globalized digital age, a seemingly minor tweak to a product design can **spiral into global ramifications** – influencing user trust, affecting social discourse, or altering mental health patterns on a large scale. A living philosophy brings **systemic awareness**, urging us to see beyond local, immediate effects and evaluate the broader web of cultural, technological, and societal influences in which each decision resides.
3. **Phase Transitions**: Complex systems often undergo **tipping points**, where slow, incremental changes lead to sudden and dramatic shifts. Just as societies can abruptly transform due to political, economic, or environmental triggers, our moral systems may require equally swift leaps in reasoning. A living philosophy recognizes these critical junctures, remaining alert to when incremental changes are no longer sufficient and a more radical ethical recalibration is necessary.

"A minor tweak in a product design might spiral into global ramifications – altering user trust, influencing social dialogue, or impacting mental health at scale."

19.2.3 The Neuroscience of Adaptability

Humans are biologically equipped for **adaptation** – our brains exhibit remarkable **neuroplasticity**, enabling continuous learning and reorganization of neural pathways in response to new experiences. The **Default Mode Network (DMN)** plays a key role in introspection and self-reflection, both indispensable for refining values, questioning assumptions, and integrating fresh insights. This adaptive capacity is further enriched by **embodied cognition**, acknowledging that emotions, bodily sensations, and intuitive flashes significantly shape our interpretations and decisions.

A living philosophy harnesses this inherent adaptability. Rather than relying solely on detached, rational principles, it incorporates emotional intelligence, cultural sensitivity, and bodily awareness, refining ethical stances as conditions evolve. It melds **rational analysis** (cognitive evaluation) with **embodied signals** (emotions, intuitions, and physical feedback) to yield decisions that are simultaneously **thoughtful, ethical, and human-centered**.

"Our Default Mode Network supports introspection, essential for refining beliefs, values, and assumptions. A living philosophy builds on this inherent adaptability."

19.2.4 Resilience and the Living Philosophy

By repeatedly **realigning your philosophical stance**, you foster **resilience** – the capacity not just to survive disruptive changes but to emerge from them stronger and more deeply grounded in your core ethics. This concept mirrors **antifragility**, the idea that certain systems grow more robust through stressors rather than merely enduring them.

- **Personal Growth**: Each time you absorb a shock – be it a career upheaval, a social crisis, or a technological disruption – and respond with adaptive moral action, you accumulate **resilience evidences**. These evidences confirm your ability to handle future shocks with creativity and ethical clarity.
- **Societal Contribution**: As you refine your philosophy, you become a positive force in your communities – modeling how ethical adaptability can lead to solutions that benefit both individuals and the greater good.

19.2.5 A Philosophy in Motion

In essence, a **living philosophy** is not a static decree but a constant evolution – a moral and intellectual dance with the ever-shifting world around us. Central to this dance are four key principles:

1. **Continuous Learning**: Stay open to new knowledge, experiences, and cultural perspectives. Seek to question, explore, and update your ethical positions as new facts and contexts arise.
2. **Ethical Foresight**: Anticipate potential outcomes – whether they be social, environmental, or psychological – and act responsibly before harms escalate. Consider both **immediate impacts** and **long-term consequences**.
3. **Systemic Awareness**: Recognize that each decision resonates through a web of interlinked systems – cultural, technological, and ecological. Acknowledge feedback loops, second-order effects, and the interplay of local actions with global impacts.
4. **Narrative Agility**: Stay flexible in how you frame and interpret the world. As new truths emerge, be ready to **revise** your narratives, beliefs, or strategies accordingly. This agility keeps you aligned with reality, not clinging to outdated paradigms.

"Such an approach flourishes when combined with a clear method for sifting through data and developing wisdom—cue the DIKW Pyramid."

These principles lay the groundwork for a moral architecture that thrives in a rapidly evolving context. But **knowing** you need a living philosophy doesn't, by itself, show you **how** to build and apply it – especially amid information overload and shifting truths. That's where the **DIKW Pyramid** comes into play, guiding us from raw data to meaningful wisdom in a time-sensitive manner, ensuring our philosophy remains not just alive, but effectively engaged with the world's complexities.

THE ETHICS OF CONVENIENCE

THINKING BOX

"At what unseen cost do we demand one-click ease? Sometimes, as with **42**, the real complexity sits beneath a surface of simplicity."

19.3 THE DIKW PYRAMID: BUILDING A FOUNDATION FOR WISDOM

Modern life bombards us with data – tweets, headlines, research studies, and infinite scrolls of information. We risk drowning in a sea of facts, half-truths, and biased viewpoints if we don't have a **rigorous yet flexible** way to sift through it all. The **DIKW pyramid** – Data, Information, Knowledge, and Wisdom – has long served as a staple model for transforming raw facts into meaningful action.

But in an era of fast-moving innovations and shifting cultural norms, we must enhance DIKW with **quality checks**, **temporal awareness**, and **ethical foresight** to keep our understanding both reliable and relevant.

19.3.1 A Living System for Navigating Truth

Think of DIKW as a **building with four floors**:

1. **Data** (ground level)
2. **Information** (context floor)
3. **Knowledge** (practical application floor)
4. **Wisdom** (ethical and foresight floor)

Between each floor stands a **quality gateway** – a checkpoint that ensures only well-founded insights rise to the next level. This model isn't static; it's **living** and must adapt as new data emerges or old assumptions become obsolete.

Stage 1: Data (Foundation)

Raw Facts and Observations

- **What to look for**:
  - **Original sources**: Who collected this data, and are they credible or transparent?
  - **Verification**: Do other independent sources confirm it?
  - **Timestamps**: Is this recent enough to remain accurate or relevant?
- **Key question**: *"Is this from a reliable source?"*

19.3.2 Data Gateway

Here, we filter out noise and questionable material. It's not enough just to gather data; we must **test** it for accuracy, reliability, and timeliness. In a world where "yesterday's truth might be today's misconception," each piece of data should be verified and reviewed. If data fails at this gateway – maybe it's from a dubious blog or a single unverified claim – it stays behind.

Stage 2: Information (Context Floor)

Data with Context and Meaning

- **What to look for**:
  - **Patterns**: Trends, correlations, or anomalies.
  - **Relationships**: How do these data points interconnect?
  - **Broader context**: Social, historical, or technical backdrop.
- **Key question**: *"What does this mean in context?"*

19.3.3 Information Gateway

Once data passes initial checks, we convert it into **information** by placing it in a broader narrative. Missing or selective context can distort our view just as much as false data. Therefore, we ask:

- *"What's the bigger picture?"*
- *"Are there cultural biases or omitted details?"*

We also maintain **temporal awareness** here. Some info is ephemeral – today's hot take might lose relevance tomorrow. By keeping the time factor in mind, we avoid anchoring future decisions to facts that expire quickly.

Stage 3: Knowledge (Practical Application Floor)

Actionable Understanding

- **What to look for**:
  - **Practical implications**: Can we apply this to real-world challenges?
  - **Verification**: Does it align with past experience or well-established principles?
  - **Adaptation**: How might this knowledge shift if new evidence arises?
- **Key question**: *"How can I use this effectively?"*

19.3.4 Knowledge Gateway

Here, information turns into **knowledge** when we test it in practice or measure it against other insights:

- *"Does it stand up under scrutiny?"*
- *"How does it fit with other pieces of knowledge I trust?"*
- *"Can others replicate or confirm these findings?"*

The **Temporal Truth Evaluation** remains vital. A concept that works brilliantly now might grow outdated or require modification in six months. Recognizing this impermanence keeps our knowledge current and open to refinement – a hallmark of a **living philosophy**.

Stage 4: Wisdom (Ethical and Foresight Floor)

Long-Term Perspective, Ethical Action

- **What to look for**:
  - **Broader impact**: Who or what might be affected by our decisions?
  - **Moral implications**: Does this align with our values and the greater good?
  - **Systemic awareness**: Are we considering cultural, technological, and environmental ramifications?
- **Key question**: *"What's the right thing to do?"*

19.3.5 Wisdom Gateway

At this apex, we blend **knowledge** with **ethical foresight**, **cultural sensitivity**, and **long-term vision**. Here's where **embodied cognition** (our emotional and intuitive responses) helps us sense potential blind spots or moral dilemmas. Wisdom acknowledges that even well-grounded knowledge can cause harm if applied without compassion or a vision for communal well-being.

19.3.6 Quality Checks and Temporal Awareness

19.3.6.1 Quality Gates

Between each DIKW stage, we have **quality checks** – stopgaps that ask:

- *"Has this data been vetted?"* (Data)
- *"Is the context sufficient?"* (Information)

- *"Does this stand under scrutiny and practice?"* (Knowledge)
- *"Have we weighed broader ethical and future consequences?"* (Wisdom)

19.3.6.2 Temporal Lens

Truth can evolve as **new data** arises or **cultural norms** shift. A living philosophy must track these changes, updating each stage in real time. A piece of knowledge that passes scrutiny today may need revisiting tomorrow.

19.3.6.3 Putting DIKW into Practice: A Concrete Example

Scenario: You're evaluating whether to deploy a cutting-edge AI tool in your company's HR department.

i. **Data**: You gather usage logs, reading articles about the tool's success elsewhere.
   - **Quality check**: Confirm the credibility of sources – industry experts, unbiased research, and up-to-date usage stats.

ii. **Information**: You integrate these raw stats with the company's context – how your HR processes work, employee feedback on existing tools, and regulatory compliance needs.
   - **Quality check**: Are you missing any perspective, like privacy concerns or user acceptance?

iii. **Knowledge**: You test the AI tool in a small pilot program to see if it actually reduces recruitment time without bias. You compare results to internal benchmarks and gather more feedback.
   - **Quality check**: Is the pilot data consistent with earlier research? Does it adapt well to your company culture?

iv. **Wisdom**: Finally, you assess the long-term ethical impact. Does the tool align with your company's values regarding fairness? Could it inadvertently discriminate against certain groups if the data sets are skewed? How does it fit into your living philosophy of fostering inclusive practices?
   - **Quality check**: Are you upholding your moral and strategic aims? Do you see potential unintended consequences or second-order effects?

If it all aligns, you decide to proceed, mindful that future updates to the AI or shifting labor laws may require re-evaluation.

Why This Model Matters

1. **Adaptable and Evolving**: By layering **temporal** checks and **quality gates**, DIKW becomes a **living** system rather than a static ladder. Each stage remains open to review, ensuring your insights remain valid in a changing environment.
2. **Ethically Grounded**: Rather than using knowledge purely for efficiency or profit, you emphasize **wisdom**, questioning moral implications and societal impact.
3. **Embodied Cognition**: Logical reasoning is crucial, but emotional and bodily cues often signal hidden risks or moral tensions. Factoring in these signals ensures truly holistic decision-making.
4. **Alignment with Your Living Philosophy**: This DIKW approach resonates with the **continuous learning**, **ethical foresight**, **systemic awareness**, and **narrative agility** that underpin a **living philosophy**. It prevents complacency by requiring that you repeatedly scrutinize your truths, moral stances, and assumptions.

A Guide to Implementation

1. **Begin at the Ground Floor (Data)**: Verify sources, cross-check claims, and ensure recency. This filters out misinformation early.

2. **Move to Information**: Seek context, identify patterns, and watch for biases or missing pieces. Evaluate how ephemeral or enduring each insight might be.
3. **Develop Knowledge**: Test theories in real scenarios. Gather user or stakeholder feedback. Stay open to revising your conclusions.
4. **Ascend to Wisdom**: Integrate ethical considerations, future consequences, and your embodied sense of rightness. Ensure alignment with your long-term vision and moral commitments.
5. **Feedback Loop**: Recognize that new information or cultural shifts may prompt a re-evaluation at any stage. Remain flexible and willing to descend the "pyramid floors" for updates.

19.3.6.4 Final Thoughts on DIKW as a Living System

In a world that changes faster than any static rulebook can accommodate, **DIKW** – enhanced by **temporal checks**, **quality gateways**, and **ethical reflection** – offers a **continually refreshing** perspective on truth. It's not a one-time climb but a cyclical process that aligns with the broader themes of **dynamic ethical emergence** and a **living philosophy**. Each iteration through these stages – data vetting, contextualizing information, cultivating actionable knowledge, and culminating in wisdom – enhances your capacity to engage with complexity responsibly and creatively. By treating DIKW as a **living approach to truth**, you ensure that your decisions remain anchored in **reality** yet guided by **moral foresight** – the ideal balance for flourishing in a rapidly shifting world.

19.4 YOUR INNER LANDSCAPE: A METAPHOR FOR CULTIVATING WISDOM

19.4.1 Introduction to the Inner Landscape

In a fast-moving, tech-driven world, it's easy to feel uprooted – tossed about by unrelenting change, new ethical conundrums, and torrents of information. **Your inner landscape** is a dynamic mental environment you can enter whenever you need to **re-center**, clarify your thinking, or reaffirm the principles of your living philosophy.

This landscape is not static. Like the living philosophy itself, it adapts and grows as you gain fresh data, insights, and ethical perspectives. Within it, you integrate the stages of the **DIKW framework** – from raw data to meaningful action – while keeping your decisions aligned with moral foresight. Think of it as a virtual realm for exploring the interplay between **complexity, emotional reflection, practical tools, integrated wisdom, and ongoing growth**. Whenever you feel overwhelmed or uncertain, mentally stepping into this landscape helps you reconnect with your resilience evidences and the guiding values at the heart of your living philosophy.

19.4.2 The Forest: Embracing Complexity and Pattern Recognition

Symbolism: The forest stands for the **intricate webs of modern life** – be they technological networks, cultural norms, or social ecosystems. At first glance, a dense forest can appear daunting: thick canopies obscure the path, and countless species intertwine in perplexing ways. Yet, as you venture in, you begin to notice patterns: certain plants thrive near water, some animals roam predictable routes, and the interplay of sunlight and soil reveals hidden logic.

Connection to DIKW:

- **Data (the "trees")**: On the ground level, the forest mirrors the endless facts and figures that flood our daily lives. Initially, they might seem random and overwhelming.
- **Information (contextual "paths")**: By spotting relationships – like which species need direct sun or which animals are nocturnal – you begin to glean meaningful context from these data points.
- **Living Philosophy**: Rather than dreading the complexity, you acknowledge it as **fertile ground** for insight. Complexity need not be chaos; it's a realm brimming with potential knowledge, waiting for you to observe patterns and extract meaning.

Action Cue: When you face an onslaught of data – be it news, analytics, or conflicting opinions – imagine yourself at the **forest's edge**. **Step in** with curiosity, seeking not to reduce the forest to one or two data points, but to find the hidden structure and relationships that turn random facts into **information**.

19.4.3 The Still Lake: Reflection, Self-Awareness, and Ethical Intuition

Symbolism: At the center of this forest, a **still, mirror-like lake** reflects both the sky above and your own inner state. As you kneel at its edge, small ripples distort the reflection before settling again, offering a clear image that blends internal and external realities.

Connection to DIKW:

- **Information to Knowledge**: The lake is where mere context (information) deepens into **knowledge** through **introspection** and **emotional intelligence**. It's not enough to see data in context; you also need to examine how it resonates with your beliefs, feelings, and ethical standards.
- **Embodied Cognition**: Just as the reflection shifts with each ripple, your emotional state and intuitive sense can color your interpretation of events. By allowing those ripples to settle, you gain **clarity** that merges rational understanding with inner alignment.

Living Philosophy: The Still Lake reminds you that **self-awareness** is indispensable for genuine wisdom. In a living philosophy, you repeatedly return to the lake to check your ethical stance, your biases, and your emotional reactions. This ensures your knowledge is **not purely intellectual** but also **deeply humane** and ethically grounded.

Action Cue: When uncertainty or inner turmoil strikes, envision yourself at the **lake's edge**. Breathe, let the internal ripples still, and ask:

- *"What biases might be influencing my interpretation?"*
- *"Are my values guiding this choice, or am I reacting from fear or habit?"*

19.4.4 The Boat: Navigating Complexity with Tools and Frameworks

Symbolism: A simple wooden **boat** rests at the water's edge, representing the **practical frameworks** that help you traverse the uncertain waters of complexity – tools like the **Integration Canvas, the Contradiction Navigator,** or the **enhanced DIKW pyramid**.

Connection to DIKW:

- **Knowledge into Action**: The boat is how you apply your carefully cultivated knowledge in real-world scenarios. While you may grasp the data, context, and meaning behind an issue, it's these practical methods that translate your insights into **purposeful, decisive movement**.
- **Living Philosophy**: Tools are flexible instruments, not dogmas. You guide the boat, adjusting your course based on shifting currents – akin to updating your approach as new data emerges or as your ethical viewpoint expands.

Action Cue: Before tackling a complex decision – like a major career choice or a team leadership challenge – visualize stepping into this **boat of frameworks**:

- *"Which tool best illuminates these conflicting priorities?"*
- *"How can the DIKW approach help me see the path forward?"*

Each successful crossing – resolving a conflict, balancing privacy vs. performance, or recalibrating a plan – yields **resilience evidences**, proof of your growing adaptability and ethical focus.

19.4.5 The Island: Integrating Knowledge into Wisdom

Symbolism: An **island** emerges in the midst of these waters, a place where diverse ideas and perspectives **coalesce** into **holistic insight**. On this island, your ethical foresight, analytical understanding, cultural sensitivity, and empathy **merge** into a unified vision.

Connection to DIKW:

- **Wisdom (the apex)**: The island symbolizes the pinnacle of the DIKW process, where **long-term impact** and **ethical considerations** crystallize into wise action.
- **Living Philosophy**: This is where your guiding principles come to life, no longer a patchwork of theories but an integrated approach that **balances rational analysis, moral responsibility, and human empathy**.

Action Cue: When you've tested your data, contextualized it, and reflected deeply, imagine arriving at **the island**. Here, you:

1. **Integrate** knowledge from different disciplines or frameworks.
2. **Consider** the broader social, environmental, and cultural ramifications.
3. **Commit** to actions that are both innovative and ethically grounded.

19.4.6 The Garden: Cultivating and Evolving Your Living Philosophy

Symbolism: Within the island's interior lies a **vibrant garden**, its plants continually adapting, blooming, or yielding fruit. This represents the **ongoing cultivation** of your living philosophy – nurturing growth through reflection, learning, and the acceptance that beliefs may need replanting or pruning as contexts shift.

Connection to DIKW:

- **Continuous Renewal**: Just as seasons change, your knowledge and wisdom undergo cycles of **updating, refining, and flourishing**. Data from last year's "harvest" may not suffice for next year's climate.
- **Resilience Evidences**: Each new insight or resolved dilemma is like fresh seeds planted in your garden – **tangible proof** of your adaptive capacity.

Living Philosophy:

- **Gardening Mindset**: You don't plant once and forget; you **weed out** outdated concepts, **fertilize** new ideas, and wait patiently for them to bear fruit.
- **A Long-Term View**: Just as a gardener plans for changing weather or shifting soil conditions, you approach ethical growth with a **flexible, future-ready mindset**, ensuring your philosophy remains **responsive and fruitful**.

19.4.7 Returning to the Landscape

Your inner landscape isn't a one-time sanctuary but an enduring **practice** – a mental space you revisit whenever you need **centering** or **renewed perspective**. Each element – forest, lake, boat, island, garden – serves a **distinct purpose** in supporting your living philosophy and the DIKW framework:

1. **Forest →** Embrace and Map Complexity (transform raw **data** into contextual **information**).
2. **Still Lake →** Reflect on Emotions and Ethics (integrate self-awareness to refine **knowledge**).
3. **Boat →** Apply Tools and Frameworks (carry that **knowledge** into **action**).

4. **Island** → Achieve Integrated Wisdom (converge insights into **wise**, long-term decisions).
5. **Garden** → Sustain and Evolve (keep your living philosophy adaptive, **continuously** iterating and welcoming new data).

Each journey through this inner world **affirms your resilience evidences**, reminding you that complexity can be navigated – indeed, **cultivated** – for deeper insight, ethical clarity, and ongoing growth. By internalizing and revisiting this landscape, you weave **DIKW** and a **living philosophy** into the core of who you are, ready for any challenge a rapidly changing world may present.

19.4.8 A Guided Journey: Revisiting Your Inner Landscape

Imagine you're stepping away from everyday noise and into a realm that is uniquely yours – a gentle, dawn-lit pathway leading toward the **forest**. The air is crisp, and a soft breeze carries a hint of possibility. **Take a slow breath** and let your shoulders drop as you begin this calming mental journey.

19.4.8.1 At the Edge of the Forest

You arrive at the forest's threshold, a place both mysterious and intriguing. It represents all the intricate information, data, and complexities swirling in your life – like a thousand different voices vying for your attention. You notice how the interlacing branches and thick foliage can feel overwhelming, much like headlines, emails, or your to-do list. Yet, instead of shrinking back, **you step forward**. In this forest, with patience and curiosity, you discern subtle patterns and hidden pathways.

- **Recall**: "I might have a flood of data or confusion, but just as in a forest, I'll look for patterns and connections, turning data into structured insight."

19.4.8.2 Deeper into the Forest, You Find the Still Lake

Following a winding trail, you emerge at a tranquil, mirror-like lake. Its surface is so calm that it reflects the sky and the forest around you. Kneeling at the water's edge, you watch your reflection wobble with each gentle ripple, reminding you of your own emotions and biases. As the water settles, the reflection grows clear – symbolizing how **self-awareness and introspection** bring clarity to your thoughts and feelings.

- **Recall**: "When I'm uncertain or emotionally tense, I can calm my internal ripples by pausing, reflecting on my ethical stance and the knowledge I've collected—thereby evolving raw information into deeper understanding."

19.4.8.3 A Wooden Boat at the Shore

Quietly docked is a simple **boat**, crafted from the tools and frameworks you rely on – like the DIKW pyramid, the Contradiction Navigator, or your favorite planning method. You step aboard, feeling the gentle sway as you push off into the lake. These frameworks keep you from drifting aimlessly; they help you chart your course.

- **Recall**: "Equipped with the right tools—my 'boat'—I can transform knowledge into effective action, be it solving a dispute at work or planning a family decision."

19.4.8.4 Reaching the Island of Integrated Wisdom

Across the water, an **island** rises. Sandy shores give way to lush greenery – a place where disparate ideas and insights converge. Stepping ashore, you sense the integration of all you've learned: your ethically grounded knowledge, your empathy, and your foresight. On this island, **wisdom** takes shape, weaving logic and intuition into a coherent, long-term perspective that respects both individual needs and the broader community.

- **Recall**: "Beyond data and knowledge, I aim for wisdom—actions aligned with my values, shaped by ethical considerations, and mindful of systemic consequences."

19.4.8.5 Venturing to the Garden

In the island's interior lies a **garden** – a living tapestry of blooming plants, each nurtured by attentive care. Sunlight filters through, and you feel a calm certainty here. This garden symbolizes the continued cultivation of your living philosophy: its growth in response to every new piece of data, every moral reflection, and every resilience evidence you earn. You plant seeds of new insights and harvest the fruits of lessons learned.

- **Recall**: "My living philosophy, like a garden, requires ongoing attention. I regularly check for weeds (outdated beliefs), provide fresh nourishment (new learning), and adapt to changing 'seasons' in my life."

Closing the Journey: As you prepare to leave this inner world, **take a moment** to look around: the forest (complex data), the lake (reflective insight), the boat (tools), the island (wisdom), and the garden (continuous growth). Each element holds a piece of the larger story – a map of how you transform raw data into caring, actionable wisdom aligned with your ethics and your evolving understanding of reality.

When you step back into daily life, carry the confidence of this landscape with you. You know where to find clarity when complexity strikes. You have a method for reflection, a set of tools to navigate, a place to form ethical perspectives, and a space for ongoing growth. Return to this mental story whenever the demands of modern life test your resilience or ethical bearings. Let each element remind you that learning, reflecting, acting, and evolving are not separate tasks but parts of one seamless, living philosophy.

By revisiting this inner landscape, **you affirm your readiness** to traverse complexity – no longer as a source of stress, but as an environment brimming with insights waiting to be discovered.

19.5 CONCLUSION: EMBRACING THE JOURNEY OF A LIVING PHILOSOPHY

19.5.1 Recap of the Three Pillars

1. **The Need for a Living Philosophy**: Static models crumble under the weight of **ever-accelerating change** – from rapid technological innovations to shifting cultural norms. In a world where yesteryear's ethical guidelines might already be obsolete today, a *living philosophy* remains the only approach flexible enough to adapt and remain relevant.
2. **Using the DIKW Pyramid**: From raw data to applied wisdom, the DIKW framework offers a structured method to **filter, interpret, and act** on information. Enhanced by **temporal sensitivity** (acknowledging that truths may evolve over time), DIKW ensures your judgments remain accurate, responsible, and ethically grounded – even as new facts emerge or cultural attitudes shift.
3. **The Inner Landscape**: You've explored a metaphorical realm – **forest, lake, boat, island, and garden** – each symbolizing a stage in how you engage with data, knowledge, and ethical reflection. This landscape is a dynamic space where you can integrate insights and continually adapt. It provides a **psychological anchor**, reminding you to embrace complexity, reflect deeply, employ practical tools, unify perspectives, and cultivate your evolving set of values.

19.5.2 Emphasis on Continuous Evolution

A living philosophy is **not a final endpoint**. Rather, it's an ongoing dialogue between you and the world around you, a commitment to:

- **Lifelong Learning**: Regularly updating and refining your mental models.
- **Ethical Foresight**: Considering potential impacts on people, systems, and the planet.
- **Systemic Awareness**: Seeing every action in the context of interconnected cultural, technical, and social webs.
- **Narrative Agility**: Remaining open to reframing stories as new information or ethical insights arise.

Cultivating this approach demands:

- **Humility**: Admitting there's always more to learn or reconsider.
- **Curiosity**: Actively seeking fresh perspectives, data sources, and cultural understandings.
- **Empathy**: Recognizing how your decisions ripple through your community and ecosystems.

Where old doctrines collapse under novelty, **a living philosophy thrives on change**, turning today's breakthroughs into tomorrow's opportunities for ethical growth.

19.5.3 The Power of Resilience Evidences

Each time you apply the DIKW approach to decode a complex challenge – each time you steer your "boat" of frameworks through ethical dilemmas – you add to your collection of *resilience evidences*. These are **tangible reminders** that you can adapt, innovate, and thrive despite uncertainty.

"These resilience evidences are not just records of past successes; they are the building blocks of your living philosophy. Each time you successfully navigate a complex situation or apply a framework effectively, you gather proof of your capacity for ethical growth."

1. **Practical Example**: Completing a project under high-pressure constraints where you used the **Contradiction Navigator** to resolve fairness vs. speed or used **DIKW** to confirm reliable data.
2. **Psychological Benefit**: Resilience evidences boost your confidence; *"If I managed that, I can manage this."*

- **Ethical Reinforcement**: They also remind you that skillful, value-based action is feasible – even in uncharted territory.

Think of storing each new resilience evidence in the **"garden"** of your inner landscape – nourishing it as a seed that will bear fruit for future endeavors.

19.5.4 Call to Action

1. **Test Your Tools**: In the coming week, pick one real challenge – personal or professional – and **apply the DIKW pyramid**. Start by verifying your data sources (Data level), build context (Information), consider real-world applications (Knowledge), and reflect on ethical ramifications (Wisdom). Document how your perspective shifts from raw data to wise, ethically informed decisions.
2. **Reflect in Your Inner Landscape**: Whenever stress or confusion strikes, **take a few minutes** to mentally revisit the forest, lake, boat, island, and garden. Ask yourself:

    - Which part of my landscape do I need most now – pattern recognition, self-reflection, tool application, or integrated wisdom?
    - Which stage of DIKW am I currently at – do I need more data, or am I missing a bigger ethical viewpoint?

3. **Evolve Consciously**: Each time you see your knowledge advancing or your ethics sharpening, *mark that as a resilience evidence*. Let each small success reinforce your capacity to handle higher-level complexities. As your garden accumulates these "growth markers," you'll find motivation to tackle even larger challenges with renewed poise.
4. **Share and Collaborate**: Spread this mindset – encourage friends, colleagues, and teams to adopt a **living philosophy**. Our global challenges (climate change, AI ethics, and mental health crises) require a **collective, adaptive ethos** that no single set of static guidelines can address. By sharing what you've learned, you become part of a wider solution, multiplying the impact of every resilience evidence you collect.

19.5.5 Final Thought

A living philosophy is a promise you make – **to yourself and to society** – to remain open, inquisitive, and ethically engaged in a world that refuses to stand still. By weaving the DIKW pyramid into your intellectual processes and grounding it in the imagery of your inner landscape, you equip yourself for complexity, not with dread, but with *curiosity*, *compassion*, and *clear purpose*.

Continue your journey confident in your tools, anchored in your evolving moral compass, and fueled by the resilience evidences you collect. In so doing, you reshape your relationship with complexity – turning it from a barrier into a **source of wisdom and ethical innovation**.

The future is not a predetermined path but a canvas awaiting your brushstrokes of thoughtfulness and integrity. Together, let us co-create a world where ingenuity meets ethics, where uncertainty sparks growth, and where human potential flourishes beneath the ever-evolving skies of a **living, adaptable philosophy**.

THE POWER OF STORYTELLING

THINKING BOX

"How do digital narratives mold our views and hopes? Their influence—like the meaning of **42**—may span far beyond the obvious."

Epilogue

The Journey Forward

Key Reflections:

1. **From Observer to Co-Creator:**
   - You no longer passively watch complexity unfold – you **engage, adapt, and shape it** through ethical frameworks and resilience evidences.
2. **Integrating Knowledge into Action:**
   - **Complex Systems Thinking:** Understanding interdependencies and leverage points.
   - **Embodied Systemic Intelligence (ESI):** Aligning intuition, emotions, and systemic awareness for wise decision-making.
   - **Narrative and Meaning-Making:** Shaping collective stories to foster unity and progress.
   - **Personal ↔ Collective Growth:** Individual insights ripple into societal transformation.
3. **Sustaining a Living Philosophy:**
   - **Continuous Learning:** Stay open to evolving truths.
   - **Practical Application:** Embed ethical habits into daily routines.
   - **Measurable Growth:** Track impact and refine approaches.

Key Takeaway:

This is **not an ending but a threshold** – your **living philosophy grows with every reflective action, ethical decision, and collaborative endeavor**. The future is uncertain, but by embracing **complexity, adaptability, and systemic intelligence**, you **co-create a world where individual and collective flourishing align**.

E.1 REFLECTING ON THE PATH

E.1.1 Stepping Out of Complexity's Shadow

As you arrive at these closing pages, pause to consider just how far you've journeyed. What began as a confrontation with overwhelming complexity – looming environmental crises, frenetic technological breakthroughs, and cultural narratives in constant flux – has slowly transformed into a deliberate reimagining of your role in this unfolding story. Once upon a time, you stood at the edge, an onlooker attempting to decode the relentless churn of events. Yet now, the complexity that seemed so intimidating has become a canvas for co-creation, calling forth your ingenuity, compassion, and resolve.

E.1.2 From Observer to Co-Creator

Earlier in your exploration, global challenges loomed as monolithic puzzles beyond your grasp. Through each page, however, you encountered frameworks. These frameworks wove scientific insights with ethical reflection, personal introspection with societal innovation, helping you see that even colossal crises are networks of interlinked factors – possible to map, understand, and influence.

This shift from observer to active co-creator is not a minor adjustment; it marks a fundamental change in how you perceive your agency – your narratives, your engagement with others, and your capacity for nuanced thought all contribute to the broader ethical ecosystem. No longer do you view moral paralysis as inevitable; you now recognize how daily choices can reshape norms, and how collective efforts can break cycles of inertia.

E.1.3 Accumulating Resilience Evidences

Along the way, you gathered resilience evidences – concrete signs of your growing ability to navigate complex moral terrains. These might range from a newly established habit of mindful reflection to orchestrating a community initiative that addresses a local ecological issue. Each proof stands as a living testament to how awareness, thoughtful engagement, and collaborative processes can reshape outcomes once assumed immutable.

Far from mere anecdotes, these resilience evidences serve as durable markers of ethical progress, reinforcing your confidence that growth is not just possible but already unfolding. They signify that you have moved beyond theoretical interest in ethics toward its practical realization, engaging with real-world dilemmas in creative, life-affirming ways.

E.1.4 Cultivating a Living Philosophy

You've also encountered concepts such as ESI, *recursive moral development*, and *narrative complexity*, each weaving insights from neuroscience, evolutionary psychology, cultural studies, and ethics into a coherent approach. The sum of these elements forms a "living philosophy." This isn't a static moral doctrine, but a fertile approach to ethics that evolves with new information, responds fluidly to emerging challenges, and thrives on reflection.

By integrating these ideas, you've begun to see how morality can remain stable yet flexible – grounded in enduring values like empathy or fairness but open to reinterpretation as social and technological landscapes shift. In this way, each lesson throughout these chapters has contributed to an adaptive moral compass, refined through reflection, experimentation, and the stories we share.

E.1.5 A Sense of Interconnectedness

Through these pages, your perspective on interconnectedness has deepened. Every aspect of your life – from intimate relationships to participation in civic or digital communities – is part of overlapping networks. This realization underscores the truth that personal well-being and social health are inseparable. Local actions can have global effects; global shifts can return to impact your day-to-day life. Your ethical growth no longer unfolds in isolation – it intersects with communities, institutions, and digital platforms, making your role as an ethical agent simultaneously personal and collective.

E.1.6 The Emerging Role of Humility and Courage

Stepping into this co-creative role demands humility – a recognition that you cannot possibly control all variables or anticipate every consequence. Yet the very chaos of global challenges implies a need for bold innovation. This requires the willingness to prototype solutions, learn from failure, and continuously refine both personal behaviors and collective policies. Only with both humility and courage can we navigate uncertainties without succumbing to fatalism.

E.1.7 An Ongoing Narrative

Though these pages conclude, the story itself continues. You possess not just new theoretical knowledge but an integrated approach to living ethically in a complex world. It's your daily engagement – through ethical experiments, dialogues, and challenging old norms – that writes the next chapters of this story. As you turn from reader to author, remember that each choice contributes to a living tapestry of ethical evolution, influencing the shared future we all inhabit.

With these reflections in mind, let's transition to a final synthesis of how a living philosophy – embodied, adaptive, and interwoven with collective narratives – can continue to guide you in the months and years to come.

E.2 THE INTEGRATION OF UNDERSTANDING

Over the course of this exploration, you have encountered four interwoven threads that shape our engagement with complexity: **Complex Systems Thinking**, **ESI**, **Narrative and Meaning-Making**, and the continual interplay of **Personal ↔ Collective Growth**. Weaving these threads together transforms insight into a cohesive, living philosophy – one that responds fluidly to new challenges, fosters well-being at both personal and communal levels, and builds on the synergy between individual action and societal transformation. Below is a deeper look at these four threads, along with more exhaustive examples and explanations to help you apply them in your daily life.

E.2.1 Complex Systems Thinking

E.2.1.1 Core Idea

Complex Systems Thinking teaches us to see beyond isolated events. It emphasizes patterns, feedback loops, emergent properties, and non-linear cause-effect relationships. This perspective unveils how small decisions can catalyze large-scale shifts and highlights that our world is interwoven by interacting systems – ecosystems, economies, cultural networks, and technological platforms – all influencing each other in sometimes unpredictable ways.

E.2.1.2 Why It Matters

- **Identification of Leverage Points**: Recognizing the nodes in a network where a modest nudge can yield outsized benefits (e.g., improving community recycling, which then galvanizes broader environmental initiatives).
- **Anticipation of Side Effects**: Spotting the potential unintended consequences of a decision – useful in personal contexts (like changing a daily routine) and larger domains (such as policymaking).
- **Enhanced Resilience**: Understanding interdependencies fosters adaptability, helping you pivot gracefully when a system experiences sudden strain.

E.2.1.3 Expanded Example

Imagine you start a simple practice: sorting your trash meticulously to reduce landfill waste. Initially, this seems trivial, but neighbors notice and begin doing the same. Within months, some advocate for better city programs – like more recycling drop-off points or composting services – making the entire neighborhood's environmental footprint smaller. By adopting one discrete habit, you have influenced your social environment to shift its broader waste management behavior, illustrating how local action intersects with systemic change.

E.2.1.4 Key Takeaway

Complex Systems Thinking reframes challenges from daunting monoliths into networks of relationships. By identifying how smaller pieces interact, you uncover powerful points for meaningful intervention.

E.2.2 Embodied Systemic Intelligence (ESI)

E.2.2.1 Core Idea

ESI contends that ethics, well-being, and insight arise from integrating signals at three levels:

- **Physical:** Bodily cues such as stress, tension, calm, or excitement.
- **Emotional:** Affective states including empathy, concern, or moral outrage.
- **Societal:** Cultural norms, collective values, and shared beliefs.

ESI bridges the gap between "hard" logic and intuitive, body-centered wisdom, showing that moral action is not just reasoned, but deeply felt.

E.2.2.2 Why It Matters

- **Immediate Emotional Feedback:** Bodily sensations (e.g., tightness in your stomach when something feels "off") can alert you to ethical or emotional red flags before logic fully kicks in.
- **Deeper Ethical Engagement:** By honoring and interpreting bodily and emotional cues, you create a richer, more responsive decision-making process that aligns with both personal integrity and societal well-being.
- **Personal-Societal Loop:** When you heed the collective emotional climate (e.g., a community's anxiety about a local crisis), you can respond compassionately and effectively, fostering trust and unity.

E.2.2.3 Expanded Example

Consider reading disheartening headlines about climate change. You notice your heart rate quickens and a wave of tension sets in. Rather than letting that anxiety spiral into despair, you pause – recognizing the tension as a physical sign prompting you to act. You decide to hold a small neighborhood workshop on practical sustainability tips. Through this bodily-emotional signal, you transform worry into empathetic outreach and collaborative innovation.

E.2.2.4 Key Takeaway

ESI integrates personal bodily cues with emotional resonance and social norms, illustrating that meaningful, ethical decisions emerge from a blend of reason, embodied feeling, and collective context.

E.2.3 Narrative and Meaning-Making

E.2.3.1 Core Idea

Narratives – personal stories, cultural myths, institutional rhetoric – are the lenses through which groups interpret reality and imagine solutions. How a community talks about a crisis, a conflict, or an innovation can determine whether it sees obstacles or opportunities.

E.2.3.2 Why It Matters

- **Shaping Collective Action**: A negative, divisive story (e.g., "us vs. them") can fracture cooperation, while a positive, unifying narrative ("we're in this together") fosters collaboration and shared purpose.
- **Moral Imagination**: Stories offer scenarios for exploring ethical dilemmas, making them tangible and relatable, and revealing new paths to conflict resolution or community-building.
- **Cultural Transformation**: Shifting the dominant narrative can realign entire systems – leading, for instance, to more inclusive policies or new social norms regarding sustainability.

E.2.3.3 Expanded Example

In a workplace where the prevailing story is "management fights with staff," morale and productivity decline. A middle manager begins circulating a new narrative: "We're a unified team pursuing a shared mission to serve our customers better." Over time, employees start referencing mutual goals in meetings, managers encourage feedback loops, and the culture gradually shifts from adversarial to collaborative. By re-authoring the narrative, the entire workplace dynamic evolves.

E.2.3.4 Key Takeaway

When you challenge worn-out or divisive stories, you free space for more empowering narratives that guide collective ethics and propel social innovation.

E.2.4 Personal ↔ Collective Growth

E.2.4.1 Core Idea

There is a dynamic feedback loop between individual development and societal change. Personal growth – your improved habits, expanded worldview, or deepened empathy – can influence social norms, just as shifting cultural contexts can propel you to grow further.

E.2.4.2 Why It Matters

- **Reciprocal Influence:** One person's courageous step toward sustainable living can spur community engagement, which in turn encourages that person to refine and amplify their efforts.
- **Scaling Up Ethics:** Positive changes in your life become visible examples, nudging others to adopt similar initiatives. When entire groups follow suit, norms shift, policies transform, and long-lasting societal impact unfolds.
- **Continual Reinforcement:** As communities and institutions adopt more ethical stances, they create environments that nurture further individual growth, reinforcing a virtuous cycle.

E.2.4.3 Expanded Example

A single household decides to switch to solar energy. Neighbors take note, curiously asking about costs and benefits. Before long, local associations provide workshops on solar installations, prompting more homes to go solar, illustrating how visible personal choices catalyze collective transformations. Meanwhile, observing such widespread adoption reaffirms the original household's commitment, inspiring them to explore further sustainability measures.

E.2.4.4 Key Takeaway

Personal ↔ Collective Growth demonstrates that meaningful progress isn't a one-way street; it's a loop where each personal victory resonates outward, and each societal shift creates supportive conditions for deeper personal transformation.

E.2.4.5 Conclusion: The Four Threads Converge

These four threads – **Complex Systems Thinking**, **ESI**, **Narrative and Meaning-Making**, and **Personal ↔ Collective Growth** – demonstrate that meaningful change arises from the dynamic interplay between personal insight and communal evolution, not from isolated ideas. Each thread enriches your capacity to perceive complexity, respond ethically, and craft resonant stories. This empowers you to co-create a world where personal flourishing and collective well-being are intertwined.

E.3 ESI: THE BRIDGE

At the heart of this evolving understanding is the principle of ESI – a pivotal concept that connects individual development with large-scale social transformation. By unifying three levels of awareness – physical (bodily states), emotional (affective responses), and systemic (collective dynamics) – ESI underscores that meaningful ethical action arises from an integration of personal insight, emotional resonance, and a clear grasp of how society's interlocking systems function.

E.3.1 What is ESI?

1. **Physical Awareness**
    - **Core Idea**: Our bodies provide real-time signals – tension, ease, heightened alertness – that can guide us toward ethical and constructive behavior.

- **Why It Matters**: Bodily cues often precede conscious rational deliberation, offering quick feedback on situations that might require moral attention. This "somatic compass" helps us detect discomfort or alignment before our intellect fully processes it.

2. **Emotional Intelligence**
   - **Core Idea**: Emotions aren't just "feelings"; they help us understand priorities, empathize with others, and sense urgency or harmony in a given issue.
   - **Why It Matters**: Emotions can either block moral insight (when left unchecked) or enhance it (when skillfully managed). By tuning in, we regulate our responses, forming genuine connections and fostering collaborative problem-solving.
3. **Systems Understanding**
   - **Core Idea**: True ethical impact requires looking beyond individual wants to see how decisions ripple through social networks, cultural narratives, and institutional frameworks.
   - **Why It Matters**: Recognizing these interconnected layers prevents narrow, short-term fixes. Instead, it encourages approaches that consider downstream impacts, long-term sustainability, and the wellbeing of diverse stakeholders.

E.3.2 ESI in Action: The Environmental Activist Example

Imagine an environmental activist deeply disturbed by the growing pollution in a local river. This visceral reaction – perhaps a knot in the stomach or a surge of stress – represents the **physical awareness** that something is fundamentally wrong. Instead of succumbing to anxiety, the activist harnesses that bodily signal as an impetus for **emotional intelligence**: they engage in empathetic dialogue with concerned neighbors, listening to their fears and forging a communal sense of responsibility.

From there, the activist applies systems understanding: they look at how city policies, corporate waste disposal practices, and community habits interlock to sustain the pollution problem. Equipped with this systems perspective, they coordinate multi-stakeholder initiatives – inviting local businesses, municipal authorities, and citizen groups to collaborate on policy changes and river clean-up drives.

E.3.2.1 Key Takeaway

This scenario shows how **ESI** translates personal concern (physical and emotional awareness) into strategic community engagement (systemic thinking). Rather than letting distress morph into cynicism, the activist leverages it for constructive activism – rallying neighbors, altering local policies, and promoting broader cultural shifts around environmental stewardship.

E.3.3 Why ESI is the Crucial Bridge

1. **Personal Flourishing Meets Collective Transformation**
   - ESI ensures that individual emotional cues and bodily states fuel ethical action, not just personal coping. By aligning personal well-being with social responsibility, ESI keeps self-care and societal care mutually reinforcing.
2. **Empathy as a Channel for Systems Change**
   - Emotional intelligence transforms raw concern into empathetic engagement. It fosters alliances across different segments of society – NGOs, government offices, private companies – facilitating the systemic changes needed to address deep-rooted issues.
3. **Responsive, Not Reactive**
   - Through ESI, ethical choices become responsive rather than reactive. Physical signals alert us early, emotional intelligence refines our motivation, and systems thinking pinpoints effective interventions – preventing hasty decisions that might overlook deeper ramifications.

4. **Scale-Independent Framework**
   - ESI applies at every scale: from a personal moment of choosing a more sustainable daily routine to orchestrating policy-level collaborations that reshape entire industries. This scalability ensures that ESI can guide action whether you're focusing on self-care, leading a team, or catalyzing community-driven change.

E.3.4 Practical Entry Points to ESI

- **Mind-Body Check-Ins**: Briefly pausing throughout the day to scan for physical sensations – tight shoulders, racing heart – can indicate emerging stressors or moral tensions.
- **Emotion Journals**: Tracking emotional states in response to local or global events. Identifying recurring patterns clarifies what truly matters to you and how you might act on those concerns.
- **Systemic Listening**: When confronted with problems, practice "mapping" their various elements. Who are the stakeholders? Which structures keep this issue in place? What cultural narratives amplify or mitigate it?
- **Community Dialogues**: Bring ESI into group discussions by intentionally acknowledging both emotional undercurrents and the complex interplay of institutional or cultural factors, ensuring decisions are holistic.

E.3.5 Conclusion: ESI as a Foundation for Ongoing Evolution

ESI is more than a technique; it's a **mindset** that respects human complexity and the interconnectedness of modern life. It offers a coherent pathway for transforming inner awareness into societal-level influence, clarifying how individual well-being dovetails with collaborative innovation and systemic reform. By emphasizing bodily attunement, emotional resonance, and multi-layered systems thinking, ESI ensures that your ethical engagements – large or small – contribute to an ever-adapting cycle of growth for yourself, your community, and the broader world.

E.3.5.1 Key Takeaway

ESI provides a clear, pragmatic framework for turning personal concerns into effective collective action, bridging self-care and community leadership. As you continue along this journey, remember that each step – each mindful breath, empathetic conversation, and systemic insight – reinforces a culture of ethical resilience, ensuring that progress is both inwardly grounded and outwardly transformative.

E.4 THE LIVING FRAMEWORK

Your insights now coalesce into a **living framework** – continuously enriched by practice, reflection, and feedback. Picture this framework like a cultivated garden: you plant seeds of understanding, water them with deliberate practice, and observe how each season yields new blossoms of ethical insight and skill. These blossoms reflect both your personal growth and its ripple effects across your community. Over time, the framework adapts to changing conditions, mirroring the dynamic nature of life itself.

Below, we examine three crucial aspects of this living framework: **Continuous Learning**, **Practical Application**, and **Measurable Growth**. Each aspect contains a rationale ("Why It Matters") and suggested ways to track progress ("How to Measure"), ensuring that your journey remains active, iterative, and grounded.

E.4.1 Continuous Learning

Why It Matters

- **Constant Adaptation:** In a world of rapid technological shifts and evolving social norms, continuous learning prevents today's best practices from becoming tomorrow's liabilities. By regularly updating your knowledge, skills, and mindsets, you ensure you're prepared to handle new dilemmas with agility and insight.
- **Neural Plasticity and Growth:** Your brain's capacity to rewire itself (neuroplasticity) allows moral understanding and cognitive flexibility to deepen over time. Each round of learning – whether through reading, mentorship, or experiential practice – helps form or strengthen neural pathways that support ethical judgment.
- **Sustained Curiosity:** A willingness to learn from diverse sources – mentors, peers, and global events – fuels both personal and communal resilience. This curiosity broadens your perspective, challenges entrenched biases, and keeps your cognitive and moral senses sharp.

How to Measure

- **Reflective Practices Log:** Keep a brief record (digital or paper) of each journaling session, mindfulness practice, or guided reflection. Note any shifts in clarity, mood, or decision-making.
  Example: After a morning meditation, jot down a "focus score" (1–5) indicating how centered you felt. Over weeks, look for patterns – do lower scores coincide with busier schedules or stressful events?
- **Decision-Cycle Reviews:** Document major decisions, noting how much analysis you applied, which mental models you used, and the outcomes.
  Example: A simple spreadsheet could have columns for the "Situation," "Tool/Framework Used," "Outcome," and "Lesson Learned." Periodically review to see which approaches correlate with better results.
- **Skill-Development Tracking:** When learning new approaches (e.g., systems mapping, narrative bridging), periodically evaluate your progress using stages like "Beginning," "Practicing," and "Mastering."
  Example: A short weekly self-assessment might show you've moved from "Beginning" to "Practicing" in conflict resolution after regular role-playing exercises.
- **Final Note:** By regularly reviewing these logs and templates, you can identify patterns, track progress, and refine your approach to continuous learning. Over time, you'll see how small, consistent efforts accumulate, enabling ever-deeper ethical, cognitive, and emotional development.

E.4.2 Practical Application

Why It Matters

- **ESI:** Ethical and personal growth requires embodiment. Practical application translates abstract concepts into tangible actions that shape daily life. Even if you grasp moral principles mentally, it is through concrete, lived practices – such as mindful pauses, structured debates, or collaborative projects – that your ethics become real.
- **Habits of Implementation:** Simple, consistent frameworks – like daily "stress checks," gratitude reflections, or ethical decision prompts – lay the groundwork for habitual ethical engagement. These routines shape your attitudes, influence relationships, and can quietly shift the norms within your social circles.

- **Tangible Improvements:** When practices result in tangible improvements – calmer interactions, greener choices, or more equitable policies – they become self-reinforcing, driving further engagement and continuous refinement of your methods.

How to Measure

- **Tool Usage Frequency:** Maintain a concise log when you apply a practice or technique, be it a mindful pause or an ethical reflection prompt.
  Example: In a small notebook, tally each instance: "Used Contradiction Navigator to resolve team conflict" or "Took 30-second breath break before responding to an email." Over time, see which practices you rely on most.
- **Outcome Reflections:** Pair usage data with notes on the effect. Did a mindful pause diffuse tension in a meeting? Did your Conflict Navigator framework guide a more balanced family discussion?
  Example: A weekly or monthly review might have a column for "Outcome Rating" (1–5) or short remarks on whether the action improved communication or alleviated stress.
- **Impact on Relationships:** Informally or formally survey those around you – friends, colleagues, family – to gauge whether they perceive improvements in communication, empathy, or conflict resolution.
  Example: A quarterly anonymous survey among team members asking them to rate the "collaborative atmosphere" of the workplace. Compare results over time to see if your individual practices contribute to broader improvements.
- **Final Note:** By correlating tool usage with observed outcomes, you can refine your choice of methods and approaches, ensuring maximum effectiveness. This cycle of experimentation, feedback, and adaptation is the engine that translates ethical knowledge into embodied daily life.

E.4.3 Measurable Growth

Why It Matters

- **Visibility of Progress:** Identifying tangible changes in emotional balance, interpersonal dynamics, and even environmental markers helps you sustain motivation. Without concrete indicators, gradual improvements might go unnoticed, hindering further commitment.
- **Adaptive Feedback:** Regular assessment reveals the health of your personal growth and its communal impact. If emotional stability is rising but teamwork remains stagnant, that discrepancy highlights where additional effort or new strategies are needed.
- **Reinforcing Success:** Each documented improvement serves as evidence of your capacity to navigate complexity ethically and creatively. These recorded achievements bolster confidence and build resilience, encouraging further ethical risk-taking and innovative solutions.

How to Measure

- **Emotional Well-Being Indicators:** Use quick rating scales (mood trackers, the PERMA model) to see if mindfulness, journaling, or other practices yield lasting shifts in emotional states.
  Example: A daily "mood check" app that tracks your sense of calm or stress. Over a month, observe patterns – does your sense of well-being consistently improve when you do certain practices?

- **Relational Harmony Metrics:** Solicit brief feedback from peers or family about conflict resolution, team unity, or trust levels. Even informal text or email polls can highlight changes in group cohesion.
  Example: If you lead a small team, conduct a short "harmony pulse check" monthly, asking for a 1–5 rating of team synergy and any optional comments. Watch trends over time.
- **Community and Environmental Markers:** Note if your lifestyle adjustments (reduced waste, local volunteering) inspire neighbors or colleagues, or if they nudge organizational or community policies.
  Example: Tally the number of neighbors who followed your lead by installing solar panels, or measure how many community members joined your local climate-action group after a campaign.
- **Final Note:** By tracking these diverse metrics, you gain a holistic view of your growth across emotional, relational, and environmental dimensions. This fosters balanced development, deeper purpose, and genuine enthusiasm for continued ethical engagement.

E.4.3.1 Concluding Thought: The Framework as an Ever-Adapting System

This **living framework** – encompassing continuous learning, practical application, and measurable growth – resembles an **ecosystem** evolving under the guidance of both internal and external feedback. A cycle of reflection, adaptation, and renewed commitment ensures that your moral journey remains vibrant, flexible, and increasingly impactful.

As conditions shift – personally, socially, or technologically – your framework adjusts, integrating fresh data, new insights, and emergent challenges. Over the long arc, you'll discover calmer emotional states, more harmonious relationships, and a heightened capacity to influence your broader environment. By embodying these dynamic principles, you uphold the spirit of a **living philosophy** – one that remains grounded yet innovative, deeply personal yet resonant with collective aims, and always ready to engage with the uncharted complexities of our world.

E.5 FACING OBSTACLES: TURNING HURDLES INTO GROWTH

Proactively recognizing and transforming these obstacles into opportunities for deeper resilience and innovation is crucial. By approaching each challenge with creativity and open-mindedness, you can reinforce the core principles of ESI – ultimately weaving ethical awareness, emotional intelligence, and systemic thinking more deeply into your daily life and community.

E.5.1 Consistency: Sustaining Daily Practices

Why It Matters

1. **Habit Formation**
   - **Core Idea:** Daily ethical or mindful micro-habits (e.g., short meditations, reflection prompts) gradually become second nature.
   - **Why This Helps:** Repeated engagement stabilizes new neural pathways involved in ethics, empathy, and holistic awareness, converting good intentions into ingrained behaviors.
2. **Building Momentum**
   - **Core Idea:** Continuous engagement fosters compounding benefits. The insights of one day build upon the next, strengthening neural pathways related to ethical, emotional, and systemic awareness.
   - **Why This Helps:** Regular reinforcement minimizes "ethical drift" – the tendency to revert to less mindful behaviors when routines lapse.

How to Manage

1. **Start Small**
   - **Approach:** Begin with just two minutes of daily mindful breathing or a brief ethical "check-in."
   - **Rationale:** Minimal time investment reduces resistance; success with small steps builds confidence for more extended practices later.
   - **Example:** A daily two-minute body scan in the morning can gradually evolve into a ten-minute reflective habit.
2. **Pair with Existing Habits**
   - **Approach:** Integrate your micro-habits with established routines. For instance, do a quick mindfulness check after brushing your teeth or take a 30-second breath before checking email.
   - **Rationale:** Attaching new habits to automatic behaviors is a proven way to reinforce consistency.
   - **Example:** Perform a short gratitude exercise immediately after finishing lunch to seamlessly embed reflection into your day.
3. **Use Gentle Accountability**
   - **Approach:** Enlist a friend or colleague for weekly progress check-ins.
   - **Rationale:** Light social reinforcement helps maintain motivation without creating excessive pressure or guilt.
   - **Example:** Share one brief reflection each day in a group chat, celebrating small successes and exchanging encouragement.

Potential Pitfalls

- **All-or-Nothing Thinking:** If you expect perfection, missing one day can spiral into quitting altogether.
- **Overlooking Cumulative Effects:** Dismissing brief habits as "too small" misses the exponential gains they can yield over time.
- **Routine Fatigue:** Without occasional changes (e.g., varying the time or type of exercise), even short practices can grow stale.

E.5.2 Biases and Blind Spots: Overcoming Limitations

Why It Matters

1. **Hidden Barriers**
   - **Core Idea:** Cultural, emotional, or cognitive biases can unconsciously skew perception and decision-making.
   - **Why This Helps:** Identifying and addressing these biases leads to more accurate judgments and fairer interactions.
2. **Inclusivity and Fairness**
   - **Core Idea:** Overcoming biases leads to more equitable and innovative solutions by leveraging the full spectrum of available insights.
   - **Why This Helps:** Cultural sensitivity and perspective-taking enrich group problem-solving and foster a more cohesive community.

How to Manage

1. **Actively Seek Diverse Input**
   - **Approach:** Incorporate viewpoints from different cultural backgrounds, professions, or age groups when tackling challenges.

- **Rationale:** Exposure to varied perspectives often illuminates blind spots that homogenous teams overlook.
- **Example:** In decision-making panels, rotate leadership roles or invite community members typically underrepresented.

2. **Practice Structured Dialogue**
   - **Approach:** Use formal methods like "active listening circles," where participants paraphrase each other's points before adding new insights.
   - **Rationale:** This ensures all voices are heard, reducing the chance that hidden biases dominate.
   - **Example:** Before finalizing a workplace policy, gather the team in a listening circle so each person can articulate the policy's potential impact on them.
3. **Reflect on Emotions**
   - **Approach:** Record these instances and later, in calmer moments, reflect on how biases may have influenced your reaction.
   - **Rationale:** Emotional triggers often signal areas where personal or cultural biases are at play, allowing for more deliberate adjustment.
   - **Example:** If you feel defensive or upset during a debate, note that emotional spike. Later, ask: "Which bias might have been activated here?"

Potential Pitfalls

- **Tokenism:** Merely inviting diverse voices without sharing real power can maintain existing biases rather than resolve them.
- **Overconfidence in Objectivity:** Believing oneself immune to bias can hinder critical self-examination.
- **Hostile Confrontations:** Confronting biases too bluntly or publicly can provoke defensive reactions, worsening conflict rather than resolving it.

E.5.3 Resource and Time Constraints: Working Within Limits

Why It Matters

1. **Realistic Planning**
   - **Core Idea:** Acknowledging limited time and resources allows for more strategic and effective interventions.
   - **Why This Helps:** Overcommitting can lead to burnout, whereas realism fosters steady progression.
2. **Long-Term Impact**
   - **Core Idea:** Strategic prioritization allows smaller pilot projects to demonstrate impact, building public trust and momentum for larger initiatives.
   - **Why This Helps:** Demonstrated success becomes a catalyst for broader endorsement, funding, or community engagement.

How to Manage

1. **Pilot Projects**
   - **Approach:** Start with a manageable, short-term initiative that addresses a local need or a specific challenge.
   - **Rationale:** Early wins validate the approach and encourage further investment and replication.
   - **Example:** Launch a one-month neighborhood recycling competition to show tangible improvements in waste management.

2. **Leverage Existing Platforms**
   - **Approach:** Adapt known tools or networks (e.g., community centers, existing apps) rather than building new ones from scratch.
   - **Rationale:** Reduces setup costs, shortens the learning curve, and integrates seamlessly into daily routines.
   - **Example:** Use an established local Facebook group to coordinate a sustainability drive instead of creating a separate app.
3. **Gradual Scaling**
   - **Approach:** Expand in phases, each step drawing on lessons from previous efforts.
   - **Rationale:** Controlled growth prevents resource overload and allows for iterative refinement.
   - **Example:** A successful pilot recycling program in one district can then expand city-wide, eventually informing state-level environmental policies.

Potential Pitfalls

- **Overextension:** Jumping into large-scale projects without foundational support can drain resources prematurely.
- **Undercommunication:** If small wins go unnoticed, you lose opportunities to build wider backing.
- **Complacency:** Initial success can create a false sense of completion, halting the systemic changes needed for lasting impact.

E.5.3.1 Concluding Note: From Obstacles to Catalysts

Each obstacle – 111 sustaining consistency, 222 navigating biases, and 333 managing resource constraints—can, with the right mindset and strategy, become a catalyst for personal growth and collective advancement. Every challenge overcome strengthens your ESI, demonstrating that genuine growth often arises from challenging experiences. By reframing these hurdles not as dead ends but as invitations to refine your approach, you embody the essence of a living philosophy – dynamic, empathetic, and responsive to a world in flux.

As you fine-tune your daily habits, engage with diverse perspectives, and make strategic use of limited resources, remember that these challenges are integral to the journey. Overcoming them fosters resilience, sharpens ethical insight, and lays a stronger foundation for sustainable, meaningful change – both for you as an individual and for the communities you help shape.

THE EVOLUTION OF ETHICS

THINKING BOX:

"Are our moral compasses evolving in lockstep with new tech breakthroughs? The path forward, like the quest for **42**, might call for deeper inquiry."

E.6 CALL TO ENGAGE: THE NEXT STEP

True transformation, however, requires translating these insights from conceptual understanding into real-world action. The following steps provide a **starting point** for integrating **ESI** into your daily life – whether at home, in the workplace, or within your community. These **initial actions** are intentionally simple yet capable of yielding profound ripple effects when approached consistently and adaptively.

E.6.1 Pick One Habit for Personal Integration

E.6.1.1 Opening Thought

Adopting even a single, small practice can anchor you in ESI's core principles. This habit becomes a daily reminder that ethical growth and systemic awareness aren't abstract ideals; they're lived realities shaped by repetitive, deliberate actions.

Why It Matters

1. **Momentum Building**
   - *Core Idea:* Each small victory – like noticing bodily tension or pausing before an impulsive choice – builds deeper ethical reflexes.
   - *Why This Helps:* Success with one habit encourages you to expand or refine other practices, creating a **positive feedback loop** that strengthens moral and systemic awareness.
2. **Sustained Focus**
   - *Core Idea:* Consistency over time transforms ephemeral insights into stable neural pathways supporting ethical, emotional, and societal sensitivity.
   - *Why This Helps:* By focusing on a single habit, you eliminate overwhelm and ensure consistent engagement, which fosters genuine behavioral change.

How to Implement

1. **Choose a Manageable Habit**
   - *Approach:* Set a simple daily "embodied check," such as scanning your body each morning to note tension, comfort, or anxious energy.
   - *Rationale:* A small, concrete action lowers resistance and allows for easy repetition, embedding ESI principles in your routine.
2. **Tie It to Existing Routines**
   - *Approach:* Link the new habit to an existing routine. For instance, do a 1-minute mindfulness scan after brushing your teeth or pause for 30 seconds before opening email.
   - *Rationale:* Linking the new habit to an existing routine promotes automaticity and reduces resistance.
3. **Check in Weekly**
   - *Approach:* At week's end, note any changes. Did you feel calmer? More aware of bodily signals?
   - *Rationale:* Brief self-reflection ensures ongoing refinement, keeping the habit adaptive and responsive to emerging needs.

Potential Pitfalls

- **Over-Ambition:** Trying multiple new habits simultaneously dilutes focus and can lead to burnout.
- **Routine Fatigue:** Repeating the same micro-habit daily can become stale. Periodically refresh your micro-habits by changing their nature or timing.
- **Neglecting Emotional Insight:** Focusing solely on physical awareness without acknowledging emotional states reduces ESI's holistic benefits.

Tracking Progress

- **Self-Logging:** Maintain a short daily note (e.g., "Completed 1-min body scan: noticed slight jaw tension, felt calmer afterward").

- **End-of-Week Reflection:** Evaluate if you felt more centered or if you managed stress better. This helps identify how one small habit can ripple into other areas of life.

E.6.2 Catalyze a Collective Conversation

E.6.2.1 Opening Thought

ESI gains momentum when shared. By sparking even a brief group dialogue, you harness **collective insights** to illuminate challenges more fully and stimulate **cooperative action**.

Why It Matters

1. **Shared Insights**
   - *Core Idea:* Hearing diverse perspectives expands your systemic understanding, identifying hidden links between personal experiences and societal trends.
   - *Why This Helps:* Collaborative reflection can reveal new narrative angles or practical solutions that individual contemplation might overlook.
2. **Narrative Influence**
   - *Core Idea:* Group discussions shape the stories communities tell about their problems and potential futures.
   - *Why This Helps:* Adjusting these collective narratives – from scarcity to abundance, from helplessness to proactive solutions – can catalyze community transformation.

How to Implement

1. **Identify a Pressing Issue**
   - *Approach:* Pick a concrete topic – a local recycling challenge, a workplace communication gap, or a rising community concern.
   - *Rationale:* Specificity grounds the discussion, ensuring participants can see and feel the direct relevance.
2. **Frame the Discussion**
   - *Approach:* Host a concise 30-minute session. Prompt participants to notice bodily sensations (tension, excitement), emotional responses (frustration, hope), and possible systemic impacts (interdepartmental, neighborhood-wide).
   - *Rationale:* This prompts participants to connect personal experiences with broader systemic issues, ensuring the conversation stays constructive and multi-layered.
   - *Example:* "Take a moment to note any physical reaction (like a tightness in your chest) when you hear this issue. Now, share one feeling (anger, concern) and one idea about how it affects our wider community."
3. **Steer Toward Action**
   - *Approach:* Conclude with a single group commitment – testing a solution or scheduling a follow-up.
   - *Rationale:* Without actionable steps, engaging discussions risk fading into inaction.

Potential Pitfalls

- **Dominant Voices:** One or two outspoken members can overshadow quieter participants. Use round-robin formats or breakouts to ensure balanced input.
- **Emotional Overwhelm:** Complex or sensitive topics may surface strong emotions. Provide gentle facilitation and time-outs if tension escalates.
- **No Follow-Up:** Without clear follow-up actions, the conversation's momentum will be lost, and lessons learned might never manifest in real-world outcomes.

Tracking Progress

- **Discussion Notes:** Keep a communal record of main talking points, emotional currents, and proposed actions.
- **Action Check-Ins:** Revisit the group's agreed-upon step within a set time (e.g., two weeks). Did the group complete it, and what changed? Are there emerging breakthroughs or new obstacles?

E.6.3 Join or Form a Support Network

E.6.3.1 Opening Thought

ESI doesn't thrive in isolation. Collective accountability, resource-sharing, and the synergy of diverse expertise can **amplify** your efforts and **sustain** long-term engagement.

Why It Matters

1. **Collective Accountability**
   - *Core Idea:* Joining a circle of peers – whether for ecological activism, mental health advocacy, or community empowerment – ensures mutual motivation.
   - *Why This Helps:* Shared commitments push members to stay consistent, offering resources and moral support, which is especially crucial when confronting systemic challenges.
2. **Cumulative Expertise**
   - *Core Idea:* A group setting merges specialized knowledge (technical, cultural, managerial) and fosters creative tension that fuels innovation.
   - *Why This Helps:* Individuals in different professions or from distinct cultural backgrounds can spot blind spots and propose complementary strategies, enriching the entire group's systemic insight.

How to Implement

1. **Identify Existing Groups**
   - *Approach:* Search for local neighborhood alliances, workplace committees, online communities on sustainability, mental health, or social justice.
   - *Rationale:* Tapping into established groups accelerates your learning curve, letting you contribute skills quickly.
2. **Form a New Circle**
   - *Approach:* If you spot a gap – for instance, no local "well-being hub" or no mental health support group – initiate one with a few like-minded people.
   - *Rationale:* Starting small often fosters tight-knit collaboration that can later expand, attracting members who share ESI-aligned values.
3. **Document Progress Together**
   - *Approach:* Maintain a shared digital tool (e.g., a Slack channel, Google Sheet) for tasks, goals, and achievements.
   - *Rationale:* Documenting progress visually reinforces motivation and clarifies next steps, ensuring no valuable insight gets lost.

Potential Pitfalls

- **Group Inertia:** Without regular check-ins or a clear vision, groups can drift. Scheduling monthly "progress huddles" helps maintain momentum.
- **Conflict or Burnout:** High passion can spark disagreements or exhaust members. Establish respectful dialogue norms and emphasize rest, recognition, and self-care.
- **An Insular Group:** An insular group can limit the influx of fresh perspectives; remain open to new members or rotating outside expertise.

Tracking Progress

- **Shared Logs:** Post weekly tasks completed, upcoming targets, and personal reflections on the group's shared platform.
- **Periodic Reviews:** Every few months, jointly review metrics like membership growth, project milestones, or community responses to your efforts.

E.6.4 Measure and Reflect

E.6.4.1 Opening Thought

Periodic evaluation transforms **small gestures** into cumulative breakthroughs. By systematically tracking data – both qualitative (personal stories) and quantitative (statistics) – you gain **insight** into what's **working, what's faltering, and how to pivot**.

Why It Matters

1. **Continuous Improvement**
   - *Core Idea:* Assessing progress ensures that your evolving practices remain aligned with real-world needs and your evolving sense of purpose.
   - *Why This Helps:* Regular evaluation highlights which habits or group initiatives yield tangible benefits, steering you toward more effective strategies.
2. **Accountability and Celebration**
   - *Core Idea:* Recognizing concrete achievements galvanizes ongoing commitment, while transparent review fosters trust within a network.
   - *Why This Helps:* Highlighting successes, even small ones, boosts morale and encourages continued experimentation.

How to Implement

1. **Pick Focus Areas**
   - *Approach:* Determine which metrics matter most – personal stress levels, conflict resolution success, local environmental indicators, or policy changes.
   - *Rationale:* Clarity of focus ensures you don't drown in data. Choose a handful of relevant, meaningful markers.
2. **Collect Data or Observations**
   - *Approach:* Both qualitative (narratives, anecdotes) and quantitative (statistics) data provide valuable insights. Use simple surveys, logs, or community feedback to gauge changes.
   - *Rationale:* Combining narratives with numbers deepens understanding and can reveal patterns otherwise missed.
3. **Reflect and Share**
   - *Approach:* Schedule a quarterly or monthly assessment session, alone or with your network. Review any shifts in emotional well-being, group cohesion, or environmental impact.
   - *Rationale:* Public discussions of findings encourage collective adaptation. If certain strategies aren't effective, pivot promptly; if they are, expand them.

Potential Pitfalls

- **Data Overload:** Monitoring too many metrics complicates analysis. Start modestly, then expand if needed.
- **False Attribution:** External factors (e.g., a new local policy or economic shift) might skew results, so remain critical and open-minded.
- **Inconsistent Data Collection:** Inconsistent data collection can lead to inaccurate assessments of progress, masking true trends.

Tracking Progress

- **Shared Dashboards:** In group contexts, visualize progress in a common digital space. Individuals might keep personal metrics in a mindfulness app or journal.
- **Narrative Logs:** Complement charts or graphs with short reflective statements, capturing **qualitative** impressions of growth, hurdles, or unexpected breakthroughs.

E.6.4.2 Closing Note: Embracing an Active Role

None of these steps demands perfection or grand-scale heroics. **ESI** shines precisely when **small, consistent actions** build toward impactful shifts – both in your own neural wiring and across the networks you inhabit. By **starting modestly**, forming supportive alliances, creating new dialogues, and measuring impacts, you lay the **foundation** for broader transformations yet to come.

Each personal habit you adopt, **each conversation you spark**, **each connection you foster**, and **each metric you track** forms part of an evolving tapestry of ethical awareness and systemic literacy. As you **adapt, learn, and share** these experiences, you continuously strengthen the cycle of personal ↔ collective flourishing – a feedback loop that ultimately **shapes not just your own life, but the shared world** in which we all strive to thrive.

E.7 FINAL REFLECTION: EPILOGUE AS A BEGINNING

As you arrive at these final pages, remember that while this book closes, your ethical journey continues. **Everything you have explored – complex systems, embodied wisdom, and personal-collective synergy – provides a toolkit** for navigating life's challenges with creativity, compassion, and a deep sense of interconnectedness. Rather than standing still, this learning remains alive in how you perceive, respond to, and ultimately shape the evolving moral and societal landscapes around you.

E.7.1 Embracing Complexity as an Ongoing Teacher

Throughout this journey, you've come to see **complexity not as a static state, but as a dynamic process**. Every novel situation – be it rapid technological innovation, changing social norms, or unforeseen global events – calls for adaptive thinking and heartfelt engagement.

- **Continuous Unfolding:** Complexity unfolds daily in myriad forms. Policies, relationships, technologies, and cultures all interact in ways that can be surprising yet offer fresh insights if observed carefully.
- **Opportunity in Ambiguity:** What feels uncertain or chaotic can be recast as a stimulus for learning and moral discovery. By engaging complexity proactively, your insights remain **relevant and applicable**, avoiding stagnation.

E.7.1.1 Key Takeaway

Complexity is not static; it is a continuous teacher. Your ongoing engagement ensures that your understanding expands in both depth and utility.

E.7.2 Harnessing ESI in Daily Life

From personal routines to large-scale projects, **ESI** connects mind, body, and collective awareness, guiding actions that merge empathy and reason.

- **Responsive to Change:** ESI keeps your moral compass flexible. Whether navigating a sudden global shift or a personal dilemma, it prompts you to integrate bodily cues, emotional states, and social factors for balanced decisions.

- **Relevance Across Scales:** ESI isn't confined to personal well-being. It extends to workplace leadership, community initiatives, and policymaking. Wherever decisions carry ethical weight, ESI can nurture more humane, grounded outcomes.

E.7.2.1 Key Takeaway

Let ESI guide you through future uncertainties, combining reason, emotion, and collective awareness to create **balanced, humane solutions**.

E.7.3 Personal Growth as Collective Potential

Your individual growth has never been an isolated project; it's interwoven with the well-being of your community and society at large.

- **Reciprocal Influence:** The personal insights you gain today can shape how you approach family, teams, or civic groups. In turn, supportive communities empower you to refine your moral imagination.
- **Ripples of Innovation: Small shifts – mindful communication, eco-conscious habits, or inclusive leadership – can spark significant communal changes**, creating a cycle of mutual growth. By living your values, you set visible examples that others can adapt and multiply.

E.7.3.1 Key Takeaway

Personal flourishing and social progress nurture each other. **Your ethical evolution contributes to collective resilience**, just as collective movements fuel your ongoing growth.

E.7.4 The Power of Consistent, Adaptive Practice

Insight alone cannot transform your world. **Genuine change** emerges through repeated, iterative engagement that refines your habits and deepens your resolve.

- **Iteration Over Perfection:** Ethical mastery is not a destination but **an ongoing process** of learning from both successes and setbacks. Each day offers a testbed to recalibrate your framework, turning lessons into actionable wisdom.
- **Learning from Missteps:** When a plan falters or a relationship strains under ethical conflict, treat that not as failure but data. Adjust your approach, integrate new knowledge, and step forward with renewed clarity.

E.7.4.1 Key Takeaway

Ethical mastery is not a final state – it's an unfolding journey enriched by the lessons of both achievements and challenges.

E.7.5 Facing Tomorrow's Challenges with Confidence

The **fast-paced world** ahead – whether it's technological leaps, environmental shifts, or social tensions – requires the adaptive capabilities you have cultivated.

- **Technological Transitions:** Emerging Artificial Intelligence (AI), data privacy issues, and biotech innovations may demand nuanced moral reasoning and community consensus-building.
- **Ecological and Social Pressures:** Climate emergencies, resource allocation problems, and social justice concerns underscore the need for calm, systemic action, grounded in empathy and evidence.
- **Cultural Shifts:** Evolving norms around identity, equity, and collaborative governance can be embraced when approached with an open mind, humility, and ESI's integrative lens.

E.7.5.1 Key Takeaway

View each new challenge as an opportunity to practice ESI, integrating bodily, emotional, and systemic awareness to develop **compassionate, pragmatic solutions**.

E.7.6 Cultivating Flourishing Communities

Your continued engagement helps co-create **environments where individuals and communities thrive**:

- **Modeling Ethical Action:** Consistent application of reflective dialogue, mindful communication, or equitable policies shows others that these principles are workable and beneficial.
- **Strengthening Social Fabric:** As networks see moral integrity in practice, trust and cooperation often flourish. When members experience empathy and mutual understanding, collective intelligence grows, fueling cultural evolution.

E.7.6.1 Key Takeaway

Your sustained ESI-based approach not only evolves you personally but **shapes communities** that value empathy, adaptability, and shared progress.

E.7.7 The Invitation Forward

Think of this epilogue as **an open threshold rather than a closed door**. Your living framework – consisting of complexity insight, embodied wisdom, and collaborative action – remains vital as you face uncharted territories.

- **Maintain Your Practices:** Continue (and periodically refine) your habits of reflection, micro-checks, dialogue, and measurement. Adapt them to shifting circumstances, whether personal or societal.
- **Share and Collaborate:** Join forums, workshops, or discussions that build ethical literacy in your community or professional sphere. Seek out new perspectives; incorporate them into your growing repertoire.
- **Evolve Consciously:** Apply each new lesson, from minor revelations to major shifts, refining your approach in a cycle of continuous improvement.

E.7.7.1 Key Takeaway

Accept this epilogue as an invitation to sustained growth – one that you shape day by day, in tandem with evolving realities and shared aspirations.

E.7.7.2 Conclusion

Ultimately, your **living philosophy** grows each moment you choose reflective engagement over passive acceptance. **Every mindful decision**, each collaborative initiative, and every instance of narrative bridging extends the tapestry of ethical wisdom that weaves through families, workplaces, and societies.

- **You Are Not Alone:** Others around the world are on a parallel path, exploring ways to align personal well-being with collective good.
- **You Are Well-Equipped:** Your toolkit – comprised of complexity thinking, ESI, and narrative awareness – can address crises and nurture communal flourishing.
- **You Are Ongoing:** This epilogue signals **a beginning, not an ending** – a living, dynamic process that evolves with each new insight, challenge, and collective endeavor.

Step forward with confidence and empathy. Let your cultivated **ESI** guide your interactions, your pursuits, and your innovations. As you do, you will find that **each moment of awareness and each act of care** shapes a future where individual fulfillment and societal well-being rise together.

COUNTERARGUMENT

THINKING BOX:

"Are you open to challenging your own cherished assumptions? Cultivating that openness might be as fundamental – and maybe as cryptic – as **42**."

A Hitchhiker's Guide Through Complexity

42 Stops on the Digital Road

A.1 INTRODUCTION

"Don't Panic." – Douglas Adams

The digital landscape, like the universe itself, is vast, bewildering, and often absurd. This chapter is not a map, for the terrain is constantly shifting. Instead, it's a guidebook – a collection of 42 thoughts, observations, and "what if" scenarios to help you navigate the complexities of our increasingly technological world. Think of it as a pocket companion for the digital hitchhiker, offering insights into the ethical dilemmas, philosophical paradoxes, and unexpected consequences that arise when human nature collides with rapid technological advancement.

Each entry is a brief stop on this journey – a moment of reflection designed to spark further thought and discussion. So, grab your towel, keep your wits about you, and remember: the most important thing is not to panic.

A.2 THE 42 THOUGHTS

1. **Communication as a Weapon**
   - **Core Idea:** Communication tools, from spoken words to social media posts, can be immensely powerful and are often underestimated. Because every tool can be weaponized, even our simplest interactions can shape or distort reality – and sometimes wreak havoc.
   - **Context:** From ancient rhetoric to modern social media manipulation, words can both persuade and devastate. For instance, deepfakes have emerged in political disinformation campaigns, illustrating the raw power of weaponized speech.
   - **What If?:** What if advanced "deepfake" technology makes it almost impossible to discern truth from fiction, turning every online interaction into a potential minefield?
   - **Ethical Angle:** Who bears responsibility when communication becomes a weapon – should we blame the speaker, the platform, or the algorithm?
2. **Interdependence**
   - **Core Idea:** True independence is largely an illusion because we depend on countless systems and people for survival, stability, and innovation. Recognizing this interconnectedness can reshape how we approach collaboration and accountability.
   - **Context:** From the air we breathe to the software we rely on, our existence depends on intricate, intertwined systems.
   - **What If?:** Imagine digital networks worldwide suddenly shutting down. How would our illusion of self-reliance fare?
   - **Practical Takeaway:** Embracing our interdependence fosters collaboration and shared accountability for the digital commons we inhabit.
3. **Understanding Proxies**
   - **Core Idea:** Proxies are surface-level indicators – like wealth, likes, or followers – that often stand in for deeper aspirations, such as belonging or autonomy. When we confuse

the symbol for the substance, we risk chasing illusions of success rather than meaningful goals.

- **Context:** Social media metrics (likes, followers) often become proxies for self-worth, blurring our sense of genuine value.
- **What If?:** What if we could measure genuine connection or meaningful impact directly, instead of chasing superficial numbers?
- **Added Note:** Quantifying "meaningful impact" is ethically tricky and can introduce new distortions.
- **Ethical Angle:** Do tech platforms exploit our hunger for status by emphasizing these proxies?

4. **Discipline vs. Motivation**
   - **Core Idea:** Motivation is like a spark – it can ignite our energy but fades fast under real-world pressures. Discipline, on the other hand, provides the steady fuel that keeps us going when excitement wanes.
   - **Context:** Motivation ebbs and flows, while discipline underpins consistent effort in mastering complex technologies or skills.
   - **Practical Takeaway:** Rely on well-structured habits – like daily coding practice – rather than short bursts of enthusiasm.
5. **Cognitive Bias**
   - **Core Idea:** Cognitive biases can protect our sense of identity and sanity, yet they simultaneously distort our grasp of objective truth. In a hyperconnected era, these biases often get magnified by feedback loops and filter bubbles.
   - **Context:** In the digital age, confirmation bias is amplified by algorithmic feeds, funneling us into echo chambers of our own making.
   - **What If?:** Imagine Artificial Intelligence (AI) tools that identify and correct personal biases in real time. Could they steer us toward more objective thinking?
6. **Reality's Infiltration**
   - **Core Idea:** Virtual platforms can simulate entire worlds, but they can't seal out our physical, emotional, or cultural realities. No matter how immersive a digital environment, real-life factors – like hunger or outside events – intrude and reshape our experience.
   - **Context:** Even immersive VR can't eclipse real-world factors – like emotional needs or unforeseen interruptions. Current VR systems increasingly blur the line between physical and digital realms.
   - **What If?:** If VR tech keeps advancing, might the physical and virtual merge so completely that "reality" becomes subjective?
7. **Science as Inquiry**
   - **Core Idea:** Science isn't a set of static truths but a continuous quest to better understand our world and ourselves. Its strength lies in curiosity, skepticism, and the willingness to revise beliefs as new evidence emerges.
   - **Context:** Mistaking scientific consensus for dogma erodes trust, especially in tech debates such as climate science or vaccine efficacy.
   - **Practical Takeaway:** Hone critical thinking by focusing on evidence-based evaluations, remembering that science evolves as new information unfolds.
8. **Fulfillment vs. Happiness**
   - **Core Idea:** Happiness can be a fleeting emotion triggered by external events, whereas fulfillment grows from sustained purpose or meaning. Mixing them up leads to chasing dopamine hits rather than building a richer sense of self and community.
   - **Context:** Social platforms often provide quick bursts of pleasure, but seldom yield long-term fulfillment. Gamification can be designed for positive ends if it prioritizes deeper engagement over instant gratification.

- **What If?:** What if social networks prioritized meaningful collaboration over fleeting validation?

9. **Dreaming Big and Resilience**
   - **Core Idea:** Ambitious goals can inspire monumental achievements, but they can also overshadow the importance of resilience and adaptability. When we idolize big dreams without a plan for inevitable setbacks, we set ourselves up for greater disappointment.
   - **Context:** Tech culture glorifies rapid disruption and grand visions but often neglects the importance of recovery from failure.
   - **Practical Takeaway:** Combine lofty ambitions with pragmatic strategies and clear contingency planning.
10. **Curiosity Against Biases**
    - **Core Idea:** Genuine curiosity is the antidote to many biases, driving us to explore perspectives that challenge our assumptions. By confronting new ideas, we disrupt our own echo chambers and foster intellectual flexibility.
    - **Context:** Actively seeking out diverse viewpoints counters the digital echo-chamber effect.
    - **Practical Takeaway:** Regularly engage with opinions that clash with your own – step outside your comfort zone.
11. **It Is Easier to Fool People**
    - **Core Idea:** Once misinformation takes root, it's far more difficult to uproot it than to plant it in the first place. Our pride and sense of identity often prevent us from admitting we've been deceived.
    - **Context:** Misinformation morphs swiftly; deepfakes and propaganda tactics make debunking harder.
    - **What If?:** Picture an AI-based fact-checker that flags misinformation in real time. Could it keep pace with evolving disinformation?
12. **You Are Not Your Thoughts**
    - **Core Idea:** There is a distinction between the mental chatter that runs through your head and the deeper sense of self that observes it. Recognizing this gap can help you navigate digital-era stress and social comparisons more gracefully.
    - **Context:** This mindfulness principle mitigates anxiety, envy, and FOMO in a world teeming with social media highlights.
    - **Practical Takeaway:** Practice detachment and awareness – notice recurring mental scripts without automatically believing them.
13. **Compliance and Risk Assessment as a Gateway to Plausible Deniability**
    - **Core Idea:** Formal compliance or risk assessments can serve as a veneer of responsibility, masking deeper ethical failings. When used as liability shields rather than genuine protective measures, they can harm users while absolving organizations.
    - **Context:** The Cambridge Analytica scandal exemplifies how formal procedures can fail to stop data misuse in practice.
    - **What If?:** What if companies shifted from box-ticking to truly ethical design, prioritizing user well-being from inception?
14. **The Blue Dot Effect**
    - **Core Idea:** We have a built-in tendency to fixate on negativity or problems, even when things are generally stable or positive. The mind can magnify minor issues, overshadowing the broader reality of progress or normalcy.
    - **Context:** Overexposure to grim headlines on 24/7 news cycles can warp our perspective of how the world is actually doing.
15. **Countering the Blue Dot Effect**
    - **Core Idea:** While negativity bias is strong, it can be tempered by deliberately training ourselves to notice positives and moments of calm. Shifting focus from a constant problem-seeking mode allows for a more balanced view of reality.

- **Practical Takeaway:** Limit exposure to toxic media, practice daily gratitude, and intentionally seek out encouraging stories.

16. **AI Singularity**
    - **Core Idea:** Advances in AI – driven by disparate fields like the military, industry, and academia – might collectively push us toward a tipping point of machine intelligence. Some call it a "singularity," envisioning AI surpassing human cognition in unpredictable ways.
    - **Context:** Experts fiercely debate whether a true singularity is imminent, far off, or even a possibility.
    - **What If?:** Could superintelligence emerge before humanity is ready?
    - **Ethical Angle:** Who "owns" an AI that outperforms human intellect?
17. **Balancing AI Regulation**
    - **Core Idea:** Striking a balance between fostering AI innovation and preventing harm is a delicate dance – regulations can protect society but can also choke creativity if too heavy-handed. This tension underscores the complexity of shaping AI policy in real time.
    - **Context:** Over-regulation may stifle breakthroughs; under-regulation could unleash serious risks.
    - **Practical Takeaway:** Push for global, multi-stakeholder frameworks that ensure AI uplifts rather than undermines human welfare.
    - **Ethical Angle:** How do we ensure equitable outcomes and long-term benefits while mitigating AI's potential pitfalls?
18. **Alan Watts' Key Insights**
    - **Core Idea:** Philosopher Alan Watts spotlighted the interconnectedness of all things, advocating for a fluid approach to life rather than a rigid quest for control. His teachings invite us to see technology as part of a broader ecosystem, not merely a tool we dominate.
    - **Context:** Embedding a sense of "oneness" into digital networks could foster more collaboration and empathy online.
    - **What If?:** Could letting go of our obsession with control – central to Watts' philosophy – reduce friction in tech-driven lives?
    - **Practical Takeaway:** Explore Watts' writings on dissolving the ego's hold, an approach that might offset modern digital anxiety.
19. **Dopamine Inequality**
    - **Core Idea:** Digital platforms often provide quick bursts of pleasure through notifications and likes, creating an uneven "dopamine economy." While some people thrive on constant stimulation, others feel left behind or unfulfilled in offline settings.
    - **Context:** This imbalance can make real-world experiences seem dull, further drawing users into incessant app-checking or social media loops.
    - **What If?:** What if these platforms nudged us toward occasional disconnection and richer offline experiences?
    - **Practical Takeaway:** Consider digital detox strategies – regular intervals without phone notifications or social media.
20. **The Paradox of Choice**
    - **Core Idea:** Having seemingly infinite options may look like freedom but can breed anxiety, second-guessing, and endless searching for "the best." The more we can choose from, the more paralyzed we can become.
    - **Context:** Streaming services or online stores exemplify this overload – an unending buffet of items, leaving us overwhelmed. Predicting personal satisfaction compounds the challenge.
    - **Practical Takeaway:** Try "satisficing" – selecting a "good enough" option – rather than chasing an elusive, perfect choice.

21. **The Law of Unintended Consequences**
    - **Core Idea:** Every new technological development, no matter how well-intentioned, can spin off results nobody predicted – sometimes beneficial, sometimes harmful. Innovations aimed at solving specific problems can inadvertently create new ones.
    - **Context:** Social media bridged communication gaps but opened the door to misinformation and cyberbullying. Accurately forecasting all effects is notoriously difficult.
    - **What If?:** Mandating "impact assessments" for emerging tech could reduce negative outcomes – provided they're enacted thoroughly and not just as lip service.
22. **The Concept of Anti-Fragility**
    - **Core Idea:** Anti-fragility is about thriving under uncertainty and stress, rather than merely surviving. By viewing failures and disruptions as opportunities for growth, systems and individuals can become stronger through adversity.
    - **Context:** A rapidly changing tech world favors designs that adapt and improve under stress. The internet's distributed structure, for instance, resists localized failures.
    - **What If?:** Could the same concept guide personal development, allowing each setback to spark innovation instead of defeat?
    - **Practical Takeaway:** Incorporate redundancy, adaptability, and a willingness to experiment – both in technology and personal skill sets.
23. **The Illusion of Control**
    - **Core Idea:** We often seek complete control over our digital and physical environments, forgetting that chaos is a natural part of life. Letting go of the need for absolute control can lead to reduced stress, better creativity, and healthier acceptance of unexpected outcomes.
    - **Context:** Micromanaging digital systems – whether through excessive notifications or constant monitoring – frequently backfires.
    - **What If?:** Embracing unpredictability could unleash innovative solutions that rigid control might stifle.
24. **The Concept of Time Affluence**
    - **Core Idea:** Having ample free time is often more valuable to our well-being than material wealth. By managing technology usage thoughtfully, we can "buy back" the hours lost to endless scrolling or digital distractions.
    - **Context:** Technology can save time through automation, yet it can also consume our days with infinite feeds and notifications.
    - **Practical Takeaway:** Use tech intentionally – reserve pockets of uninterrupted time for relationships, creativity, and rest, thereby achieving greater "time affluence."
25. **The Influence of Digital Minimalism**
    - **Core Idea:** Trimming digital excess – like unnecessary apps, notifications, or mindless online habits – can greatly enhance clarity and peace of mind. Just as a cluttered desk can impede efficiency, a cluttered digital life can fragment focus and well-being.
    - **Context:** Constant updates, pings, and alerts fragment our attention, reducing productivity and elevating stress.
    - **Practical Takeaway:** Set boundaries for tech usage – consider notification-free hours, device-free zones, or a periodic digital "cleanse."
26. **The Power of Daily Rituals**
    - **Core Idea:** Simple, consistent routines anchor us in a world of rapid change, offering stability and mental grounding. Whether it's morning meditation or evening journaling, these small acts can help us reclaim control amid digital overload.
    - **Context:** Amid the chaos of instant communication and shifting work-life boundaries, daily rituals provide comforting structure.
    - **Practical Takeaway:** Embed short, meaningful habits – like an early walk or a bedtime reflection – to buffer against digital overwhelm.

27. **Weaponized Empathy**
    - **Core Idea:** Empathy is a potent force that can unite or divide. In the wrong hands, emotional appeals and personalized triggers can be harnessed to manipulate opinions, recruit followers, and even radicalize communities.
    - **Context:** Social media algorithms can push content designed to provoke strong emotional reactions, potentially steering user behavior toward specific outcomes.
    - **Ethical Angle:** Should platforms guard against exploitative emotional targeting? And how open should they be about such manipulative tactics?
28. **The Illusion of Independence**
    - **Core Idea:** While we pride ourselves on being autonomous, our daily reality depends on intricate networks – technological, social, economic – that we seldom understand fully. Recognizing this illusion can spur collaborative thinking and shared responsibility.
    - **Context:** We rely on a global web of devices, software, and infrastructure that few individuals can maintain alone.
    - **What If?:** If the internet vanished abruptly, would society's routines grind to a halt? How prepared are we to handle such a disruption?
    - **Added Note:** Fostering digital literacy can help reduce overreliance and empower us to cope with unexpected breakdowns.
29. **The Simplicity Bias**
    - **Core Idea:** Sometimes we chase complexity for a sense of control or sophistication, overlooking elegant, straightforward solutions. Complexity can be alluring, but it can also introduce unnecessary costs, delays, and failure points.
    - **Context:** High-tech fixes often overshadow simpler, equally effective alternatives.
    - **Practical Takeaway:** Whenever possible, opt for minimalistic answers before investing in elaborate, resource-intensive ones.
30. **The Anxiety Feedback Loop**
    - **Core Idea:** Anxiety, especially in an always-online culture, can spiral out of control as we repeatedly check for updates, bad news, or potential threats. This cycle amplifies perceived dangers and locks us into a state of vigilance.
    - **Context:** Endless doomscrolling on social platforms fosters a self-reinforcing loop of fear and stress.
    - **Practical Takeaway:** Break the loop by limiting exposure to anxiety-inducing content and embracing mindfulness techniques to recalibrate emotional balance.
31. **The Proxy Trap**
    - **Core Idea:** We might mistake symbolic achievements (like follower counts or "likes") for real, fulfilling accomplishments. This disconnect can drive addictive behaviors and leave us emptier than before.
    - **Context:** Online validation via likes and followers often substitutes for genuine relationships or professional milestones.
    - **Cross-Link:** Ties back to "Fulfillment vs. Happiness" by highlighting deeper, more lasting rewards beyond mere numbers.
32. **The Paradox of Progress**
    - **Core Idea:** Each technological solution tends to bring new challenges. Progress can solve old problems while inadvertently spawning fresh ones, urging us to question whether we're truly moving forward or just sideways.
    - **Context:** The internet connected the planet yet birthed privacy nightmares and misinformation at scale.
    - **What If?:** Encouraging "long-term thinking" in research and development could yield more sustainable innovations with fewer harmful side effects.
    - **Added Note:** Ethical reflection at every design stage is crucial to minimizing unforeseen risks.

33. **The Filter Bubble Dilemma**
    - **Core Idea:** Personalized content can feel empowering yet insulate us from diverse opinions, locking us into echo chambers. As algorithms refine our feeds, we risk intellectual stagnation and polarized communities.
    - **Context:** Many users prefer these bubbles, finding comfort in repeated affirmations of their views.
    - **Counterpoint:** True growth often arises from uncomfortable exposure to conflicting ideas.
    - **Practical Takeaway:** Actively seek out contradictory sources and engage with varied perspectives.
34. **The Attention Economy Conundrum**
    - **Core Idea:** Our focus has become a monetizable commodity, with platforms vying for every spare second of our attention. This endless competition can fragment our thoughts, leaving us with reduced capacity for deep work or genuine social connection.
    - **Context:** Social networks and media outlets optimize for engagement, often at the expense of our mental well-being.
    - **Ethical Angle:** Should there be guidelines or limits on how companies profit from human attention?
35. **The Meaning Crisis**
    - **Core Idea:** In an age that celebrates individualism and endless consumer choices, many people struggle to find shared purpose. While technology can isolate us, it can also help us discover communities aligned with deeper values.
    - **Context:** Platforms that unify people around common ideals could help counter rising feelings of emptiness or aimlessness.
    - **What If?:** Building digital spaces focusing on collaboration and shared goals might restore a sense of collective meaning.
36. **The Death of Expertise**
    - **Core Idea:** Easy access to information can lead us to assume we're as knowledgeable as actual experts, undermining respect for specialized skills and research. This overestimation can result in misguided decisions and weaker societal discourse.
    - **Context:** The internet's egalitarian nature sometimes dilutes legitimate expertise with noise and overconfidence.
    - **Practical Takeaway:** Seek credible, specialized sources; remain humble about your own knowledge gaps and be wary of superficial certainty.
37. **The Ethics of Convenience**
    - **Core Idea:** Modern conveniences often hide significant moral trade-offs – like exploitative labor or environmental harm. By demanding frictionless services, we may unknowingly perpetuate unjust systems.
    - **Context:** One-click shopping, rapid deliveries, and on-demand services can come at a high human and ecological cost.
    - **Practical Takeaway:** Investigate supply chains, choose sustainable options where possible, and accept minor inconveniences if they align with higher ethical standards.
38. **The Power of Storytelling**
    - **Core Idea:** Stories shape how we perceive reality, influencing cultural norms, political opinions, and our sense of identity. Digital platforms magnify storytelling power, allowing narratives to circle the globe within minutes.
    - **Context:** Whether used for social change or propaganda, narratives can unite or polarize communities.
    - **Ethical Angle:** Can we foster media literacy so people discern manipulative stories from genuinely informative ones?

39. **The Age of Loneliness**
    - **Core Idea:** Ironically, in our hyper-connected era, many feel more isolated than ever. Superficial online connections may lack the depth and warmth of real-world relationships, exacerbating loneliness.
    - **Context:** Despite (or perhaps because of) near-constant digital interaction, people often struggle to form lasting bonds offline.
    - **Practical Takeaway:** Consciously invest in face-to-face engagements and communities to combat the stark sense of isolation.
40. **The Nature of Consciousness**
    - **Core Idea:** Consciousness remains one of humanity's greatest mysteries – how do neural processes give rise to subjective experience? Some anticipate that AI might eventually mirror or even surpass human awareness, posing philosophical and ethical quandaries.
    - **Context:** Research into AI self-awareness sparks debate about whether machines can truly "think" or merely simulate cognition.
    - **What If?:** If an AI became conscious, would it hold rights or responsibilities akin to human beings?
41. **The Evolution of Ethics**
    - **Core Idea:** Traditional moral codes may prove insufficient for emerging technologies like gene editing or highly autonomous AI. Ethics must evolve alongside our inventions, requiring ongoing dialogue and adaptation.
    - **Context:** Rapid innovation can outpace older frameworks – leading to confusion, loopholes, or moral blind spots.
    - **Practical Takeaway:** Encourage continuous ethical discussions, readiness to update norms, and collective participation in shaping moral guidelines.
42. **Counterargument**
    - **Core Idea:** Without a robust counterargument, beliefs risk becoming dogmatic or entrenched in personal bias. Openness to dissenting views fosters humility, adaptability, and more rigorous thinking.
    - **Context:** Echo chambers make it easy to avoid disagreements, but intellectual growth demands exposure to contrary evidence and perspectives.
    - **Practical Takeaway:** Actively pursue credible opposing viewpoints, remain open to revision, and cultivate respectful debate for richer discourse.

A.3 FINAL OBSERVATIONS

This collection of 42 entries has traveled widely – from the weaponization of communication to the mysteries of consciousness. Several tensions recur: the delicate balance between chaos and control in a rapidly shifting world; the search for purpose and meaningful human connections in our digital age; and the importance of ethical reflection when technologies evolve faster than our cultural norms.

These thoughts are not definitive answers. Rather, they're prompts for deeper reflection and conversation. The digital road is often winding and unpredictable, and the journey matters as much as any destination. Embrace the detours, question the shortcuts, and carry a healthy curiosity – your best guide on any hitchhiking trip through complexity.

FSC
www.fsc.org
MIX
Papier | Fördert gute Waldnutzung
FSC® C083411

Kolibri 360 GmbH
Ferdinand-Jühlke-Straße 7
99095 Erfurt, Deutschland
produktsicherheit@kolibri360.de